Statics, structures and stress

Statics, structures and stress

A teaching text for problem-solving in Theory of Structures and Strength of Materials

W. Fisher Cassie
C.B.E., LL.D., F.R.S.E., M.S., Ph.D., C.Eng.
Emeritus Professor of Civil Engineering
University of Newcastle upon Tyne

Longman

Longman Group Limited
London

Associated companies, branches and representatives throughout the world

© Longman Group Limited 1973

All rights reserved. No part of this publication may be reproduced, stored in a retrieval system, or transmitted in any form or by any means, electronic, mechanical, photocopying, recording, or otherwise, without the prior permission of the Copyright owner.

First published 1973

ISBN 0 582 44912 X (Cased)
      0 582 44913 8 (Paper)

Made and printed in Great Britain by offset by William Clowes & Sons, Limited London, Beccles and Colchester

Contents

Preface   ix

PART I RIGID STRUCTURES

GROUP A The Working Conditions

UNIT 1 HOW TO USE THIS BOOK
  1.1 The GROUP and UNIT classification   4
  1.2 Rules and injunctions   6
  1.3 The use of the INDEX   7

UNIT 2 QUANTITIES AND SYMBOLS
  2.1 Quantities used in these Units of Study   8
  2.2 Multiple measurement   8
  2.3 Multiplying quantities   9
  2.4 Dividing quantities   9
  2.5 Keeping the number of digits small   11

UNIT 3 FORCES, COUPLES AND MOMENTS
  3.1 Surface tractions and body forces   13
  3.2 Relationship between mass and weight   14
  3.3 Directions and senses of forces   15
  3.4 Moments   16
  3.5 Couples   18

GROUP B The Equilibrium of Rigid Structures

UNIT 4 LIMITATIONS OF STUDY FOR RIGID STRUCTURES
  4.1 First limitation   23
  4.2 Second limitation   23
  4.3 Third limitation   24
  4.4 Fourth limitation   24

UNIT 5 THE ACHIEVEMENT OF EQUILIBRIUM
  5.1 Conditions for balanced loading   26
  5.2 Equilibrium of a structure under two or under three forces   28
  5.3 Equilibrium under more than three forces   30
  5.4 The effect of applied couples on the Conditions of Equilibrium   32
  5.5 Expression of the three Conditions of Equilibrium   33

UNIT 6   FREE STRUCTURES AND THEIR SUPPORTS
    6.1   Defining the FREE STRUCTURE   36
    6.2   Loading the Free Structure   37
    6.3   Types of support   38
    6.4   Relationships between the types of support   44
    6.5   Couples as part of the supporting system   45
    6.6   The senses of the supporting forces and couples   46

UNIT 7   THE GEOMETRY OF THE STRUCTURE
    7.1   The right-angled triangle   51
    7.2   Components of forces   53
    7.3   The geometry of curves   55

GROUP C   Techniques and Problems

UNIT 8   SIX STEPS TO SOLUTION
    8.1   The mnemonic – S-U-R-E-T-Y   60
    8.2   S   The free structure and supports   60
    8.3   U   The unknown values   61
    8.4   R   The centre of rotation (COR)   61
    8.5   E   Evaluation (writing down the quantities)   63
    8.6   T   Total is zero (the equations of equilibrium)   64
    8.7   Y   Yes or No? (checking the senses of the unknowns)   65

UNIT 9   RIGID STRUCTURES IN NUMERICAL SOLUTION
    9.1   The guiding rules   67
    9.2   Arrangement of problems   68
    9.3   THE PROBLEMS   69

UNIT 10   DIAGRAMS FOR DESIGN
    10.1 Representation for the effects of forces and couples   185
    10.2 Diagrams for horizontal beams with vertical loads   186
    10.3 Deflected shapes   189
    10.4 Couples and eccentric loads   189
    10.5 Influence lines   191
    10.6 THE PROBLEMS   194

PART II   DEFORMABLE STRUCTURES

GROUP D   The Working Conditions

UNIT 11   THE GEOMETRY OF THE CROSS-SECTION
    11.1 The geometrical calculations   252
    11.2 The tabular method of recording operations   253

11.3 First moment of Area	255
11.4 Second moment of Area	257
11.5 Polar Second moment of Area	258
11.6 The three categories	259
11.7 THE PROBLEMS	262

UNIT 12 STRESS, DEFORMATION AND ENERGY

12.1 The three effects	285
12.2 The elastic condition	286
12.3 Three origins	287
12.4 Fundamental relationships	288
12.5 Mnemonics or G-U-I-D-E-S	292

GROUP E Stresses and Deformations

UNIT 13 STRESS AND DEFORMATION CAUSED BY DIRECT FORCE

13.1 Direct stress and deformation	296
13.2 Longitudinal strain	297
13.3 Lateral strain	298
13.4 Change in volume	299
13.5 Controlled or restrained strain	300
13.6 THE PROBLEMS	302

UNIT 14 STRESS CAUSED BY BENDING

14.1 Translation of bending moment into stress	329
14.2 The significance of the six quantities	330
14.3 Sections unsymmetrical about the neutral or centroidal axis	332
14.4 Deformation due to bending	333
14.5 Solving the relationships	333
14.6 THE PROBLEMS	335

UNIT 15 STRESS CAUSED BY TRANSVERSE SHEAR

15.1 Evaluation of the relationship	353
15.2 The influences of variations in individual quantities	354
15.3 Suggestions and the mnemonic	355
15.4 THE PROBLEMS	356

UNIT 16 TORSION

16.1 The relationship in torsion	375
16.2 The influences of variations in individual quantities	376
16.3 THE PROBLEMS	378

GROUP F Synthesis of Stress and Deformation

UNIT 17 DEFORMATION CAUSED BY BENDING
17.1 Deflection and deflected shapes 393
17.2 Sketching the deflected shapes of bent beams 393
17.3 Relationship between curvature and deflection 398
17.4 The double-integration method of finding deflections 400
17.5 Macaulay's variation 401
17.6 Method of Area-moments 401
17.7 THE PROBLEMS 405

UNIT 18 DIRECT AND BENDING STRESS COMBINED
18.1 Bending combined with direct force 454
18.2 Overturning moments 456
18.3 Eccentric loading 457
18.4 Quantities and equations 458
18.5 THE PROBLEMS 459

UNIT 19 DIRECT AND SHEARING STRESS COMBINED
19.1 The stressed block 481
19.2 Principal stresses 482
19.3 Graphical calculation by Mohr's Circle 484
19.4 Lengths in the Mohr's Circle diagram 485
19.5 Angles in the Mohr's Circle diagram 486
19.6 Drawing the Mohr's Circle 488
19.7 Relationships between stresses and directions of planes 490
19.8 THE PROBLEMS 492

UNIT 20 STRAIN ENERGY
20.1 Representation of the magnitude of energy stored 511
20.2 The use of strain energy in the determination of deformation 514
20.3 Calculation of displacement under bending couples 514
20.4 Laying out the calculation for displacement due to bending 515
20.5 The unit-load variation 518
20.6 The unit-load method applied to the deformation of pin-jointed frames 519
20.7 THE PROBLEMS 521

Index 553

Preface

The assumption is often made that, if students understand basic principles, and the theories developed from them, they will be well equipped to solve numerical problems. These solutions, after all, depend only on the application of the proved theories. In the subjects covered in this book – long known as elementary theory of structures and strength of materials – this conception has been shown, by experience, to be unsound. The average student of these topics is always sufficiently versed in mathematics to have little difficulty in following the developed arguments leading to the establishment of a principle.

He shows himself, nonetheless, to be at a loss in 'problem solving'. Many of the failures and inaccuracies evident in the results of any test or 'problem session' are due, not to a lack of comprehension of basic principles, but to an unfamiliarity with the techniques needed in the solutions demanded by the problems.

This book is not merely an *ad hoc* collection of old examination questions. It is a teaching text with the object of teaching techniques in problem solving. The problems are arranged in ascending order of complexity, each being designed to show a particular feature or variation. The problems are grouped in UNITS, sometimes with introductory UNITS dealing with various aspects of technique. Through each UNIT runs a consistent pattern, simplifying the solution to a series of defined steps. There is much repetition, for, as any musician will confirm, certainty and smoothness in performance depends on repeated practice of the same techniques. It is not enough to read this book.

From this numerical approach, the principles and the important concepts underlying development of the subjects are more clearly understood.

Waterhouse & Partners W. FISHER CASSIE
Rowlands Gill
Co. Durham
1972

PART I
RIGID STRUCTURES

GROUP A
The Working Conditions

Part I: Group A

1 How to use this book

Don't read this book. It is not intended to be read. It is a working tool which you should learn to use for the purpose for which it is intended – that of linking the theory of elementary Statics, Structures, and Strength of Materials with its efficient application to numerical problems.

Most students find little difficulty in eventually understanding at least the mathematics of the theory underlying these fundamental topics. They do not, however, except perhaps the most brilliant of them, find the translation of the theory into numerical evidence and practice at all easy. This book tries to show how that translation can be accomplished by systematic application of the principles so well described in most books on the subjects under review.

The names of the subjects – Theory of Structures and Strength of Materials, with Statics thrown in – are far from being accurately descriptive of the real distinctions which occur between the different types of numerical problem, and of the assumptions underlying them. In this book there are two PARTS – PART I and PART II – indicating that there are two distinct and non-overlapping subjects. These can be further defined as the *study of rigid structures*, and the *study of deformable structures*.

1.1 The Group and Unit classification

The book is broken into twenty UNITS OF STUDY. These each deal with a distinct and separate portion of the subjects concerned. In turn the UNITS are linked into GROUPS each GROUP dealing with a broader but well-defined aspect of work. Finally, the GROUPS form the two PARTS. There happen to be ten UNITS in each PART, and also three GROUPS in each PART: The pattern is as follows. Study it carefully so that you can see ahead how the plan of attack on these numerical problems can be broken into manageable portions. You ought to be really familiar with each portion of the work, first a UNIT and then a GROUP before going on to the next. PART II can be read independently of PART I, but the work of PART I, if well understood, will certainly assist in the labours of PART II.

As you read at the beginning, this book is not meant to be read through in a comfortable armchair. You should never work from the book without being seated at a desk and having a pencil and paper ready. Work through each problem *yourself*. Personal experience of solving problems is very important in the training you are receiving. No amount of reading can replace benefit obtained from working out the problems afresh.

The Plan of the Book

PART I There are three UNITS at the beginning of PART I which merely set out *The Working Conditions*. These apply to the whole book, and should be referred to when you are working through both PARTS.

> From UNIT 4 to UNIT 7 you have GROUP B:
> *The Equilibrium of Rigid Structures*
> The limitations imposed in this GROUP, and the step-by-step preparation for solving numerical problems are of vital importance: don't skip this.
>
> From UNIT 8 to UNIT 10 you have GROUP C:
> *Techniques and Problems*
> Here, you are introduced to the routines which should be followed in the solution of all problems in Planar Rigid Structures (or Bodies). Stick to the S-U-R-E-T-Y routine.
>
> At this point you have come to the end of the study of the simpler type of rigid structure.

PART II There are two UNITS at the beginning of PART II which, together with the UNITS 1 to 3, form the basis for, and present the working tools of the study of deformable structures.

> From UNIT 13 to UNIT 16 you have GROUP E:
> *Stresses and Deformations*
> Here the stresses caused by Direct Force, Bending, Transverse Shear and Torsion are treated as a Group, showing similar relationships between the loading applied and the

stresses produced; the deformations produced by bending are not considered, for they require separate study, but the deformations produced by the other three loadings are included in this GROUP.

From UNIT 17 to UNIT 20 you have GROUP F:
Synthesis of Stress and Deformation
Here, the simpler forms of stress studied in the previous GROUP are combined as they are in structures, the joint effects being studied. The deformation caused by bending is also calculated by several methods, and the use of strain energy in elucidating other types of deformation and deflection ends the study of deformable structures.

Remember the INDEX – see Section 1.3
Where *you* work:
In UNITS 2, 3, 5, and 6 there are Tests which show you, if you don't cheat, how much you have absorbed of the important fundamentals.
In UNITS 9, 10, 11, 13, 14, 15, 16, 17, 18, 19, and 20 there are many worked examples and problems which you must try again for yourself, and supplement by attacking similar problems obtained from other sources.

1.2 Rules and injunctions

The mnemonics, rules, and injunctions presented may seem childish and fussy to the confident young man or woman. But each of these simple guides is based on long experience of the mistakes and misconceptions of the student population. They are insurances against the common errors – errors which would not be made if the plan of work were logical and methodological. Accurate solutions of problems is possible if the challenge is met by a pre-determined sequence of operations and an attention to detail which is not generally vouchsafed. The individual moves, each simple in itself, add up to the presentation of an inevitable and easily reached result.

Do not, therefore, scorn to use the insurances against inaccurate handling of numerical problems which are built into the problems. Do not feel superior and decide that you, at least, are quite capable of taking short cuts. The evidence of thousands of examination papers is that quick methods seem to lead almost inevitably to mistakes when the student is confronted with the apparently complex circumstances presented in examination questions, wrapped up as they often are in tortuous language. Apply the rules of procedure even to the simple problems, and the so-called 'difficult' problems become like 'miracles' – they take a little longer.

1.3 The use of the INDEX

You must never read this book from beginning to end. Each piece of reading is merely a preliminary to the working out of problems from this or another textbook. You will, therefore, need to be able to find any fact, method, or principle at a moment's notice. The INDEX at the end of the book is very full, and there is never any need to thumb through the pages to find what you want to know. Keep using the index; a reading of it will show you how much information you may glean and how directly you may be led to the piece of advice or information which you seek.

Don't READ this book: WRITE it again.

Use pencil and paper as you work.

Part I: Group A

2 Quantities and Symbols

The system of measurement used throughout this book is the *Système Internationale d'Unités* or the SI system as it is usually known. The system has six *basic quantities* but, of these, most are not required for the problems of statics, structures, and stress. By combining basic quantities you will arrive at others known as *derived quantities*.

2.1 Quantities used in these Units of Study

The three quantities basic and derived, used in the study of rigid structures, and of deformable structures are:

Mass measured by the *kilogram* (kg)
Force measured by the *Newton* (N)
Length measured by the *metre* (m)

Note carefully, and remember:

(a) There are no full stops after contractions.
(b) Both kilogram and metre use small letters (kg, m) in the contractions, while the Newton uses a capital letter (N).
(c) Singular and plural units have the same contraction; there is no letter 's' to denote the plural.

2.2 Multiple measurement

The units described above may be made larger or smaller in multiples of 10^3.

    To multiply by 10^3 use *kilo*; small k
    To multiply by 10^6 use *Mega*; capital M
    To divide by 10^3 use *milli*; small m

Again, be sure to use the small letters and the capital letters correctly. Their misuse is a common mistake.

2.3 Multiplying quantities

Quantities can be squared or cubed or raised to any power, and become, say m² (an area) or m³ (a volume).

When two different quantities are multiplied together, the result is a quantity which may have a physical and readily appreciated meaning. Sometimes, though, such products have no recognizable state, and are used only as mathematical tools.

When two quantities are multiplied by each other, their symbols are placed side by side, with a short space between them. There is no other mark or line, merely the two symbols. Many mistakes are made by students in writing products and quotients of symbols, so obey the simple rules. *Newtons* multiplied by *metres* has the symbol N m or *Newton-metres*.

2.4 Dividing quantities

Symbols can be raised or lowered by a power as for multiplication, but when two different units are divided, one into the other, a *solidus* is used. The *solidus* is a sloping line, and always represents a division. This is an important difference from the complete absence of any mark for multiplication.

A force of *Newtons* applied over a length of *metres* gives a distribution of *Newtons per metre* (N/m).

A force of *Newtons* applied over an area gives a distribution of *Newtons per square metre* (N/m²).

A mass of material measured in Megagrams (Mg) has a known volume. Its *density* or mass per unit volume is known as (Mg/m³).

An accepted rule in the use of SI units is that the quantity shown to the right of the solidus (the denominator in a division) should always be in basic units. These, in our case are the units of *mass* and *length*. The powers of these units can be used in the denominator, but not their multiples. Thus, you would not be strictly correct to write N/mm² instead of MN/m². *Metre* is a basic unit and *millimetre* is not. In many writings, and even in official publications, there are faulty uses of units. You will find calculation much simpler if the above rule, of using only basic units in the denominator is followed. However, the British Standards Institution has admitted N/mm² into 'respectability' as an alternative to MN/m².

2.4

Changing from one unit to another

Not many of the calculations in this book are complex but, whether complex or not, they must always be accomplished with a regard for the units in which the results are to be expressed. Unless a number is dimensionless, never write it down without giving it its nomenclature in the form of the SI units it represents.

Students are often confused about the units resulting from the multiplication or division of known quantities. This is because they concentrate on obtaining the numerical values without a parallel regard for the manipulation of the units. The techniques you must acquire are: to change from one unit to another, both being of the same type, and also to mix units of different types in calculation so that the final numerical result is obtained with its appropriate unit attached. Here are examples:

1. Multiply 7·2 MN/m² by 6 N/m² and divide by 10 kN/m². Here, the denominator is the same for all three, so the multiplication becomes (7·2 × 6)/10 and is parallel with (MN × N)/kN. Since the result must be in one unit, choose, say, kN. The unit then becomes

$$\frac{(10^3 \text{ kN}) \times (10^{-3} \text{ kN})}{\text{kN}}$$

which is kN. The unit is thus kN/m² and the numerical value is 4·32.

2. Multiply 6·8 MN/m² by 8 m and divide by 4 kN/m. Express the result in kN and m. The calculation goes in two parallel phases, one for the numerical values, and one for the units:

$$\frac{6\cdot 8 \times 8}{4} \frac{\text{MN/m}^2 \times \text{m}}{\text{kN/m}} = 13\cdot 6 \frac{(10^3 \text{ kN}) \times \text{m} \times \text{m}}{\text{m}^2 \times \text{kN}} = 13\,600$$

which is a dimensionless figure, all the units having cancelled out.

3. Multiply the weight of a mass of 12 kg by 7 m² and divide by 15 kN/mm², stating the result in N and mm.

$$\frac{12(9\cdot 81 \text{ N}) \times 7 \text{ m}^2}{15 \text{ kN/mm}^2} = 54\cdot 94 \frac{\text{N} \times \text{m}^2 \times \text{mm}^2}{\text{kN}}$$

$$= 54\cdot 94 \frac{\text{N} \times (10^3 \text{ mm})^2 \times \text{mm}^2}{(10^3 \text{ N})} = 54\,940 \text{ mm}^4$$

4. Divide 15 kN/mm² by 4 N/m and multiply by 6 MN/m², expressing the results in kN and m.

$$\frac{15 \times 6}{4} \frac{kN \times MN \times m}{N \times mm^2 \times m^2} = 22{\cdot}5 \frac{kN \times (10^3 \, kN) \times m}{(10^{-3} \, kN) \times (10^{-3} \, m)^2 \times m^2}$$

$$= 22{\cdot}5 \times 10^{12} \, kN/m^3$$

These examples are somewhat far-fetched, and you are not likely to meet anything as complicated. The method, however applies also to simple problems: replace any unit by its equivalent in the required units, and take note that the new bracketed unit must be raised to the same power as the original.

2.5 Keeping the number of digits small

One of the complaints made about the use of SI units is that the numbers are sometimes very large or very small. For example, the danger to accuracy of handling 0·000 030 75 MN/m² or 7654·321 kN m is clearly great. The method you should use is to include in the number the terms 10^3, 10^6, 10^9, 10^{-3}, 10^{-6}, etc. Multipliers of this kind should go in steps of 10^3 only. In addition, the three symbols m, k, M given earlier in this Unit, can be incorporated. You thus have numerous alternatives to choose from. For the two numbers given above, depending on the number of significant figures with which you are working, the choices available can keep the number to an acceptable 'shape'.

0·000 030 75 MN/m² could be 30·75 N/m², 30·75 × 10^{-3} kN/m², or even 30 750 × 10^{-9} MN/m².

7654·321 kN m could be 7·65 MN m, 7·654 × 10^3 kN m, 7654·3 × 10^3 N m or 7·7 × 10^6 N m.

The choice is made by considering the units in which the problem is being solved. Always define these before you start.

Symbols multiplied have a gap between.
Symbols divided are parted by a solidus.

2.5

TEST 2A

The answers to this test are printed a page or two further on. First *write down your answers* before checking. If you have made errors, revise the preceding units of study. Never move on to the next unit until you are thoroughly familiar with the work already covered.

Pencil in the missing words or symbols:

2A/1   Kilograms are a measure of the amount of _____ in a structure, and have the symbol _____.

2A/2   Newtons are quantities of _____ and have the symbol ____.

2A/3   The symbol mm represents one _____ of a _____ and is a measure of _____.

2A/4   The unit Mg represents one _____, and can be called a _____. It is a measure of _____.

2A/5   The symbol m^2 means _____ multiplied by _____ and represents an _____.

2A/6   There is a space between N and m because this _____ quantity represents the _____ of N and m.

2A/7   The quantity kg/m^3 means _____ _____ and is a measure of _____.

2A/8   The ratios of the four terms N/mm^2, N/m^2, kN/mm^2, MG/m^2 are _____, _____, _____, _____. You may assume the first one to be unity and give the relative values of the others.

2A/9   In 2A/8 some of the terms are not written in the best and most accepted style. Decide which are correct and which are not according to recommended practice. Explain why. _____
_____
_____.

2A/10   Lower case or small letters are used for the basic quantities of _____
_____
_____, and for fractional or multiple quantities
_____.

Capital letters are used for the quantity of _____ and for fractional or multiple quantities_____.

12

Part I: Group A

3 Forces, couples and moments

It can be assumed that you know intuitively what a force is. We all experience forces of various kinds – pushes, pulls, and weights. But if you are to solve the problems in this book, you must be able to deal with forces formally, exactly, and in order.

3.1 Surface tractions and body forces

There is only one variety of *mass* (kg) and only one kind of *length* (m), but there are two types of *force* (N). The first is a force applied by physical contact – the pull of a friend's hand as he helps you to your feet; the force you experience in a rugger tackle. Such forces can be called *external forces*, although, technically, they are known as *surface tractions* because they occur through the contact of the surfaces of two bodies. On the sketches accompanying numerical problems, an *external force* is indicated by a thin arrow shown in the direction of the force, and acting through the point at which contact is made between two surfaces.

The second type of force is applied without physical contact and is called a *body force*. This category of force acts on all parts of a body or structure simultaneously and is caused by an acceleration. The sinking feeling you experience when you are suddenly dropped or raised, or the unfamiliar sensations you purchase on a scenic railway are examples of *body forces* resulting from mechanically produced accelerations. In this book only one type of body force will be encountered, and only one triggering acceleration. This is the force produced on a mass by the acceleration due to gravity – the *weight* of the *mass*. If the acceleration applied by gravity is unresisted and allowed to act, the mass falls with increasing speed. If its effect is opposed and prevented, the latent acceleration produces a resistance called *weight*. This is the only body force used in this book. A body force is represented in the diagrams by a broad arrow, to distinguish it from a surface traction and to indicate that it does not need contact in order to act.

13

3.2

3.2 Relationship between mass and weight

For the problems of this book, a numerical relationship must be established between mass and force, when the force is a weight and the mass is known. The simple rule is that, very approximately, a kilogram of mass has a weight of 10 Newtons. The exact figure is 9·806 65, or 9·81.

For example:
Forces produced by gravity acting on various masses

Mass(kg)	Approximate force; Weight in Newtons or kilonewtons	More exact force: Weight in Newtons or kilonewtons
1·00 kg	10 N	9·8 N
5·66 kg	56 N	55·6 N
25·34 kg	253 N	248·5 N
315·70 kg	3160 N	3096·0 N or 3·1 kN
1·00 Mg or 1·00 tonne	10 kN	9·81 kN

Test 3A

The answers to this test are printed a page or two further on. First, write down your answers before checking.
If you have made errors revise the preceding units of study.

Pencil in the missing words or symbols:

3A/1 A surface traction is a _____. It is measured in _____ and given the symbol _____.

3A/2 There are _____ types of force. They are called _____ and _____.

3A/3 When a surface traction is applied to a body and is unresisted, the body _____.

3A/4 When a body has mass it is acted on by a _____ which is called its _____.

3A/5 When an unresisted body force is the weight of the body, the result is called _____.

Decide whether these statements are true or false (answer on a separate sheet):

3A/6   All body forces produce weight.
3A/7   A surface traction acts on all parts of the body.
3A/8   The only body force which can be found from the mass of a body by multiplying by 9·81 is its weight.
3A/9   There are six basic units in the SI system.
3A/10   If a body is not to move under the action of forces their effects must balance.

TEST 2A   The answers

2A/1   mass; kg
2A/2   force; N
2A/3   thousandth; metre; length
2A/4   million grams; tonne; mass
2A/5   metres; metres; area
2A/6   derived; product
2A/7   kilograms per cubic metre; density
2A/8   unity; 10^6; 10^3; unity
2A/9   First and third not strictly correct, but the first is considered acceptable. Second and fourth correct.
Only basic units and powers of basic units in the denominator.
2A/10   kilogram; metre
kilo (multiply by 10^3), and milli (divide by 10^3)
Newton
Mega (multiply by 10^6)

3.3   Directions and senses of forces

In order to provide a guiding framework for the solution of most of the problems in later UNITS, you must select each time two co-ordinate axes at right angles to each other in order to give a frame of reference for the definition of forces and other quantities. For the numerical solutions you must choose the guiding axes in the most appropriate direction for the conditions presented for solution. There is no need for the axes always to be *up-and-down* and *across* the paper, or *vertical* and *horizontal*. The only forces which are invariably vertical are the applied weights. Other forces may be at different angles and the most convenient guiding axes may be at some angle to the vertical and horizontal.

3.4

It is convenient, since the axes are chosen at right angles to each other, to use the convention of the bridge table and speak of *North–South* and *East–West directions*. However, for each *direction* there are two *senses*. The force may act in the *northerly sense of the N–S direction* or in the southerly sense. Similarly a force may act in the *westerly sense of the E–W direction*, and another in the easterly sense. Senses such as N and S, being directly opposed to each other can be given positive and negative signs. It is not important which of the senses is chosen as positive, but the signs adopted must be used consistently through the calculations. In this book you will use the following convention:

	Directions	
	N–S	*E–W*
Positive sense	North	East
Negative sense	South	West

3.4 Moments

The combination of force (N) and length (m) gives a rotating or turning effect. A force applied to the handle of a door, combined with the distance from the handle to the hinge causes the door to rotate and open. Similarly if you take a small card (sketch 3A) and pin it lightly to a board, a force applied by a match at some distance from the pin causes a turning of the card.

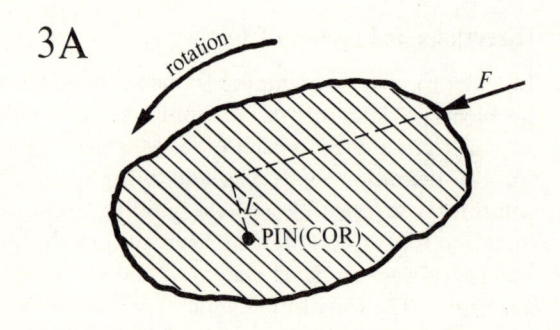

3A

3.4

The value or intensity of the turning effect is measured by multiplying together the force (N) and the distance (m) between the line of the force and the centre of rotation (COR). In the instance shown in the sketch, the rotational effect has the value *FL*. The numerical value is measured in Newton-metres, and given the symbol N m with a space between the two basic symbols to indicate multiplication. To avoid using the term 'rotational effect', use the technical term *moment*. There is no need to understand why this term is used. Use it as a shorthand description of a rotational effect caused by forces acting around centres of rotation (COR). The force and the distance concerned are multiplied.

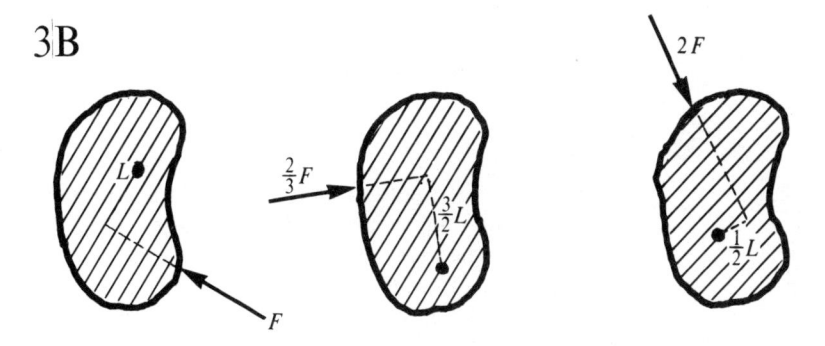

3B

The same value of the turning effect or *moment* is obtained with any combination of force and length which gives the same product. The three sketches (sketch 3B) all show the same numerical value of moment (*FL*), clockwise (or positive).

You must remember that a *centre of rotation* (COR) must be selected and defined. The moment of a force *about that centre* is then the product of the force and the shortest distance between the line of the force (produced if necessary) and the chosen centre of rotation. The COR used must always be identified and stated when moments are calculated.

Structures are acted on by more than one force, if they are to be maintained in equilibrium, for one force alone, causes movement. The various forces act simultaneously, all causing their own turning effects or moments about the same chosen centre of rotation (COR). These individual moments can be combined into one nett turning effect. Just as forces can act positively or negatively, so moments can cause rotations in one sense or the other. This time there is no North or South, East or West, but the sense is concerned with rotation, and is either *clockwise* (positive) or *anticlockwise* (negative). The individual moments

3.5

added, with due regard to *plus* and *minus*, gives the combined effect of all the forces together. A little chart or rubric must be sketched at the start of each solution to remind you of the conventions. It should be drawn parallel to the chosen N and E directions (sketch 3C).

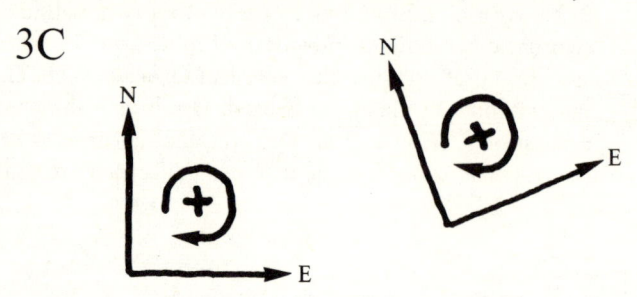

3C

3.5 Couples

When forces act simultaneously, there is one combination which often occurs, and which is worthy of attention as a special case. This arrangement occurs when two forces act on a structure and they are
(*a*) equal in magnitude
(*b*) parallel to each other
(*c*) opposed in sense.
This system is called a *couple* (sketch 3D).

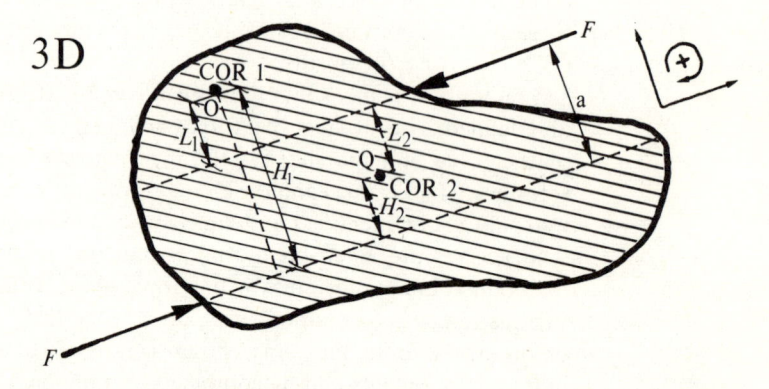

3D

The sketch shows a planar structure (piece of card) acted on by two forces which are equal, parallel and opposite in sense. This is a couple, and the *moment* or turning effect caused by this arrangement can be studied about the COR defined by the letter O.

3.5

Clockwise turning is positive

The moments of the two forces about the centre of rotation (COR) defined by the letter O, are
$$+FL_1 \text{ (clockwise)} - FH_1 \text{ (anticlockwise)}$$
This is equal to
$$+F(L_1 - H_1)$$
but $+L_1 - H_1 = -a$ so the moment is of a value Fa and anticlockwise. Similarly if you take another point Q as the centre of rotation, and measure its distance from each of the forces, you can obtain the combined moment of the forces about this point.
$$-FL_2 - FH_2 = -F(L_2 + H_2)$$
$$= -Fa \text{ once more}$$

The rule emerging from this is that, if the forces take up the special condition of a *couple*, the same moment is applied to the structure wherever on the structure the couple is situated. The value of the *moment of a couple* is calculated as the product of *one* of the forces and the distance between them. So long as the couple is acting, this moment is applied to all points on the structure and can be added to the moments of the forces which have been taken as acting about particular centres of rotation (COR). *There is thus no need to define a centre of rotation for a couple.*

TEST 3B

The answers to this test are printed a page or two further on. First, write down your answers before checking. If you make errors, revise the preceding units of study.

Pencil in the missing words or symbols:

3B/1 A moment of a force is applied by a _____ acting at some distance from a _____ (_____).

3B/2 A moment measures a _____ _____ and is evaluated in _____ _____.

3B/3 To find the moment of a force requires knowledge of the position of the _____; to find the moment of a couple requires _____ _____.

3.5

3B/4 A force acting in the N–S direction is reckoned to be positive in effect if it has a _____.

3B/5 A moment is reckoned to have a positive turning effect if it acts _____.

State whether these statements are true or false (answer on a separate sheet):

3B/6 The centre of rotation for a couple must be located midway between the forces. True or false?

3B/7 A force acting at 2 m from the centre of rotation gives half the moment of the same force acting at 4 m from the same centre of rotation. True or false?

3B/8 A force has a positive moment of 20 N m about a given centre of rotation (COR). A couple, acting on the same planar structure has a negative moment of 40 N m. The centre of the couple is twice as far from the centre of rotation as is the force. The combined moment is, therefore, $+20 - 2 \times 40$ N m $= -60$ N m. True or false?

3B/9 If a couple of 5 N m is added to a moment of a force of 5 N m the result is 10 N m. True or false?

3B/10 Forces which are parallel and of opposite senses form a couple, and the word 'couple' means 'moment of a couple'. True or false?

Learn and remember

(a) Surface tractions are forces acting on the surface of a structure by contact with something else.

(b) Body forces are caused by accelerations and act on all parts of the structure simultaneously.

(c) The body force used in this book is weight. The weight in Newtons of a mass is obtained by multiplying the mass in kg by 9·81, to obtain a value in N.

(d) There is a difference between *direction* and *sense*. Two forces in the same direction may have different senses (positive and negative).

(e) The *sense* of moments is clockwise (positive) and anticlockwise (negative). Moments have nothing equivalent to the *direction* of a force. They are not directional.

(f) To find the *moment of a force* the Centre of Rotation (COR) must be defined.

3.5

(g) To find the *moment of a couple*, no COR is required. The moment of a couple is the same at any part of the structure, and its effect is the same wherever it is applied.

(h) To add the moments of forces they must all have the same COR. The moments of any applied couples can then be added, but these couples need not have a centre of rotation.

(i) When writing the values of forces or moments *always* precede them with the sign which indicates their sense.

(k) The word 'couple' describes a physical arrangement of forces. The turning or rotational effect of this arrangement must always be called the 'moment of the couple' and not abbreviated to 'couple'.

**A line for DIRECTION:
An arrow for SENSE.**

Sketches are your best friends.

Define the S.I. UNITS in which you are to work and write everything in these units.

GROUP B
The Equilibrium of Rigid Structures

Part I: Group B

4 Limitations of study for rigid structures

This is the first unit of study for rigid structures. To be successful in the solution of problems in this field, you must have a clear appreciation of the conditions controlling the work. In the next six units of study, the problems are all based on the limitations defined below. You must be conscious of these limitations, and remember that the routines used in the solutions are not applicable outside the range defined.

4.1 First limitation:

The structures studied are rigid

The body (or structure) – the terms are synonymous here – under the action of loading must change in shape. However, when the loading is easily supported and is not close to the limit at which collapse takes place, the changes in shape are very small. If account were to be taken of this tiny alteration in the shape of the structure, at this stage, an unnecessary complication would be introduced. Thus, you may assume that the deformation of the structures are so small that it makes no difference to the loading conditions. This is equivalent to assuming that the structures are rigid and continue to be of the same form, whatever loads are applied. Thus, for the purpose of studying the system of forces acting on a structure, the structure may be looked upon merely as a rigid geometrical shape which locates and supports the various forces concerned. Since it is rigid, the internal construction of the structure is of no significance; only the forces externally acting on it are relevant to the solution of the problem.

4.2 Second limitation:

All parts of the structure lie in one plane

Structures which depend on their third dimension for their resistance to applied loads are called *space structures*, and lie outside the present

4.4

field of study. Domes, pylons, shell roofs, television masts, aircraft fuselages, and other space structures are not considered.

Thus, all the rigid structures studied in PARTS I and II may be sketched on a sheet of paper; they are flat. Such structures are known as *planar structures*. You may think this is somewhat far-fetched, for there are few structures which are truly flat and in one plane. However, many full-scale structures consist of a number of planar structures connected together. A roof truss, or a beam supported on columns, lies sufficiently in one plane to fall into the *planar* category. A number of such planar structures erected parallel to each other and connected together, form the type of building with which you are, no doubt, familiar. Cylinders and prisms can also be imagined as a connected series of slices, each of which can be studied as a planar body or structure.

4.3 Third limitation:

The loading on each body is co-planar with the body (or structure)

It would be illogical to simplify the problems by assuming structures to lie in one plane if the loading applied to them were to act from the third dimension. Thus, the loading applied to the structure being studied is assumed to lie in the plane which contains the outline of the rigid body. This is known as *co-planar* loading.

4.4 Fourth limitation:

The solution required to any problem is a full description of the forces acting externally on the rigid structure

This is a limitation not of conditions, but of requirements. Students are often confused in answering problems set in class or in examination papers because they are not certain of what is required. Once you have *defined all the forces and couples* acting on the rigid structure, you have done all that is required. Usually you will be presented with details of some of the forces and loadings and you will have to find the others.

Learn and remember

(*a*) The structures or bodies (the terms mean the same) you work on are rigid.

(*b*) The internal construction of the structures does not concern you: how they became rigid does not matter.

4.4

(c) Since the internal construction does not matter, any forces you think of as acting internally do not come into the picture.

(d) Only forces acting externally on the rigid structure are your concern.

(e) Your task is to take the values and directions of the forces given you, and the geometrical conditions presented by the shape of the rigid structure, and find the unknown forces. Everything is in the same plane.

(f) These known and unknown forces hold the rigid structure in equilibrium; it does not move. You start every problem with this condition held firmly in mind.

Test 3A The answers

- 3A/1   force; Newtons; N
- 3A/2   two; surface tractions; body forces
- 3A/3   moves
- 3A/4   force; weight
- 3A/5   falling
- 3A/6   False: only gravity acting on a mass produces the force called weight.
- 3A/7   False: it acts only where the contact is made. A body force acts on all parts of the body.
- 3A/8   True: other body forces are caused by other accelerations and are not considered in this book.
- 3A/9   True: four of them are not used in this book.
- 3A/10   True: this is the subject of a later Unit of Study.

KNOW each Unit before going on.

Part I: Group B

5 The achievement of equilibrium

The paramount and over-ruling condition applying to all the structures in these chapters is that they are postulated as being held in equilibrium while acted upon externally by surface tractions and internally by their own weight. The forces acting on the structure must be such as to keep it motionless. You start every problem with the assumption that the structure is held stationary, and then determine the conditions which cause this assumption to be true. This procedure results in the determination of the values of all the forces and couples acting on the structure. This is your sole aim; once you have made these definitions and decided on the magnitude, direction, and sense of each force or couple, your task is accomplished.

5.1 Conditions for balanced loading

Once again, as in Unit 3, cut a piece of card to an irregular shape to represent a rigid planar body or structure. Lay it on a board. Hold a match in each hand and get a friend to do the same. Apply all four matches to the edges of the card, and exert pushes in the plane of the board. The nett effect of these pushes can cause the card to move across the board. If your pushes are adjusted so that the card remains stationary even while the matches are exerting their surface tractions, the forces have achieved a complete balance. This represents the condition of equilibrium which you study in all the problems.

In Unit 4 it was agreed that all forces acting in one direction would be called the *North–South* forces, while those acting at right angles are the *East–West* forces. Remember the difference between 'direction' and 'sense' and revise Unit 4.

Since an unopposed or unbalanced force – one match only on the card – causes the structure to move in the direction of that force, the structure remains stationary in a N–S direction only if the N–S forces are balanced and add up to a nett force of zero. The total effect of the forces in the *North sense* (+) must oppose, exactly, the effect of the

5.1

forces in the *South sense* (−). Similarly, if the structure is to remain stationary in the E–W direction, all the forces in the *East sense* (+) must counteract all those in the *West sense* (−), the total effect being zero.

The two conditions described are known as the
Force-conditions of Equilibrium
They can be stated in words:

> *Condition 1*: If a structure remains in equilibrium, the nett value of the forces acting on the structure in the *North–South* direction must be zero.

> *Condition 2*: If a structure remains in equilibrium, the nett value of the forces acting on the structure in the *East–West* direction, must be zero.

If these *Conditions of Equilibrium* are fulfilled, the structure does not move either in the East–West or in the North–South directions. If this is so, it cannot move in any direction. If you ensure that the forces are balanced in the two co-ordinate directions, there can be no *translation* – movement of the structure as a whole, each part moving parallel to all the other parts.

Test 3B   The answers

- 3B/1   force; centre of rotation (COR).
- 3B/2   turning effect or rotational effect; Newton-metres.
- 3B/3   centre of rotation; no centre of rotation, only the value of the distance between the forces.
- 3B/4   northerly sense.
- 3B/5   clockwise.
- 3B/6   False: there is no centre of rotation for a couple. The moment of a couple is the same at all parts of the structure.
- 3B/7   True: the value of a moment of a force is the value of the force multiplied by the moment arm from the line of the force to the COR.
- 3B/8   False: the distance of the couple from the centre of rotation is irrelevant. The combined moment is 20 − 40, or −20 N m about the chosen centre of rotation. The moment of the couple is merely added or subtracted as it is.
- 3B/9   False: a couple is an arrangement of forces and cannot be added to a moment. If the *moment of the couple* is 5 N m it can be added to

27

5.2

the moment of a force about a chosen COR and the final result will be 10 N m. This is not 'quibbling'! A couple is not a moment.

3B/10 False: they must also be of equal numerical value (in Newtons) before the arrangement can be called a couple. The word 'couple' describes a physical arrangement of forces and is NOT the same as 'moment of a couple' which is a 'rotational' effect.

5.2 Equilibrium of a structure under two or under three forces

All the structures considered in these pages are in equilibrium – they are motionless. Put each of your forefingers on a match lying on the board. Bring the matches together to touch the edges of the rigid structure represented by the card. Keep your fingers at the ends of the matches. Adjust the matches as you push so that the card does not move.

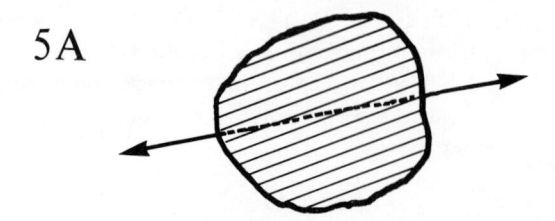

5A

You find that the two matches point towards each other and are in one straight line. This applies also if the forces pull (sketch 5A).

Ask a friend to bring another match to bear on the edge of the card so that there are now three matches pushing the edges of the rigid body.

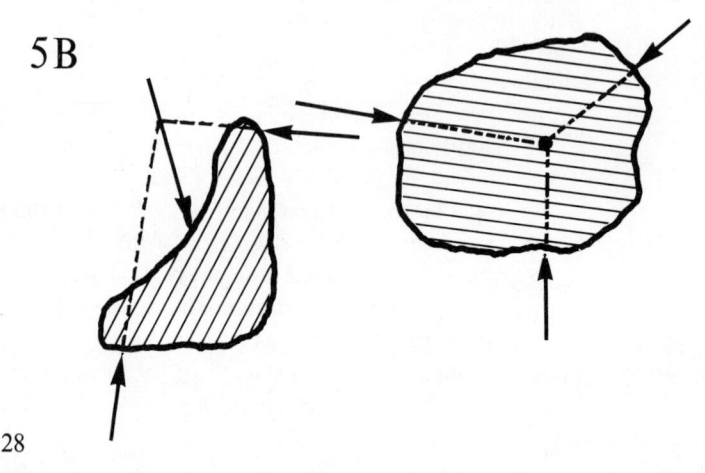

5B

5.2

Adjust the pushes of the three matches so that the card remains in equilibrium. If you now let go carefully, so that you do not disturb the matches, you find that they all point so that their lines of action pass through one point (sketch 5B). The point through which the lines of force pass need not necessarily lie within the area of the planar body.

Test 5A

The answers to this test are printed a page or two further on. First write down your answers; then check. If you have made errors, revise this Unit of Study. Never move to the next until you are thoroughly familiar with the work already covered.

Decide which of these statements are true and which false:

5A/1   When two forces act on a rigid structure they keep it in equilibrium.

5A/2   When two forces act on a structure and keep it in equilibrium, they may point in any direction.

5A/3   When three forces keep a planar body in equilibrium they sometimes pass through one point.

5A/4   When three forces act on a body and keep it in equilibrium the application of a fourth force will not alter the conditions.

5A/5   If a body is in equilibrium under the action of two forces there is no translation.

5A/6   If a body is in equilibrium under the action of three forces there may be translation in the direction of the greatest force.

5A/7   When two forces have the same value in Newtons and act in the same line the body is in equilibrium.

5A/8   When two forces act in opposite senses along the same direction, the body is always in equilibrium.

5A/9   When three forces acting on a body pass through one point, the body is in equilibrium.

5A/10 One of the *conditions of equilibrium*, studied above, must be true if the body is to be in equilibrium.

5A/11 When three forces act on a body and keep it in equilibrium they all pass through one point.

5A/12 If a body is to be in equilibrium the forces in the North direction must be zero, and in the East direction also zero.

5.3

5.3 Equilibrium under more than three forces

The two cases in the last section – equilibrium under two forces and under three forces – are special cases, for, generally many more forces than three are acting on a structure. The general case of four or more forces brings in another aspect of equilibrium. Go back to your card lying on the board, and apply equal forces through the two matches, one in each hand. Unless the directions of the forces are very carefully adjusted to comply with the conditions in section 5.2 (sketch 5A), there will be no equilibrium, but the card will move. The forces are either

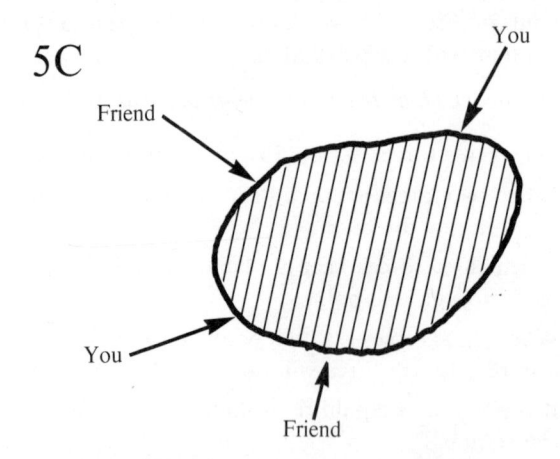

5C

[Sketch 5C: an irregular shaded shape with four arrows labelled "You", "Friend", "You", "Friend" pushing inward from different directions]

acting to apply moments about any chosen centre of rotation (sketch 5C) or they are in the special state of being a couple (sketch 5D). In either event, there is a rotation of the structure.

Assume that the two forces act as in LEARN AND REMEMBER (*a*) (see end of Unit) except that condition (ii) is not fulfilled, and the forces are not in one straight line. The force conditions of equilibrium are thus complied with, for the two forces are equal in magnitude, opposite in sense, and act along the same direction. They are not, however, in one line, and thus form a couple which causes a pure rotation. The compliance with the force-condition of equilibrium means that there is no translation as there would be in sketch 5C.

Now get a friend to apply two equal parallel forces to the condition in sketch 5D, so that four are now acting (sketch 5E). They need not be of the same magnitude – you could never be sure that they were, and it is not important. But the two moments exerted by these couples,

5.3

which act in opposite senses, should be the same. If they are, then the card remains stationary; equilibrium has been established once again. Sketch 5E shows merely one of an infinite number of possible arrangements of balanced couples.

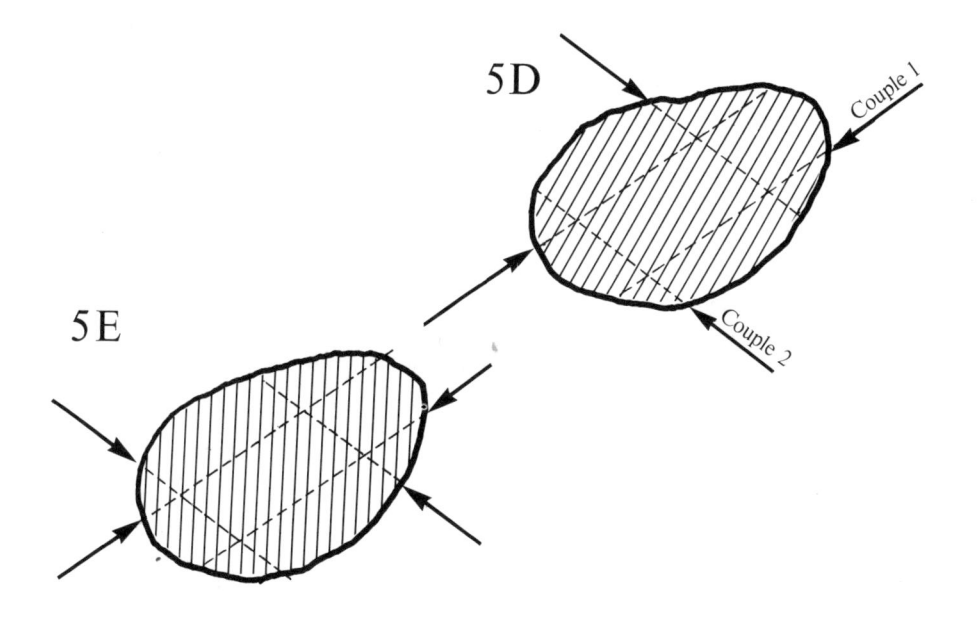

Thus, to achieve equilibrium in a planar structure under the action of more than three forces, it is not enough to say that the N–S and the E–W forces each add up to zero. They may balance numerically, but they may also be applying a moment which is unbalanced and could cause movement. The moment is applied either by forces acting round a centre of rotation, or by couples acting anywhere on the structure and having no reference to any centre of rotation.

A third *Condition of Equilibrium* is thus required. It can be stated in words as:

Condition 3: If a structure remains in equilibrium, the nett value of the moments applied by forces acting round any centre of rotation, together with the moments of any applied couples, must be zero.

5.4

5.4 The effect of applied couples on the Conditions of Equilibrium

The two forces forming a couple are equal in *magnitude* and opposite in *sense*, and act in the same *direction*. They thus balance exactly in their translational effect and fulfill one of the force-conditions of equilibrium. Because they are not in the same line, however, they apply a *moment of a couple* to the structure – a purely rotational effect.

Thus, you need not be particularly interested in the exact formation of the couple – what value are the forces, or how far apart they are – all you need to know is the effect the couple produces, which is a moment. Thus, on sketches showing systems of forces and couples acting on a structure, couples are indicated by a curved arrow, and alongside is written the *moment of the couple*. Remember not to confuse *couple*, which is a physical arrangement of forces, and *moment of a couple* which is the rotational effect of forces in the configuration of a couple. The curved arrow then, indicates the sense of the couple, and the figure alongside indicates the moment of the couple in Newton-metres.

Test 5B

The answers to this test are printed a page or two further on. First write down your answers; then check. If you have made errors, review this Unit of Study, especially the LEARN AND REMEMBER entries.

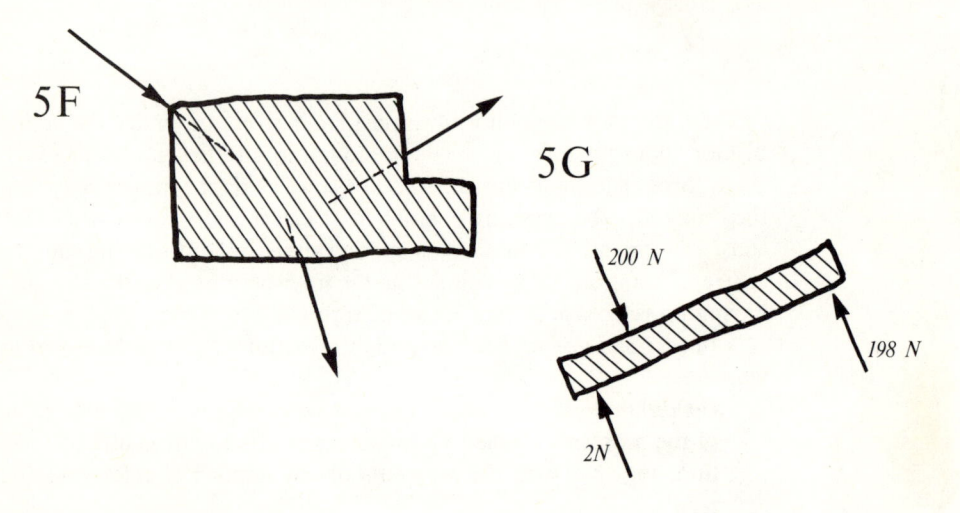

5F

5G

5.5

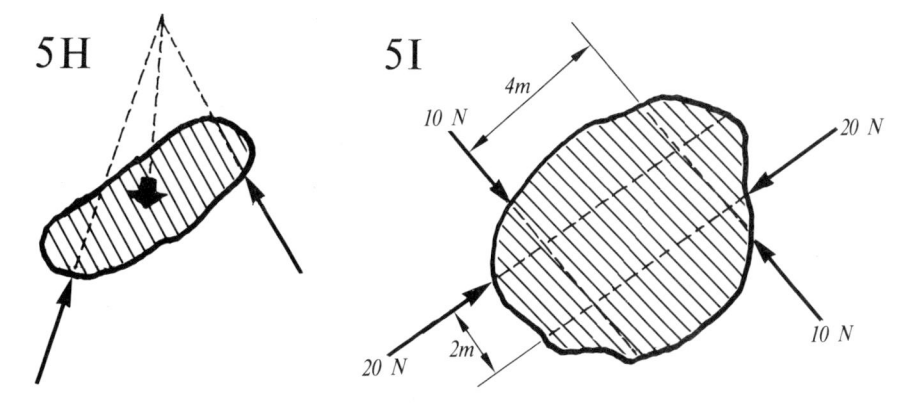

5H

5I

5B/1 Examine carefully, the sketches 5F, 5G, 5H and 5I and put against each
5B/2 a letter indicating one of the following states:
5B/3   E means that the structure is undoubtedly in equilibrium.
5B/4   C means that the structure could be in equilibrium, but not enough information has been given.
        N means that the structure cannot possibly be in equilibrium in the state shown, regardless of any further information.

5.5 Expression of the three Conditions of Equilibrium

The *Conditions of Equilibrium* form the basis of all the solutions of problems in rigid structures – so far as this book goes. They must be constantly kept in mind and in view, but it would be tedious to write out each one time and again. A symbolic method of expression is used to shorten the work, and these symbolic statements of the *Conditions* must be written down at the start of any calculation. Much of the difficulty encountered by students in applying theory to problems is that they are not clear in their minds what their detailed intentions are. The symbolic conditions make these intentions clear.

$\sum$ (sigma) means 'the algebraic sum of' (i.e., taking account of + and − signs)

F_{NS} means 'forces in the North–South direction'

F_{EW} means 'forces in the East–West direction'

5.5

M_O means 'moments of both forces and couples' (the letter O, or any other chosen, indicates the position of the centre of rotation about which the moments of forces are measured. The moments of couples are independent of a centre of rotation)

(A) *The algebraic sum of all the forces in the North–South direction is zero*

$$\sum F_{NS} = 0$$

(B) *The algebraic sum of all the forces in the East–West direction is zero*

$$\sum F_{EW} = 0$$

(C) *The algebraic sum of the moments of all forces and couples is zero*

$$\sum M_O = 0$$

All these three conditions must be fulfilled for a planar body under co-planar loading if it is to remain in equilibrium. Since you start every problem with the assumption that the body or structure *is* in equilibrium, these three equations must be true when, instead of symbols, the applied forces and couples are fed into the equations.

5.5

Learn and remember

(*a*) If *two* forces act on a structure and keep it in equilibrium, they must
  (*i*) comply with one of the force conditions of equilibrium (equal values and opposite senses)
  (*ii*) act along the same line.

(*b*) If *three* forces act on a structure and keep it in equilibrium, they must
  (*i*) comply with two conditions of equilibrium – the force-conditions
  (*ii*) act through the same point, which need not be within the structure.

(*c*) If *more than three* forces act on a structure and keep it in equilibrium they must
  (*i*) comply with all three conditions of equilibrium.

(*d*) If a system of loading includes any applied couples, all three conditions of equilibrium must be fulfilled if equilibrium is to be maintained.

(*e*) Commit all *three* Conditions of Equilibrium to memory, (especially in their symbolic form), for the whole of the succeeding study is built on these statements.

Define the NORTH and draw the axis-rubric (Sketch 3C)

Moments do not appear in the FORCE equations.

Read the INDEX and use it.

35

Part I: Group B

6 Free structures and their supports

In each problem you tackle, the *structure* to which you must devote your attention (and which is assumed to be in equilibrium) must be precisely defined. Unless you are quite clear in your mind to what structure the Conditions of Equilibrium apply, you will find difficulty.

6.1 Defining the FREE STRUCTURE

The term *free structure*, which has been given capitals in the title of this Section, is a very important one. It means exactly what it says; the structure you are considering must be entirely free of everything except the forces and couples acting on it externally. It must be cut off from all contact with other structures, with the ground, or with its surroundings. It is floating in a planar space, maintained in equilibrium by the applied forces and couples.

To obtain this condition, the structure you have chosen must be cut off from everything except itself. In the beginning, you are advised to carry out this cutting-off physically by drawing, on scrap paper, the arrangement of physical conditions given in the problem, and then cutting off a portion of this arrangement with a pair of scissors. The portion cut off is a 'structure' in equilibrium, since the whole physical arrangement was in equilibrium. Later you will achieve the same end by sketching a *free-structure diagram*.

In defining the *free structure* it makes the problem seem much simpler if you remember that all the structures you deal with are like the cards you used in earlier units. They are flat, and the internal arrangement – or how they are constructed – does not have any significance. The structure, when it is freed from its surroundings, can be shaded to show that internal structure is not your concern, and, as a corollary, internal forces are also to be neglected; only external forces and couples come into the picture.

Sketches 6A show, on the left, the physical arrangements presented by the problems and, on the right, the *free structures* which can be made

6.2

by taking a pair of scissors and cutting away pieces of the arrangement. Note that the supports are also cut away and vanish completely. The free structure must be really free and float in a planar space.

6.2 Loading the Free Structure

Since your aim (Unit 2) is to define for each structure every force and couple acting externally on it, the free structures shown in sketches 6A are incomplete. These sketches form the geometrical shape on which the various forces act, but it is these forces in which you are interested, so they must be added to the free structure, and added in such a way that the structure remains stationary.

In all the problems presented to you, some forces and couples are already defined, and you are expected to define the rest. You can calculate only three unknown items, for there are only three equations defining the conditions of equilibrium, and thus only three values can be determined.

The first step is to sketch the *free-structure diagram*. This consists of the sketches on the right of sketches 6A, together with the external

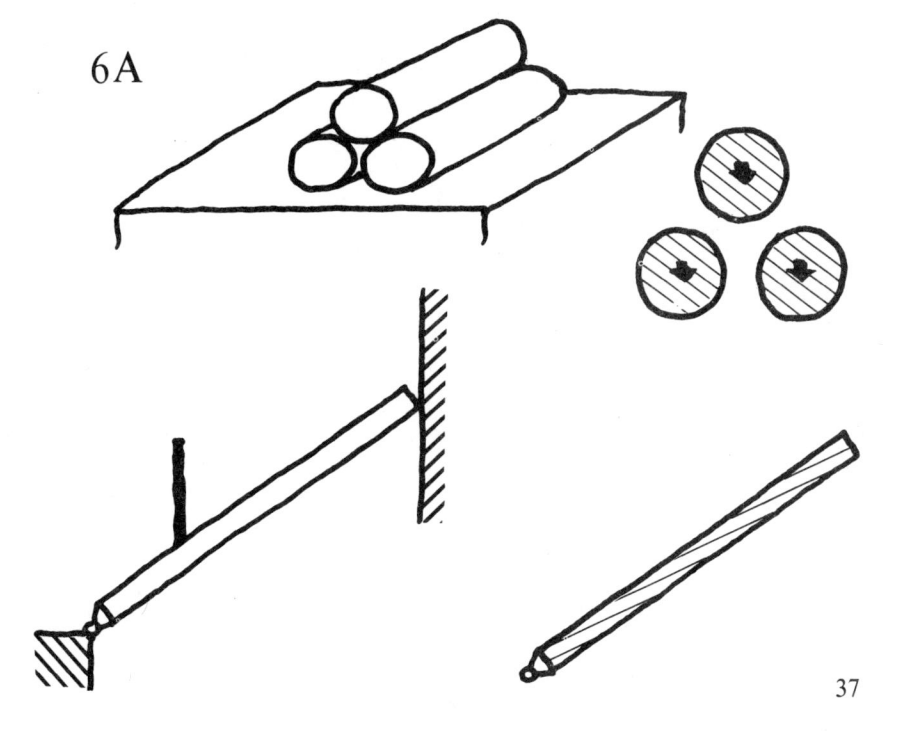

6A

37

6.3

forces and body forces (in this book only *weight*). This *free-structure diagram* is the basis of the routines of solution, and unless you sketch it correctly, you will be likely to make an error. Sketches 6B show some physical arrangements and the corresponding *free-structure diagrams* which can be built up from them by cutting off portions.

The effect of an adjoining part of a structure in supporting the part cut off is shown by adding *external forces* and *couples*. How these are determined is explained later; in the meantime merely realize that one portion of a structure cannot be discarded unless its effect is shown by forces and couples. If you were leaning against a fence which collapsed, you would start moving unless a friend applied an external force by seizing your elbow and holding you up.

6.3 Types of support

You will see, by examining sketches 6A and 6B, that when *free-structure diagrams* are formed, the loading shown in the original physical arrangement is merely repeated, but the supports are cut away and replaced by forces which were not shown on the general arrangement. The supports are translated into forces and couples, so that a total force-system can be applied to the free structure.

In the problems encountered in Statics and Structures there are five different, although related types of support. You must be able to recognize these and to translate them into forces and couples acting on the structure at the former cut-away point of support. These types of support can be designated as:

Simple, Hinged, Direct, Frictional, and Held

The Simple Support

The word *simple*, here, is used as a technical term to indicate that the structure is merely laid on its support, as on a sliding or roller bearing. Sometimes – for large bridges, for example – complete rollers are specially fabricated and the ends of the bridge laid on them, so that the designers are certain that a *simple support* has been provided. For lighter structures, rollers may be replaced by greased plates, one resting on the other. The use of durable plastics has also been developed for this type of support. For small structures you may assume that if one portion is merely laid on another, without fastening, you have a *simple support*.

6.3

The characteristic of a *simple support* is that it moves by sliding if a force is applied in any direction except at right angles to the surfaces in contact. The direction of the force acting on all *simple supports* can be visualized if you think of standing on a pair of roller skates. If you

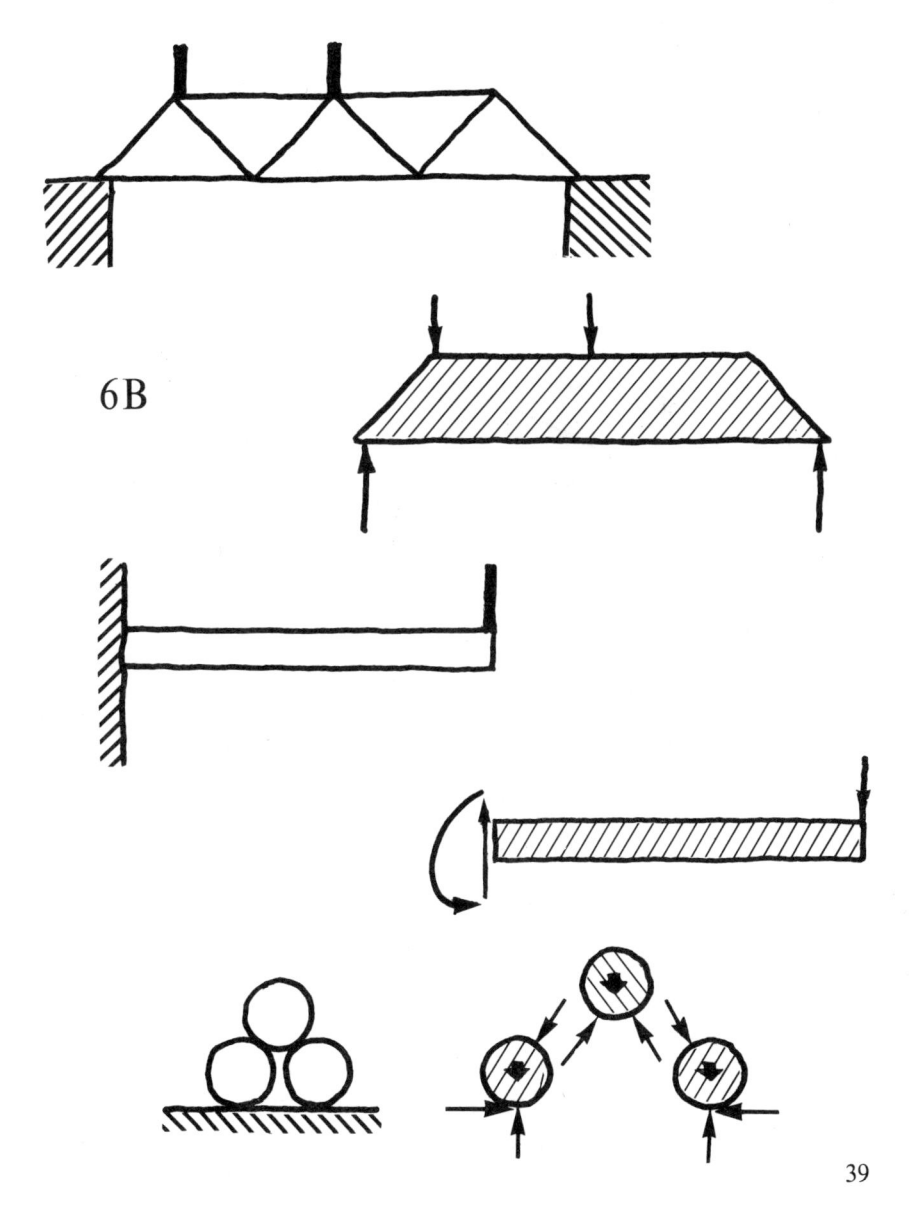

6B

6.3

lean backward or forward, the rollers move and equilibrium is destroyed! If you wish to stand motionless (in equilibrium) all the load from your body (body force) must act at right angles to the floor. The floor is pushing up to support you through the stationary rollers. A *simple support*, therefore, when removed in the formation of the FSD, can be replaced by one force *at right angles to the surfaces in contact*. This force is known as a *normal force* – another important term, meaning *at right angles to the surface*. Note that the supporting force need not be vertical; inclined supports can be *simple* if they are on rollers or greased plates, or have similar non-fastened conditions. Study sketches 6C carefully.

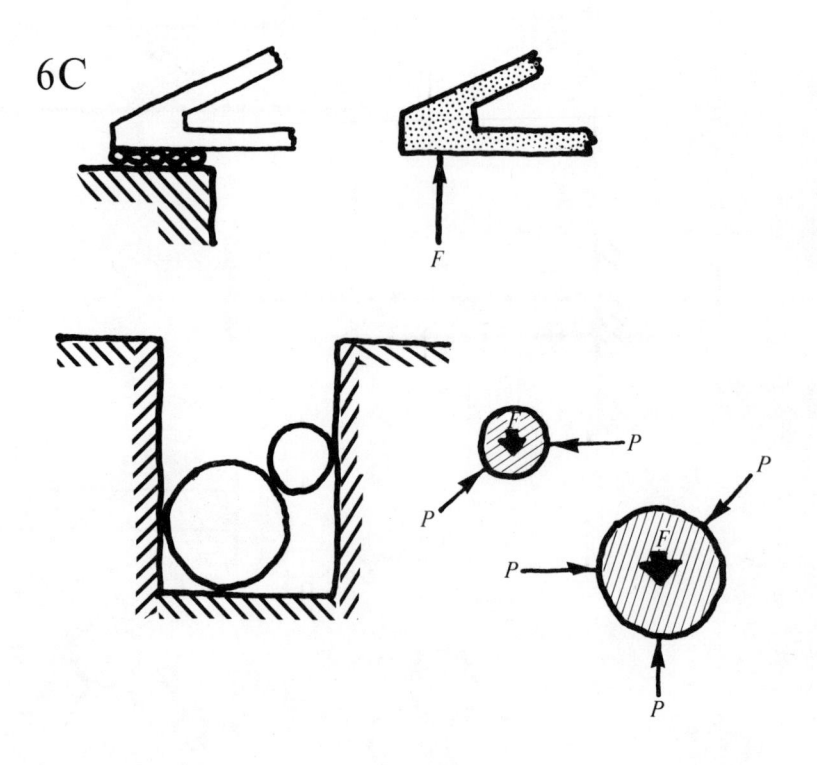

The Simple Support

6.3

The Hinged Support

This type of support (sometimes called *pinned*) is used to attach a structure to a solid base or to hinge one portion of a structure to another. The hinge can be specially formed of metal, like the hinge of a door, or it may merely be a portion of the structure deliberately made weak. A hinge in a reinforced-concrete structure or the rivetted joint in a light roof truss are examples of the weak-spot hinge.

The important characteristic of a *hinged support* parallels that of a *simple support*. In a simple support any inclined force causes motion, and, since you know that there is no motion in a structure in equilibrium, there can be no inclined force acting on a simple support. In a *hinged support* the application of a *couple*, or of a *force producing a moment* about the hinge causes motion. Since you know that there can be no motion in a structure in equilibrium, no moment (whether produced by a force or a couple) can be in action at a hinged support. Indeed, hinges are designed for the sole purpose of producing motion when moments are applied, or to be known points of zero moment when equilibrium is established. The hinge of a door has the first of these functions, but a moving door is not in equilibrium and does not come into the ambit of your studies here. In structures, the hinged support is in equilibrium like the rest of the structure, and is inserted where it is convenient or important to provide a simplified and precisely defined type of supporting condition (sketch 6D). The single force representing the reaction at the hinge usually has an unknown direction, and it is usually split into two 'components' in the co-ordinate directions.

6D

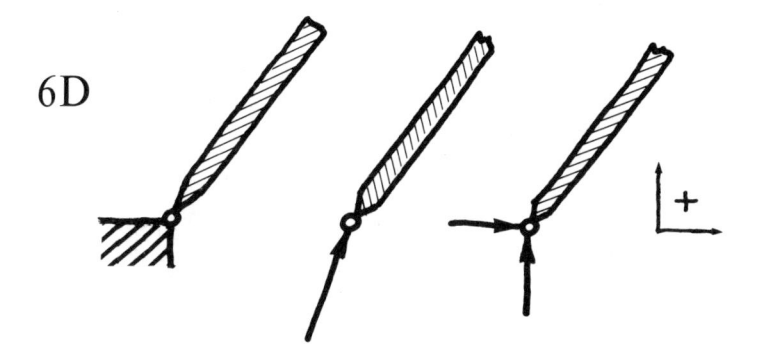

The Hinged Support

6.3

The Direct Support

This is a variation of the hinged support when a bar (straight or curved) is provided, not only with one hinge, but with one at each end. To understand subjectively the action of such a double-hinged member of a structure, open a door and place your palm on its edge. Your hand, not gripping the door but merely laid on it, represents one hinge, and the door also has a mechanical hinge.

You will find that if you push at an angle even slightly inclined to the line of the door, the door moves, and your hand rotates like a hinge, while the door hinge also moves. By experiment, you will find that the only way to maintain equilibrium (essential in all structures) is to direct the line of push from one hinge to the other – down the line of the door. The door is put into compression, but the same result would be obtained if the door were pulled instead of pushed. The line of action of the push or pull, for a state of equilibrium, must be directly in a straight line from one hinge to the other.

This applies even if the member, or portion of structure, between the two hinges is curved. A good example of this is a strung bow, which is in equilibrium when not in use. The ends form hinges, the material is curved, but the tensile force maintaining equilibrium (the bowstring) acts directly from one hinge to the other.

Structures are often built in this way, with hinges (or weak-spot hinges) in order that the designer can be sure that the forces acting are pure tension or compression between pairs of hinges. This time, the direction of the unknown force at the hinge is defined by a straight line between the hinges.

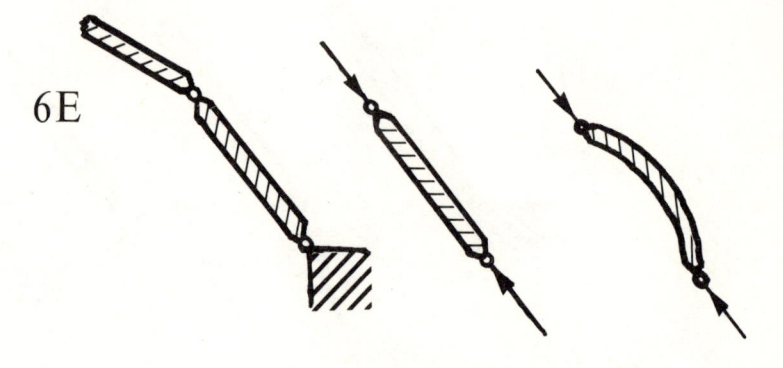

6E

The Direct Support

6.3

The Frictional Support

If the foot of a ladder were placed on a smooth sheet of glass or on ice, you would be very doubtful of the advisability of climbing. You know from experience that the foot of the ladder would be likely to move outward away from the wall against which it was leaning. If, however, the ground is rough, you know that the ladder will not be likely to slip. Why? Because the ground, through its roughness, exerts a force opposing the incipient motion. For a ladder, the force exerted by the ground is inward towards the wall, since the incipient motion is outward.

The forces acting where the support is frictional are conveniently represented by a *normal force* (at right angles to the surfaces in contact) and a *frictional force* opposing the incipient motion. In all problems dealing with structures where there are frictional supports, you assume that slipping is just about to take place, but has not yet started. This ensures that the *frictional force* is as large as possible, and also that you

6F

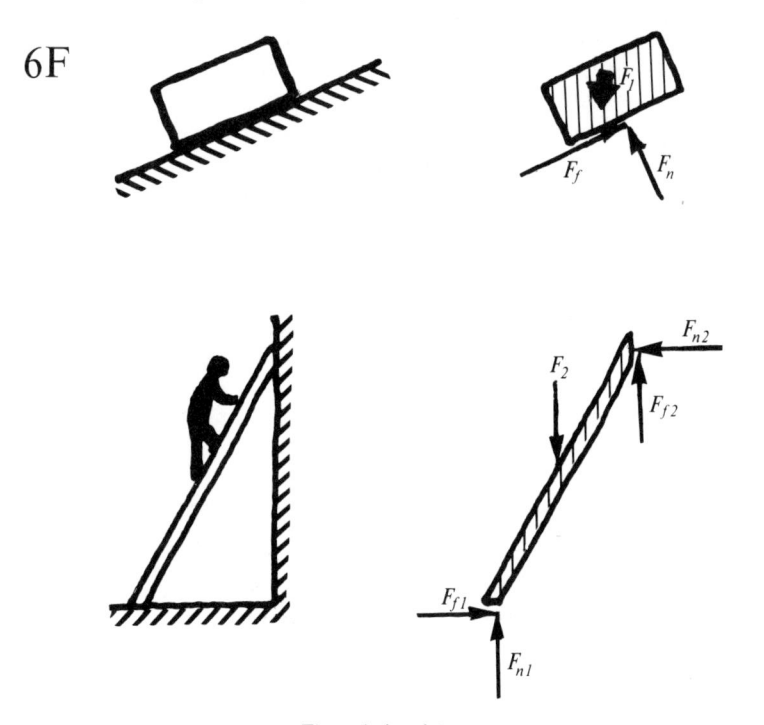

The Frictional Support

6.4

know its value in relation to the normal force. In sketches 6F the frictional force is given the symbol F_f and the normal force, F_n. Just as slipping is about to take place, these two have a known relationship, and thus, if one can be evaluated, the other is also defined.

The relationship between F_f and F_n is a figure known as the *limiting coefficient of friction*. Its value depends on the types of surfaces in contact. If the interface between the two surfaces is rough, the coefficient is large, if the contact is smooth, the coefficient is small.

Remember the ratio
$$\mu = F_f/F_n \quad \text{(pronounced 'mew')}$$

6.4 Relationships between the types of support

The fifth type of support – the *Held Support* – is discussed in the next Section because it is of a distinctive character and needs special attention. At this stage, therefore, it is well to review the four types already studied. They can all be represented by one force passing through the point of contact – the point at which the cut was made. It is true that *two* forces were defined above for the *frictional support*, but this was a matter of convenience for calculation and the solution of problems. Any force can be broken down into its effects in various directions – its *components*. The single force representing the other supports – simple, hinged, and direct – can also be broken into two directional effects or components. Note that when the ratio of the two component forces is found, or known in advance, the angle of the single force which they replace (their *resultant*) is defined by that ratio. This angle is not dependent on the magnitude of the force forming the support. You should now study carefully the sketches 6G which show supporting conditions and their components mutually at right angles in directions chosen at random. Later you will find how to determine the best directions to choose for N–S and E–W.

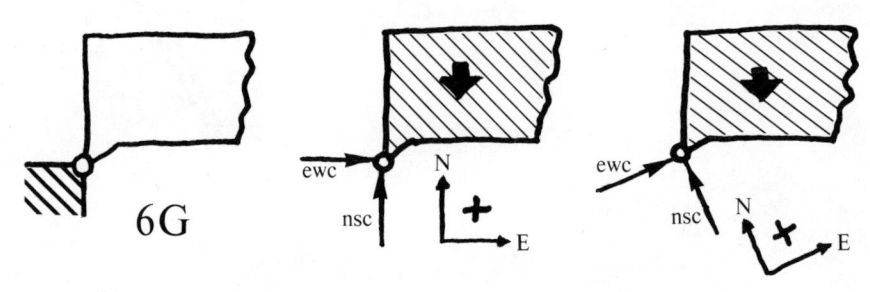

6G

The following table summarizes the findings made so far about the supporting conditions, and shows how they are related:

	SIMPLE	HINGED	DIRECT	FRICTIONAL
Direction of the single force	*Known* Normal to surfaces in contact	*Unknown* Passes through hinge but in an unknown direction	*Known* Passes through both hinges and is thus defined in direction	*Known* Acts at angle to normal, as defined by the limiting coefficient of friction
Resolution into two components N–S and E–W	*Not needed* Single normal force: no second component	*Necessary* Components in both directions unknown	*Not Needed* Direction of force is known	*Necessary* The ratio of F_f to F_n is known but not their values.

6.5 Couples as part of the supporting system

A couple always applies a pure rotational effect. Hold up a scale or a piece of stick or a pencil in one hand, with the thumb above, at the extreme end, and the forefinger a short distance along underneath. Now, press down on the free end of the pencil or rule with your other hand. You find that you have to exert a downward pressure with the thumb and an upward pressure with the forefinger if equilibrium is to be maintained. You are not only pressing upward with your forefinger to balance the pressure downwards of the other hand at the end of the pencil, but you are applying a moment by means of a couple formed of your finger and thumb.

If your forefinger is applying both a supporting force upwards, and one of the equal forces required in the couple, it ought to be greater than the pressure exerted by your thumb. This can readily be experienced if you press down really hard with the other hand. The difference between the larger upward force of the finger and the smaller downward force of the thumb is then readily apparent.

6.6

The Held Support

What is described above is one example of a *held support*. This type of support has within it one or two component forces as for the first four supporting conditions, but, in addition, it contains a couple which applies a moment to prevent rotation. In the first four conditions of support no attempt was made to prevent rotation, and thus no couple could act and still maintain equilibrium since a couple requires resistance if it is to act while maintaining equilibrium. No moment can act at the well-oiled hinge of a door, but if the hinge rusts solid, then a moment can be applied through the handle. No motion occurs, but a couple causing a resisting moment develops at the solid hinge. The door cannot be moved.

Thus, when a supporting condition, where there is a solid physical connection, is cut away it must be replaced by a force, or by two components of a force, and *also* by a couple. This condition appears if a structure is built into a wall, for example (a *cantilever*) or when the cut is made through the solid mass of a continuous body, like cutting through a log. One portion of a structure may be *held* by the adjacent one, and when a clean cut is made, the force and couple come into play. Sketches 6H show examples of *held supports* and of the forces and couples replacing them.

6.6 The senses of the supporting forces and couples

Students often try to imagine the senses of the forces and couples acting where the support has been moved. This can be very puzzling indeed, and is quite unnecessary. The more you can replace puzzlement by routine the faster and more accurate you will be. When putting in the forces and couples in the FSD as in sketches 6D and 6E, assume or guess senses for the forces and for the curved arrows of the couples. These indicate the *senses* of *directions* already determined. There is no need to know the 'proper' senses; merely as a matter of routine, put the arrows in either direction, at random. It is quite likely that some of the arrows in sketches 6D and 6E are wrongly directed, but so long as you work from the sense shown in the FSD, the final calculation will show clearly whether you guessed the sense correctly or not. So forget about trying to think out the senses of the supporting forces and couples. Put in the arrows at random.

6.6

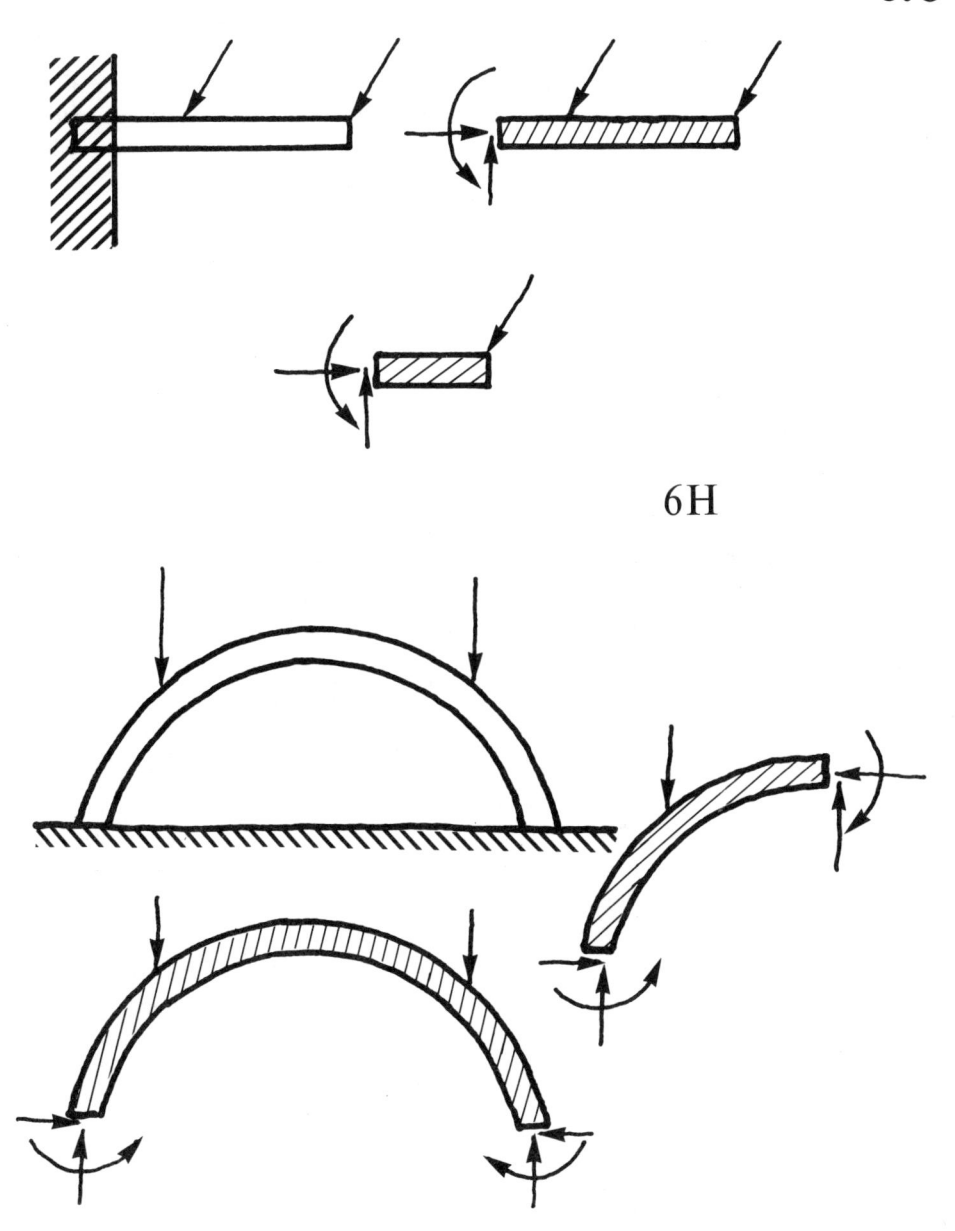

6H

The Held Support

6.6

Test 6A

The answers to this test are printed a few pages further on. First write down and sketch your answers and then check. If you have made errors review this unit of study. You cannot go further without a thorough knowledge of the free-structure diagram.

6A/1 A horizontal beam is in equilibrium and is laid on rollers at each end. There is a concentrated load at one-third of the way from one end. How many unknown forces are there if
(a) the concentrated load is known and is vertical?
(b) the concentrated load is known and is inclined at 60° to the horizontal?

6A/2 Wind blows on a tall building at right angles to its length. The building is supported on a row of columns on the windward side and another row on the leeward side. Draw the FSD.

6A/3 A solid cylinder lies on a smooth horizontal table. The cylinder does not roll. Draw the FSD for the cylinder. Draw a second FSD for the same cylinder and conditions if the table were rough.

6A/4 A rectangular stiff card has one corner on a smooth china plate. It is lightly held vertically and prevented from falling over sideways. A string tied to the opposite corner is pulled horizontally in the plane of the card. Draw the FSD for this condition, and for a rough tablecloth.

6A/5 A crane, in the form of an oblique triangle, is simply supported on two corners, the side being parallel to the ground, which is horizontal. At the other corner, forming the jib, the crane carries a weight. Draw the FSD for the crane and also for the weight being lifted just before it begins to move, and is still in equilibrium.

6A/6 A horizontal beam is simply supported, and carries two vertical concentrated loads, one at the middle and the other at a quarter of the span from one end. The beam is cut through at the other quarter-point, and is assumed to remain in equilibrium. Draw the FSD for each of the two parts.

6A/7 You are about to enjoy a spoonful of soup. Look at the spoon and your hand as it is poised above the plate. Draw the FSD for the spoon.

6A/8 You lean nonchalantly with one elbow on a rough mortared wall well built into a gravel soil. Your feet are on the gravel about a pace away from the wall. Draw two FSD's – one for you, and one for the wall.

6.6

Learn and remember

(a) Always define the structure to which the *Conditions of Equilibrium* apply.

(b) The sketch forming this definition is called the *Free-structure Diagram* (FSD, for short).

(c) The FSD consists of the geometrical outline of the structure. The internal construction is irrelevant.

(d) The FSD is formed by cutting off the appropriate part of the general arrangement presented, and replacing the supporting conditions by forces and couples according to one of the five types of support.

(e) The five types of support are *Simple, Hinged, Direct, Frictional,* and *Held*.

(f) The first four of these are each represented by one force. It is sometimes convenient to break this force into two components parallel to the chosen co-ordinate directions.

(g) The *held* support is represented by a force, also, but it has, in addition a *couple*. When the force is broken into two components, the held force is shown as two forces in the co-ordinate directions together with a couple acting round the cut portion.

(h) The senses of unknown forces are sometimes obvious, but sometimes unknown. There is no need to puzzle over this. Mark the arrows on the unknown forces and couples (curved arrow) at random. Later calculation decides whether the sense chosen is the correct one.

(i) NEVER start a problem without constructing the FSD, which must show ALL forces and couples (whether components are used or not), and each must carry an arrow showing the sense, known or guessed.

6.6

Test 5A The answers

5A/1 False: They must fulfil the conditions given in unit 5 before they bring equilibrium.

5A/2 False: They may point in any direction but cannot achieve equilibrium except as in 5A/1.

5A/3 False: They ALWAYS pass through one point if they keep the structure in equilibrium.

5A/4 False: The fourth force would be unbalanced and would cause movement.

5A/5 True: If any body is in equilibrium there is no 'translation' (which describes movement).

5A/6 False: If a body is in equilibrium there is no movement.

5A/7 False: They might have the same senses. For equilibrium they must have opposite senses.

5A/8 False: They must also be in the same line, otherwise they become a couple which causes rotation.

5A/9 False: The forces may pass through one point, but have such values or senses that the conditions of equilibrium are not satisfied. See the answer to equilibrium under three forces in 5A/11.

5A/10 False: All of the conditions must be complied with.

5A/11 True: The significant phrase here is 'and keep it in equilibrium'. There can be conditions when three forces acting through one point cause motion.

5A/12 False: First, there is no 'North direction'. The word 'north' refers to a *sense* in the *North–South direction*. Similarly, there is an East *sense* in the *East–West direction*. The true statement about the force conditions of equilibrium is given above, in unit 5.

Remember the limitations: keep inside them.

Part I: Group B

7 The geometry of the structure

The stationary structures which you study, form geometrical shapes delineating locations, directions and senses of the forces and couples acting. It is important, therefore, that you are skilled in the geometry of simple shapes. For complex shapes, the superposition of several simple triangles or quadrilaterals may give all the facts required. For still more complex arrangements, it may be necessary to draw the structure to scale and measure off the lengths and areas required. This can be quite accurate if done well, but it is time-consuming and requires drawing equipment. In this book, only simple shapes are used, and the lengths and areas calculated from the geometry.

7.1 The right-angled triangle

The lengths of the sides of triangles are often required for calculation, and can be obtained by comparing similar triangles. Normally, one of the angles of a triangle, in problems dealing with structures, is a right-angle. This occurs because one of the lengths frequently required is measured from a *centre of rotation* (COR) to the line of action of a force. The line defining this length is at right angles to the force.

It is well, therefore to be familiar with the geometry of right-angled triangles. Work, as far as possible from the lengths and slopes of the

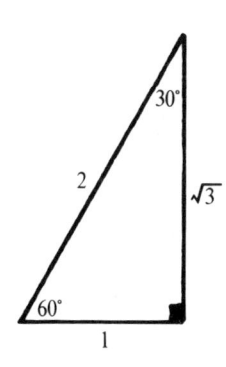

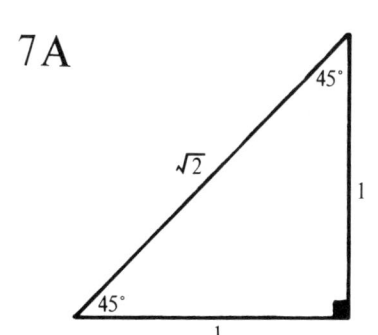

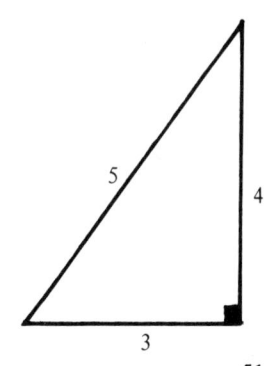

7A

7.1

sides; only as a last resort should you use books of trigonometrical tables. Some special cases of right-angled triangles should be committed to memory. The most useful are shown in sketch 7A with the proportions of the lengths of their sides. Remember these, for they appear frequently in problems.

Often, the appropriate right-angled triangles must be selected from a network of lines forming the free-structure diagram. When these triangles are isolated, they often form such comparisons as are shown in sketch 7B.

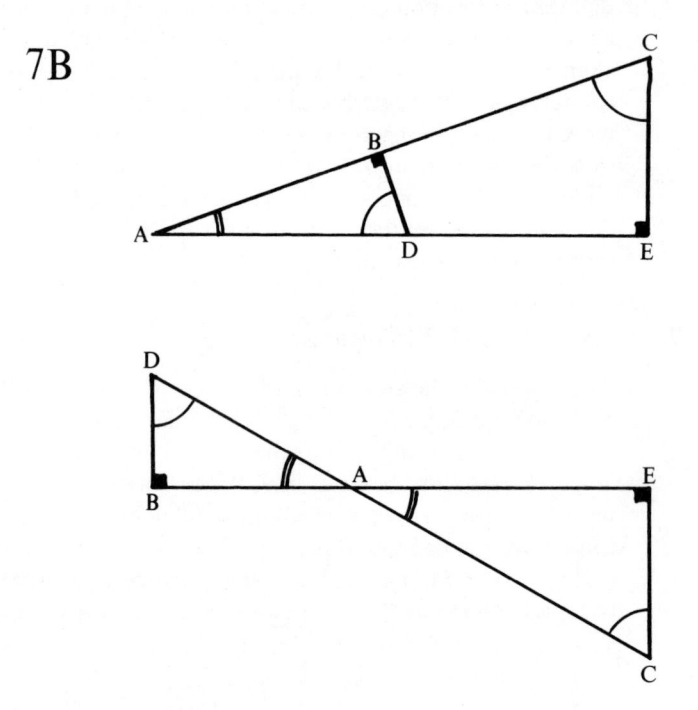

7B

In both figures, compare triangles ABD and AEC. They are similar triangles, each pair having the same angles. The relationships in both are:

$$\frac{AD}{AC} = \frac{AB}{AE} = \frac{DB}{CE}$$

If two of these lengths are known in one triangle, and one in the other, the remainder can be determined by substituting the known values in the above equation.

7.2 Components of forces

When a force is not acting parallel to one of the chosen co-ordinate directions, its effect in these two directions can be found by sketching right-angled triangles and comparing them. In sketch 7C a force is shown alongside the defined co-ordinate directions. To determine the effect of the force in the two co-ordinate directions, the following procedure should be adopted until you are familiar with the technique:

(a) Draw a line representing the force, at some distance from the FSD. The diagram to be drawn forms a graphical calculation, so the length of the line can be scaled equal to the magnitude of the force; the arrow shows its known or assumed sense. The line is lettered AB.

(b) Draw from A a line parallel to one of the co-ordinate directions, and, from B, another line parallel to the other co-ordinate direction. It does not matter which is chosen to be drawn from A or from B. These lines meet at C forming a right-angled triangle.

(c) The length AC, in the diagram of sketch 7C represents the east-west component of the force represented by AB. The length CB similarly represents the value of the north-south component of the force AB (these components are shown as *ewc* and *nsc*).

(d) As always in these problems the senses of the forces are of prime importance. The sense of AB, as obtained from the FSD is shown by an arrow pointing from A to B. The two components, together, therefore, must act from A to B. Thus on the FSD you can show force AB, *or* the two forces AC and CB acting together. They have the same effect as AB. These are the two components of the force, and their values can be found from the geometry of this right-angled triangle.

To become familiar with this technique, used red pencil for the *nsc* (line CB) and blue pencil for the *ewc* (line AC). If all inclined forces are resolved into components by this method, all the red lines will be parallel to the NS direction and can be used in the $\sum F_{NS} = 0$ equation, and all the blue lines brought into the $\sum F_{EW} = 0$ equation. In these sketches the dotted line represents the blue lines and the full line represents the red.

7.2

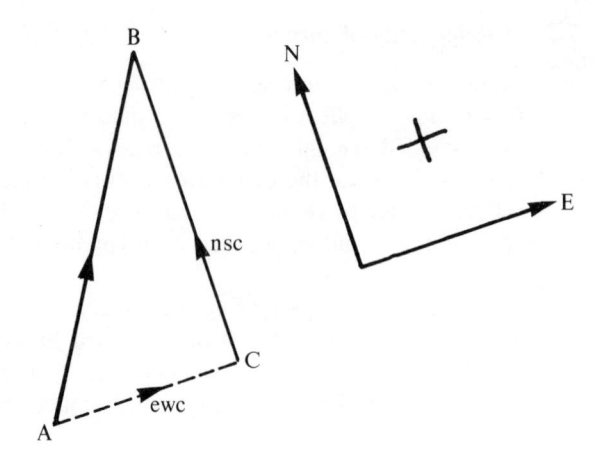

7C

[Figure: Triangle with vertices A (bottom-left), B (top), C (bottom-right). Arrow from A to B labeled along AB; arrow labeled "nsc" near C pointing up along CB; dashed arrow from A toward C labeled "ewc". To the right: compass showing N (up) and E (right).]

Test 5B   The answers

5B/1   The structure under load in sketch 5F cannot ever be in equilibrium, even if you know the values of the forces in Newtons. Although they pass through one point, they have an aggregate force towards the right, and the structure would move in this direction no matter how small are the forces. The answer is: N.

5B/2   The structure under load in sketch 5G supports a system of three parallel forces. The nett sum of these in the N–S direction shown, is zero. However, wherever a centre of rotation is assumed – say on the line of the right-hand force – it is clear that the moments are not zero, and the structure would rotate about the centre of rotation. Imagine a pin inserted where the right-hand force touches the structure. The turning effect of the 200 N about this centre of rotation is obviously much greater than the turning effect or moment of the 2 N force about the same centre. The structure is thus not in equilibrium but rotating counter-clockwise. The answer is: N.

5B/3   The structure under load in sketch 5H has a body force which acts vertically downward, since it is a weight. The other two forces acting meet on the line of action of the weight. Examination shows that the N–S effects of the three forces could be balanced if they were of the right values, and the same applies to the E–W forces. The answer is: C.

5B/4   The structure under load in sketch 5I is under the action of two couples. The 20 N forces act to cause rotation in a clockwise sense, and the 10 N forces have a moment of the opposite sense. The values of the moments

7.3

of each couple are given. The one is 20 × 2 N m and the other is 10 × 4 N m. These are equal and the moments therefore balance. In a couple the forces are always balanced, being equal in value and opposite in sense, so the whole structure is in equilibrium. The answer is: E.

7.3 The geometry of curves

The curves which you will find are the most important in the study of structures are the CIRCLE and the PARABOLA. You must be able to find your way about within the geometry of these curves, deciding co-ordinates, lengths, and slopes. These are of most use in the study of curved structures such as arches (Unit 10). The circle is more complex than the parabola, but the dimensions usually required, and used extensively in Unit 10, are given in this section. Other variations of these

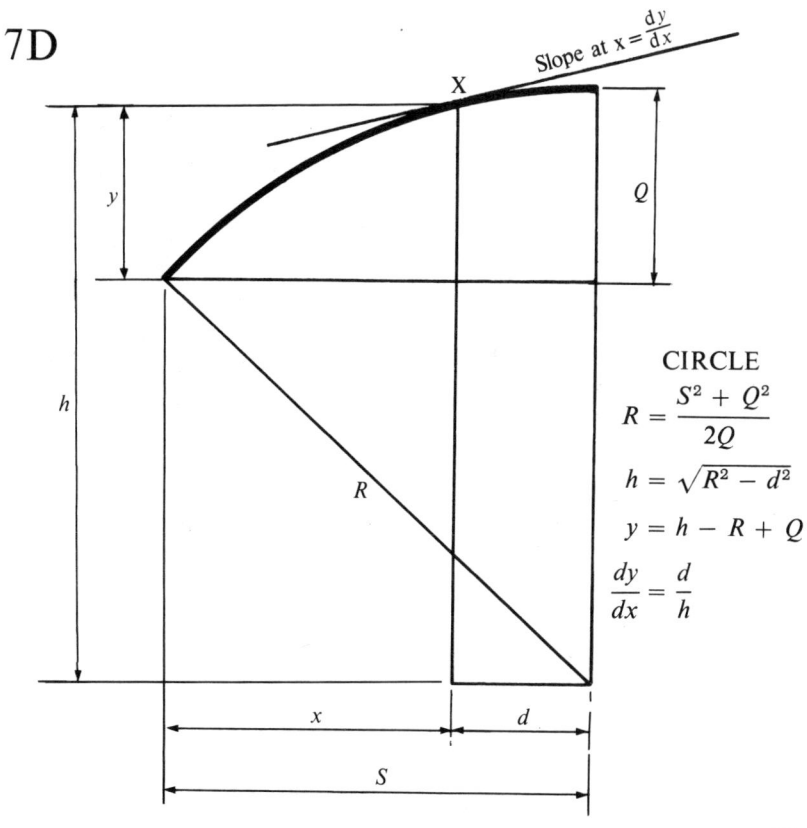

7D

Slope at $x = \dfrac{dy}{dx}$

CIRCLE

$$R = \frac{S^2 + Q^2}{2Q}$$

$$h = \sqrt{R^2 - d^2}$$

$$y = h - R + Q$$

$$\frac{dy}{dx} = \frac{d}{h}$$

55

7.3

formulae are possible, but these are put in the form and in the order most usually needed in calculations of arch structures. The data given in arch problems are most commonly span and rise – S and Q – and the derivations are placed in the order which is of use in such a situation. Only one part of the curve is shown, coming to a horizontal slope at the top. The other portion can be different. In the problems of this book a hinge is inserted somewhere on the arch rib, and thus the portions on either side of the hinge can be treated individually, even if the one is not a mirror image of the other.

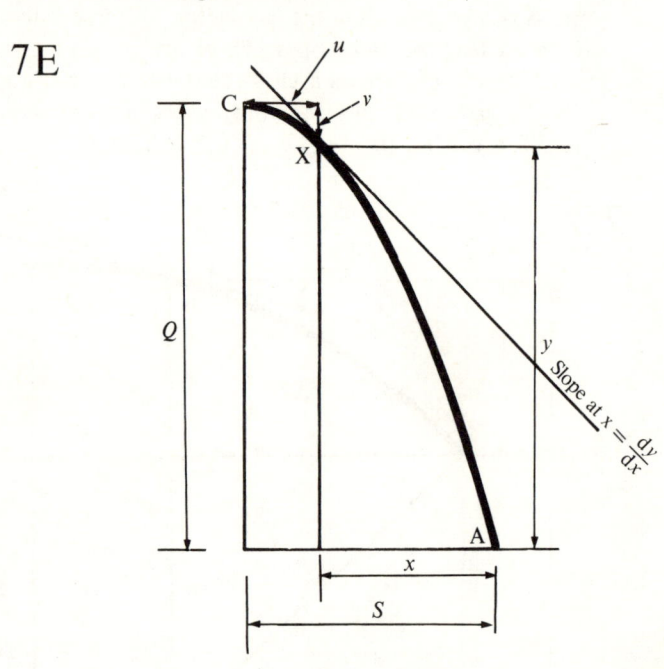

7E

PARABOLA

Curved length $AC = S + \dfrac{2Q^2}{3S}$

$$y = \dfrac{Qx}{S^2}(2S - x)$$

$$\dfrac{dy}{dx} = \dfrac{2Q}{S}\left(1 - \dfrac{x}{S}\right)$$

$v = Cu^2$ where C is a constant for each parabola

GROUP C
Techniques and Problems

Part I: Group C

8 Six steps to solution

In textbooks on statics and structures the various types of problem encountered are usually treated separately, under separate chapter headings. The inference to be drawn from this division is that these problems differ sufficiently and fundamentally to require specific and individual methods of attack and solution. The impression this gives to the student is often daunting – so much seems to need learning.

Such a complex approach is unnecessary, for all problems dealing with forces on rigid, planar structures with only three unknown quantities, can be solved by a single application of the fundamental principles. If you are to achieve a confidence in the solution of the problems of statics and structures, it is suggested that you follow the six steps to solution, implicitly until you are familiar with the results achieved. Later, you may wish to develop your own short-cuts, but, since you are reading this book you clearly need help, so do not diverge from the simple steps of 'the method'. There is nothing new in this method; it is merely presented as a formal routine. This routine eliminates guesswork and mental strain; it represents a controlled advance to an accurate and confident solution.

Test 6A The answers

6A/1   (*a*) Two; the reactions at each end.
      (*b*) This is an impossible situation for the beam would move on the rollers and there would be no equilibrium.

6A/2   The supports are assumed to be just resting on a fixed block, perhaps with weak spot hinges.

6A/3   The FSD for both conditions is the same. Even if there is a rough base, the cylinder is applying the same single downward weight, and is resisted by the same upward reaction from the table. This is an instance of only two forces keeping a body in equilibrium.

6A/4   This is impossible for a china plate, since there is no chance of a frictional resistance at the corner. The FSD is drawn as if the corner were on a rough tablecloth.

6A/5   The weight shows a similar FSD to the cylinder in 6A/3. All the forces on the crane are parallel and vertical, since no horizontal load is applied to the structure.

6A/6   The senses of the force and couple at the cut portions (held condition) can be guessed by using a scale and finger and thumb. Do this for both portions.

6A/7   The end of the spoon is a held condition of support. Without the couple applied by finger and thumb, the soup would not be in equilibrium.

6A/8   You have two frictional supports at elbow and feet. The wall is in the held condition when it is built into the ground. You apply a downward force and one to the right with your elbow.

Test 6A   Sketches illustrating answers

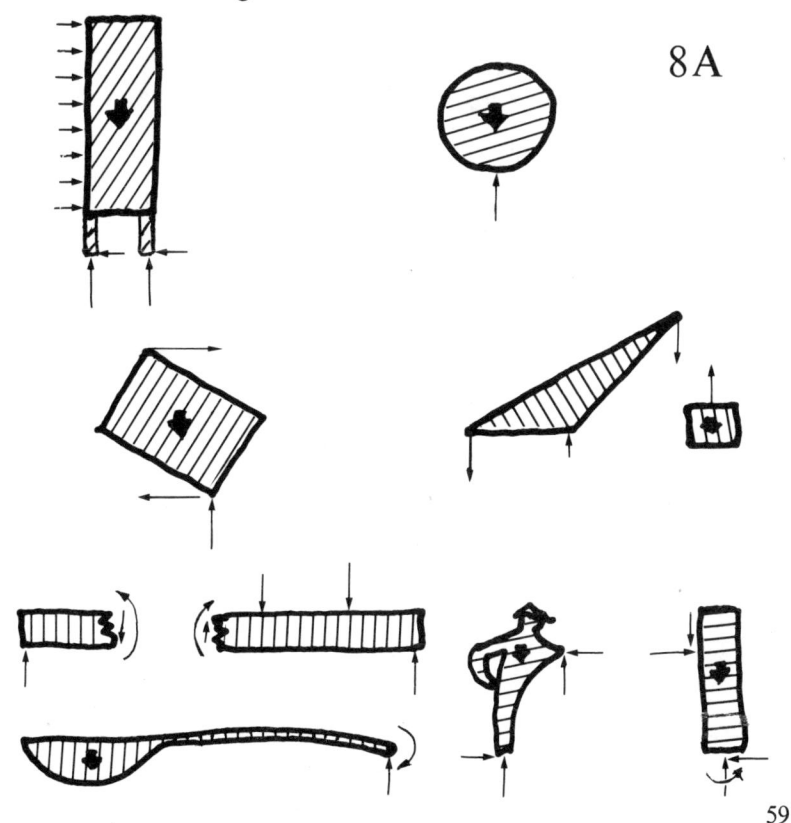

8.1

8.1 The mnemonic – S-U-R-E-T-Y

The word SURETY means 'the state of being sure'. In these studies your sole purpose, for the moment, is to be quite sure, without any doubt, that you are working towards a correct solution of a problem in *statics* or *theory of structures*. The word is used to remind you of the six steps towards an inevitably accurate solution. The six letters have the following meanings, which you should memorize:

S represents the *Structure and Supports*. A sketch of the FSD is the very first step to solution.

U represents the *Unknowns*. Always define these clearly. Their directions can be defined, but their senses are put in *at random* on the FSD.

R represents *Rotation*. Any point on the plane of the structure will do as a COR, but a little thought and ingenuity in selecting the centre of rotation reduces the work of solution.

E represents *evaluation*. Algebraic sums of *all the forces* and *all the moments* are required. Work from the FSD.

T represents *Total*. The sums obtained in the evaluation are equated to zero.

Y represents *Yes or No*? It reminds you to check by a very simple test at the end of the solution whether the senses of the unknowns, *chosen at random*, were correct.

In the remaining sections of the unit, rules of procedure are presented to you for each of the six steps. Follow these implicitly. Later, you can shorten the work by cutting corners, but not until you are quite familiar with all types of problem.

8.2 S The free structure and supports

A whole unit of study has been devoted to this, so revise it. The steps are:

(*a*) Decide which part of the whole arrangement you wish to study. Cut it off from everything else, draw round its outline and cross-hatch it so that you are not tempted to worry about internal arrangements, which don't concern you.

(b) Sketch the rubric showing the directions of chosen co-ordinates, and the positive senses. The word 'rubric' strictly interpreted indicates a printing or writing in red, so it is really important. This little sketch must appear with every FSD.

(c) Draw in on the FSD, any body force, with a broad arrow across the cross-hatching since body forces act internally. The only one body force which appears is weight. Sometimes weight is small, and is neglected, but sometimes it is important.

(d) Draw in the known surface tractions, (which always act externally on the outer limits of the FSD). Show arrows on these known forces to indicate the known senses. Draw in the applied and known couples with senses indicated. Although the senses of 'the unknowns' are shown at random, those of the applied loading are, normally, known.

8.3 U The unknown values

The object of all the solutions is to determine the unknown forces and couples. Usually, the unknowns are the supporting forces and couples, although this is not invariable. The steps are:

(e) Sketch in on the FSD the unknown forces and couples. The direction may be known, or two components of the unknown force can be sketched in the co-ordinate directions [(6)].

(f) Complete the FSD by putting arrows on the unknown forces and couples to indicate senses. Put these arrows in at random. There is no need to try to imagine the correct senses.

(g) Make a list of the unknowns so that you are quite sure what you are looking for in the solution.

8.4 R The centre of rotation

The discussion necessary on this point was not long enough to be made into a full unit of study, but careful planning of the choice of COR is worthwhile. Such planning can considerably reduce the amount of calculation required for the final solution.

Any point in the plane of the structure may be chosen as a centre of rotation for the purpose of defining the moments of forces. A solution is

8.4

quite possible with such a random selection. However, a careful choice pays dividends, and is planned on the following lines. If a force passes through a point, it has no moment about that point, for there is no distance between the point and the line of action of the force. Such forces disappear from the moment equations of equilibrium. If, then, we can choose centres of rotation through which some unknowns pass, these unknowns disappear in the moment-equation of equilibrium and do not complicate the solution.

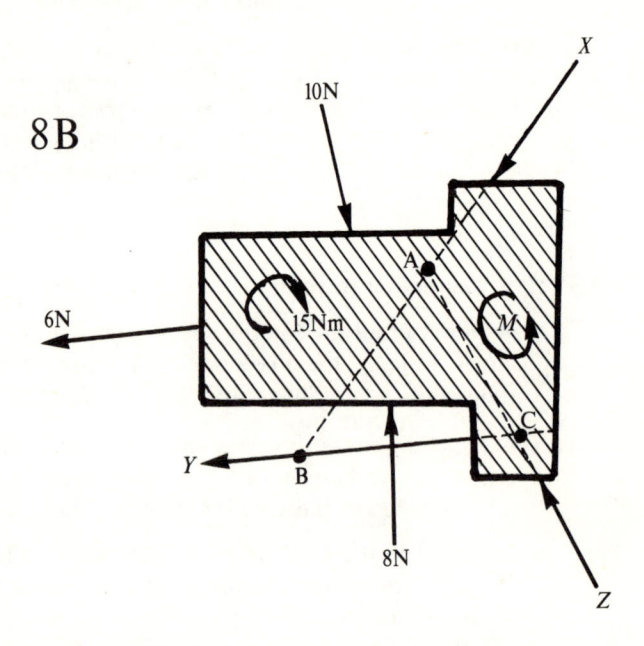

8B

The first step, therefore, is to produce the lines of action of the unknowns so that they all intersect each other. In sketch 8A, for example, the FSD has been completed by applying the known forces (6, 10, 8) and couples (15), and by showing the unknown forces (X, Y, Z) and couples (M) with senses guessed at random. If the lines of action of the unknown forces X, Y, and Z are drawn so that three intersections take place (A, B, and C) you have selected the COR's required. In writing a moment equation of equilibrium to determine the value of the unknown Y, you choose the COR where X and Z intersect (A); these two forces have no moment about A and disappear from the moment equation which is left with one unknown – the force Y. In writing a moment equation to determine X, the COR is at the intersection of Y and Z – the point C. In writing a moment equation to determine the force Z, the COR is at the intersection of X and Y (the point B).

The rules for the selection of the *centre of rotation* in all likely circumstances are:

(*h*) IF THERE ARE THREE UNKNOWN FORCES, the COR for any one of these is chosen at the intersection of the other two. There are three choices (A, B, and C in the sketch).

(*i*) IF THERE ARE THREE UNKNOWN FORCES OF WHICH TWO ARE PARALLEL, there are only two choices of COR – where the inclined unknown force cuts the two parallel ones.

(*j*) IF THERE ARE TWO UNKNOWN FORCES, the COR is anywhere on the line of action of one of them. There is usually a 'best' position on that line, normally easily selected.

These rules must be well remembered, but there are one or two negative points to keep in mind. (1) The presence of couples, known or unknown, has no influence on the selection of the COR. The COR is concerned with forces only. (2) If all the forces acting on the structure pass through one point (known forces and unknown) there is no need for a centre of rotation (COR). (3) If the three unknown forces are all parallel, the problem cannot be solved by the methods dealt with in this book. (4) If there are more than three unknowns, the problem cannot be solved by the methods given in this book, either, but you must be careful to be quite sure that the forces are really independently unknown before abandoning the solution. Sometimes one apparently unknown force is related to one of the others. For example, the frictional force at a frictional support is always a known proportion of the normal force. Also, four or more apparently unknown forces may sometimes be found by using several FSD's made from different parts of the whole arrangement.

8.5 E Evaluation (writing down the quantities)

You now have your FSD drawn in front of you. It is cross-hatched to ensure that you do not worry about the internal arrangements. It has marked on it *all* the forces and couples acting, whether known or unknown. The known values usually are given numerically, while the

8.6

unknowns are marked by a symbol. All the forces and couples have an arrow showing the sense selected. The most appropriate co-ordinate directions are marked in a little sketch alongside.

From now on, you work entirely from this FSD. The signs (plus or minus) for each of the quantities are taken from the FSD by inspection. You will have no difficulty in deciding whether a force has a *North-and-an East* effect, a *North-and-a-West* effect, a *South-and-an-East* effect or a *South-and-a-West* effect. Similarly it is clear whether the couples are acting clockwise or anticlockwise.

The *evaluation* consists merely of writing down in three statements, (sometimes not all three are required, but three are possible) *all* the values of forces in the E–W direction, *all* the values of the forces in the N–S direction, and *all* the values of the moments of forces and couples. The word *all* has been emphasized, for many students omit a force or a couple by mistake. The sketching of the FSD makes this impossible, for you write down every force and every couple *you can see* on the FSD. The steps are:

(k) Determine the N–S and E–W components of each force (unit 7).

(l) Write down in two lines, the values of the forces in the N–S and in the E–W directions. Each *must* carry its sign even if it is the first in the line and has a positive sense.

(m) Write down in one line the values and senses (signs) of all the moments. Moments of forces about the COR plus the moments of all the couples added algebraically.

8.6 T Total is zero (the equations of equilibrium)

This step is simple, for it consists merely of equating each of the three evaluations to zero, and so writing in physical terms the three equations of equilibrium which you have, so far, known only in symbolic form. The solution of these three equations, simultaneously if necessary, gives you the values of the three unknowns. It is very important if you wish to be sure of a correct solution that this routine should be followed – first the evaluation, and then equation to zero.

The solution of these problems in statics and structures is often attempted by writing the equations with the positive forces or moments on one side of the equation, and the negative forces or moments on the other side. If done correctly every time, the student must try to think out which are opposed to which. This very often leads to mistakes in signs, for, in writing such an equation, not only must imagination be strained, but the signs of negative values must be changed to positive before the solution takes place, and the signs of the unknowns must be correct if confusion is not to prevail. Apart from the fact that the conditions of equilibrium say nothing about the balancing of forces or moments, but merely that they are, in the aggregate, zero, this common method brings in a great risk of mistakes, and your surety is shaken. Always carry out the complete evaluation before completing the equation with the statement 'equals zero for a condition of equilibrium'. The step is:

(*n*) Form the evaluation into an equation by equating it to zero, and then solve all three equations.

8.7 Y Yes or No? (checking the senses of the unknowns)

In drawing the FSD, the senses of the unknowns were put in at random and the directions of these arrows used in the subsequent steps as if they were correct. It is, of course extremely important to know whether a force or a couple whose value you have determined by the solution of equations in section T, has a positive or a negative sense. If you have followed the rules above, and made sure that all the values carry with them their sign (plus or minus) the rule is quite simple:

(*o*) Write the value of the unknown with a positive sign for the symbol of the unknown on the left of the equation, and the value with its appropriate sign on the right of the equation, using the rules of algebra carefully. If then there are two plus signs, the sense selected was the correct one. If the symbol for the unknown is positive and the value negative, the senses must be reversed on the FSD or a new FSD drawn.

$+F_3 = -6\text{kN}$ means that the unknown force has a value of 6 kN but should have been pointed in the other direction.

$+F_7 = +8\text{N}$ means that the unknown force has a value of 8 N and that the sense shown in the FSD is correct.

8.7

There is a point which must be noted here. All through the writing of the evaluation and the totalling to zero, the signs indicate the senses as seen on the carefully drawn FSD. In the final equation giving the value of the unknown, however, the sign merely indicates if the sense shown in the FSD is correct or not. A statement that $+F_6 = +10$ N does not necessarily mean that the force is acting in the positive sense, but merely that the sense given in the FSD is the right one. It might well be acting in the negative sense, despite the positive sign. Remember then, that when the solution is found, the signs have a somewhat different meaning. To get the correct results, of course, the left-hand side of the final statement must always be written with a plus sign.

Learn and remember

(a) All the six operations S-U-R-E-T-Y.

(b) All the fifteen steps within these operations.

(c) Signs used in the equations of equilibrium, up to operation T, refer to senses *as seen* on the FSD, whether these senses are correct or not.

(d) Signs in the final statement of the value of an unknown refer to the correctness of the interpretation of sense. The left-hand side of the final equation must always be positive. Be sure to apply operation Y every time: it is very important to the designer of a structure to have the sense correct.

Direction lines in FSD are:
Straight for FORCES: Curved for MOMENTS OF COUPLES

COUPLES shown in the FSD do not appear at all in the FORCE EQUATIONS.

Part I: Group C

9 Rigid structures in numerical solution

This is the unit of study to which the earlier ones have led. Here you have the solution of problems; work through them independently. It is not enough to read them. Use the six operations until you are thoroughly familiar with any problem dealing with rigid planar structures with co-planar loading. Even then you may well find this division of the task into six steps helps efficiency.

9.1 The guiding rules

Some of the guiding rules to be remembered, if there is to be a diminished chance of error, are:

1. The most likely equation of equilibrium to be used first in any problem is the moment equation.

2. It is important, therefore, that the rules for choosing the centre of rotation (COR) should be well known (Unit 8).

3. Choose the co-ordinate directions to suit the majority of the forces acting. Since most forces on structures are weights acting vertically, the vertical and horizontal directions are the most likely to be chosen.

4. Remember that applied and induced couples do not appear in the force equations of equilibrium, but must appear in the moment equations.

5. There is a serious risk of error if signs are not written with every quantity. The signs are those shown on the FSD, whether they are 'correct' or not. The final solution shows whether the correct sense has been chosen.

6. Never equate a number of terms to another set. The terms of the equation should be evaluated with their signs, and then equated to zero.

9.2

7. When a hinge is used as a COR, there is no *M*-term in the equation. If the COR is a section cut through a beam, the supporting conditions show a couple whose moment must be added in to the moment equation around the COR.

9.2 Arrangement of problems

In the following pages there is little distinction made between the different types of problem, although they naturally fall into categories. These categories do not require different treatments; the method of attack is consistent and need not be altered for any of the problems of this Unit. The problems are set out approximately in the order of increasing complexity, and of increasing difficulty in geometrical shapes. The final equations of even a complex problem do not present difficulties. It is the setting up of the equations which offers a challenge. A knowledge of the geometry of the shape concerned, and a clear technique in sketching FSD lead to a correct solution of even the most exacting problem.

Remember to USE THE INDEX at the end of the book. You will find there a direction to any aspect of the work in which you are interested, and also links with the conventional classifications of such groups as 'method of sections' 'bending moments', three-hinged arches' 'concurrent forces', 'suspension cables', and others, which are often treated as if they were separate and different problems, whereas all are soluble by the one method outlined in the preceding units.

Go no further with UNIT 9 unless you are well versed in the earlier UNITS.

9.3 THE PROBLEMS

9/1 ONE EQUATION: One FSD: Four concurrent forces

It is known that the system of forces shown in the Sketch is in equilibrium. The 50 N force and the force F_2 are in the same straight line. Only the sense and magnitude of F_2 remains to be determined to complete the solution.

S *Structure*

The only addition to the sketch as supplied in the problem, to make it into a free-structure diagram is an arrow on the line of F_2 to show the sense (chosen at random). The co-ordinate directions are then most conveniently chosen parallel to and at right angle to F_2, since three out of the four forces act in these directions. Any other direction for the co-ordinates would result in a much more complex calculation.

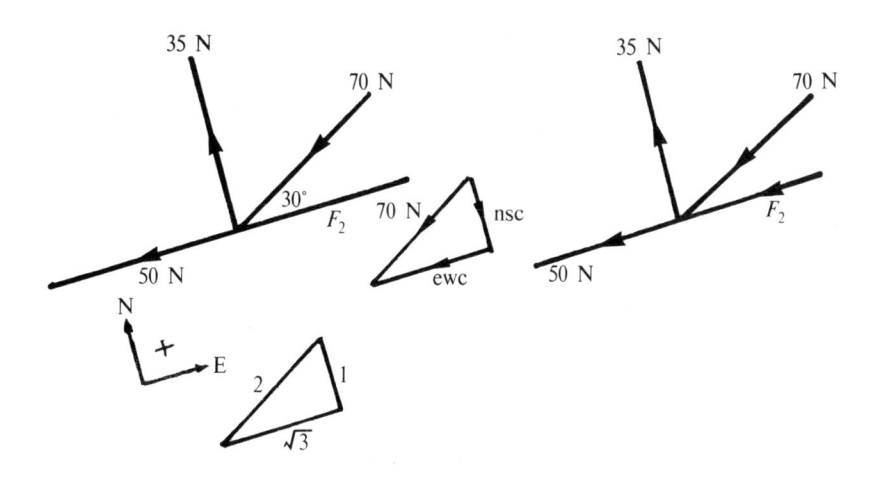

U: *Unknowns and Units*

The only unknown is F_2. Use N.

R: *Rotation*

COR not required as these are concurrent forces.

9/2

E: *Evaluate forces*

The *ewc* (east–west component) of 70 N is calculated from comparing right-angled triangles. All E–W forces visible in the FSD are written down with the senses visible in the FSD.

$$\Sigma F_{EW}$$

$-50 - F_2 - \text{ewc } 70$ (N)
$-50 - F_2 - \sqrt{3}/2 \times 70$ (N)

T: *Total is zero*

$$\Sigma F_{EW} = 0$$

$-50 - F_2 - 60\cdot6 = 0$ $+F_2 = -110\cdot6$ N

Y: *Yes or No?*

The negative value of $+F_2$ shows that it really acts in the opposite sense. This is important. Sketch the correct FSD.

9/2 TWO EQUATIONS: One FSD: Simple supports: Four concurrent forces

A cylinder lies in a Vee-block and is partially supported by two cords giving pulls of 120 N at 37° to the horizontal, and 240 N at 42° to the horizontal. The cylinder has a mass of 81·5 kg. What loads are applied to the two faces of the Vee block by the weight of the cylinder? The two faces of the Vee block lie at 28° and 73° respectively to the horizontal.

S: *Structure and Supports*

The FSD required is the cylinder, represented by its cross section. The supports are *simple*, and are thus at right-angles to faces of the Vee-block.

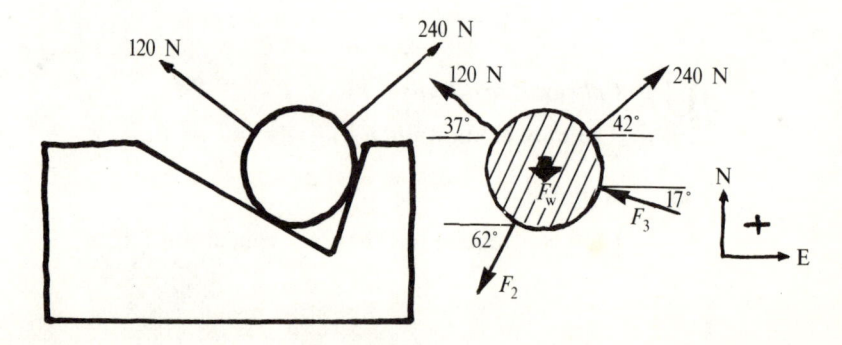

9/3

U *Unknowns*

These are F_2 and F_3, senses chosen at random.

R *Rotation*

Since the forces are all directed through the centre of the cylinder, no COR is required.

E : T *Evaluate and equate to zero*

$$\Sigma F_{NS} = 0$$

$+ \text{nsc } 120 + \text{nsc } 240 - F_W - \text{nsc } F_2 + \text{nsc } F_3 = 0$
$+ 120 \sin 37° + 240 \sin 42° - F_W - F_2 \sin 62° + F_3 \sin 17° = 0$
$- 568 - 0·88 F_2 + 0·29 F_3 = 0$

$$\Sigma F_{EW} = 0$$

$- \text{ewc } 120 + \text{ewc } 240 - \text{ewc } F_2 - \text{ewc } F_3 = 0$
$- 120 \cos 37° + 240 \cos 42° - F_2 \cos 62° - F_3 \cos 17° = 0$
$+ 82 - 0·47 F_2 - 0·96 F_3 = 0$
Solving these simultaneously,
$- 721 + 2·09 F_3 = 0$

$+ F_3 = +35 \text{ N}$
$+ F_2 = -53 \text{ N}$

Y *Yes or No?*

The sense of F_2 has been chosen wrongly, as was fairly obvious. Re-draw the FSD.

9/3 **ONE EQUATION: One FSD: Simple supports: Vertical forces on a horizontal beam**

A horizontal beam carries three vertical loads of 300, 600, and 800 Newtons, and is simply supported. The left-hand supporting reaction is known to be 575 N. Complete the solution.

The solution of this and of the other problems in this Unit is complete when all the forces and couples acting are known. Here, only one force is left unknown. In this and the other problems, apply the six operations: S-U-R-E-T-Y.

9/3

S *Structure and supports*

The supports are simple, and therefore act at right angles to the beam (i.e., vertical). The forces are all parallel and the N–S direction can, conveniently, be made parallel to them.

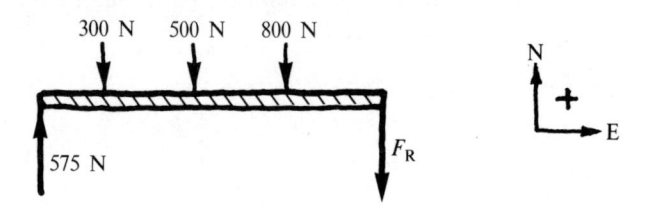

```
      300 N   500 N   800 N
        ↓       ↓       ↓
   ┌─────────────────────────┐              N
   │▓▓▓▓▓▓▓▓▓▓▓▓▓▓▓▓▓▓▓▓▓▓▓▓▓│              ↑  +
   └─────────────────────────┘              └──→ E
    ↑ 575 N                ↓ F_R
```

U *Unknowns and Units*

There is one unknown, F_R. Sense shown at random. Use N.

R *Rotation*

COR not required. Forces are parallel, and a force-equation of equilibrium may be used.

E *Evaluate forces*

$$\sum F_{NS}$$
$$+575 - 300 - 600 - 800 - F_R \quad (N)$$

These are *all* the forces acting on the FSD in the N–S direction and the signs used are taken from the senses shown by the arrows in the FSD.

T *Total is zero*

$$\sum F_{NS} = 0$$

$$-1700 + 575 - F_R = 0 \qquad\qquad +F_R = -1125\ N$$

Y *Yes or No?*

The numerical value of $+F_R$ is negative. This means that the guess as to its sense was wrong. The force F_R must act in the upward direction.

This was really quite obvious at the start. A support for downward forces in these simple circumstances is clearly upwards. However, this routine treatment shows no thought need be taken about the senses of unknowns in the FSD. Put an arrow on at random, and treat it as if it were correct. The final signs show whether the guess was right or not.

9/4

9/4 ONE EQUATION: One FSD: Hinged support: Three forces on tilting block

A rectangular block slides on a rough surface, being pushed as shown in sketch 9.4A. The mass of the block is 50 kg. What horizontal force F_H would cause the block to tilt over when it encounters a small ridge R?

S *Structure and Supports*

When the block tilts over the obstruction R acts as a hinge. You must draw the FSD for the condition that the corner A is just clear of the supporting surface, which is removed. The mass of 50 kg produces a force of (Unit 3)

$$50 \times 9.81 \text{ N} = 49.05 \text{ N} = F_7$$

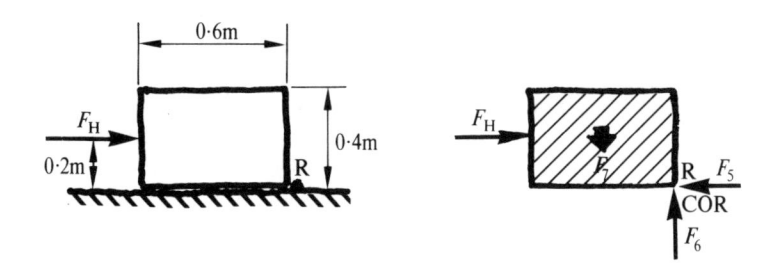

U *Unknowns and Units*

There are three; F_6, F_7, and F_H, but only F_H is required. Use N and m.

R *Rotation*

The COR must be taken where F_6 and F_5 intersect, if they are to be eliminated from the equation.

E *Evaluate moments about R*

It is clear that a moment equation is required, since this problem deals with the rotation of a block.

$$\sum M_R$$

Forces: F_H F_7 F_5 F_6 (N)
Forces × moment arms: $+ F_H \times 0.2 - F_7 \times 0.3$ (Nm)
The signs represent incipient clockwise or anticlockwise rotations.

73

T *Total is zero:*

$$\Sigma M_R = 0$$

$$+0.2F_H - 0.3 \times 490.5 = 0 \qquad\qquad +F_H = +735.8 \text{ N}$$

Y *Yes or No?*

The double plus signs checks the calculation, for the direction of F_H was given as a known condition. This is the horizontal force at half the height of the block, which would just begin to cause it to tip over on meeting an obstruction.

9/5 ONE EQUATION: One FSD: Hinged support: Three forces on circular roller

A roller of mass 170 kg, is to be pulled up over a step, 30 mm high. The roller is 200 mm in diameter, and the handle is held at 30° to the horizontal. What pull is required?

The only difference between this problem and the last is that the geometry of the circle is concerned. In other respects, the procedure is the same. Many students of Statics are more confounded by geometry than by the study of forces.

S *Structure and Supports:*

When the roller is about to mount the step, its contact with the ground has just been broken, and it is supported on the edge of the step as on a *hinged support*. The single force which represents the action of the hinge can be subdivided into two components in the co-ordinate directions, both known in magnitude and having senses chosen at random. Sketch 9.5A shows the FSD.

$$F_2 = 170 \text{ kg} \times 9.81 = 1670 \text{ N}$$

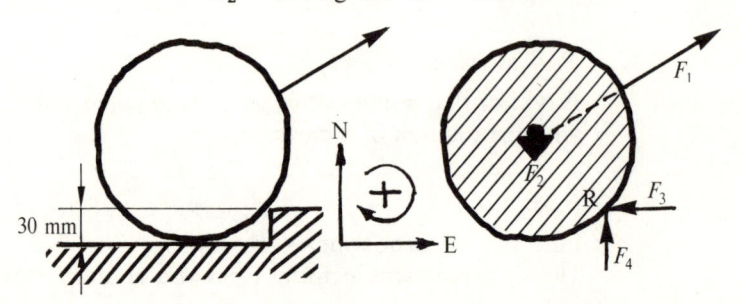

9/5

U *Unknowns and Units:*

There are three – F_1, F_2, and F_3, but only one value is needed – F_1. Use N and mm.

R *Rotation:*

To eliminate the two unwanted forces, the COR must be taken at R, the point where the forces F_2 and F_3 intersect. Only two forces are left acting as shown in sketch 9.5B.

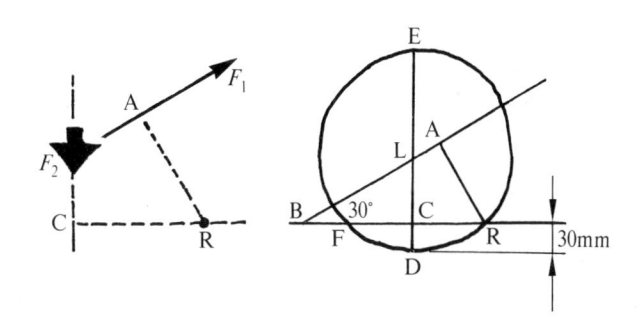

E *Evaluate moments about R:*

From sketch 9.5B:

$$\sum M_R = +F_1 \times AR - F_2 \times CR \quad \text{(N mm)}$$

$+F_1 \times AR - 1670 \times CR \quad$ (N mm)

Before you can go on to equate this to zero in order to find F_1 the values of AR and CR must be found. This can be done as a last resort, by drawing the cross-section of the roller to scale and measuring them off. If this is done accurately, the result can be quite acceptable, but it is more satisfactory to be able to calculate the dimensions required.

From the geometry of the circle:

$CR \times CF = CR^2 = CE \times CD = (DE - CD)CD$
$CR^2 = $ (diameter $-$ 30 mm step) $\times$ 30 mm
$CR^2 = 170 \times 30$ **CR = 71·4 mm**
$CL = 100 - 30 = 70$ mm
$BC/CL = \cot 30° = \sqrt{3}$
$BC = CL\sqrt{3} = 70\sqrt{3} = 121·2$ mm   BR = BC + CR = 192·6 mm
$AR = BR \sin 30°$ $AR = 0·5 \times 192·6$ **AR = 96·3 mm**

T *Total is zero:*

$$\sum M_R = 0$$

$+F_1 \times AR - 1670 \times CR = 0$ (N mm)
$+96 \cdot 3 F_1 - 1670 \times 71 \cdot 4 = 0$ $\qquad +F_1 = +283 \cdot 5 \text{ N}$

Y *Yes or No?*

The two positive signs show that the sense of the pull F_1 was correct in the FSD. This is only a check on the working, since the sense of this unknown was defined by the problem.

9/6 ONE EQUATION: One FSD: Hinged and simple supports: Three forces on bent rod

A cranked rod is pinned to a solid support at X and bears against a vertical stop at Y. What is the force exerted on this vertical surface?

S *Structure and Supports:*

In the FSD, there must be a force acting through the hinge. As you do not know its direction it can be replaced by two components in the direction of the chosen co-ordinates. In this problem the force at Y (simple support) is at right angles to the surface (given as vertical), and the co-ordinates may then be horizontal and vertical.

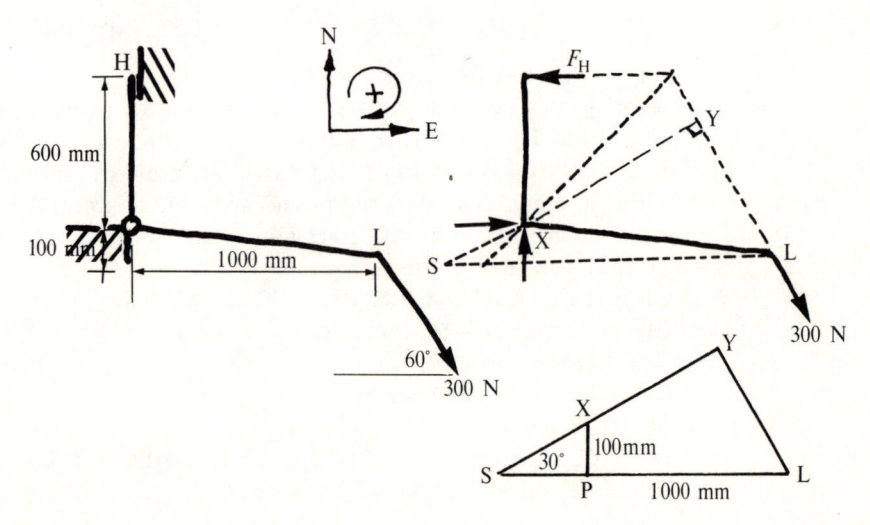

76

U *Unknowns and Units:*

There are three but only one is required (F_H). Use N and mm.

R *Rotation:*

Choosing the COR at X eliminates the two components of the force at the hinge, in the moment equation.

E *Evaluate moments about X:*

Clockwise rotations are positive, as shown by the rubric sketch.

$$\sum M_X = -F_H \times 600 + 300 \times XY \qquad \text{(N mm)}$$

This again is a problem in geometry. To determine the value of XY, produce it until it meets the horizontal through L.
The known dimensions are PL = 1000 mm, XP = 100 mm. The line SY lies at 30° to SL.
Thus the other lengths can be found as:

$$SP = 100\sqrt{3} = 173 \cdot 2 \text{ mm}$$
$$SX = 100 \times 2 = 200 \text{ mm}$$
$$SL = SP + PL = 1173 \cdot 2 \text{ mm}$$

In triangles SYL and SPX,
SY/SL = SP/SX = 173·2/200   Thus SY = 1016 mm
XY = SY − SX = 1016 − 200,   **XY = 816 mm**

T *Total is zero:*

$$M_X = 0$$

$-F_H \times 600 + 300 \times 816 = 0$ $\qquad\qquad +F_H = +408$ **N**

Y *Yes or No?*

Inadequate geometry is a principal source of error and delay.

9/7

9/7 ONE EQUATION: One FSD: Hinged and simple supports: Parallel forces on curved rod

A curved rod in the form of a quarter of a circle, is held by a hinge and carries a vertical force of 25 N. It is supported on a smooth horizontal ledge and has the dimensions shown in the diagram. Find the load on the ledge.

S *Structure and Supports:*

The left-hand support is a hinge (one force inclined at an unknown angle), and the right-hand support is *simple* since the ledge is smooth. The left-hand support can be shown by using two components in the co-ordinate directions.

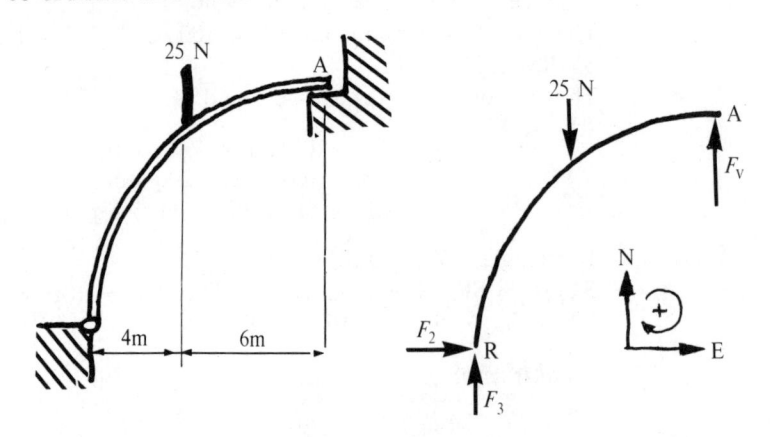

U *Unknowns and Units:*

One only is asked for: F_V. Use N and mm.

R *Rotation:*

To eliminate the other two unknown forces from the moment equation, take the COR at the hinge.

E *Evaluate moments about R:*

$$\sum M_R$$

All forces:	F_V	25	F_2	F_3	(N)
Distances to COR: point R:	10	4	0	0	(m)

Moments (+clockwise)
$$-F_V \times 10 + 25 \times 4 + F_2 \times 0 + F_3 \times 0 \quad (\text{N m})$$

T   *Total is zero:*
$$-10F_V + 100 = 0 \qquad\qquad +F_V = +10 \text{ N}$$

Y   *Yes or No?*

The sense of F_V is correct (two plus signs). Inspection of the FSD shows that F_2 must be zero since there are no other forces in the E–W direction, and the equation of equilibrium says that the total of the forces in the E–W direction is zero. Check this by writing the E–W force equation.

Note, that the curve of the support is a 'red herring'. You are dealing with force systems, and exactly the same result would be obtained by placing 25 N at 4 m from the end of a horizontal beam, simply supported at each end.

9/8   ONE EQUATION: One FSD: Hinged supports: Three forces on curved rod

This curved rod is hinged at both ends. The object is to find the value of the horizontal thrust at the lower hinge. Loading and dimensions are shown in the diagram. The shape of the curve, from lower hinge to upper is, in this instance immaterial to the solution, since dimensions are given and you are dealing with a system of forces defined by these dimensions. The force in a vertical direction at A (which is one of the components of the force through the hinge) is known, from an earlier calculation, to be 58 kN.

S   *Structure and Supports:*

The hinges are each replaced by two components in the co-ordinate directions, and the FSD drawn.

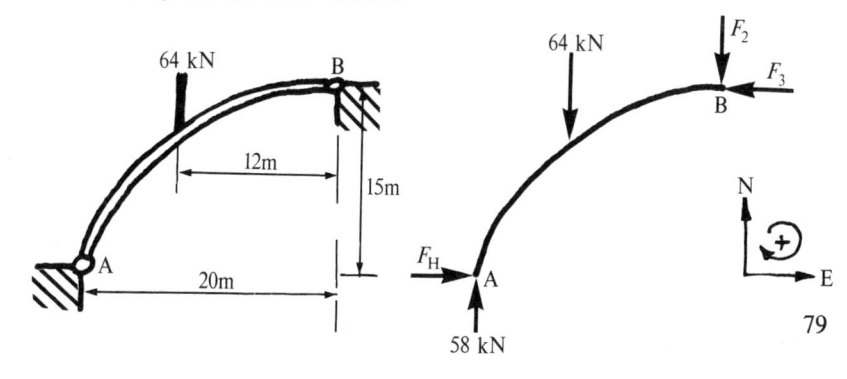

9/9

U *Unknowns and Units:*

Is F_L; others not asked for. Use kN and m.

R *Rotation:*

To eliminate F_2 and F_3 from the calculation take the COR where they intersect – at B.

E *Evaluate moments about B:*

| All forces: | F_H | 58 | 64 | F_2 | F_3 | (kN) |

| Distances from B: | 15 | 20 | 12 | 0 | 0 | (m) |

Moments with senses from the incipient rotation: Take signs from senses seen on the FSD.

$$-15F_H + 58 \times 20 - 64 \times 12 \quad \text{(kN m)}$$

T *Total is zero:*

$$-15F_H + 1160 - 768 = 0 \qquad\qquad +F_H = +26 \text{ kN}$$

Y *Yes or No?*

It is now becoming clear that the double plus sign means that the direction of F_H is correct. In many of the later problems the sense may not be so obvious, so this check should always be applied. If a negative sign appears, the sense of the unknown has been guessed incorrectly.

9/9 ONE EQUATION: One FSD: Hinged and direct supports: Inclined forces on inclined beam

A straight beam, hinged at the left-hand end and supported by a link at the right-hand end, lies at 14° to the horizontal. A complete solution is not required, but determine the force in the link.

S *Structure and Supports:*

Since two of the forces – the link support and the 300 N both lie at 45° to the horizontal, the co-ordinate directions may conveniently be taken parallel and at right angles to this direction.

9/9

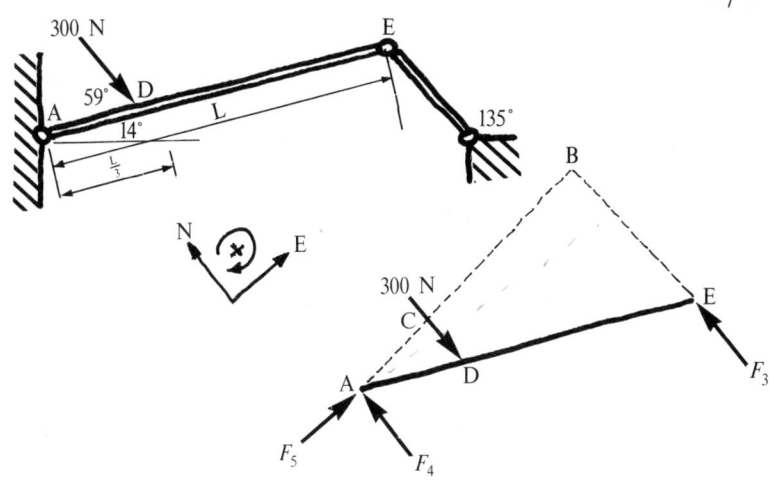

U *Unknowns and Units:*

There are three unknowns, F_3, F_4, and F_5. Only one is required; F_3, so only one equation need be used. Later, after completing this unit, come back and find F_4 and F_5. Use N.

R *Rotation:*

The COR must be taken at the point where the other two unknowns intersect, or they will appear in the equation and complicate the solution. COR is at A.

E *Evaluate moments about A:*

$$\sum M_A$$

All visible forces: F_4 F_5 300 F_3

Moment arms of forces about A: 0   0   AC   AB

Geometry: AD is given as one-third of AE. Thus, AC is bound to be one-third of AB.

Moments: $+300 \times AB/3 - F_3 \times AB$

T *Total is zero*

$-F_3 AB + 100 AB = 0$ $\hspace{4cm} +F_3 = +100 \text{ N}$

81

9/10

Y *Yes or No?*

F_3 has been assumed in the correct sense. The angle of 14°, although factually correct, is not relevant to the solution. Most students, unless solving by a system such as SURETY, would find this angle puzzling. It is a 'red herring'. This problem underlines the fact that you are dealing with force-systems. The same result would have been obtained with any shape of supporting structure between A and E. Even the length of AE is not required. The same result would be obtained with a long beam or with a short one, provided the proportions remained the same.

9/10 ONE EQUATION: One FSD: Simple supports: Horizontal beam with applied couple

A beam is loaded with a couple as shown. What is the effect on the values of the supporting forces as the couple is placed at different points along the beam?

S *Structure and Supports:*

Three FSD's are shown to make clear what happens when a couple is applied, as it often is in structural engineering.

U *Unknowns and Units:*

Both F_1 and F_2 are unknown, but only one will be determined in each instance, the other being assumed to be known, or to be found later. Use kN and m.

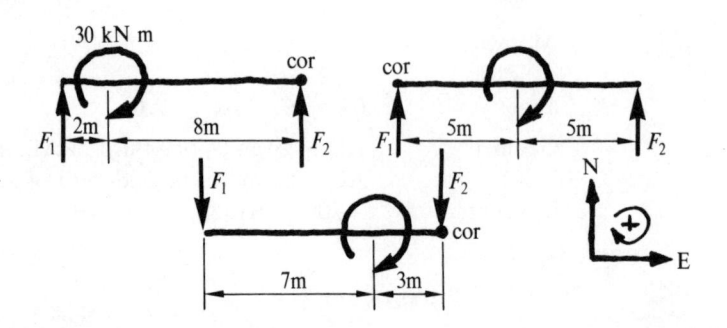

R   *Rotation:*

The two reactions, F_1 and F_2 are parallel because they are simple supports and the supporting surfaces are horizontal. COR is taken on the line of action of one of them.

E : T   *Evaluate moments and equate to zero:*

First FSD. First write down moments of the forces and, later, add the moment of the couple. Remember the moment arm is measured from the force to the COR.

$$\sum M_{COR} = 0 \quad (+ \text{ clockwise})$$

$+F_1 \times 10 + F_2 \times 0$ (moments of forces)
$+10F_1 + 30 = 0$ (add applied moment: kN m) $+F_1 = -3\,\text{kN}$

Second FSD:

$$\sum M_{COR} = 0 \quad (+ \text{ clockwise})$$

$-F_2 \times 10 + F_1 \times 0 + 30 = 0$ $+F_2 = +3\,\text{kN}$

Third FSD:

$$\sum M_{COR} = 0 \quad (+ \text{ clockwise})$$

$-F_1 \times 10 + F_2 \times 0 + 30 = 0$ $+F_1 = +3\,\text{kN}$

Y   *Yes or No?*

In the first FSD, F_1 should be in the opposite direction. In the second FSD, F_2 has the correct sense. In the third FSD, F_1 has the correct sense. You have not calculated the other reaction, but it will clearly also be 3kN. The two reactions form a couple and are always of the opposite sense, whatever the position of the applied couple. You can prove this by using a scale, and the two forefingers of a friend. None of the above FSD's is correct. Re-draw them.

Any quantity written down must carry its sign—plus or minus.

9/11

9/11 ONE EQUATION: One FSD: Hinged and simple supports: Applied couple on bent rafter

The ridged rafter shown in the diagram has a couple applied at the ridge, and two vertical and one horizontal forces. The right hand end is simply supported. Find the value of this support.

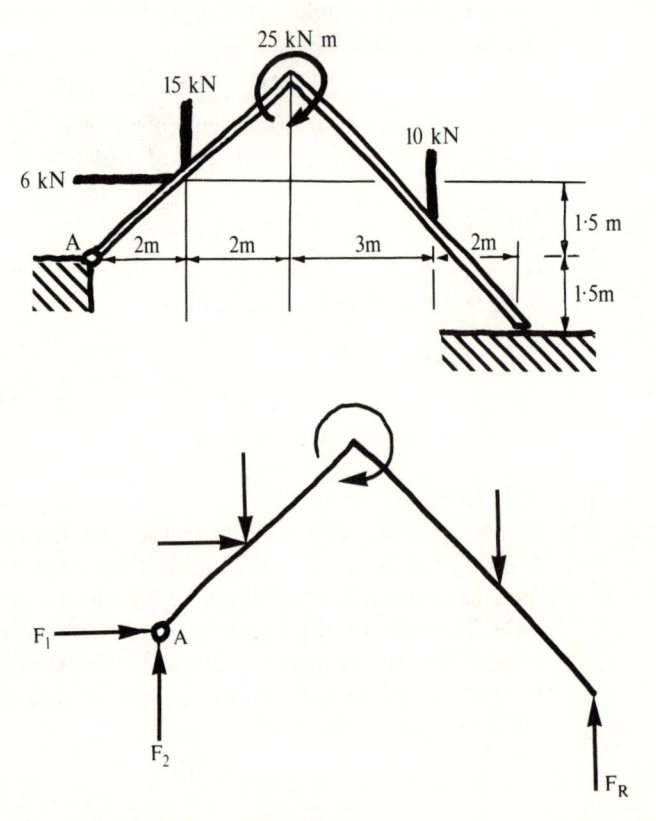

S *Structure and Supports:*
The left-hand support is by a single force through the hinge, but this can be broken into two components in the co-ordinate directions. The FSD is drawn.

U *There is one unknown:*
$$F_R$$

84

Units:

Use kN and m.

R *Rotation:*

The COR is at A. If you are not aware of the reason for this choice, go back and study the earlier problems in this unit.

E *Evaluate moments about* A: $\sum M_A$

| All forces | 6 | 15 | 10 | F_1 | F_2 | F_R | (kN) |

| Distances to A: | 1·5 | 2 | 7 | 0 | 0 | 9 | (m) |

| Moments of forces with senses: | +9 | +30 | +70 | 0 | 0 | $9F_R$ | (kn m) |

| Add moment pf applied couple: | +9 | +30 | +70 | $-9F_R + 25$ | | | (kN m) |

Note the procedure; first calculate the moments of the *forces* about the chosen COR. Then, and only then add, as a separate operation, the moments of any *couples* acting. Remember to include the appropriate sign – positive for clockwise rotation of the couples, and negative for anticlockwise.

T *Total is zero:*

$$\sum M_A = 0$$

$+9 + 30 + 70 - 9F_\$ + 25 = 0$ $\qquad\qquad +F_R = +14\cdot9\,\text{kN}$

Y *Yes or No?*

The sense of the force F_R has been chosen correctly in the FSD.

Students are often puzzled and confused about the action of couples applied to a structure or acting as unknowns. The procedure is always the same. Use the moment equation of equilibrium, find the moments of the forces about a chosen COR. Then add the moments of the couples either as numerical quantities or as symbols for an unknown. The position of the couple on the structure is irrelevant; the moment is merely added algebraically.

9/12

9/12 ONE EQUATION: One FSD: Simple and hinged supports: Two applied couples

Determine what vertical force must be applied at point X in order to bring this pin-jointed structure into equilibrium. Assume the moment equation of equilibrium to be satisfied.

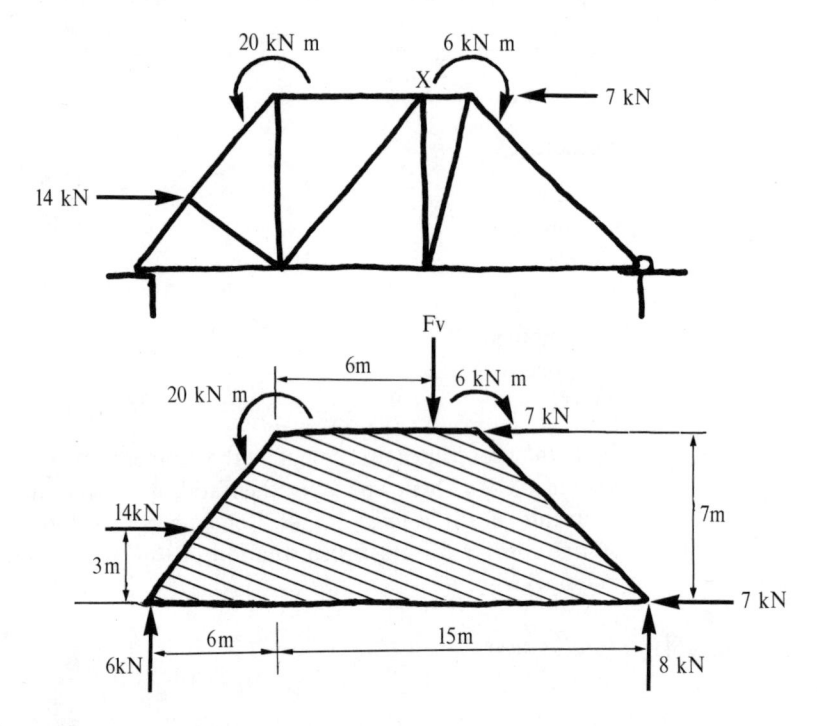

S *Structure and Supports:*

Draw the FSD as in sketch 9.12B. Note that the articulation of this open-web girder does not concern you. It is a 'red herring'. You assume the structure to be rigid; how it is constructed is irrelevant. Its weight is assumed negligible in comparison with the forces expected. Cross-hatch the outline so that it is evident that the distribution of members is not significant.

U *Unknowns and Units:*

There is only one: F_v drawn with a random sense. Since the moment equation is satisfied, no lengths need to be drawn. Use kN.

86

E:T   *Evaluate forces in the N–S direction: Equate to zero:*
$$\sum F_{NS} = 0$$
Forces 14 and 7 kN have no nsc and thus, do not appear.
$$+6 - F_V + 8 = 0 \qquad\qquad +F_V = +14\text{ kN}$$

Y   *Yes or No?*

The point to be noticed here is that in the force equations of equilibrium, the moments of couples do not appear. The forces which cause these moments are in a couple, already balanced and in equilibrium as a pair, and need be considered only in the moment equation. A complex problem can, therefore be much simplified by remembering this rule.

If the structure is in equilibrium it should not rotate. The forces and couples should be balanced about any chosen COR. Try this check, which is a useful insurance against having made a mistake.

Chose the COR at random – let us say at Y where the left-hand couple is acting.

CHECK:

All forces: 6   14   8 F_V 7   7   (kN)

Moment arms
about Y: 6   4   15   6   6   0   (m)

Moments of forces
with senses
(clockwise about
COR positive):

$+36 - 56 - 120 + 84 + 42 + 0$ (kN m)

Add moments
of couples
with senses
clockwise positive:

$+162 - 176 - 6 + 20 = +162 - 162 = 0$

This shows the solution is correct.

9/13

9/13 ONE EQUATION: One FSD: Simple and hinged supports: Unknown couple

The somewhat improbable structure shown in the figure is not in equilibrium. Assuming that the forces at the hinge are known, what is the value of the moment of a couple which must be applied to bring the structure into equilibrium? Where should this couple be applied? The body force may be assumed to act at the intersection of the diagonals.

S *Structure and Supports:*

The problem here is the complexity of the structure, so it is well to draw two diagrams, one giving the forces applied – the FSD – and the other showing the relevant dimensions once you have decided on a COR. Do not try to put everything on one diagram, or mistakes will result. Draw the FSD – sketch 9.13B, but without dimensions. One support is hinged and the other simple. The reaction at P is known to be 6 kN.

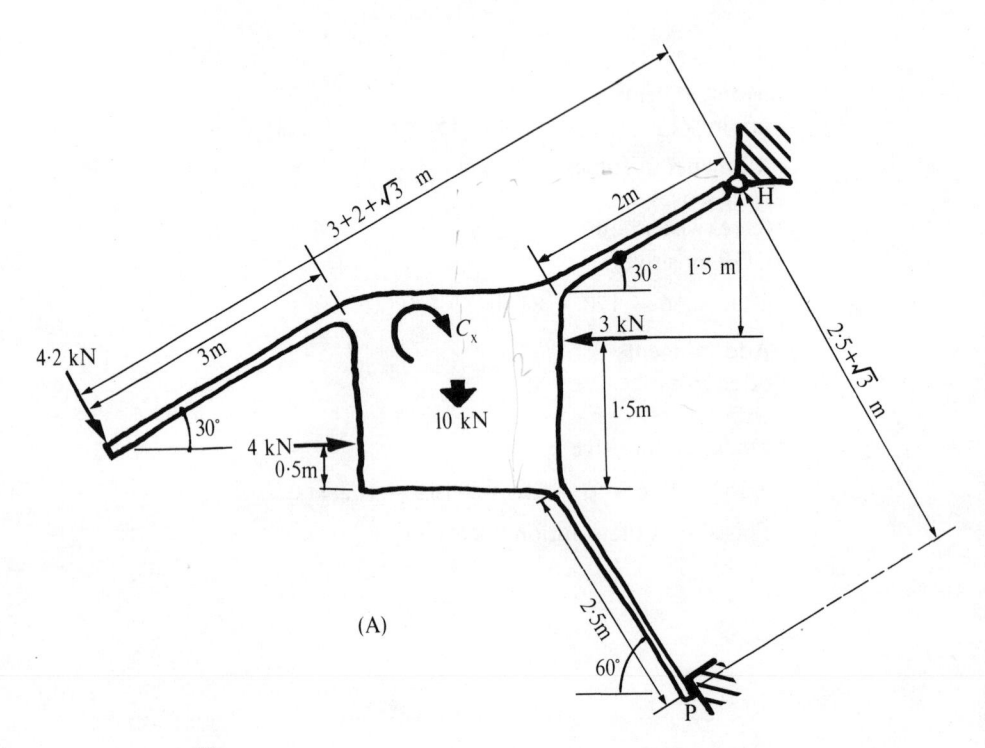

(A)

88

U *Unknowns and Units:*

The two unknown forces at the hinge can be eliminated in the moment equation, and you are left with the one unknown – the value of the

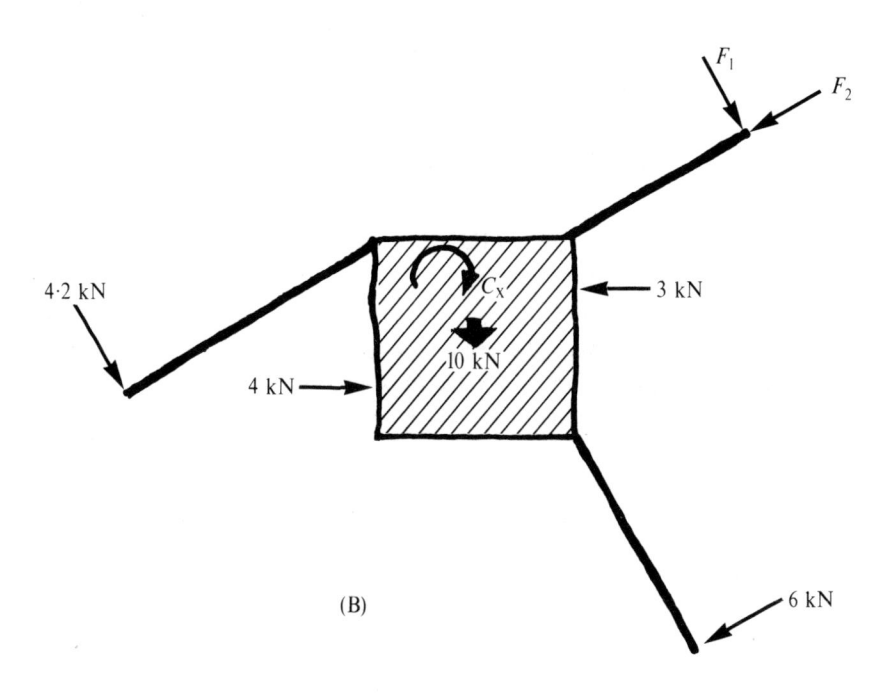

(B)

moment of the unknown couple. It has been shown at random and with a random sense. Use kN and m.

Rotation:

The centre of rotation can be taken at the hinge (H) where the two unwanted and unknown forces intersect. This eliminates them from the equation. Make the dimensioned sketch.

E:T *Evaluate moments about* COR:

$$\sum M_H = 0$$

All forces:	4·2	4	10	3	6	(kN)
Moment arms:	6·7	2·5	2·7	1·5	4·2	(m)

89

9/14

Moments of forces
with senses
(clockwise positive):

$-28·1 - 10·0 - 27·0 + 4·5 + 25·2$ (kN m)

Add the values of any couples acting

$-65·1 + 29·7 + C_x = 0$ $\qquad +C = +35·4$ kN m

There is no particular part of the structure where this couple need be applied. Its moment adds in wherever it is placed.

9/14 **TWO EQUATIONS: One FSD: Frictional support: Three forces on block**

A rough board carries a block having a mass of 10 kg. The limiting coefficient of friction between board and block is 0·2. The board lies at a gradient of 3:4. What horizontal force must be applied to keep the block from sliding down the slope?

S *Structure and Supports:*

The support is frictional and the force is divided into two components F_n and F_f. From the general arrangement, the FSD for the block can be drawn. There is one body force, one horizontal force F_h and the components of the single frictional force.

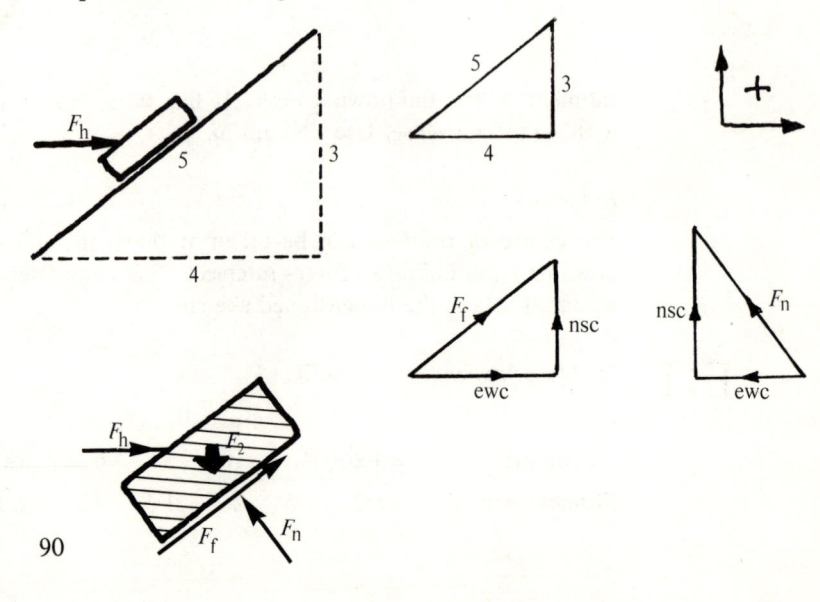

U *Unknowns:*

Only one unknown is asked for in the question, but both the force F_h and the frictional supporting force must be found, for they are interdependent, both having components in the horizontal direction (E–W).

R *Rotation:*

No COR is needed as there are only three forces acting. These must all pass through one point and only the force equations are needed.

E *Evaluate:*

$$\Sigma F_{EW}$$

$+F_h + \text{ewc } F_f - \text{ewc } F_n$
$+F_h + \tfrac{4}{5} F_f - \tfrac{3}{5} F_n$

But just as slipping is about to take place, $F_f = 0.2 F_n$

$+F_h + (0.8/5) F_n - \tfrac{3}{5} F_n$
$+F_h - 0.44 F_n$

$$\Sigma F_{NS}$$

$-F_2 + \text{nsc } F_f + \text{nsc } F_n$
$-10 \times 9.81 + \tfrac{3}{5} F_f + \tfrac{4}{5} F_n$. Substitute $F_f = 0.2 F_n$
$-98.1 + (4.6/5) F_n$

T *Totals are zero:*

1. $+F_h - 0.44 F_n = 0$ $\qquad +F_h = +0.44 F_n$
2. $-98.1 + (4.6/5) F_n = 0$ $\qquad +F_n = +106.6 \text{ N}$
1. $+F_h - 0.44 \times 106.6 = 0$ $\qquad +F_h = +46.9 \text{ N}$

Y *Yes or No?*

The forces have all been chosen with the correct senses, which is not difficult to do with a simple problem of this kind. However, keep to the rules of procedure and the more advanced problems will not seem difficult.

Try this problem again using the co-ordinates at right angles to and parallel to the slope. The same values should be found.

9/15

9/15 TWO EQUATIONS: One FSD: Frictional support: Three forces only

A skip, full of concrete, is pulled up a rough plank. The limiting coefficient of friction between skip and plank is 0·4. The mass of the skip and concrete is 815 kg, and the slope of the plank is 60° to the horizontal. Find the pull required on the rope and the pressure exerted by the skip on the plank. The skip is just on the point of moving.

S *Structure and Supports:*

From the general arrangement, extract the skip and apply the forces acting on it; this gives the FSD to be used. The supporting force is frictional, and is resolved into two components F_n and F_f.

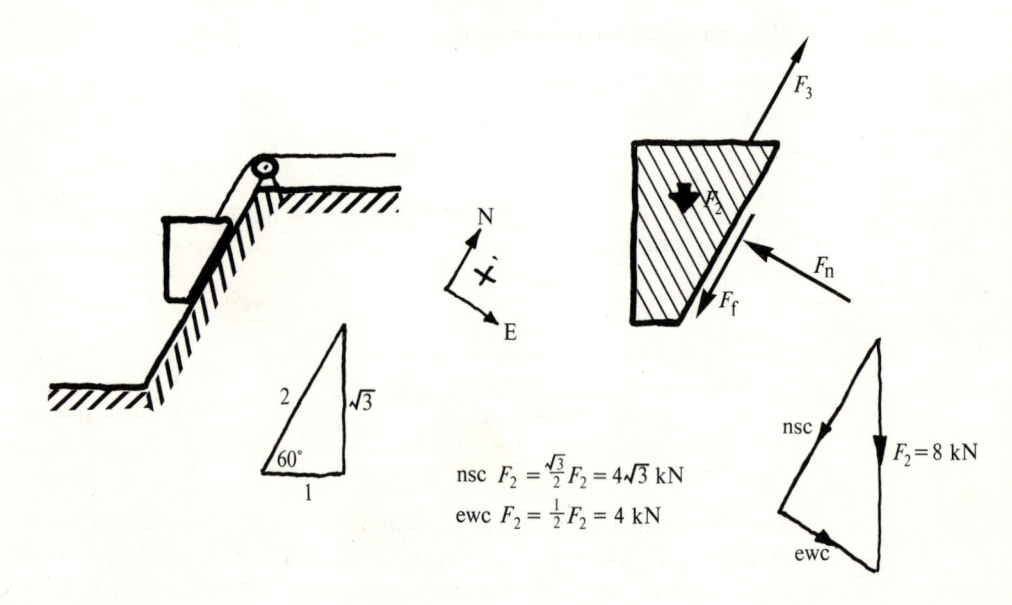

nsc $F_2 = \frac{\sqrt{3}}{2} F_2 = 4\sqrt{3}$ kN

ewc $F_2 = \frac{1}{2} F_2 = 4$ kN

$F_2 = 8$ kN

U *Unknowns:*

There are two unknowns; F_n and F_3. From F_n, the force F_f can be found from the limiting coefficient of friction.

R *Rotation:*

There are only three forces acting on the skip, and these must, therefore, pass through one point. COR is not required.

E *Evaluate:*

ΣF_{NS}
$+F_3 - F_f - \text{nsc } F_2.$ But $+F_f = +0.4 F_n$
$+F_3 - 0.4F_n - 4\sqrt{3}$

ΣF_{EW}
$-F_n + \text{ewc } F_2$
$-F_n + 4$

The components of F_2 (8 kN) are found from the triangle and the known values of cos 60° and sin 60°.

T *Totals are zero:*

1. $+F_3 - 0.4F_n - 4\sqrt{3} = 0$
2. $-F_n + 4 = 0$ $+F_n = +4 \text{ kN}$
1. $+F_3 - 0.4(+4) - 4\sqrt{3} = 0$ $+F_1 = +8.5 \text{ kN}$

Y *Yes or No?*

Both F_3 and F_n were assumed in the correct directions.

9/16 THREE EQUATIONS: One FSD: Simple supports: Bent rod

The diagram shows a bent rod. It rests on a smooth vertical wall at X, and on a smooth horizontal floor at Z. A horizontal cord, attached to the midpoint Y, where the rod is bent, is fastened to the wall to keep the rod in equilibrium. If the total mass of the rod is 3 kg, what is the force in the cord? ZY = YX = 400 mm.

S *Structure and Supports:*

Since the wall and floor are smooth, the supporting conditions are *simple*. The body force representing the weight of each portion of the rod may be considered as acting at its centre. The force applied by the weight of 3 kg is 3 × 9·81 N, or 29·4 N. The supporting conditions having been recognized, the FSD can be drawn. The wall is entirely removed.

U *Unknowns:*

These are F_1, F_2, and F_3.

R *Rotation:*

COR is chosen at X which is on the line of action of F_1 and so eliminates it from the moment equation. Use N and mm.

9/16

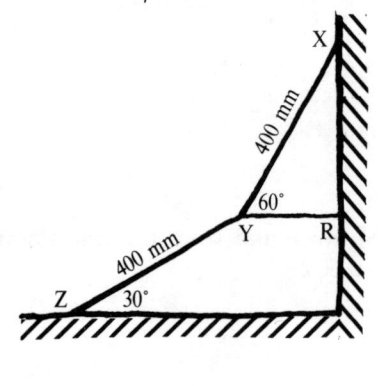

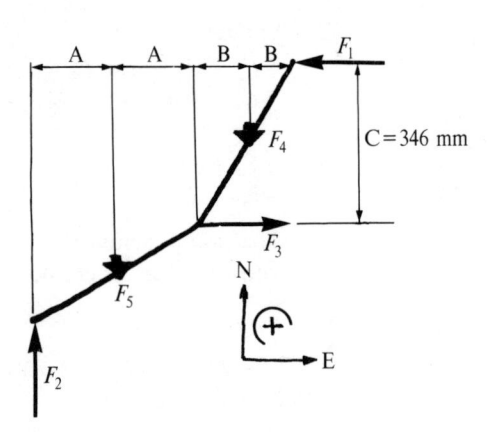

E *Evaluate:*
$$\sum M_X = 0$$

Moments:
The first step is to be sure of the various distances required. From the geometry of a 30/60° triangle the lengths in the co-ordinate directions can be determined. (A, B, and C).

All forces: F_2 F_4 F_3 F_5 F_1 (N)

Moment arm
(distance from X):   2A + 2B   B   C   A + 2B   0
                        546   100   346   373   (mm)

Moments: $+546F_2 - 14\cdot 7 \times 100 - 346F_3 - 14\cdot 7 \times 373$
               $\sum M_X = +546F_2 - 346F_3 - 6952$

N–S Forces
$$\sum F_{NS}; \quad +F_2 - 29\cdot 4$$

E–W Forces
$$\sum F_{EW}; \quad -F_1 + F_3$$

T *Totals are zero:*
The three equations are:
1. $+F_2 - 29\cdot 4 = 0 \quad +F_2 = +29\cdot 4$ N
2. $-F_1 + F_3 = 0 \quad +F_1 = +F_3$
3. $+546F_2 - 346F_3 - 6952 = 0$
    $+546(+29\cdot 4) - 6952 + 346F_3 = 0 \quad\quad +F_3 = +26\cdot 2$ N
                                                     $+F_1 = +26\cdot 2$ N
                                                     $+F_2 = +29\cdot 4$ N

Y *Yes or No?*

The directions in which forces were assumed to act are correct.

9/17 FOUR EQUATIONS: Two FSD'S: Frictional supports: Two sliding blocks

Two blocks are tied together with a cord, and rest on a board. The board is slowly tilted up. The upper block is heavier than the lower, and has a rougher base. The cord between the blocks, therefore, remains taut. At what angle of the board to the horizontal will the blocks begin to slip?

S *Structure and Supports:*

The FSD's for the two blocks show a common force, F_1, pulling upwards on the lower block and downwards on the upper block.

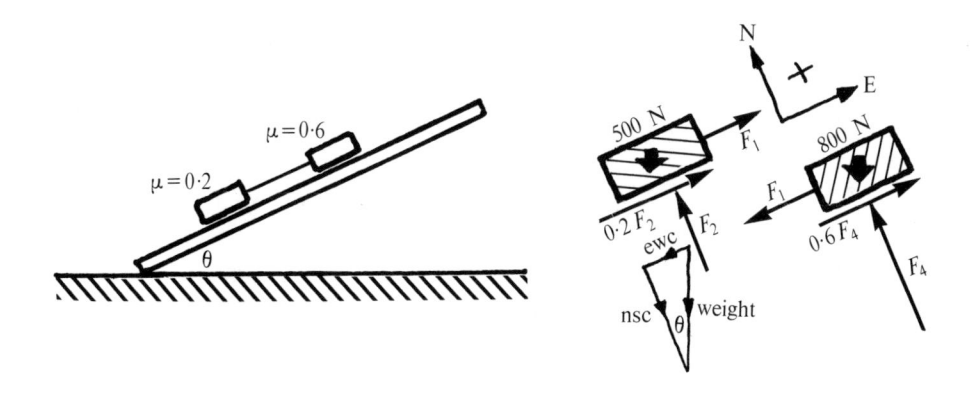

U *Unknowns:*

There are four unknowns, F_1, F_2, and F_4, and the angle of the block to the horizontal (θ). This indicates the need for two FSD's, since one FSD gives a solution for only three unknowns.

R *Rotation:*

COR is not required; only force equations need be used. This conclusion comes from the fact that each block is in equilibrium under the action of three forces (weight, cord pull, and frictional support), so forces act through one point.

95

9/17

E *Evaluate:*

FIRST FSD: UPPER BLOCK

N–S forces

$$\sum F_{NS} \qquad +F_4 - 800 \cos \theta$$

E–W forces

$$\sum F_{EW} \qquad -F_1 - 800 \sin + 0{\cdot}6 F_4$$

SECOND FSD LOWER BLOCK

N–S forces

$$\sum F_{NS} \qquad +F_2 - 500 \cos \theta$$

E–W forces

$$\sum F_{EW} \qquad +F_2 - 500 \sin \theta + 100 \cos \theta$$

T *Totals are zero:*

FIRST FSD: UPPER BLOCK

1. $+F_4 - 800 \cos \theta = 0$ $\qquad\qquad +F_4 = +\mathbf{800 \cos \theta}$
2. $-F_1 - 800 \sin \theta + 0{\cdot}6 F_4 = 0$
   $-F_1 - 800 \sin \theta + 480 \cos \theta = 0$ (inserting value of F_4)

SECOND FSD: LOWER BLOCK

3. $+F_2 - 500 \cos \theta = 0$ $\qquad\qquad +F_2 = +\mathbf{500 \cos \theta}$
4. $+F_1 - 500 \sin + 0{\cdot}2 F_2 = 0$
   $+F_1 - 500 \sin \theta + 100 \cos \theta = 0$ (inserting value of F_2)

Since both *equation* 2 and *equation* 4 show quantities equal to zero, these quantities must together equal zero. Thus:

$$-F_1 - 800 \sin \theta + 480 \cos \theta + F_1 - 500 \sin \theta + 100 \cos \theta = 0$$

$$\tan \theta = \mathbf{580/1300}$$

This represents a gradient of vertical/horizontal of 580/1300 or a gradient of $1:2\frac{1}{4}$. Only very occasionally need an angle in degrees be evaluated.

Use the INDEX to trace the type of problem you find difficult

9/18

9/18 **FOUR EQUATIONS: Three FSD'S: Frictional and hinged supports: Suspended weight**

A horizontal rod XY, of 600 mm length, has a wire firmly fastened to it at Y. Both rod and wire may be assumed to be so light as to be weightless in comparison with the other forces acting. The wire is 1000 mm long, and has a vertical downwards force of 16 N applied at the centre of the length of the wire (Z). The other end of the wire (X) is fastened to a ring threaded on to the rod, and free to move along it. The weight of the ring exerts a vertical force of 4 N. The limiting coefficient of friction between rod and ring is 0·4. When the load has been allowed to hang, and the ring has stopped slipping on the rod, what is the length of XY?

S *Structure and Supports:*

Since the load at Z is large, the light wire may be considered to be straight from X to Z and also from Z to Y. Throughout the calculations XZ can be taken as equal to ZY, as the weight does not slip on the wire, but is tied to it. The support at X is *frictional* and that at Y *hinged*.

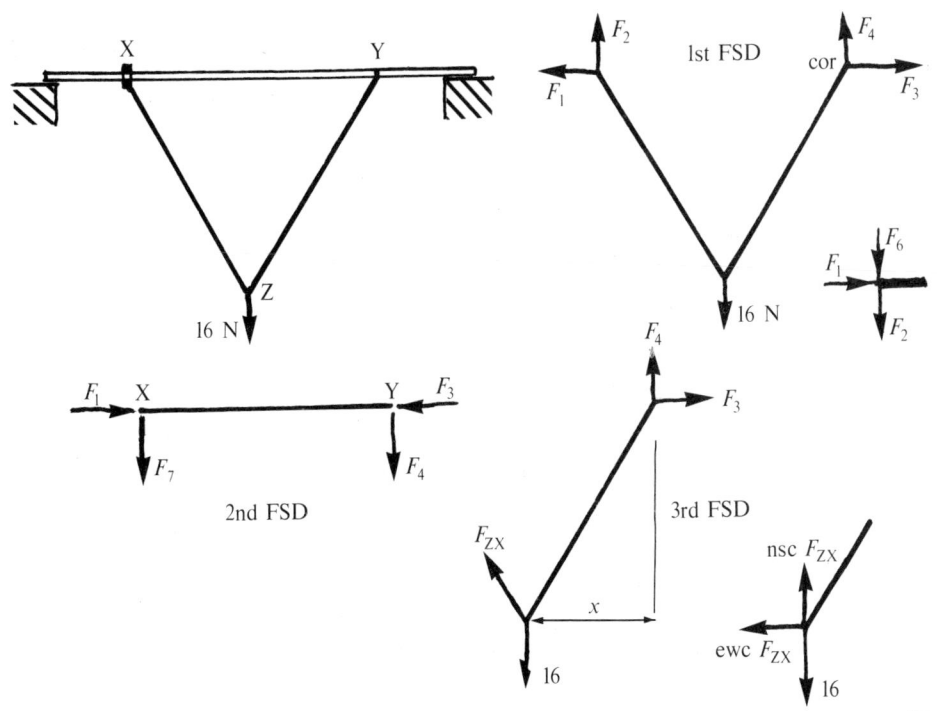

97

9/18

U *Unknowns:*

There are, apparently six unknowns, as seen in the FSD's: F_1, F_2, F_3, F_4, and F_7 and XY. Although some of these are related, more than one FSD will be required.

FIRST FREE-STRUCTURE DIAGRAM (The wire)

R *Rotation:*

Taking the wire as giving the first FSD, two of the unknown forces intersect at Y which can be taken as the COR.

E:T *Evaluate and equate to zero:*

$$\sum M_Y = 0$$

All forces acting:	F_1	F_2	F_3	F_4	16	(N)
Moment arm about Y:	0	XY	0	0	0·5 XY	(mm)
Moments:	$+F_2 \times$ XY $- 16 \times 0.5$XY $= 0$					$+F_2 = +8$ N

This result would be expected from everyday experience. Half the load is supported at each end.

$$\sum F_{NS} = 0 \qquad +F_2 + F_4 - 16 = 0 \qquad +F_4 = +8\,\text{N}$$

SECOND FREE-STRUCTURE DIAGRAM (The rod)

E:T *Evaluate and equate to zero:*

$$\sum F_{EW} = 0$$

All forces acting: $+F_1 - F_3 = 0$.

But F_1 is a frictional force, and is thus related to F_7 by the limiting coefficient of friction. $F_1 = 0.4F_7$. But F_7 is not known yet. It is made up of the downward pull of the weight of the ring (F_6). The total is 12 N. F_1 is, therefore, of the value $F_1 = 0.4 \times 12 = 4.8$ N.

$$+F_1 - F_3 = 0 \qquad +4.8 - F_3 = 0 \qquad +F_3 = +4.8\,\text{N}$$

All the forces are now known, but XY must still be determined. This cannot be done with the two FSD's already used – try it. A third FSD must be cut off.

THIRD FREE-STRUCTURE DIAGRAM (Right half of wire)

There are only two choices left for the third FSD – one of the two halves of the wire. The right half is chosen; it is a straight body supported at hinges at each end – *direct supports*.

R *Rotation:*

Since we do not wish to get involved with the left-hand half, the COR can conveniently be at Z where there is a pull from the left-hand half of the wire, and a further load from the 16 N. COR at Z.

E: T *Evaluate and equate to zero:*

$$\sum M_z = 0$$
$$+F_3[\sqrt{(500^2 - x^2)}] - F_4 x + F_5 \times 0 + 16 \times 0 = 0$$
$$+4{\cdot}8[\sqrt{(500^2 - x^2)}] - 8x = 0 \qquad +x = +257 \text{ mm}$$
$$+XY = +514 \text{ mm}$$

This is the length of rod exposed between X and Y when the ring has stopped slipping and equilibrium has been established. From the geometry of the arrangement (XY = 514 mm, and XZ = ZY = 500 mm) the finished result can be sketched.

9/19 HORIZONTAL BEAM: Vertical loads: Concentrated

In the beam shown in the sketch, the values of the reactions which are 'simple' are given. Determine the shearing force and bending moment at C, D, and B.

S *Structure:*

This is the simplest of the structures which show a *held*-support condition when a section is made. The thrust force in the held support is zero, since all the forces are parallel and vertical.

U *Unknowns and Units:*

There are six unknowns and three FSD's are required. Use kN and m.

R *Rotation:*

Normally, the first FSD would be the whole structure, with COR on the line of one of the reactions, but since the reactions are given in the

9/19

problem, there is no need for this calculation. The COR for each of the other sections is the cut section itself, the discarded part of the beam being replaced by a held support.

E:T *Evaluate: Total is zero:*

FIRST FREE-STRUCTURE DIAGRAM

$$\Sigma M_C = 0$$

$+2 \times 3 - M_C = 0$ $\qquad\qquad +M_C = +6 \text{ kN m}$
$\Sigma F_{NS} = 0; +2 - S_C = 0$ $\qquad\qquad +S_C = +2·0 \text{ kN}$
or, if section taken to right of 3kN,
$\Sigma F_{NS} = +2 - 3 + S_C = 0$ $\qquad\qquad +S_C = +1 \text{ kN m}$

SECOND FREE-STRUCTURE DIAGRAM

$$M_D = 0$$

$2 \times 6 - 5 \times 3 + M_D = 0$ $\qquad\qquad +M_D = +3 \text{ kN m}$

$$\Sigma F_{NS} = 0$$

$-2 + 5 - S_D = 0$ $\qquad\qquad +S_D = +3 \text{ kN}$
or if section cut to left of D,
$+5 - 2 - 2 - S_D = 0$ $\qquad\qquad +S_D = +1 \text{ kN}$

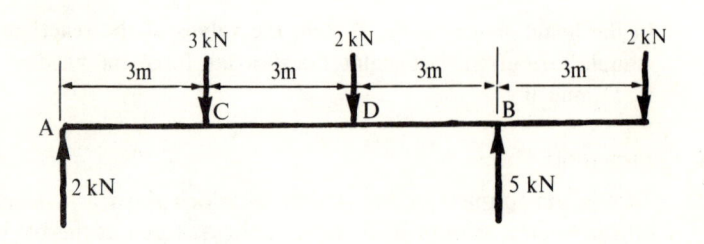

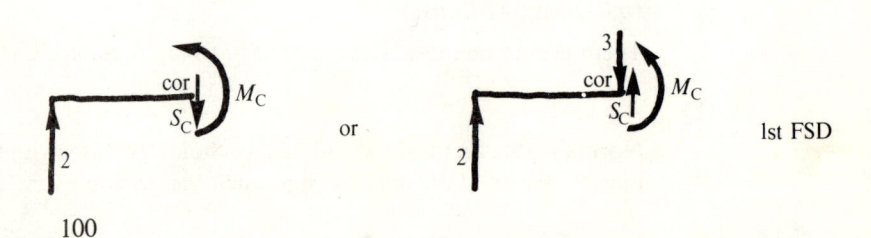

1st FSD

100

2nd FSD

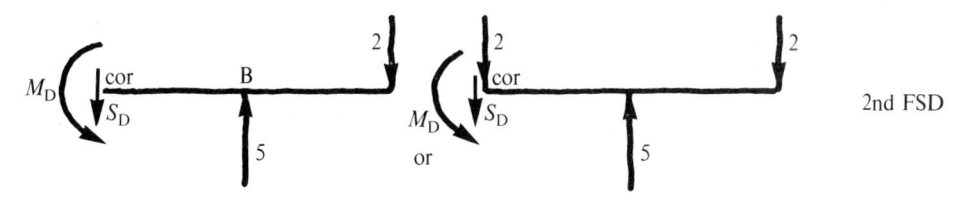

3rd FSD

THIRD FREE-STRUCTURE DIAGRAM

$\sum M_B = 0$

$+2 \times 3 - M_B = 0$ $+M_B = +6$ kN m

$\sum F_{NS} = 0:$ $-2 + S_B = 0$ $+S_B = +2$ kN

or when 5 kN reaction included,

$\sum F_{NS} = 0:$ $+5 - 2 - S_B = 0$ $+S_B = +3$ kN

Y   *Yes or No?*

All forces and couples in the FSD have the correct senses.

9/20   **HORIZONTAL BEAM:** Vertical loads: Concentrated and uniformly distributed

Determine the shearing force and bending moment at C and at D, on the beam sketched.

S   *Structure:*

The uniformly distributed load can be considered as acting at its centre of gravity for the purpose of calculating the bending moments. More problems of this kind are given in Unit 10.

101

9/20

U   *Unknowns and Units:*

The two reactions are given, so there is no need to use the whole structure as a FSD. There are four unknowns, two from each FSD. Use kN and m.

R   *Rotation:*

The **COR** for these points is taken at the cut section which ensures that the unknown value of the shearing stress is cut out of the moment equation.

E:T   *Evaluate and equate to zero:*

FIRST FREE-STRUCTURE DIAGRAM

$$\sum M_C = 0$$
$-3 \cdot 8 \times 5 + 2 \times 3 + M_C = 0 \qquad\qquad +M_C = +\mathbf{13 \cdot 0 \text{ kN m}}$
$$\sum F_{NS} = 0$$
$+3 \cdot 8 - 2 \cdot 0 - S_C = 0 \qquad\qquad +S_C = +\mathbf{1 \cdot 8 \text{ kN}}$

If the section were taken just to the left of 3 kN, the bending moment would be unaltered, but S would be $1 \cdot 2$ kN and of the opposite sense.

SECOND FREE-STRUCTURE DIAGRAM

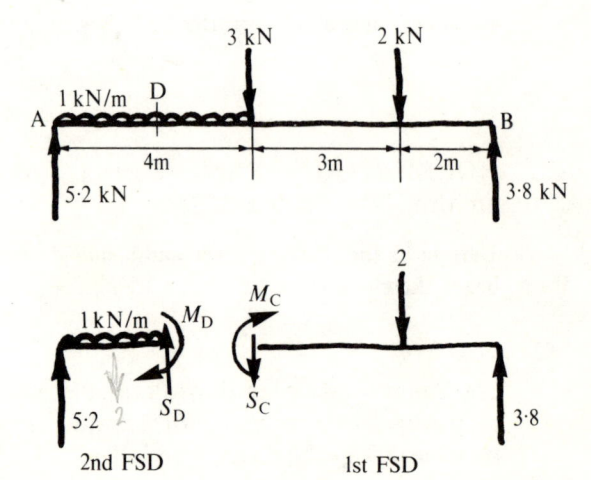

$$\sum M_D = 0$$
$+5 \cdot 2 \times 2 - 2 \times 1 + M_D = 0 \qquad\qquad +M_D = -\mathbf{8 \cdot 4 \text{ kN m}}$
$$\sum F_{NS} = 0$$
$+5 \cdot 2 - 2 + 2 + S_D = 0 \qquad\qquad +S = -\mathbf{3 \cdot 2 \text{ kN}}$

Y *Yes or No?*

The senses of both M_D (clockwise) and of S_D (up) have been taken in the wrong sense. M_D, acting on the broken end, should be anticlockwise, and S_D acting down on the cut section.

9/21 **HORIZONTAL BEAM: U.D.L. With cantilevers: Shear and bending moment**

Find the shearing force and bending moment at the two supports of the beam shown, and also define the values of the greatest positive and greatest negative bending moment on the beam.

S *Structure:*

Horizontal beams with vertical loads are the simplest of the structures in which the shearing force and bending moment must be found in order to provide data for design. The drawing of numbers of FSD is tedious, but necessary until the techniques are thoroughly assimilated. Later, the method of drawing an opaque card across the diagram can be used, as it is in unit 10.

U *Unknowns and Units:*

There are four unknowns, two shearing forces and two bending moments at the two supports. In addition, the maximum values of positive and negative bending must be found. Use kN and m.

R *Rotation:*

COR for first FSD is on the line of the unknown reaction (B).
COR for second FSD is just to the left of the support A.
COR for the third FSD is just to the right of the support B.
COR for the fourth FSD is where the shear in AB passes through zero.

E:T *Evaluate and equate to zero:*

FIRST FREE-STRUCTURE DIAGRAM

$$\sum M_B = 0$$
$$+R_A \times 18 - (1 \times 32)(\text{centre of gravity of load to B}) = 0$$
$$+18R_A - 32(16 - 12) = 0 \qquad\qquad +R_A = +7\cdot1 \text{ kN}$$
$$+R_B = +24\cdot9 \text{ kN}$$

103

9/21

SECOND FREE-STRUCTURE DIAGRAM

$\sum M_A = 0 \qquad \sum F_{NS} = 0 \qquad\qquad +S_A = +2\,\text{kN}$
$\qquad\qquad\qquad\qquad\qquad\qquad\qquad\qquad +M_A = +2\,\text{kN m}$

THIRD FREE-STRUCTURE DIAGRAM

$+F_{NS} = 0 \qquad +12 + S_B = 0 \qquad\qquad +S_B = -12\,\text{kN}$
$+\sum M_B = 0 \qquad +(12 \times 6) - M_B = 0 \qquad +M_B = +72\,\text{kN m}$

FOURTH FREE-STRUCTURE DIAGRAM

The greatest negative bending is the 72 kN m at B.

The greatest positive bending occurs where the shear passes through zero, which is at 5·1 m from A, at which point the total load on the beam to the left of the section is equal to the reaction R_A and the shear is zero.

$$\sum M_X = 0$$
$+5\cdot1 R_A - (7\cdot1 \times 1)3\cdot55 - M_X = 0 \qquad +M_X = +11\cdot0\,\text{kN m}$

The greatest positive bending is thus 11·0 kN m.

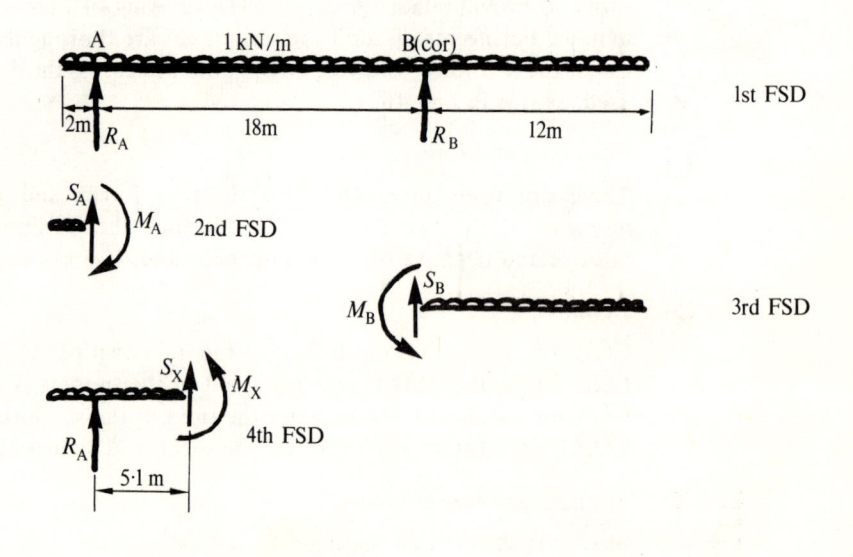

1st FSD

2nd FSD

3rd FSD

4th FSD

9/22

9/22 **HORIZONTAL BEAM: Vertical loads: S.F. and B.M.**

Determine the shearing force and bending moment at the centre of the beam shown, taking into account its own weight of 0·2 kN/m.

S *Structures:*

There must be two FSD's. The first is the whole structure, used to find the reactions (this is standard practice unless the values of reactions are given in the problem). The second is one half of the beam with the effect of the other half replaced by a shearing force and bending couple.

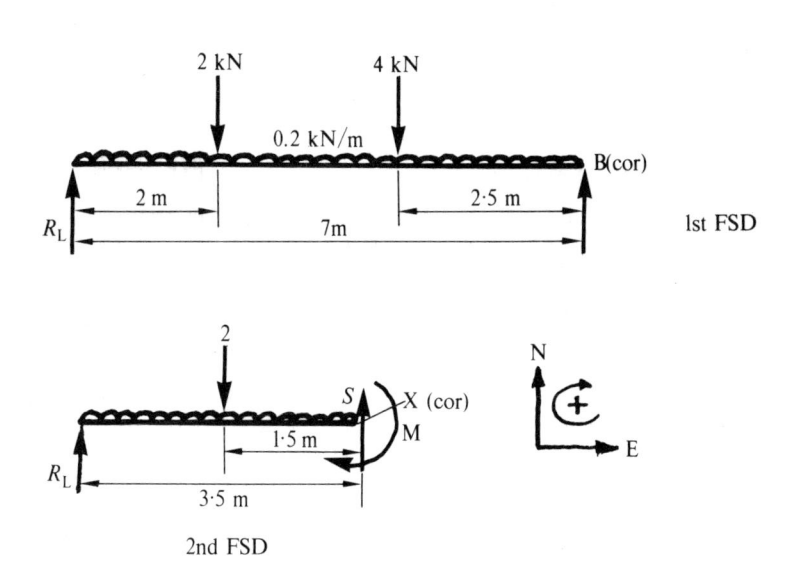

1st FSD

2nd FSD

U *Unknowns and Units:*

Apart from the reactions, there are two unknowns; S and M. The value of thrust, T, is zero since the forces are vertical and the beam horizontal – the simplest condition.

R *Rotation:*

The COR for the reaction R_L is at B on the line of action of the other unknown reaction.

The COR for the second FSD is at X where the bending couple acts.

105

E *Evaluation:*

FIRST FREE-STRUCTURE DIAGRAM

$$\sum M_B = 0$$
$$+R_L \times 7 - 2 \times 5 - 4 \times 2\cdot 5 - (0\cdot 2 \times 7)3\cdot 5 = 0$$
$$\sum F_{NS} = 0 \qquad +R_L + R_R - 7\cdot 4 = 0 \qquad\qquad +R_L = +3\cdot 6 \text{ kN}$$
$$+R_R = +3\cdot 8 \text{ kN}$$

SECOND FREE-STRUCTURE DIAGRAM

$$\sum M_X = 0$$
$$+R_L \times 3\cdot 5 - 2 \times 1\cdot 5 - (0\cdot 2 \times 3\cdot 5)1\cdot 75 + M = 0$$
$$+M = -8\cdot 4 \text{ kN m}$$
$$\sum F_{NS} = 0$$
$$+R_L - 2 - (0\cdot 2 \times 3\cdot 5) + S = 0 \qquad\qquad +S = -0\cdot 9 \text{ kN}$$

Y *Yes or No?*

The senses of S and M have been marked at random on the second FSD. Using the senses marked, the equations show, by the negative sign, that the senses have been marked wrongly. S should act downwards on the left half of the beam, and the bending couple has an anticlockwise moment, rather than the clockwise moment shown.

9/23 **HORIZONTAL BEAM: Inclined loads: Shear, thrust, bending moment**

Calculate the thrust, shear, and bending moment in this beam at E, halfway between C and D.

S *Structure:*

The reaction at B must be inclined, as there is a hinged support. Substitute a vertical and a horizontal component at B. The inclined force must be broken also into horizontal and vertical components according to the shape of the 60° triangle.

U *Unknowns and Units:*

There are three unknowns at the section cut through E. These are M, T, and S. Use kN and m.

106

9/23

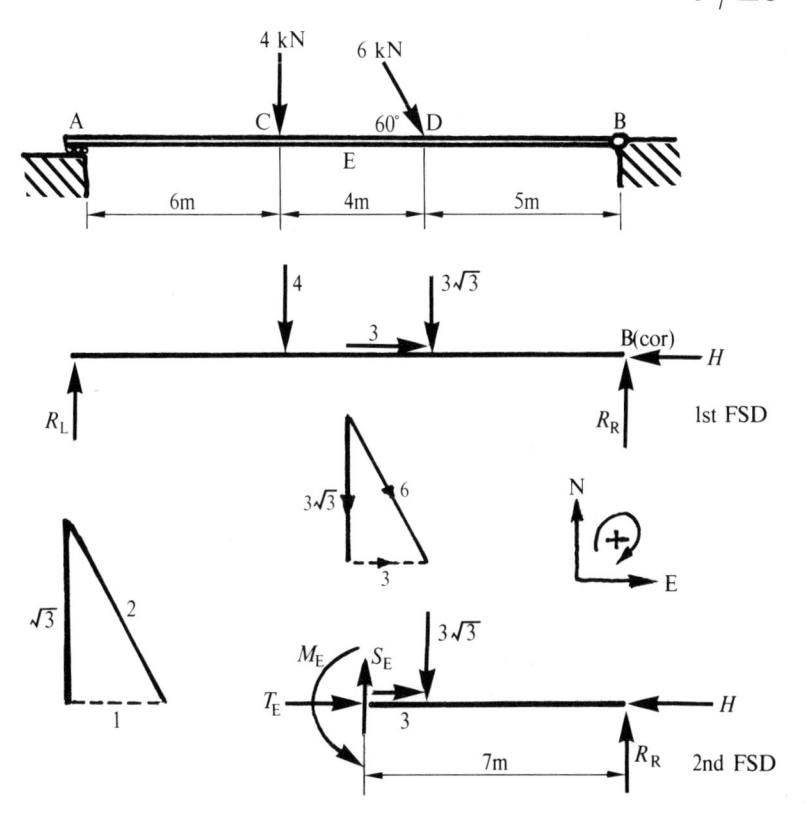

R *Rotation:*

COR for the first FSD is at B, where two unknowns intersect.
COR for the second FSD is at E, the cut section.

E:T *Evaluate: Total is zero:*

FIRST FREE-STRUCTURE DIAGRAM

$$\sum M_B = 0$$
$$+R_L \times 15 - 4 \times 9 - 3\sqrt{3} \times 5 = 0 \qquad +R_L = +4\cdot13 \text{ kN}$$
$$\sum F_{NS} = 0$$
$$+R_L - 4 - 3\sqrt{3} - R_R = 0 \qquad +R_R = +5\cdot07 \text{ kN}$$
$$\sum F_{EW} = 0$$
$$+3 - H = 0 \qquad +H = +3\cdot00 \text{ kN}$$

107

9/24

SECOND FREE-STRUCTURE DIAGRAM

$$\sum M_E = 0$$
$$-R_R \times 7 + 3\sqrt{3} \times 2 - M_E = 0 \qquad +M_E = +25 \cdot 1 \text{ kN m}$$
$$\sum F_{NS} = 0$$
$$+S_E - 3\sqrt{3} + R_R = 0 \qquad +S_E = +0 \cdot 13 \text{ kN}$$
$$\sum F_{EW} = 0$$
$$+T + 3 - H = 0 \qquad +T = \text{zero kN}$$

Y *Yes or No?*
All the unknowns have been given correct senses in the FSD.

9/24 **HORIZONTAL BEAM**: Uniformly distributed load: Hinges inserted

This beam has two hinges at C and D. The loading over the whole length of 25 m is at the rate of 1 kN/m. Find the values of the bending moments and shearing forces at B, at the centre of the suspended span CD, and at the point of maximum bending moment.

S *Structure:*
Hinges are points where no bending moment can be developed, so the moment at hinges is zero. Such a span can be treated as a separate suspended span, simply supported.

U *Unknowns and Units:*
There are six values of shearing force and bending moment asked for. Two FSD's are required. Mark senses of unknowns at random. Use kN and m.

R *Rotation:*
COR taken at points where the values are asked for.

E: T *Evaluate: Total is zero:*
FIRST FREE-STRUCTURE DIAGRAM
$$R_C = R_D = \tfrac{1}{2} \times 1 \times 6 = 3 \text{ kN} \qquad +R_C + R_D = +3 \text{ kN}$$

9/24

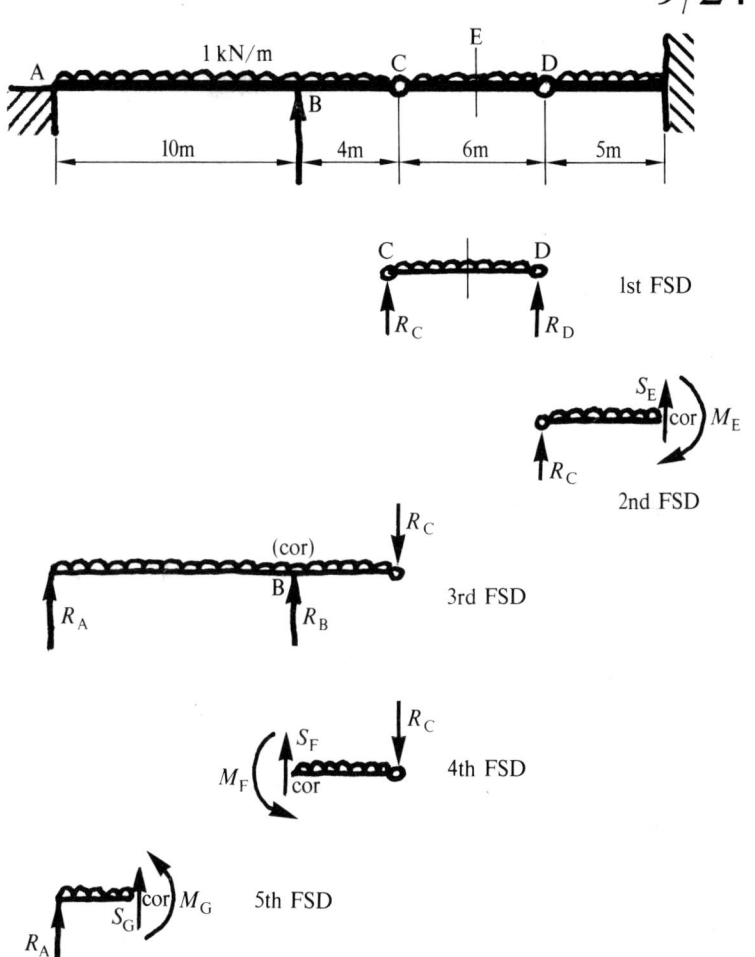

SECOND FREE-STRUCTURE DIAGRAM

$\sum F_{NS} = 0 = 3 - 3 + S_X = 0$ $\qquad +S_X = $ **zero**

$\sum M = 0 + R_C \times 3 \times 1\tfrac{1}{2} + M_E = 0$ $\qquad M_E = -4\tfrac{1}{2}$ **kN m**

THIRD FREE-STRUCTURE DIAGRAM

$$\sum M = 0$$
$+10R_A + 4R_C - 1 \times 14 \times 3 = 0$
$+10R_A + 4 \times 3 - 42 = 0$ $\qquad\qquad +R_A = +3$ **kN**

109

9/25

FOURTH FREE-STRUCTURE DIAGRAM

$$\Sigma M = 0$$
$$+4R_\text{C} + 4 \times 2 - M_\text{F} = 0 \qquad\qquad +M_\text{F} = +20 \text{ kN m}$$
$$\Sigma F_\text{NS} = 0$$
$$-R_\text{C} - 4 + S_\text{F} = 0 \qquad\qquad +S_\text{F} = -7 \text{ kN}$$

FIFTH FREE-STRUCTURE DIAGRAM

Maximum bending moment occurs where the shearing force passes through zero. Since the reaction at A is 3 kN and the loading is 1 kN/m, there is a zero shear at 3 m from A, and this is the point of maximum bending moment.

$$\Sigma M = 0$$
$$+3R_\text{A} - 3 \times 1\tfrac{1}{2} - M_\text{G} = 0 \qquad\qquad M_\text{G} = +4\tfrac{1}{2} \text{ kN m}$$
$$\Sigma F_\text{S} = 0$$
$$+R_\text{A} - 3 + S_\text{G} = 0 \qquad\qquad +S_\text{G} = 0 \text{ as a check}$$

9/25 **INCLINED BEAM**: Vertical load: Shear, thrust, bending moment

The beam shown is resting on rollers at the bottom and is held by a hinge at the top. Find the thrust, shear and bending moment just to the right of the load and just to the left.

S *Structure:*

The first FSD allows you to find the values of the reactions, and the others to determine the values sought. Co-ordinates vertical and horizontal.

U *Unknowns and Units:*

Apart from the reactions, there are six unknowns; the thrust shear and bending moment at two sections close together. Use kN and m.

R *Rotations:*

COR for the first FSD is at the hinge at B.
COR for the second and third FSD is at the cut section X, in each instance.

110

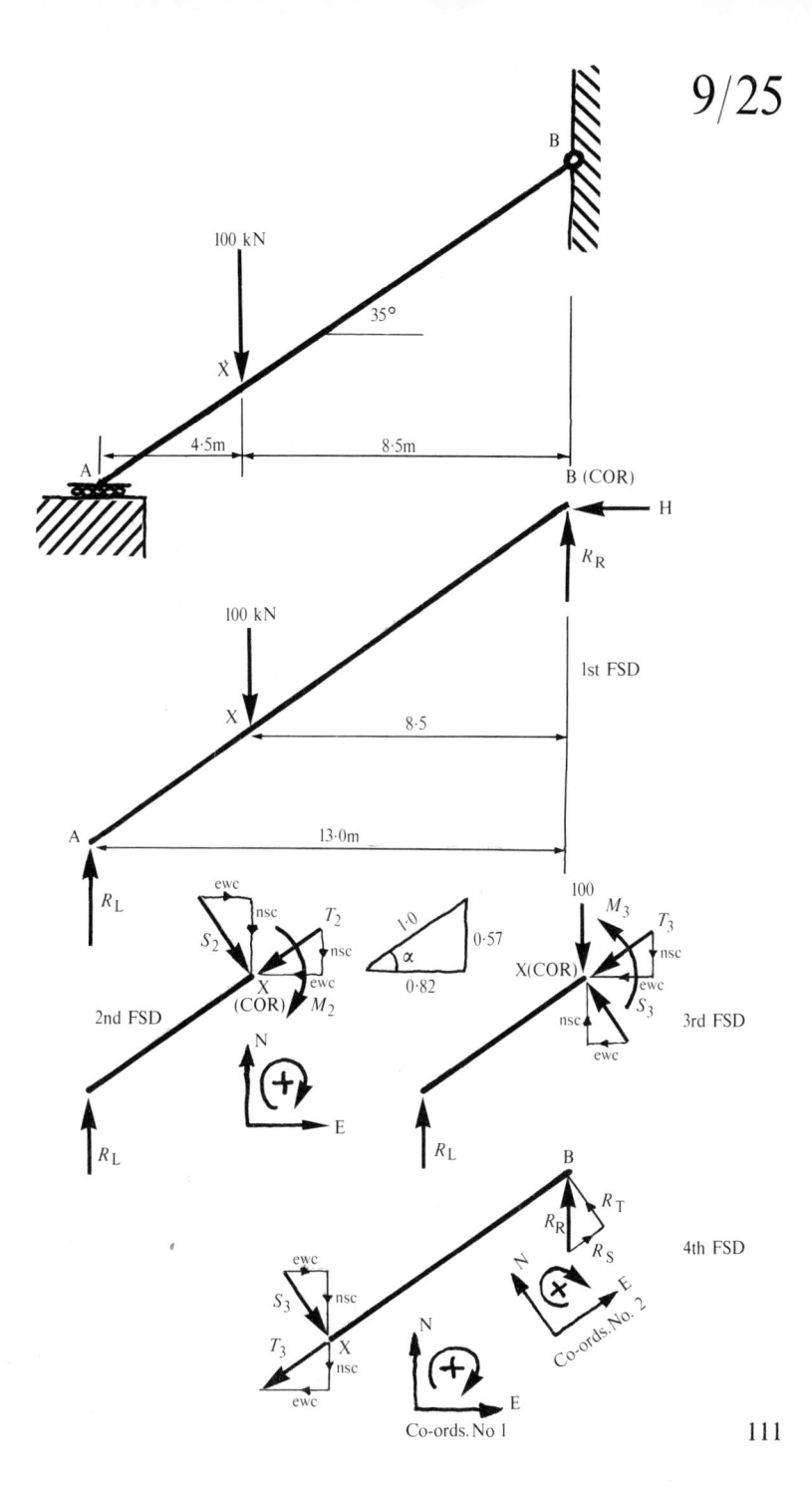

9/25

111

9/25

E : T   *Evaluate: Total is zero:*

FIRST FREE-STRUCTURE DIAGRAM

$$\sum M_B = 0$$

$+R_L \times 13 - 100 \times 8 \cdot 5 = 0$ $+R_L = +65 \cdot 4 \text{ kN}$

SECOND FREE-STRUCTURE DIAGRAM

Before proceeding with calculation, the geometry must be studied. The angle of the beam to the horizontal is given in degrees, and

$$\left. \begin{array}{l} \cos 35° = 0 \cdot 82 \\ \sin 35° = 0 \cdot 57 \end{array} \right\} \text{Ratio: } 1 \cdot 438$$

with these ratios in mind we can obtain the components of S and T, the shearing force and the thrust.

In the FSD both the forces S and T and their components are shown. This is, strictly, wrong for *either* the forces act *or* the components act. However, showing the little triangles resolving the main forces parallel to the two co-ordinate directions, helps in setting up the equations. Remember to accept the sense arrows on the FSD. These have been put in at random, but the final calculation will show whether they are correct or not.

$$\sum F_{NS} = 0$$

$+R_L - \text{nsc } S_2 - \text{nsc } T_2 = 0$
$-0 \cdot 82 S_2 - 0 \cdot 57 T_2 + 65 \cdot 4 = 0$

$$\sum F_{EW} = 0$$

$+\text{ewc } S_2 - \text{ewc } T_2 = 0$
$+0 \cdot 57 S_2 - 0 \cdot 82 T_2 = 0$

Multiply this by the ratio of cos to sin, and solve simultaneously with the NS equation.

$\left. \begin{array}{l} +0 \cdot 82 S_2 - 1 \cdot 18 T_2 = 0 \\ -0 \cdot 82 S_2 - 0 \cdot 57 T_2 + 65 \cdot 4 = 0 \end{array} \right\}$ $+T_2 = +37 \cdot 4 \text{ kN}$

$+0 \cdot 82 S_2 - 1 \cdot 18 \times 37 \cdot 4 = 0$ $+S_2 = +53 \cdot 8 \text{ kN}$

$$\sum M_X = 0$$

$+R_L \times 4 \cdot 5 + M_2 = 0$ $+M_2 = -294 \cdot 3 \text{ kN m}$

Y   *Yes or No?*

The sense of M_2 has been wrongly shown but the other values (S and T) are correct on the FSD. The value of M_2 is correct, but its sense should be anticlockwise.

THIRD FREE-STRUCTURE DIAGRAM

The only difference here is that the 100–kN load is included in the FSD, since the section is cut just to the right of the load. Sense arrows have been chosen at random. It is convenient to be able to do this without puzzling how the forces ought to act, for in the FSD the two forces of 100 kN and S kN can be separated without any disadvantage to the calculation.

$$\sum M_X = 0$$
$+R_L \times 4\cdot 5 + 100 \times 0 - M_3 = 0 \qquad +M_3 = +294\cdot 3 \text{ kN m}$
$$\sum F_{NS} = 0$$
$+R_L + \text{nsc } S_3 - \text{nsc } T_3 - 100 = 0$
$+0\cdot 82 S_3 - 0\cdot 57 T_3 - 34\cdot 6 = 0$
$$\sum F_{EW} = 0$$
$-\text{ewc } S_3 - \text{ewc } T_3 = 0$
$-0\cdot 57 S_3 - 0\cdot 82 T_3 = 0$

Multiply this equation by the ratio of cos to sin (1·438) and solve simultaneously with the NS equation.

$\left. \begin{array}{l} -0\cdot 82 S_3 - 1\cdot 18 T_3 = 0 \\ +0\cdot 82 S_3 - 0\cdot 57 T_3 - 34\cdot 6 = 0 \end{array} \right\}$

In these equations the coefficients of S and T are the same as in the second FSD. The signs are different for you chose the signs for sense quite at random. The numerical value, however, is quite different, and this is more significant. The result of solving the simultaneous equations is $\qquad +T_3 = -19\cdot 8 \text{ kN}$

In substitution remember T_3's sign goes with it.
$-0\cdot 57 S_3 - 0\cdot 82(-19\cdot 8) = 0 \qquad +S_3 = +28\cdot 5 \text{ kN}$

Thus, on one side of the load the thrust and shear are quite different from the values on the other side, but the bending moment remains the same. This is worth remembering in Unit 10.

Y *Yes or No?*

Surprisingly enough, we find that T_3 is not a Thrust but a tension. If this is so, then the two forces at the hinge have the wrong senses. Returning to the first FSD, you can check the results by the use of a fourth FSD, after evaluating H and R_R.

In the first FSD: $\qquad \sum F_{NS} = 0$
$+R_R - 100 + R_L = 0 \qquad\qquad +R_R = +34\cdot 6 \text{ kN}$
$$\sum F_{EW} = 0$$
$-H = 0$ since there is no other horizontal force.

Check FSD: The fourth FSD is obtained by making two cuts, one just to the right of the 100-kN load, and one just to the left of the hinge, and replacing these supports by forces and a couple. At the hinge there is only one vertical force R_R but at the lower end there is a force S, a force T, and a couple whose moment is M. These have been sketched in with the senses shown to act through the previous calculations. Since the forces and couple are acting on the upper part XB, they must have opposite senses to the same forces in the third FSD where they are acting on the lower portion XA. If the calculations have been correct, then all equations should come out to zero for equilibrium. Remember *either* forces parallel to the No. 1 co-ordinates, *or* those parallel to No. 2 co-ordinates are used. (T_3 has been corrected to the proper sense):

With co-ordinates No. 1

$$\sum F_{NS} = 0$$
$-\text{nsc } S_3 - \text{nsc } T_3 + R_R = 0$
$-0·82 \times 28·5 - 0·57 \times 19·8 + 34·6 = 0$
$$\sum F_{EW} = 0$$
$+\text{ewc } S_3 - \text{ewc } T_3 = 0$
$+0·57 \times 28·5 - 0·82 \times 19·8 = 0$

With co-ordinates No. 2
$$\sum F_{NS} = 0$$
$-S_3 + R_S = 0$
$-28·5 + 0·82 \times 34·6 = 0$
$$\sum F_{EW} = 0$$
$-T_3 + R_1 = 0$
$-19·8 + 0·57 \times 34·6 = 0$

All equate to zero, which checks the accuracy of the work.

9/26 **INCLINED BEAM:** Applied couple: Thrust, shear, and bending moment

In the beam shown, the vertical load of 20 kN is combined with an applied couple of -10 kN m. Find the values of the thrust, shear, and bending moment at each side of the load and couple.

S *Structure:*

The direction taken by the chosen co-ordinates must be such as to give an easy and smooth solution to the problem. The choice should be made separately for each problem. Usually, since loads acting as weights are concerned, the best directions are likely to be vertical and horizontal.

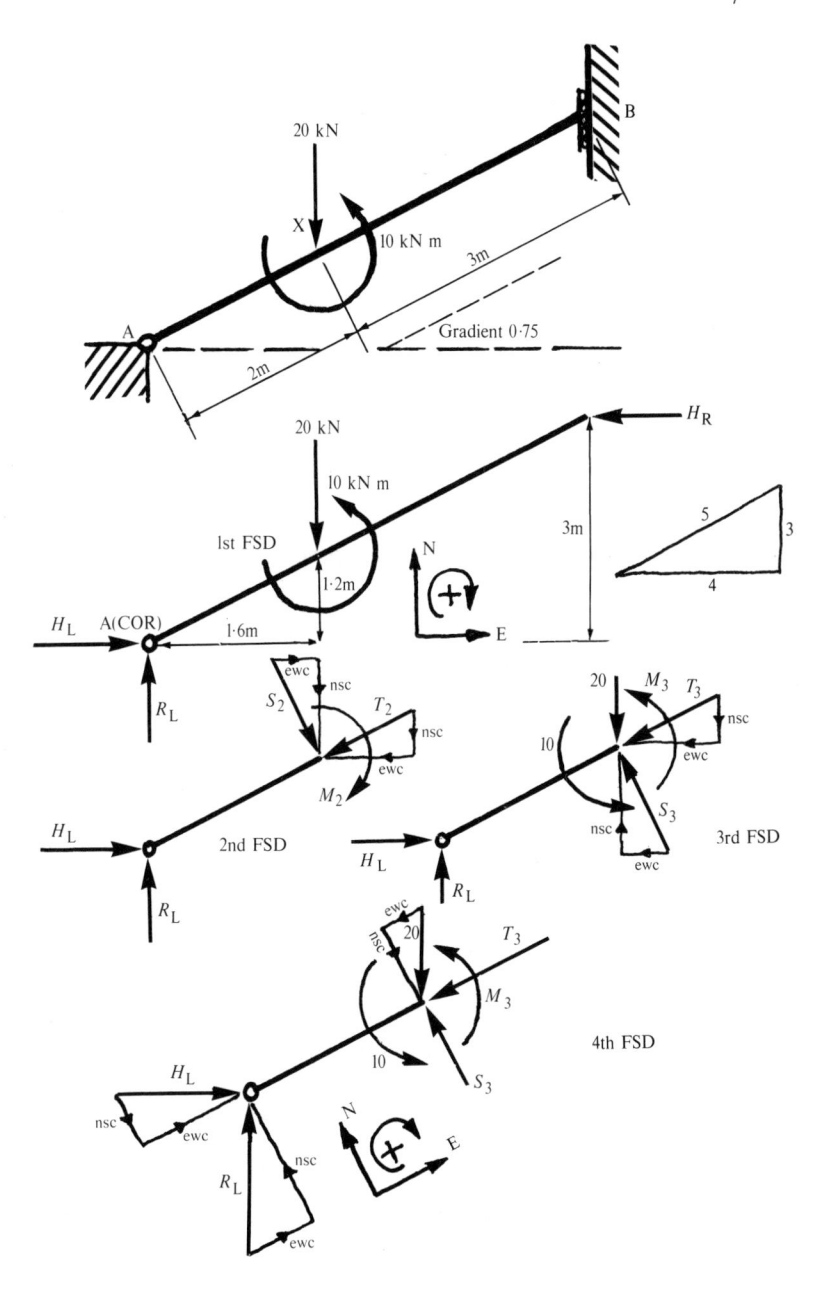

9/26

9/26

U *Unknowns and Units:*

After the components of the reactions have been found, there are six unknowns. Use kN and m.

R *Rotation:*

COR for the first FSD must be where two unknowns intersect if an easy solution is sought for (at A).
COR for the other two FSD's are at X, the cut section.

E: T *Evaluate: Total is zero:*

FIRST FREE-STRUCTURE DIAGRAM

$$\sum M_A = 0$$
$+20 \times 1\cdot6 - H_R \times 3 - 10 = 0$ $+H_R = +7\cdot33$ kN
$\sum F_{NS} = 0 \quad \sum F_{EW} = 0$ $+H_L = +7\cdot33$ kN
$+R_L - 20 = 0: +H_L - H_R = 0$ $+R_L = +20$ kN

SECOND FREE-STRUCTURE DIAGRAM

$$\sum M_X = 0$$
$+R_L \times 1\cdot6 - H \times 1\cdot2 + M_2 = 0$
$+32 - 8\cdot8 + M_2 = 0$

$$\sum F_{NS} = 0$$
$+R_L - \text{nsc } S_2 - \text{nsc } T_2 = 0$
$+20 - 0\cdot8 S_2 - 0\cdot6 T_2 = 0$

$$\sum F_{EW} = 0$$
$+H_L - \text{ewc } S_2 - \text{ewc } T_2 = 0$ $+M_2 = -23\cdot2$ kN m
$+7\cdot3 + 0\cdot6 S_2 - 0\cdot8 T_2 = 0$ $+T_2 = +17\cdot8$ kN
     $+S = +11\cdot6$ kN

THIRD FREE-STRUCTURE DIAGRAM

$$\sum M_X = 0$$
$+R_L \times 1\cdot6 - H_L \times 1\cdot2 - 10 - M_3 = 0$
$+32\cdot0 - 8\cdot8 - 10 - M_3 = 0$

$$\sum F_{NS} = 0$$
$+R_L - 20 + \text{nsc } S_3 - \text{nsc } T_3 = 0$
$+0\cdot8 S_3 - 0\cdot6 T_3 = 0$

$$\sum F_{EW} = 0$$
$+H_L - \text{ewc } S_3 - \text{ewc } T_3 = 0$ $+M_3 = +13\cdot2$ kN m
$+7\cdot3 - 0\cdot6 S_3 - 0\cdot8 T_3 = 0$ $+T_3 = +5\cdot9$ kN
     $+S_3 = +4\cdot4$ kN

Alternative evaluation with other co-ordinate directions

For those who find the solution of simultaneous equations unwelcome, the alternative of choosing the co-ordinate directions parallel to and at right angles to the beam can be used. There is usually more labour in resolving forces in this method, since most loads on structures are vertical. However, you ought to solve other problems in this unit by this method and decide which attack is preferable to you. As an example, a fourth FSD for this problem has been drawn, and forces resolved parallel to and normal to the beam. The solution is given below.

FOURTH FREE-STRUCTURE DIAGRAM

$$\sum F_{NS} = 0 \quad \text{(forces normal to the beam)}$$
$$+\text{nsc } R_L - \text{nsc } H_L - \text{nsc } 20 + S_3 = 0$$
$$+0.8 \times 20 - 0.67 \times 7.3 - 0.8 \times 20 + S_3 = 0$$
$$\sum F_{EW} = 0 \quad \text{(forces along beam)}$$
$$+\text{ewc } R_L + \text{ewc } H_L - \text{ewc } 20 - T_3 = 0$$
$$+0.6 \times 20 + 0.8 \times 7.3 - 0.6 \times 20 - T_3 = 0$$
$$\sum M_X = 0$$
$$-10 + \text{nsc } R_L \times 2 - \text{nsc } H_L \times 2 - M_3 = 0$$
$$-10 + 0.8 \times 20 \times 2 - 0.6 \times 7.3 \times 2 - M_3 = 0$$

$$+M_3 = +13.2 \text{ kN m}$$
$$+T_3 + 5.9 \text{ kN}$$
$$+S_3 = +4.4 \text{ kN}$$

9/27 FIVE EQUATIONS: Two FSD'S: Simple and direct supports: Pin-jointed frame

A bridge, constructed to carry a pipe across a stream, is shown in the sketches. The loading is applied to the two girders by cross-girders at the panel points. The loading shown in the sketches is what is applied to one of the two main girders, one on each side of the pipe. All the members of the truss are equal in length. Find the forces in the members Z, X, and Q.

S *Structure and Supports:*

In a girder of this kind, the assumption is that the members are pin-jointed at the ends, so the supporting conditions for each member are *direct* (see Unit 6). The supporting conditions for the truss itself are *simple*.

9/27

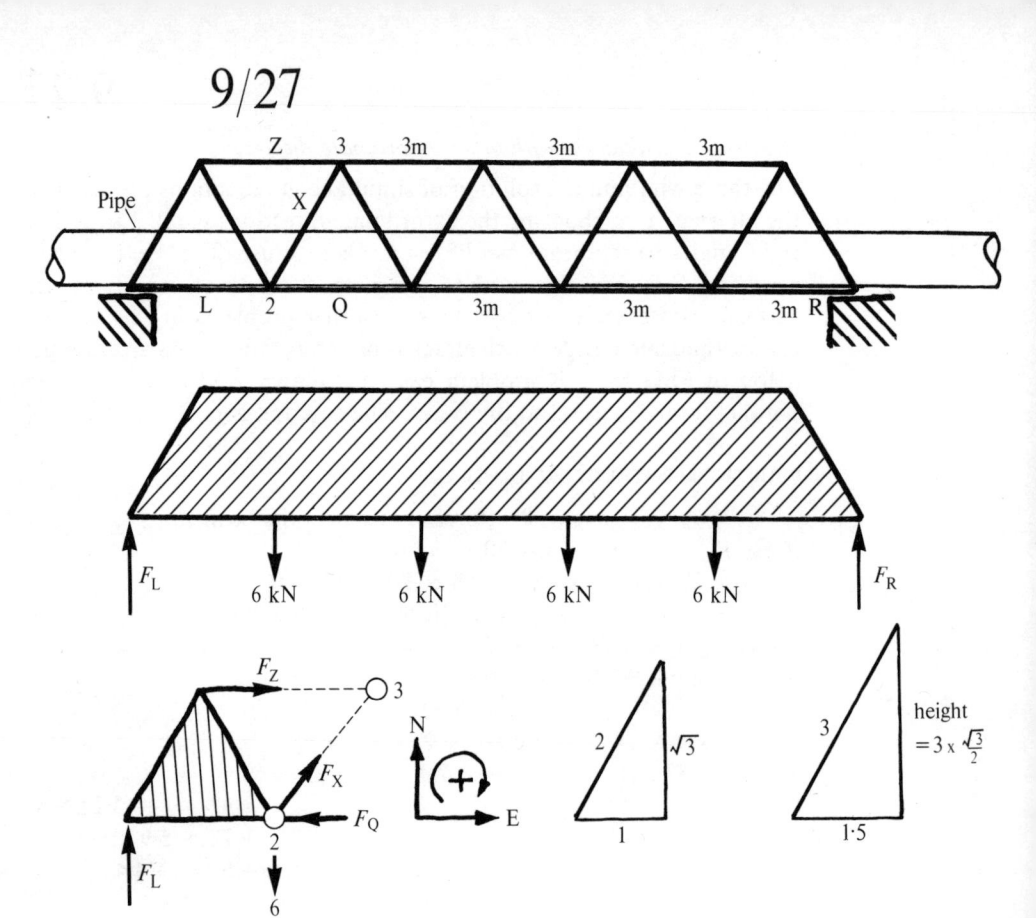

U Unknowns:

There are five unknowns, F_L, F_R, F_Z, F_X, F_Q, the first two being the supporting conditions of the truss, and the last three the forces in Z, X, and Q which are required to be found.

Since one FSD can result in only three unknowns being determined, it is clear that two FSD's are required. The first one to be studied is the whole structure. Use kN and m.

FIRST FREE-STRUCTURE DIAGRAM

R Rotation:

The loading on the truss is symmetrical, and all the forces are parallel, so no COR is required.

118

E: T *Evaluate and equate to zero:*

$$\sum F_{NS} = 0$$
$$+F_L - 6 - 6 - 6 - 6 + F_R = 0 \qquad\qquad +F_L + F_R = +24\,\text{kN}$$

Strictly speaking, another equation would be required to obtain the values of F_L and F_R, but from everyday experience it is clear that F_L and F_R must be equal. However, from

$$\sum M_L = 0: \qquad\qquad +F_R = +12\,\text{kN}$$

Both are of the correct sense, since there are two plus signs.

SECOND FREE-STRUCTURE DIAGRAM

The second FSD must have the forces in Z, X, and Q shown as external forces, since the procedure used here deals only with forces acting externally on the structure. A cut, as with a pair of scissors, leaves two portions, a short one and a longer one. The same result will be obtained from either end, so discard the more complex of the two and use the left-hand portion as the new FSD.

R *Rotation:*

The COR for F_Z is where the other two intersect – at panel point 2.
The COR for F_X is where the other two intersect – at infinity, since they are parallel. No moment equation is, therefore, possible. Use force equation.
The COR for F_Q is where the other two intersect – at panel point 3.

E: T *Evaluate and equate to zero:*

Member Z: COR at panel point 2. Sense taken at random.

$$\sum M_2 = 0$$

Forces acting: $\quad F_L \quad F_Z \quad F_X \quad F_Q \quad 6 \quad$ (kN)
Moment arm
about pp2: $\qquad\quad$ 3 $\quad$ height $\quad$ zero $\quad$ zero $\quad$ zero (m)
Moments about pp2: $+12 \times 3 + F_Z \times [3 \times \sqrt{(3)}/2] = 0$
$\qquad\qquad\qquad\quad +(12 \times 3)/3\sqrt{(3)}/2 + F_Z = 0$
$$+F_Z = -8\sqrt{3}\,\text{kN}$$

Member Q: COR at pp3 (this is outside the FSD). Sense taken at random.

$$\sum M_3 = 0$$

Forces acting: $\quad F_L \quad F_Z \quad F_X \quad F_Q \quad 6 \quad$ (kN)

9/27

Moment arm about
pp 3:   4·5   zero   zero   height   1·5
Moments about pp 3: $+12 \times 4·5 + F_Q \times [3 \times \sqrt{(3)}/2] - 9$
$$+F_Q = -10\sqrt{3} \text{ kN}$$

Member X: Use force equation since other two unknown forces are parallel
$$\sum F_{NS} = 0$$
Forces acting: $+F_L + \text{ncs } F_X - 6 = 0 \text{ (kN)}$
(the other two have no component in the N–S direction)
$+6 + \text{nsc } F_X = 0$
$+6 + \sqrt{(3)}/2 F_X = 0$ $+F_X = -4\sqrt{3} \text{ kN}$

Y   *Yes or No?*

All of the senses have been guessed wrongly, so it is best to draw a new FSD showing the correct senses. The meaning of these arrows is important.

Draw a third FSD, that of the member X, cutting off everything beyond the end hinges. If the arrow is carried with this FSD, it now becomes an internal force. FSD's do not deal with internal forces, so it is replaced by the opposite and equal external force (F_Y). But the new FSD is now unbalanced and not in equilibrium. Since in a directly supported member the supporting forces act from hinge to hinge (see Unit 6), there must be an equal and opposite force at the other hinge. The FSD is now (c) in equilibrium, and the forces are external forces, as they should be.

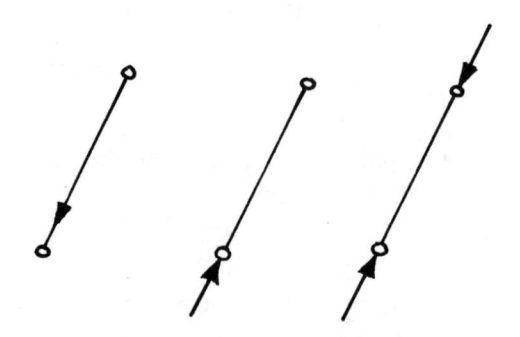

The conclusion is that an internal force, found from a second FSD, as here, to be acting outwards, shows that the external forces at the hinges must be acting inwards. This indicates that the member in question is in *compression*. If two friends pull on your arms they are putting you in *tension* by outwardly acting external forces. Your internal resistance, however, is inwards. You can feel yourself to be in *tension*.

The rule, therefore is:

Internal forces acting *inwards* within a member in a pin-jointed truss indicate *tension*.

Internal forces acting *outwards* within a member in a pin-jointed truss indicate *compression*.

It is much more difficult to design and construct a pin-ended member to resist compression than it is to design a member in which the force is tensile. You can try this with a lath or a slender rule. Tension keeps the lath straight, whereas a push inwards at each end, to put it into compression tends to cause buckling.

You must never, therefore, merely give the values of the forces in members of a pin-jointed frame. Each value must be accompanied by your decision as to whether the force is putting the member into tension or compression. For this problem the answer must be stated as shown below. Note that no signs are required. These were used to guide you on your way. It is now of paramount importance to state *tension* or *compression*. Signs do not matter.

$$F_Z = 8\sqrt{3} \text{ kN} \quad \text{(compression)}$$
$$F_X = 8\sqrt{3} \text{ kN} \quad \text{(compression)}$$
$$F_Q = 12\sqrt{3} \text{ kN} \quad \text{(tension)}$$

9/28 ONE FSD: Pin-jointed truss: Parallel booms

Determine the forces in the members 1, 2, and 3 of the truss shown. All vertical and horizontal members have a length of 4 m.

S *Structure:*
One FSD is sufficient to obtain all the values required. Use section XX.

U *Unknowns and Units:*
There are three unknowns. Units are kN and m.

9/28

R *Rotation:*

COR for F_1 is at intersection of F_2 and F_3 (P).
COR for F_2 is at infinity as F_1 and F_3 are parallel.
COR for F_3 is at intersection of F_1 and F_2 (C).

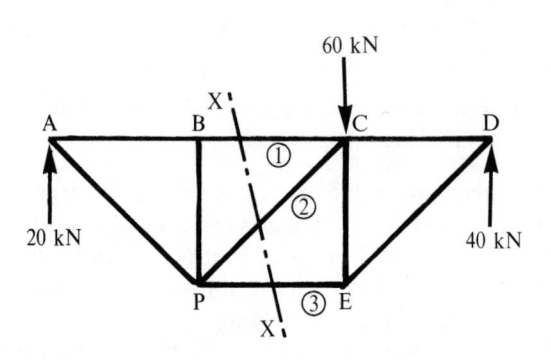

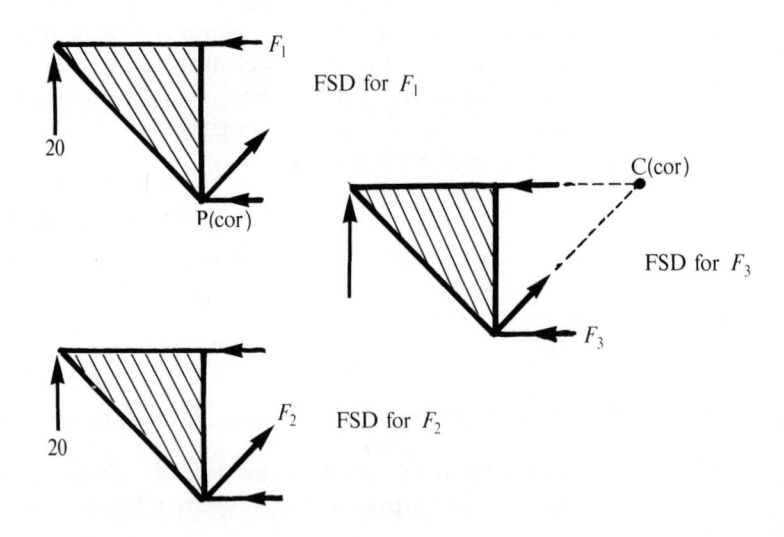

FSD for F_1

FSD for F_3

FSD for F_2

E:T *Evaluation: Total is zero:*
Derivation for F_1

$$\sum M_P = 0$$

$-F_1 \times 4 + 20 \times 4 = 0$ $\qquad +F_1 = +\mathbf{20}\text{ kN}$

Derivation for F_2

Since the intersection of F_1 and F_3 is at infinity, one of the other equations of equilibrium must be used.

$$\Sigma F_{NS} = 0$$
$$+20 + \text{nsc } F_2 = 0 \qquad\qquad +F_2 = -20\sqrt{2} \text{ kN}$$

Derivation for F_3

$$\Sigma M_C = 0$$
$$+F_3 \times 4 + 20 \times 8 = 0 \qquad\qquad +F_3 = -40 \text{ kN}$$

Y *Yes or No?*

Senses are wrongly marked for F_2 and F_3. F_2 should be in compression, and F_3 in tension. This could have been guessed at the start, but there is no advantage in trying to be 'correct'. The correct result comes automatically if the sense arrows are put in at random.

9/29 THREE FSD'S: Pin-jointed frame: Cantilever

Determine the forces in the members 1, 2, and 3 in the frame sketched. All vertical and horizontal members are 5 m long.

S *Structure:*

The placing of the members whose loading is required are such that three FSD's are required. There is no need to find the reactions. Work from the free end of the cantilever.

U *Unknowns and Units:*

There are three unknowns. The units are kN and m.

R *Rotation:*

COR for F_1 is at A where the other two members cut have their intersection. This selection removes these unknowns from the equation.
No other COR is required as members 2 and 3 can each be solved by using a force equation.

123

9/29

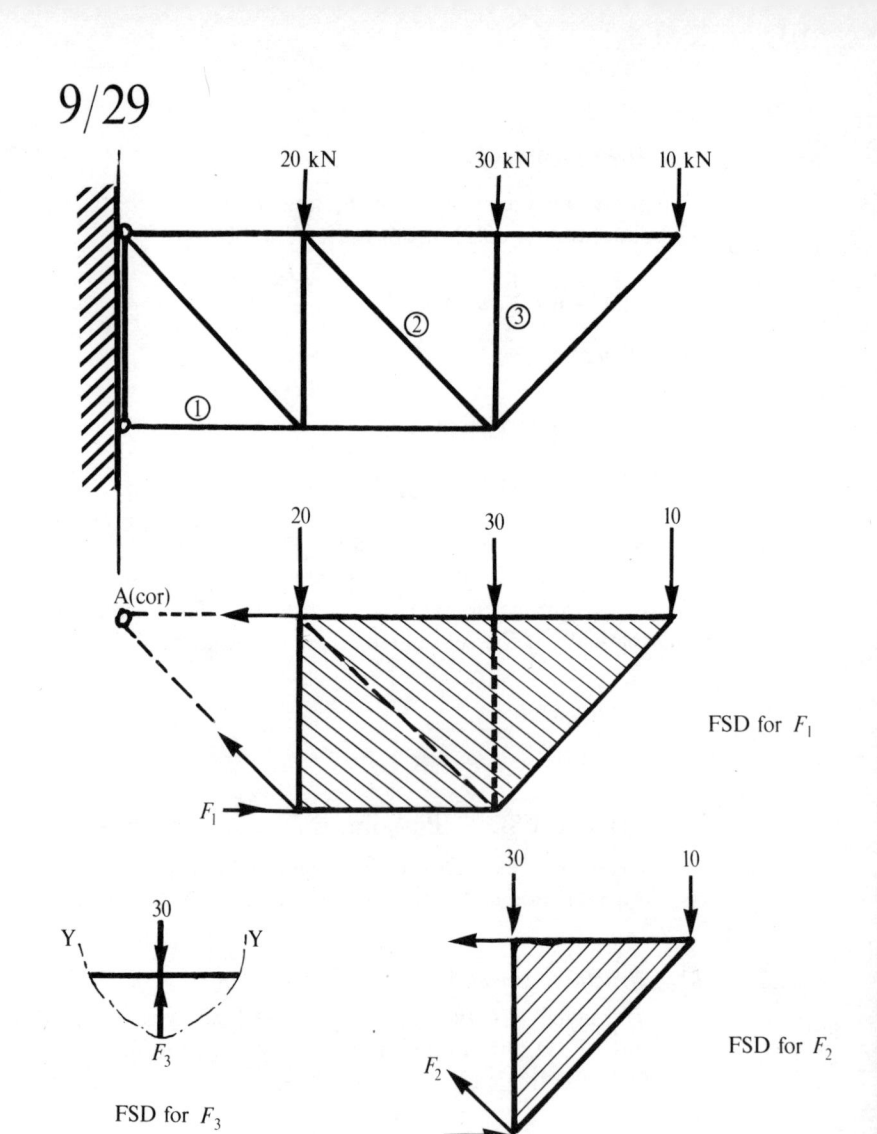

FSD for F_1

FSD for F_3

FSD for F_2

E:T *Evaluation: Total is zero:*
Derivation for F_1

$$\sum M_A = 0$$
$-F_1 \times 5 + 20 \times 5 + 30 \times 10 + 10 \times 15 = 0$
$+5F_1 = +550$ $\qquad\qquad +F_1 = +\mathbf{110}$ **kN**

124

Derivation for F_2

No moment equation is possible since the upper and lower booms do not meet. Use a force equation:
$$\sum F_{NS} = 0$$
$-30 - 10 + \text{nsc } F_2 = 0$
$+\text{nsc } F_2 = +40$ $\qquad +F_2 = +56·6 \text{ kN}$

Derivation for F_3

Sections cutting through frames need not be straight. Here, a straight section through F_3, the unknown could be used with a moment equation about B. However, in this instance and, sometimes when a straight section cuts more than three members (rendering the problem insoluble by the methods of this book) it is simpler to use a force equation of equilibrium, and a curved section line YYY.
$$\sum F_{NS} = 0$$
$-30 + F_3 = 0$ $\qquad +F_3 = +30 \text{ kN}$

9/30 THREE FSD'S: Pin-jointed frame: Parallel booms

Determine the forces in the three members marked 1, 2, and 3. All members of the girder are 3 m in length.

S *Structure:*

The vertical height of the girder is obtained from the geometry of an equilateral triangle. The height is $1·5\sqrt{3} = 2·60$ m.

U *Unknowns and Units:*

There are three unknowns. Units are kN and m.

R *Rotation:*

COR for F_1 is at G: COR for F_3 is at A.
Since the booms are parallel, a moment equation cannot be used for F_2, and a force equation gives the result required.

9/30

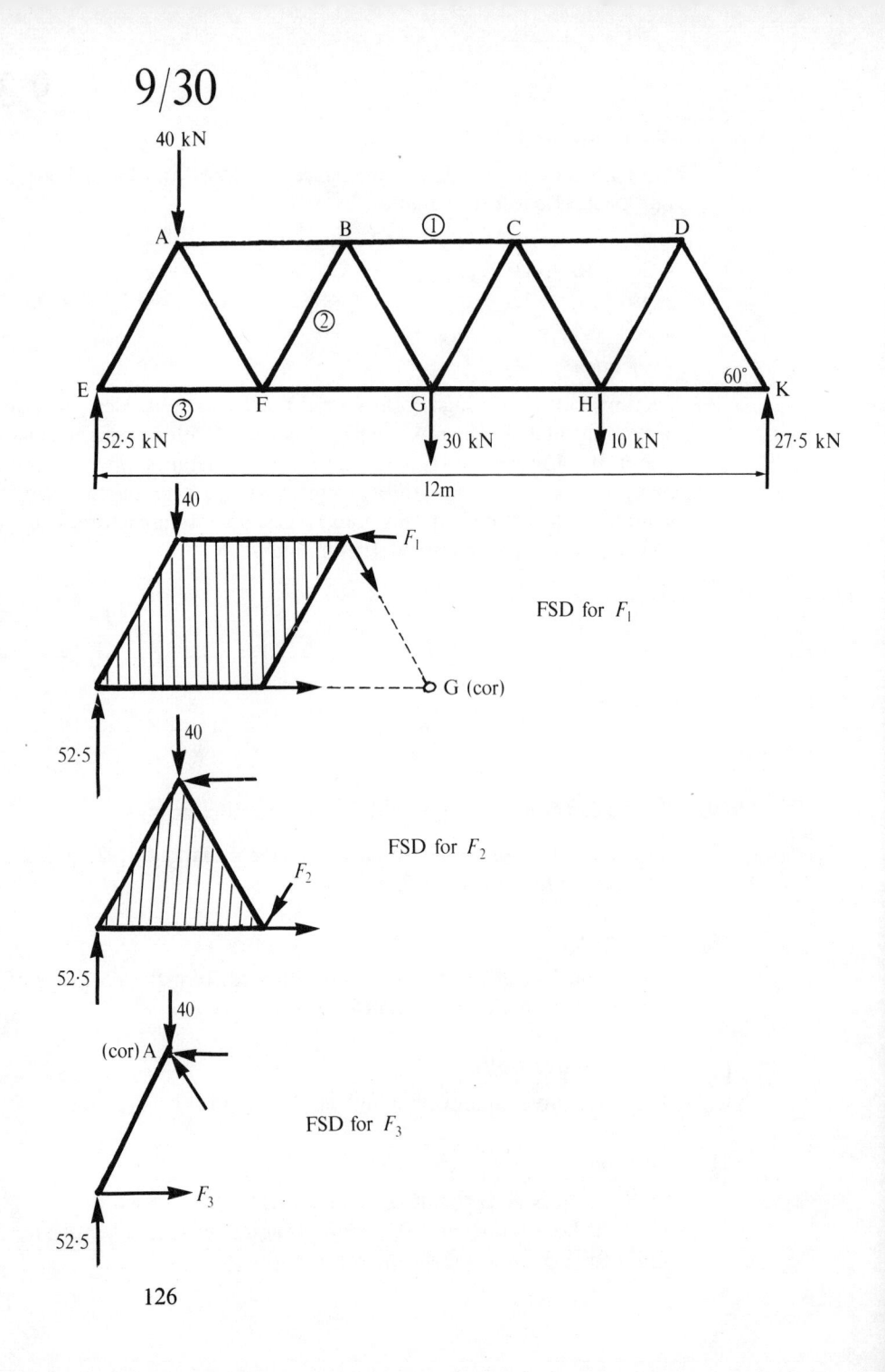

FSD for F_1

FSD for F_2

FSD for F_3

126

E : T   *Evaluation: Total is zero:*

Derivation for F_1

$$\sum M_G = 0$$

$-F_1 \times 2 \cdot 60 \text{(vertical height)} - 40(1 \cdot 5 \times 3) + 52 \cdot 5 \times 6 = 0$

$-2 \cdot 60 F_1 - 180 + 315 = 0$ $+F_1 = +51 \cdot 92 \text{ kN}$

Derivation for F_2

$$\sum F_{NS} = 0$$

$+52 \cdot 5 - 40 - \text{nsc } F_2 = 0$

$\frac{3}{2} F_2 \; 12 \cdot 5$ $+F_2 = +14 \cdot 43 \text{ kN}$

Derivation for F_3

$$\sum M_A = 0$$

$-F_3 \times 2 \cdot 60 + 52 \cdot 5 \times 0 \cdot 75 = 0$ $+F_3 = +15 \cdot 14 \text{ kN}$

9/31   ONE FSD: Symmetrical pin-jointed frame: Simple supports

Determine the forces in members 1, 2, and 3 on the symmetrical frame shown. The loading is also symmetrical.

S   *Structure:*

The FSD is obtained by the section XX.

U   *Unknowns and Units:*

There are three unknowns. Units are kN and m.

R   *Rotation:*

COR for F_1 is where F_2 and F_3 intersect (M).
COR for F_2 is where F_1 and F_3 intersect (G).
COR for F_3 is where F_1 and F_2 intersect (E).

E : T   *Evaluation: Total is zero:*

Derivation for F_1   (FM = 1·5 m)

$$\sum M_M = 0$$

$+F_1 \times 1 \cdot 5 - 30 \times 4 = 0$ $+F_1 = +80 \text{ kN}$

127

9/31

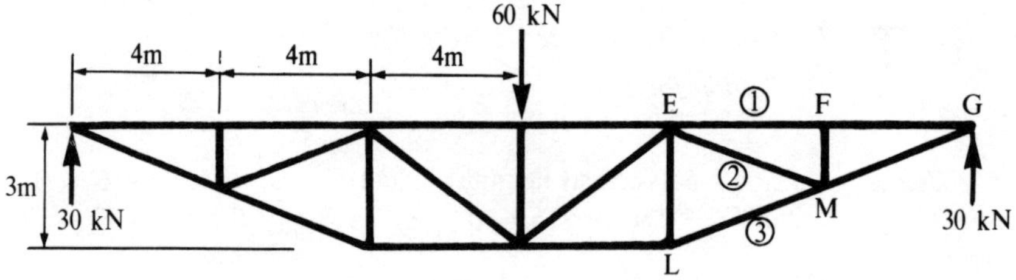

FSD for F_1

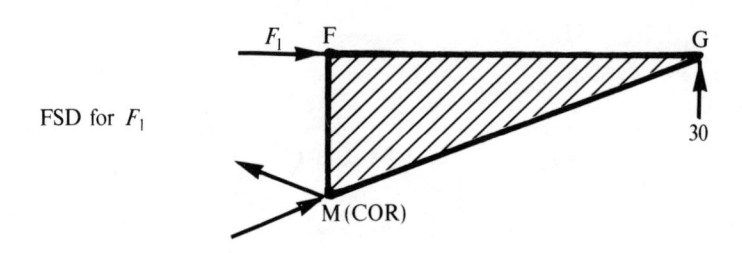

FSD for F_2

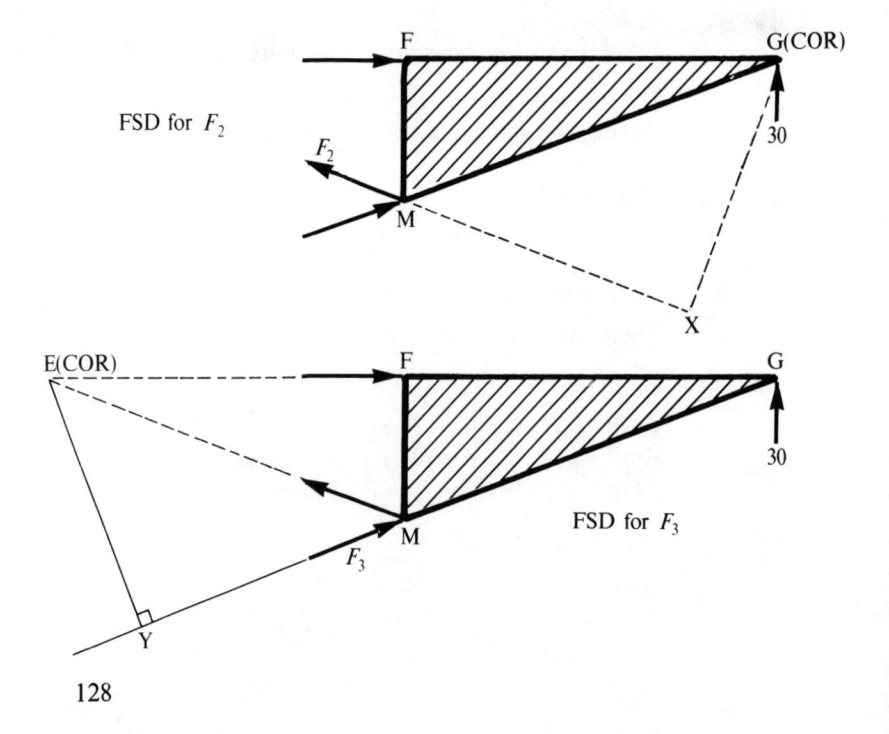

FSD for F_3

128

Derivation for F_2
$$\sum M_G = 0$$
Since there is no other force on the FSD causing a moment about G, F_2 is zero for this loading.
$$+F_2 = 0 \text{ kN}$$

Derivation for F_3
$$\sum M_E = 0$$
$-F_3 \times \text{YE} - 30 \times \text{GE} = 0$
Find value of YE:
$$\text{MG} = \sqrt{(4^2 + 1 \cdot 5^2)} = 4 \cdot 27 \text{ m}$$
Compare triangles GYE and GFM
$$\frac{\text{YE}}{\text{FM}} = \frac{\text{EG}}{\text{MG}} \quad \text{YE} = \frac{\text{EG} \times \text{FM}}{\text{MG}} = \frac{8 \times 1 \cdot 5}{4 \cdot 27} = 2 \cdot 81 \text{ m}$$
$$\sum M_E = 0$$
$-F_3 \times 2 \cdot 81 - 30 \times 8 = 0 \qquad\qquad +F_3 = -85 \cdot 4 \text{ kN}$

Yes or No?

F_1 has been correctly marked in compression, but F_3 showing a negative sign, must have the sense arrow the other way, giving tension instead of the compression shown in the FSD.

9/32 FORCES IN PIN-JOINTED FRAME: Unsymmetrical: Cantilever

Find the forces in members 1, 2, and 3 of the cantilever shown.

S *Structure:*

The FSD cut off by the section XX is sufficient to find the values of the three forces, F_1, F_2, and F_3.

U *Unknowns and Units:*

There are three unknowns. Use kN and m. Mark sense arrows at random.

R *Rotation:*

COR for F_1 is where F_2 and F_3 intersect (E).
COR for F_3 is where F_1 and F_2 intersect (B).
F_2 does not require a COR as it supports 20 kN.

129

9/32

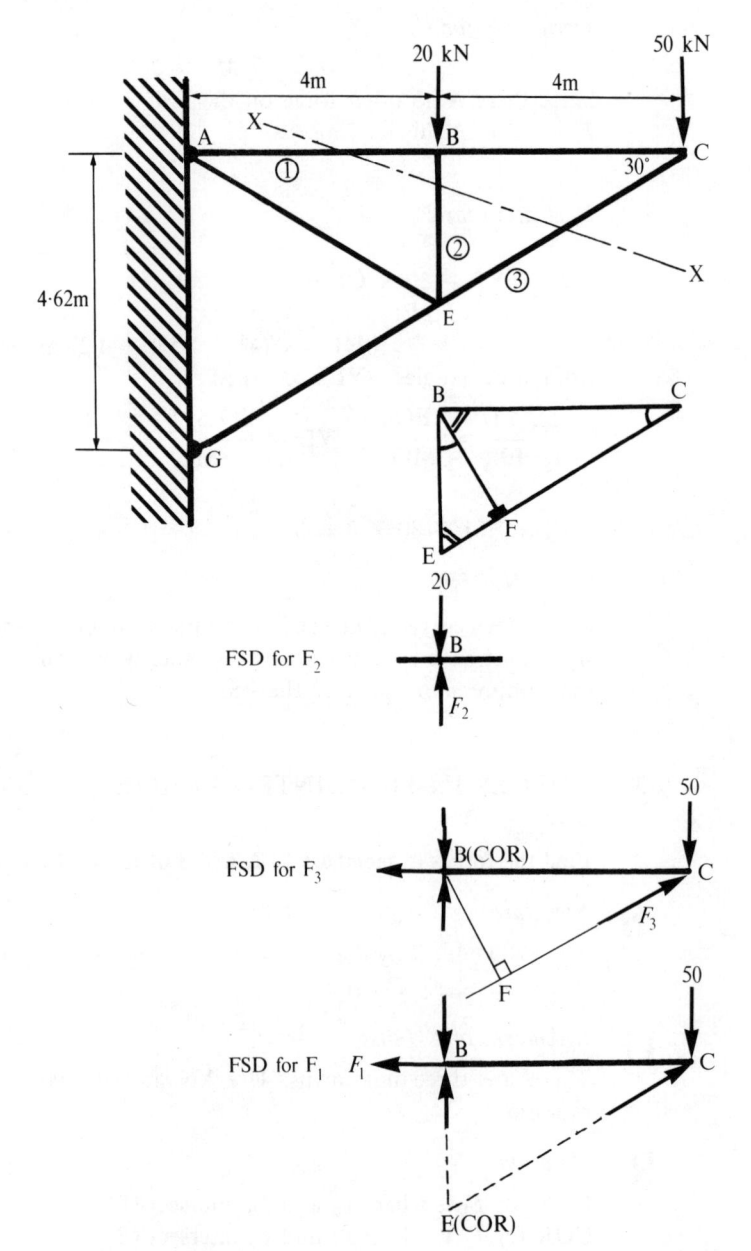

FSD for F_2

FSD for F_3

FSD for F_1

Distances required:
(a) BE = $4/\sqrt{3}$ = 2·31 m   CE = $2 \times 4/\sqrt{3}$ = 4·62 m
(b) Comparing triangles CFB and CBE

$$\frac{BF}{BE} = \frac{BC}{CE} \qquad BF = \frac{BC \times BE}{CE}$$

$$BF = \frac{4\cdot00 \times 2\cdot31}{4\cdot62} = 2\cdot0 \text{ m}$$

E : T *Evaluation: Total is zero:*

$$\sum M_E = 0$$
$-F_1 \times BE + 50 \times BC = 0$
$-F_1 \times 2\cdot31 + 50 \times 4\cdot00 = 0$ $\qquad +F_1 = +\mathbf{86\cdot6\text{ kN}}$

$$\sum M_B = 0$$
$-F_3 \times BF + 50 \times 4BC = 0$
$-F_3 \times 2\cdot0 + 50 \times 4\cdot0 = 0$ $\qquad +F_3 = +\mathbf{100\cdot0\text{ kN}}$

$$\sum F_{NS} = 0$$
$+F_2 - 20 = 0$ $\qquad\qquad\qquad +F_2 = +\mathbf{20\cdot0\text{ kN}}$

Y *Yes or No?*

All three results show double positive signs which indicates that the sense arrows were drawn correctly.

9/33   FORCES IN PIN-JOINTED FRAME: Unsymmetrical: One FSD

Find the forces acting in members 2 and 3 of the truss shown.

S *Structure:*

The FSD is obtained by cutting through AB, BE, and EF. Both forces 2 and 3 can be found by a single FSD.

U *Unknowns and Units:*

(1) Force F_2; (2) Force F_3.
Apply sense arrows at random. Use kN and m.

R *Rotation:*

COR for F_2 is where F_1 and F_3 intersect (Z).
COR for F_3 is where F_1 and F_2 intersect (B).

9/33

Distances required:
(a) From A to C the slope is 3 m in 10 m (HL). Therefore, from H to Z the distance is such as to eliminate 5 m
HZ = 50/3 = 16·67 m
EZ = 19·17 m   DZ = 14·17 m
(b) BK = mean value of 8 and 5 = 6·5 m
BE = $\sqrt{(6\cdot5^2 + 2\cdot5^2)}$ = 6·96 m
(c) Comparing triangles BEK and ZEK,

$$\frac{XZ}{EZ} = \frac{BK}{BE}$$

$$XZ = \frac{EZ \times BK}{BE} = \frac{19\cdot17 \times 6\cdot50}{6\cdot96} \qquad XZ = 17\cdot90 \text{ m}$$

E : T *Evaluation: Total is zero:*

$-40 \times DZ = F_2 \times XZ = 0$
$\qquad\qquad\qquad \Sigma M_Z = 0$
$-40 \times 14\cdot17 + F_2 \times 17\cdot90 = 0 \qquad\qquad +F_2 = +31\cdot7 \text{ kN}$
$\qquad\qquad\qquad \Sigma M_B = 0$
$+40 \times DK + F_3 \times BK = 0 \qquad\qquad +F_3 = -46\cdot2 \text{ kN}$

Y *Yes or No?*

F_2 has the correct sense as shown in the FSD. F_3 should have the opposite sense. F_2 shows compression, F_3 should show tension. Redraw FSD with correct senses.

9/34   ONE FSD: Three equations: Pin-jointed truss

Find the forces acting in members 1, 2, and 3 of the truss shown in the sketch.

S *Structure:*
The section XX cuts off the FSD required for all three members.

U *Unknowns and Units:*
There are three unknowns. Units are kN and m.

133

9/34

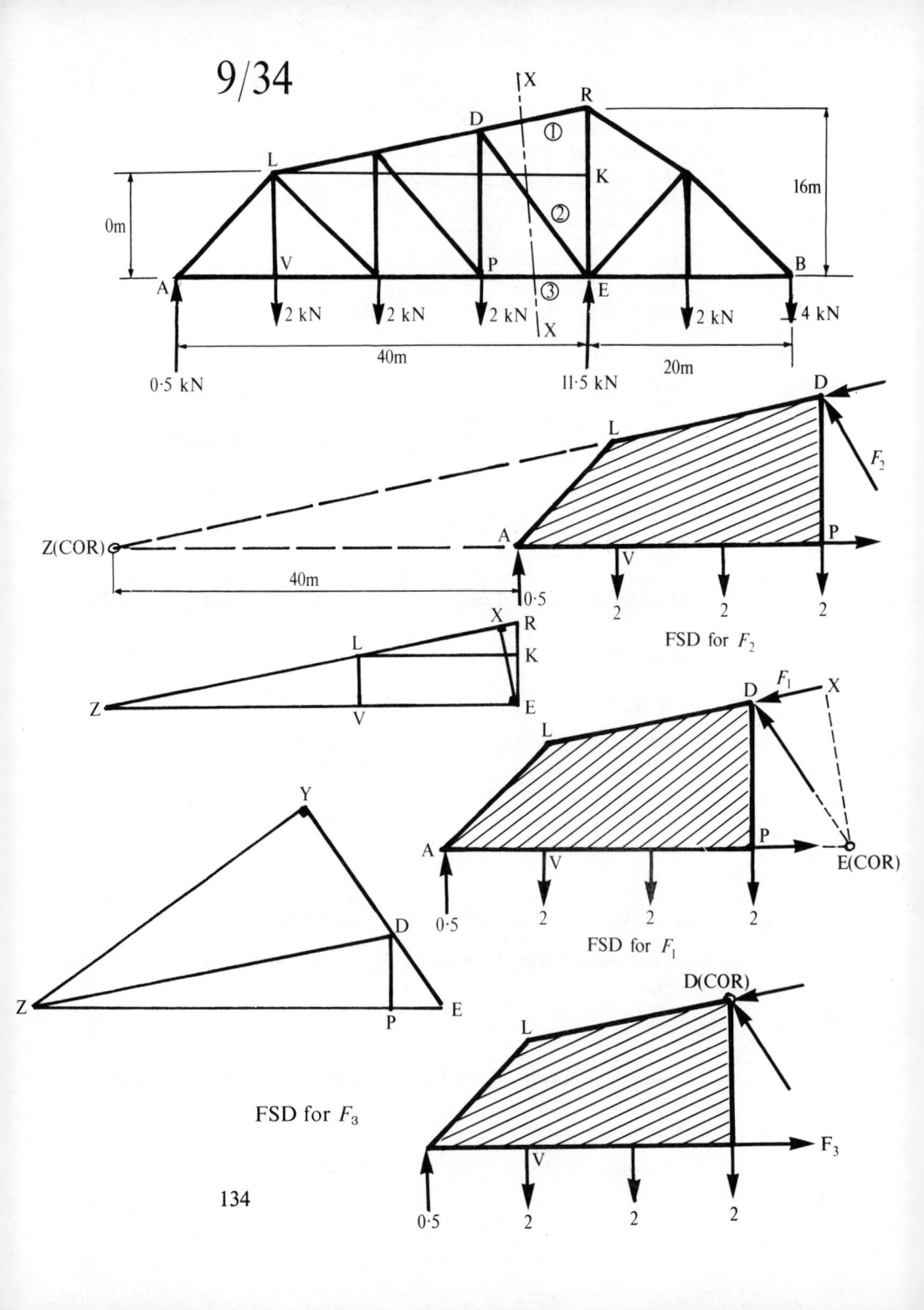

134

R   *Rotation:*
COR (centre of rotation) for derivation of F_1 (E).
COR (centre of rotation) for derivation of F_2 (Z).
COR (centre of rotation) for derivation of F_3 (D).

E: T   *Evaluation: Total is zero:*

(a) Compare triangles ZVL and LKR

$$\frac{ZV}{LV} = \frac{LK}{DK} \qquad ZV = \frac{LK \times LV}{DK} = \frac{30 \times 10}{6}$$

(b) $ZR = \sqrt{(80^2 + 16^2)} = 81 \cdot 6$ m
Compare triangles ZER and EXR

$$\frac{EX}{ER} = \frac{EZ}{ZR} \qquad EX = \frac{16 \times 80}{81 \cdot 6} \qquad EX = 15 \cdot 7 \text{ m}$$

(c) $DE = \sqrt{(14^2 + 10^2)} = 17 \cdot 2$ m
Compare triangles EYZ and EPD

$$\frac{YZ}{ZE} = \frac{DP}{DE} \qquad YZ = \frac{80 \times 14}{17 \cdot 2} \qquad YZ = 65 \cdot 1 \text{ m}$$

Derivation for F_1

$$\sum M_E = 0$$
$-F_1 \times EX - 2(+30 + 20 + 10) + \frac{1}{2}(40) = 0$
$-15 \cdot 7 F_1 - 120 + 20 = 0$ $+F_1 = -6 \cdot 37$ kN

Derivation for F_2

$$\sum M_Z = 0$$
$-F_2 \times YZ + 2(+50 + 60 + 70) - \frac{1}{2}(40) = 0$
$-F_2 \times 65 \cdot 1 + 360 - 20 = 0$ $+F_2 = +5 \cdot 22$ kN

Derivation for F_3

$$\sum M_D = 0$$
$-F_3 \times 14 - 2(+20 + 10) + \frac{1}{2}(30) = 0$
$-14 F_3 - 60 + 15 = 0$ $+F_3 = -3 \cdot 21$ kN

Y   *Yes or No?*

F_2 is in compression as shown in the FSD, but the senses of the other two have been guessed wrongly. F_1 should be in tension, and F_3 should be in compression. If you are not sure how the sense arrows indicate whether a member is in tension or compression, revise the earlier work.

9/35

The same results should be obtained by using the portion cut off as the FSD, and rejecting the portion used as the FSD in this calculation. Try this second approach, using, as FSD, the truss from B to the section XX.

9/35 **FOUR EQUATIONS: Two FSD'S: Pin-jointed frame: Simple with stay**

The pin-jointed frame shown in the sketch is stayed by the tie BH. Find the forces in BH in order to ensure that there is no vertical component of the reaction at F. When this condition is operative, find the forces in members 1, 2, and 3.

S *Structure:*

The first FSD is the whole structure. If the vertical component at F is eliminated, this leaves only the horizontal component. Three of the four unknowns can, therefore be eliminated by using E as the centre of rotation.

The forces in 1, 2, and 3 can be found from a single FSD as cut off by section XX.

U *Unknowns and Units:*

F_{BH}; F_1; F_2; F_3. Mark sense arrows at random. Use kN and m.

FIRST FREE-STRUCTURE DIAGRAM

R *Rotation:*

COR is where three unknowns intersect, in order to eliminate them from the equation ... (E).

Lengths required:

(a) $BH = \sqrt{(BF^2 + FH^2)} = \sqrt{(64 + 9)} = 8.54$ m

(b) Compare triangles HFB and HXE

$$\frac{EX}{EH} = \frac{BF}{BH} \quad EX = \frac{BF \times EH}{BH}$$

$$EX = \frac{8 \times 9}{8.54} = 8.43 \text{ m}$$

E: T   *Evaluation: Total is zero:*

$$\sum M_E = 0$$

$$-50 \times 8 + F_{BH} \times EX = 0 \qquad +F_{BH} = +\frac{400}{8 \cdot 42}$$

$$+ F_{BH} = +47 \cdot 5 \text{ kN (tension)}$$

Y   *Yes or No?*

It is highly unlikely, since the force in BH has been specified as a tie (tension), that the sense arrow would be wrongly sketched. However, the double positive shows that no mistake has been made.

SECOND FREE-STRUCTURE DIAGRAM

Derivation for F_1

R   *Rotation:*

COR is where F_2 and F_3 intersect (B).
Moment arm of F_1 about B is BY which should be found from the triangles BYA and ZBA.

E: T   *Evaluation: Total is zero:*

$$\sum M_B = 0$$

$$-F_1 \times BY + (+F_2 + F_3 + F_{BH} + 50) \times 0 = 0$$

F_1 is zero, so there is no need to find BY.
Theoretically, for this loading, the lengths CA and AB are superfluous.

Derivation for F_2

R   *Rotation:*

COR is where F_1 and F_3 intersect (Z).

E: T   *Evaluation: Total is zero:*

$$\sum M_Z = 0$$

$$-F_2 \times QZ - F_{BH} \times TZ + 50 \times BZ = 0$$

Find QZ, TZ, and BZ:
(*a*) Compare triangles BZA and DBP

$$\frac{BZ}{AB} = \frac{BD}{DP} \qquad BZ = \frac{BD \times AB}{DP} = \frac{4 \times 2}{2} = 4 \text{ m.}$$

9/35

(b) Compare triangles BQZ and BDC

$$BC = \sqrt{(BD^2 + CD^2)} = \sqrt{(4^2 + 4^2)} = 5.66 \text{ m}$$

$$\frac{QZ}{CD} = \frac{BZ}{BC} \qquad QZ = \frac{BZ \times CD}{BC} = \frac{4 \times 4}{5.66} = 2.83 \text{ m}$$

(c) Compare triangles BTZ and BFH

$$\frac{TZ}{BZ} = \frac{FH}{BH} \qquad TZ = \frac{FH \times BZ}{BH} = \frac{3 \times 4}{8.54} = 1.40 \text{ m}$$

QZ 2·83 m
TZ 1·40 m
BZ 4·00 m

E:T *Evaluation: Total is zero:*

$$\sum M_Z = 0$$
$$-F_2 \times QZ - F_{BH} \times TZ + 50 \times BZ = 0$$
$$-F_2 \times 2.83 - 47.5 \times 1.4 + 50 \times 4.0 = 0 \qquad +F_2 = +47.17 \text{ kN}$$

Y *Yes or No?*

The sense of F_2 has been assumed correctly. The member is in compression.

Derivation for F_3

R *Rotation:*

COR is where F_1 and F_2 intersect (C).

E:T *Evaluation: Total is zero:*

$$\sum M_C = 0$$
$$-F_3 \times 4 - 50 \times 4 + F_{BH} CS = 0$$

Find CS.

Compare triangles CSR and EXH:

$$\frac{CS}{CR} = \frac{EX}{EH} \qquad CS = \frac{EX \times CR}{EH} = \frac{8.43 \times 5.5}{9.0}$$

$$CS = 5.15 \text{ m}$$

$$\sum M_C = 0$$
$$-F_3 \times 4 - 50 \times 4 + 47.5 \times 5.15 = 0 \qquad F_3 = \mathbf{11.5 \text{ kN}}$$

Y *Yes or No?*

The sense of F_3 has been guessed correctly. The member is in compression.

9/35

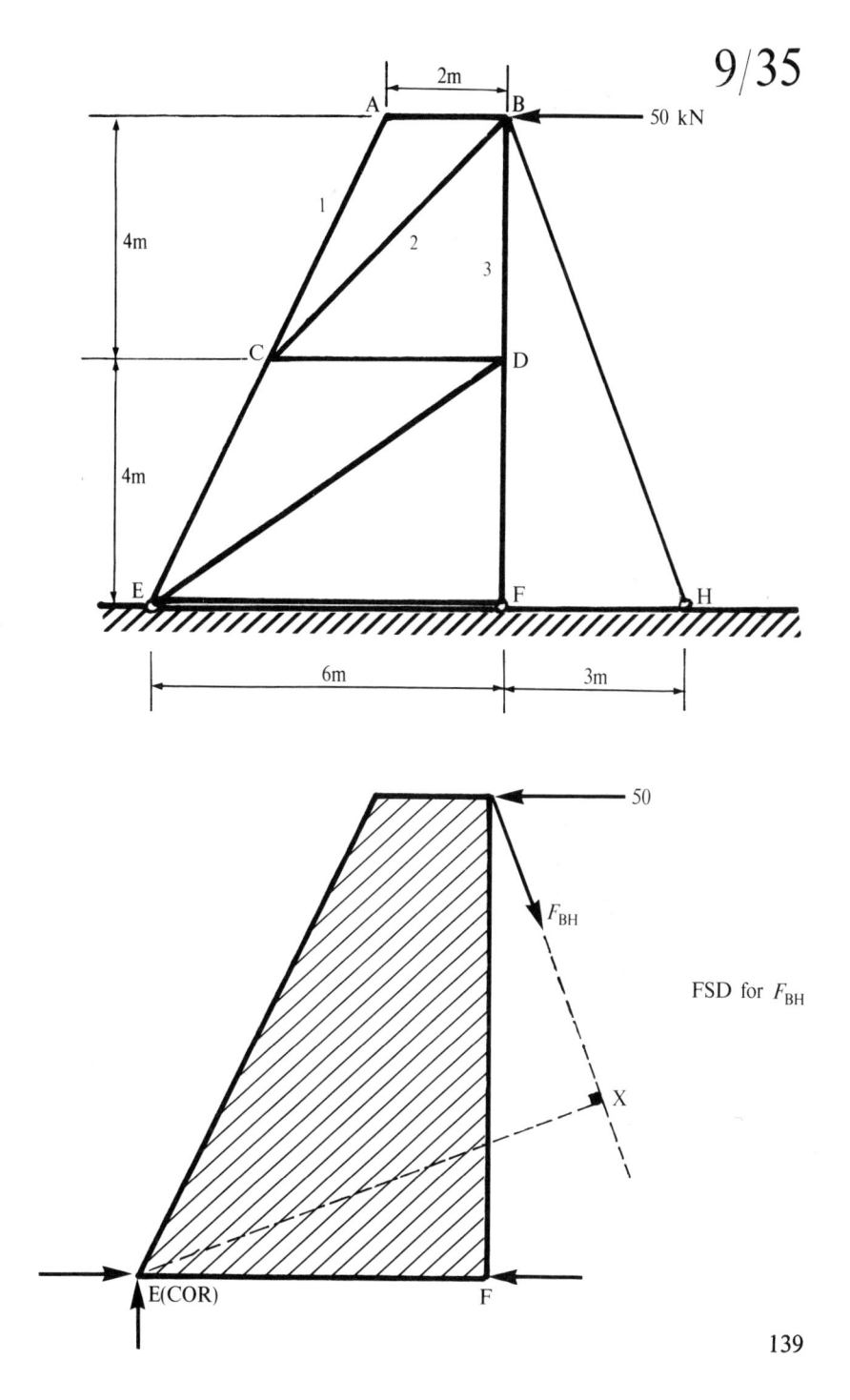

FSD for F_{BH}

9/35

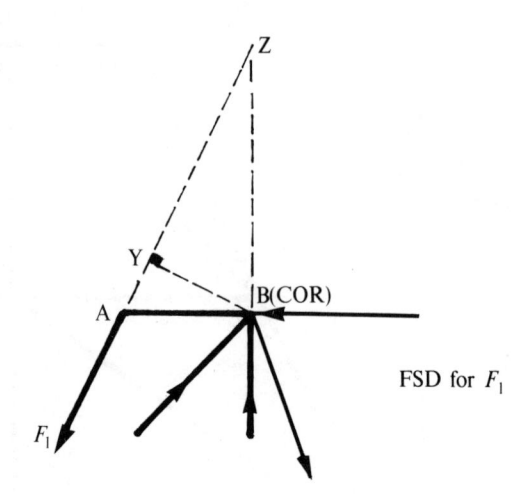

FSD for F_1

FSD for F_3

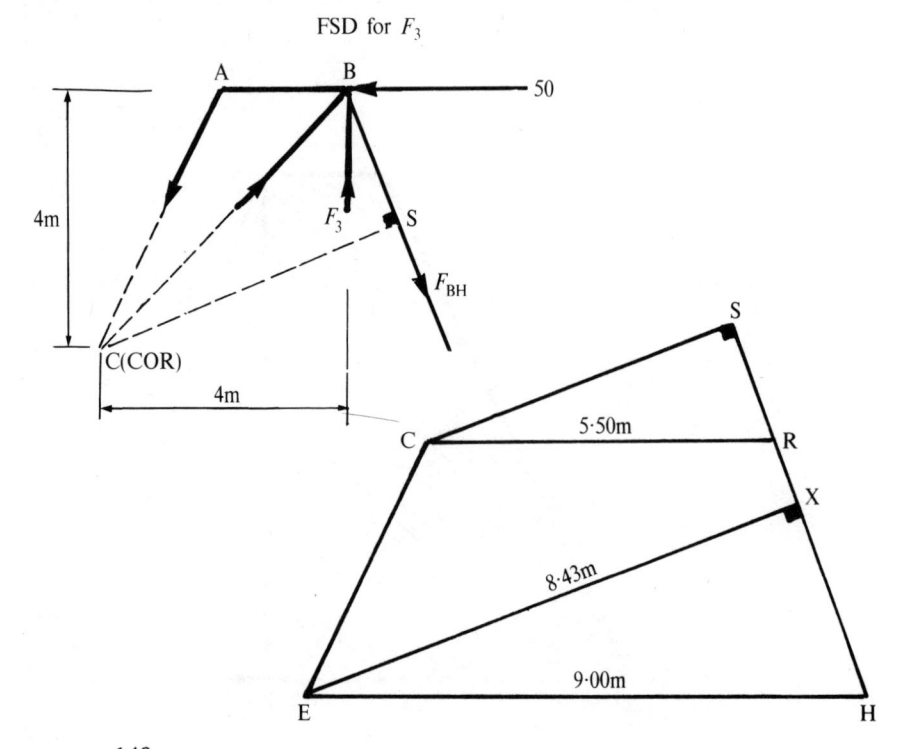

A   B   50

4m

F_3 S

F_{BH}

C(COR)

4m

C   5·50m   R

X

8·43m

9·00m

E   H

140

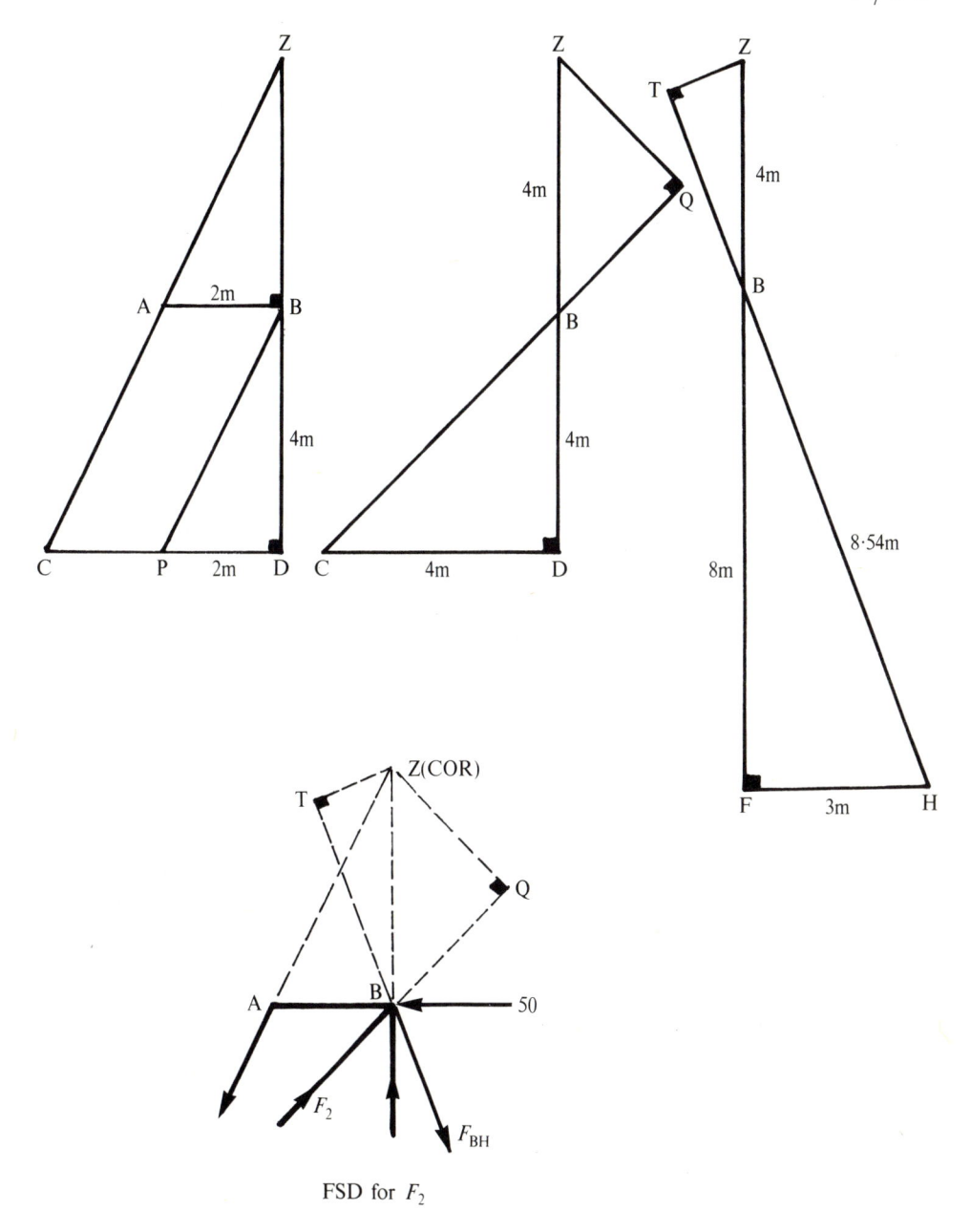

9/36

Alternative derivation

In a complex frame like this, the best solution is obtained graphically. By careful drawing in accordance with the principles of graphic statics, the forces in all the members can be obtained more readily than by working through the complex geometry required for a complete solution by computation.

But even the computation method can be made easier by an efficient order of calculation. The problem has been calculated by the full method, but it would have been better to calculate the easier F_3 first and then use a force equation of equilibrium. It should also be clear without calculation, that F_1 is zero. Suppose that F_3 has been calculated. F_2 can now be found:

$$\sum F_{NS} = 0$$

$+ \text{nsc } F_2 + F_3 - \text{nsc } F_{BH} = 0$
$+ \text{nsc } F_2 + 11 \cdot 15 - 8/8 \cdot 54 \times 47 \cdot 5 = 0$
$+ \text{nsc } F_2 + 11 \cdot 15 - 44 \cdot 50 = 0$ But F_2 is at 45° to the horizontal
$+ F_2 = + 33 \cdot 35 \times \sqrt{2}$ $\qquad\qquad +F_2 = +47 \cdot 16$ kN

This structure is badly designed for the loading. It would be much more effective with the diagonals as AD and CF. Solve the frame again with these diagonals replacing BC and DE.

9/36 THREE FSD'S: Pin-jointed frame: Simple supports

A light framework carries two loads. There is no joint between members except where shown by a small circle. At other points, the bars merely pass each other. Find the forces in members 1, 2, and 3. AB is 1·3 m, BF is 1·4 m, AF is 1·5 m, and BC is 2·8 m long. The span is 5·2 m.

S *Structure:*

The geometry of the structure is, again, the critical time-consumer, unless you can handle triangles easily and smoothly. The horizontal distances are easily obtained, but the vertical lengths are also required.

$BY = \sqrt{(AB^2 - AY^2)} = \sqrt{(1 \cdot 3^2 - 1 \cdot 2^2)} = 0 \cdot 5$ m $= PR = CZ$
Compare triangles ARX and AZE

$$\frac{RX}{AR} = \frac{ZE}{AZ} \qquad RX = \frac{ZE \times AR}{AZ} = \frac{(1 \cdot 4 - 0 \cdot 5) \times 2 \cdot 6}{4 \cdot 0}$$

$$RX = 0 \cdot 59 \quad PX = 1 \cdot 09 \text{ m}$$

9/36

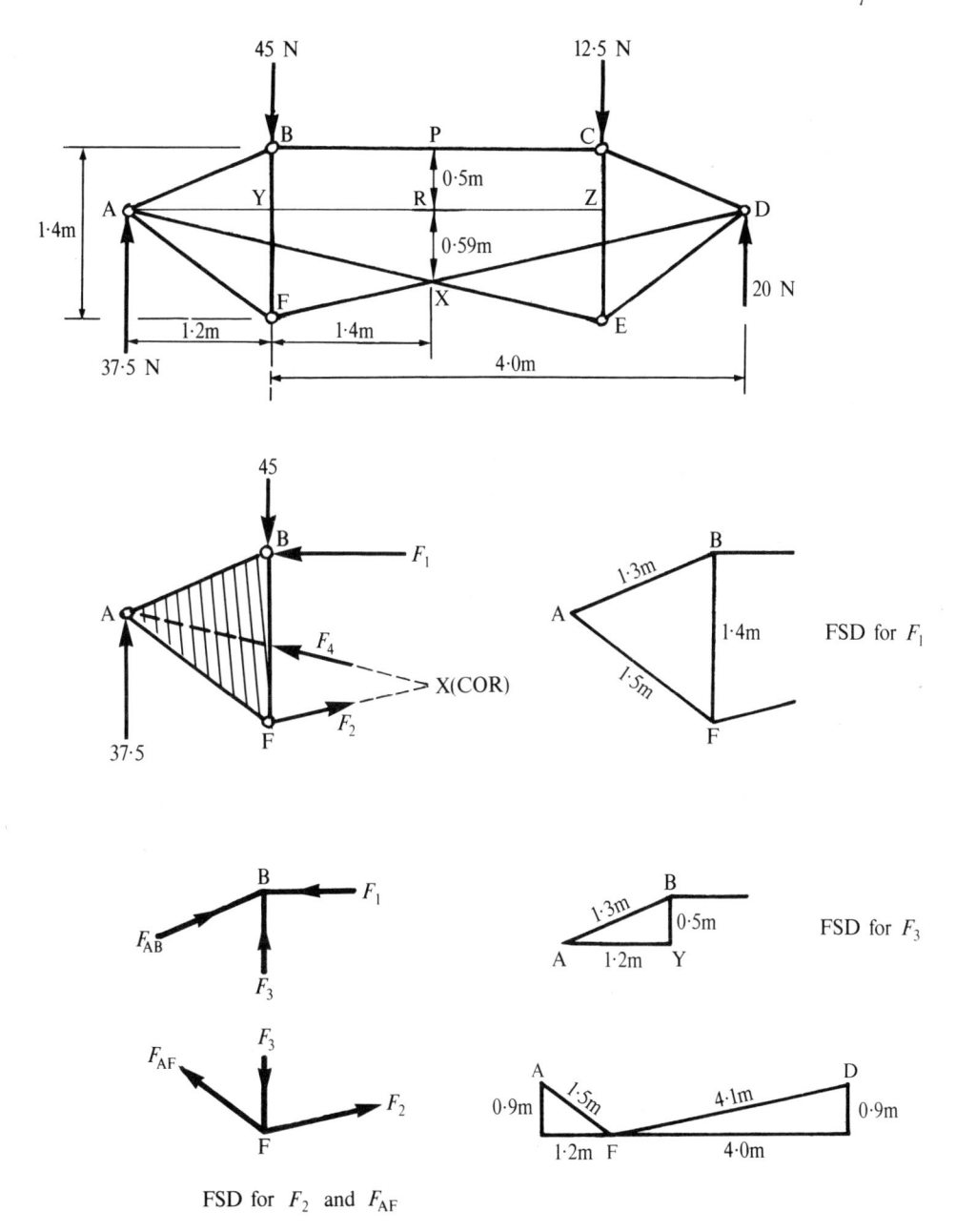

FSD for F_1

FSD for F_3

FSD for F_2 and F_{AF}

143

9/36

U   *Unknowns and Units:*

There are three unknowns. The units are N and m.

R   *Rotation:*

COR for F_1 is where F_2 and F_4 intersect (X).

E:T   *Evaluation: Total is zero:*

$$\sum M_X = 0$$
$$-F_1 \times 1 \cdot 09 + 37 \cdot 5 \times 2 \cdot 6 - 45 \cdot 0 \times 1 \cdot 4 = 0$$
$$-1 \cdot 09 F_1 + 97 \cdot 5 - 63 \cdot 0 = 0 \qquad\qquad +F_1 = \mathbf{+31 \cdot 65 \, N}$$

A second FSD can be formed at joint B, and its geometry shown in a second diagram.

$$\sum F_{EW} = 0$$
$$-F_1 + \text{ewc } F_{AB} = 0 \qquad +F_{AB} = (1 \cdot 3/1 \cdot 2) 31 \cdot 65 = +34 \cdot 29 \text{ N}$$
$$\sum F_{NS} = 0$$
$$+F_3 - 45 + \text{nsc } F_{AB} = 0$$
$$+F_3 - 45 + (0 \cdot 5/1 \cdot 3) \times 34 \cdot 29 = 0 \qquad\qquad +F_3 = \mathbf{+31 \cdot 81 \, N}$$

A third FSD can be formed at joint F, with a separate diagram showing the geometry.

$$\sum F_{NS} = 0$$
$$-F_3 + \text{nsc } F_2 + \text{nsc } F_{AF} = 0$$
$$-F_3 + (0 \cdot 9/4 \cdot 1) F_2 + (0 \cdot 9/1 \cdot 5) F_{AF} = 0$$
$$\sum F_{EW} = 0$$
$$+\text{ewc } F_2 - \text{ewc } F\alpha_F = 0$$
$$+(4 \cdot 0/4 \cdot 1) F_2 - (1 \cdot 2/1 \cdot 5) F_{AF} = 0$$

Solving simultaneously:

$$+0 \cdot 22 F_2 + 0 \cdot 60 F_{AF} = +31 \cdot 81$$
$$+0 \cdot 98 F_2 - 0 \cdot 80 F_{AF} = 0 \qquad +F_{AF} = +1 \cdot 22 F_2$$
$$+0 \cdot 95 F_2 = 31 \cdot 81 \qquad\qquad\qquad +F_2 = \mathbf{+33 \cdot 5 \, N}$$

The right-hand side of the equations of equilibrium must be written as zero.

9/37 PIN-JOINTED FRAME: Three-hinged frame: Direct forces

Determine the forces in the triangular pin-jointed frame and also those in the three-hinged frame or 'arch' which has the same dimensions.

S *Structures:*

Both these frames are really the same. In the first one, the horizontal load in the tie AB serves the same purpose as the rigid abutment supports of the second frame. The fact that one is a pin-jointed frame and the other is a three-hinged structure makes no difference to the method of attack.

U *Unknowns and Units:*

There are unknowns exterior to the frame (the various reactions) and also interior forces within the frame. Since these number more than three, there must be more than three FSD's in use during the calculation. Use kN and m.

R *Rotation:*

In the first frame, only one centre of rotation is required – to find the reactions R_L and R_R. In the three-hinged frame, two COR's will be required – one to determine the vertical reactions, and one to find the value of the horizontal reaction H (which is the same force which acts in the tie of the first frame.

E *Evaluation: Second frame*

FIRST FREE-STRUCTURE DIAGRAM
(positive clockwise)

$$\sum M_B = 0$$

$+R_L \times 5 - 10 \times 1 = 0$ $\qquad +R_L = +2\,\text{kN}$
$\qquad\qquad\qquad\qquad\qquad\qquad +R_R = +8\,\text{kN}$

SECOND FREE-STRUCTURE DIAGRAM

$$\sum M_C = 0$$

$+R_L \times 4 - H \times 3 = 0$ $\qquad +H = +\tfrac{4}{3}R_L = +2\cdot67\,\text{kN}$

THIRD FREE-STRUCTURE DIAGRAM

$$\sum F_{NS} = 0 \quad \text{(joint A)}$$

$+2 - \tfrac{3}{4}F_1 = 0$ $\qquad +F_1 = +3\cdot33\,\text{kN (comp.)}$

145

9/37

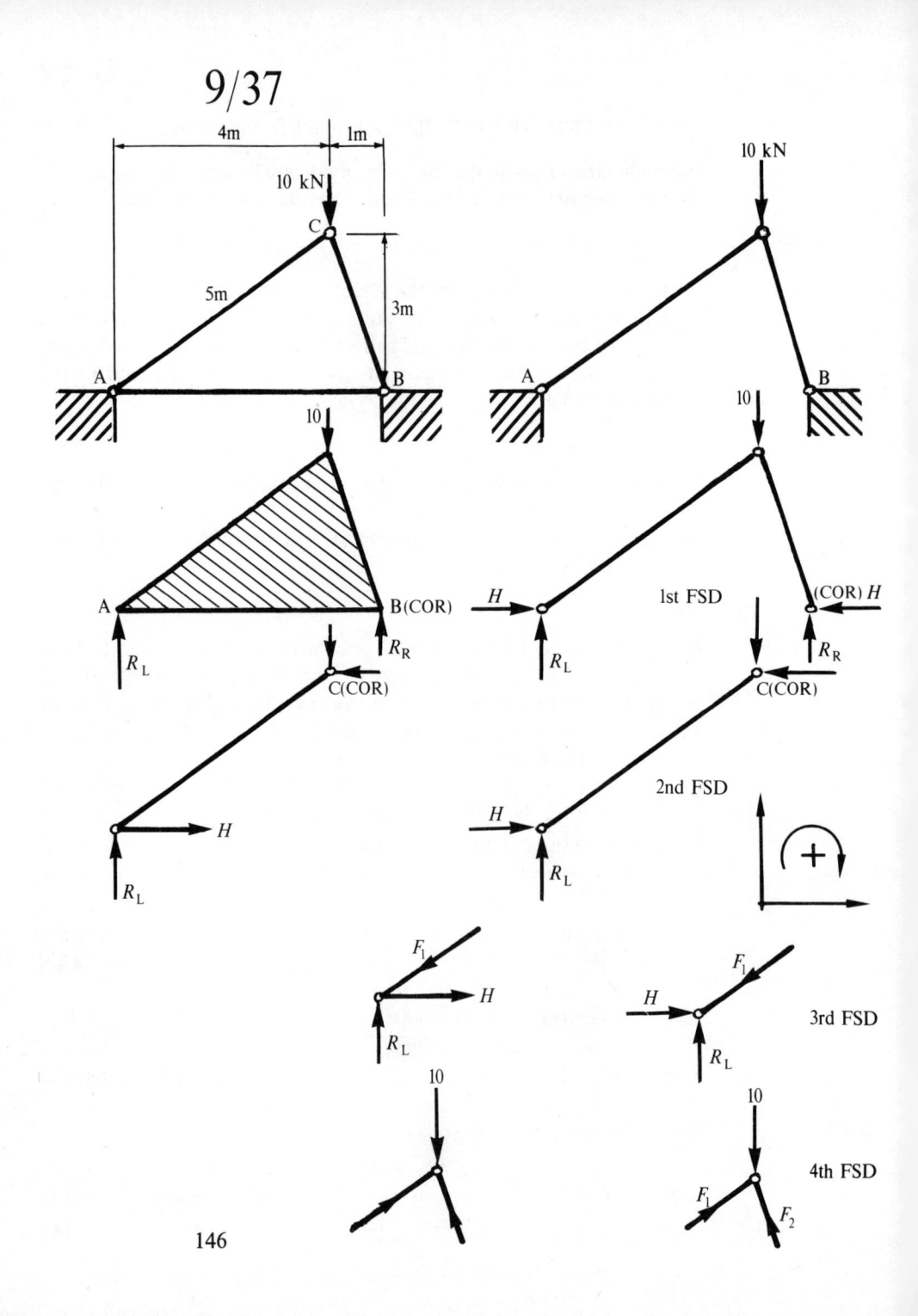

FOURTH FREE-STRUCTURE DIAGRAM

$$\sum F_{NS} = 0 \quad \text{(joint C)}$$
$$-10 + 0{\cdot}6F_1 + 3/\sqrt{10}F_2 = 0 \qquad +F_2 = +\mathbf{8{\cdot}43 \text{ kN (comp.)}}$$

E *Evaluation: First frame*

The second frame has been evaluated as an introduction to more complex three-hinged frames and arches. The first frame gives exactly the same results, except that the horizontal reaction H is now an internal force within the frame (member AB), and thus the reactions are simple supports. A tied three-hinged frame, therefore, gives fewer problems in the design of foundations at the abutments than do three-hinged arch types which require a strong support at A and B to resist the horizontal thrust. Where headroom is required and a tie across the span would be in the way, then the horizontal thrust outwards must be taken by the supporting abutments.

9/38 **THREE-HINGED FRAME**: Thrust, shear, bending moment: Unsymmetrical

Determine the thrust, shear, and bending moment at point X, just to the left of the load on the three-hinged frame shown.

S *Structure:*

This is the same shape as in the immediately previous problem, but instead of AC and BC carrying direct load only, and being either in tension or compression, the members are designed to carry bending and shear in addition. This allows loads to be placed at any point on the structure and not merely at the hinges.

U *Unknowns and Units:*

There are three unknowns within the structure (T, S, and M), and a reaction at each hinge, conveniently broken into two components. For the first FSD these components are best taken as horizontal and vertical (H and R_L).

R *Rotation:*

COR for determination of R_L is at the intersection of two of the external unknowns H and R_R (B).

147

9/38

COR for the determination of H is in the second FSD, at point C where two unwanted unknowns intersect (H and V).

COR for unknown M caused by bending couple is taken (as always for such a couple) at the cut section.

E Evaluation:

FIRST FREE-STRUCTURE DIAGRAM (COR–B)

$$\sum M_B = 0$$
$$+R_L \times 5 - 10 \times 3 = 0 \qquad\qquad +R_L = +6\text{ kN}$$

SECOND FREE-STRUCTURE DIAGRAM (COR–C)

$$\sum M_C = 0$$
$$+R_L \times 4 - 10 \times 2 + H \times 3 = 0 \qquad\qquad +H = +1{\cdot}33\text{ kN}$$

THIRD FREE-STRUCTURE DIAGRAM (COR – X)

In all determination of S, T, and M for inclined members, the best plan is to adopt the direction of the member for the E–W direction, and the direction at right angles for the N–S direction. All forces acting on the FSD should be resolved into these two directions. It is very helpful to use two different colours for the two directions. In this problem and others, dotted and full lines are used, but colours are much better. Their use helps you to avoid mistakes, for the equations are always made up

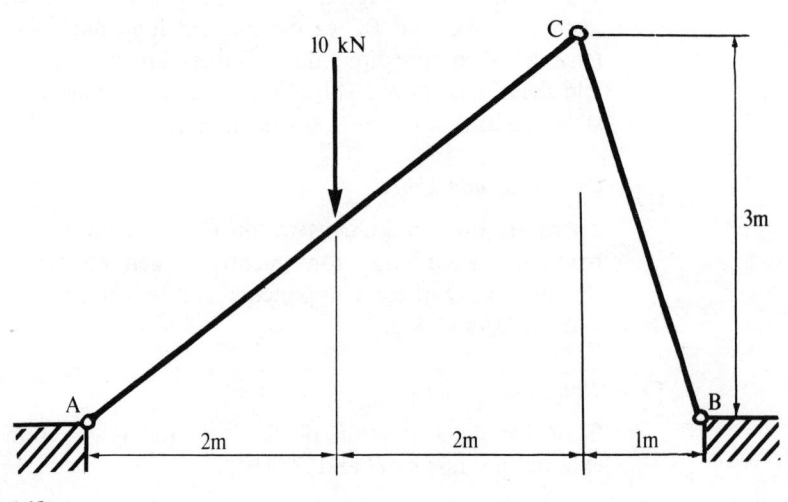

148

9/38

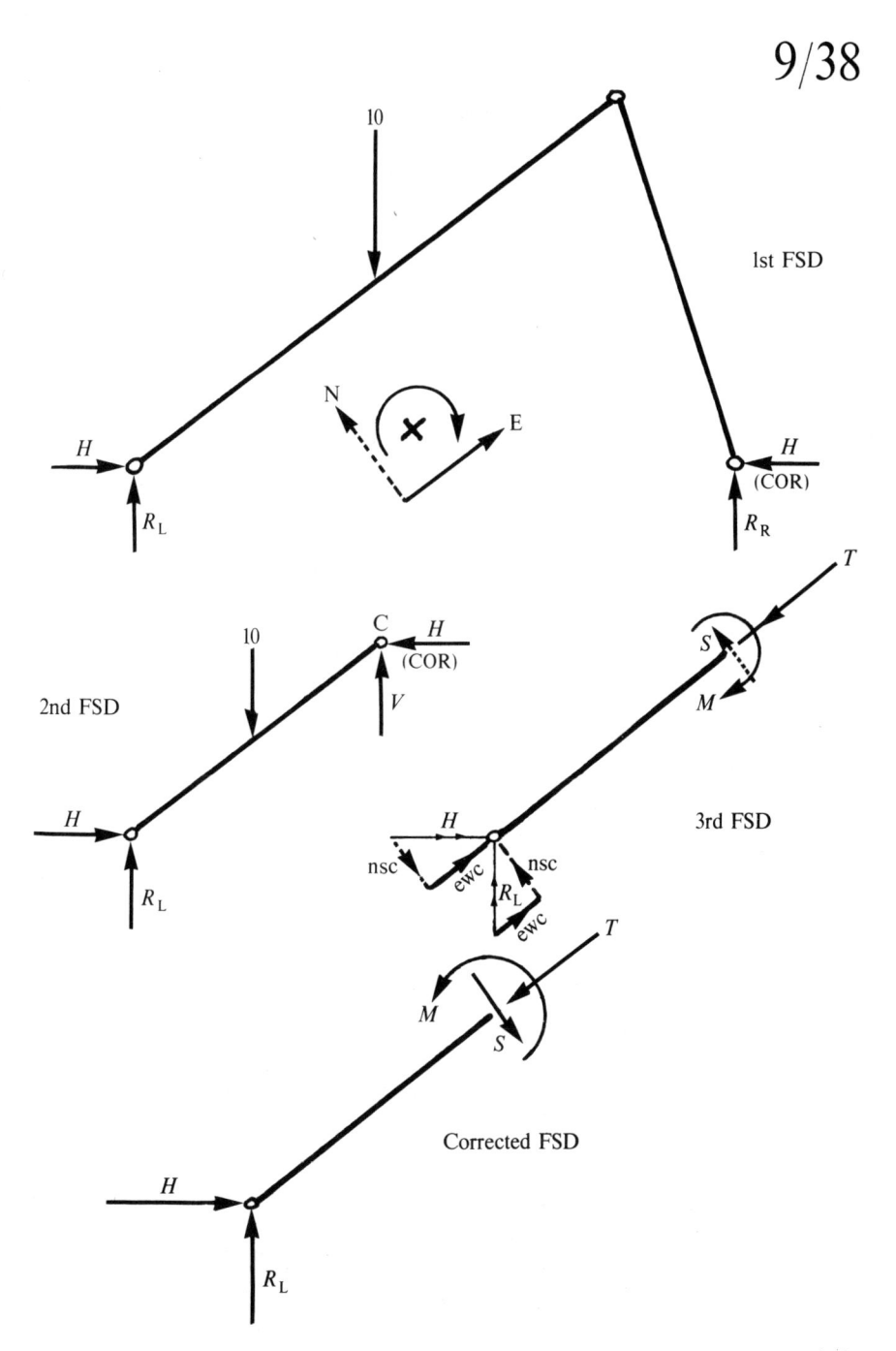

1st FSD

2nd FSD

3rd FSD

Corrected FSD

149

by inspection of directions and rotational effects on the FSD itself. It must, therefore be very clearly drawn, preferably approximately to scale and at the correct angle.

When the member is curved, as it is in an arch, the same procedure is adopted, but the slope then becomes the tangent to the curve at the point selected.

Until you are accustomed to using resolved forces, try a tabular method.

	nsc	ewc
R_L	$\frac{4}{5} \times 6 = \frac{24}{5}$	$\frac{3}{5} \times 6 = \frac{18}{5}$
H	$\frac{3}{5} \times \frac{4}{3} = \frac{4}{5}$	$\frac{4}{5} \times \frac{4}{3} = \frac{16}{15}$

$$\sum F_{NS} = 0$$
$+\text{nsc } R_L - \text{nsc } H + S = 0$
$+\frac{18}{5} - \frac{16}{15} + S = 0$ $\qquad +S = -2\cdot53 \text{ kN}$

$$\sum F_{EW} = 0$$
$+\text{ewc } R_L + \text{ewc } H - T = 0$
$+\frac{24}{5} + \frac{4}{5} - T = 0$ $\qquad +T = +5\cdot60 \text{ kN}$

$$\sum M = 0 \quad \text{(clockwise rotation positive)}$$
$+\text{nsc } R_L \times 5 - \text{nsc } H \times 5 + M = 0$
$(+\frac{18}{5} - \frac{16}{15}) 5 + M = 0$ $\qquad +M = -12\cdot67 \text{ kN m}$

Re-draw the FSD to show the forces and couple in their correct senses.

9/39 THREE-HINGED ARCH: Abutments level: Parabolic

Determine the thrust, shear, and bending moment at a point 4 m horizontally from the left abutment A. The curve of the arch rib is parabolic.

S *Structure:*

Three FSD will be required, to find, in turn, the vertical components of the abutment reactions, the horizontal components of the abutment reactions, and the three quantities required. The first two FSD's are provided with hinges at their ends and the third FSD shows a *held* support at the cut section.

9/39

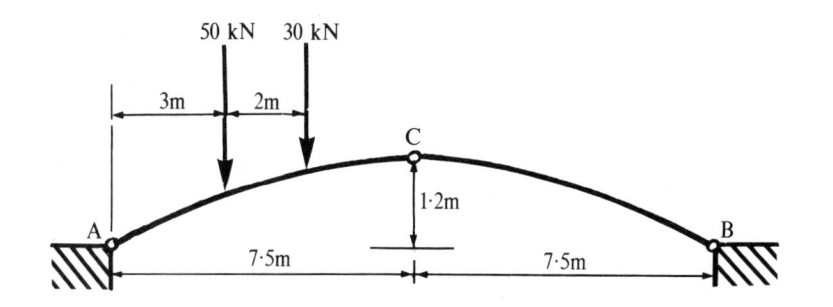

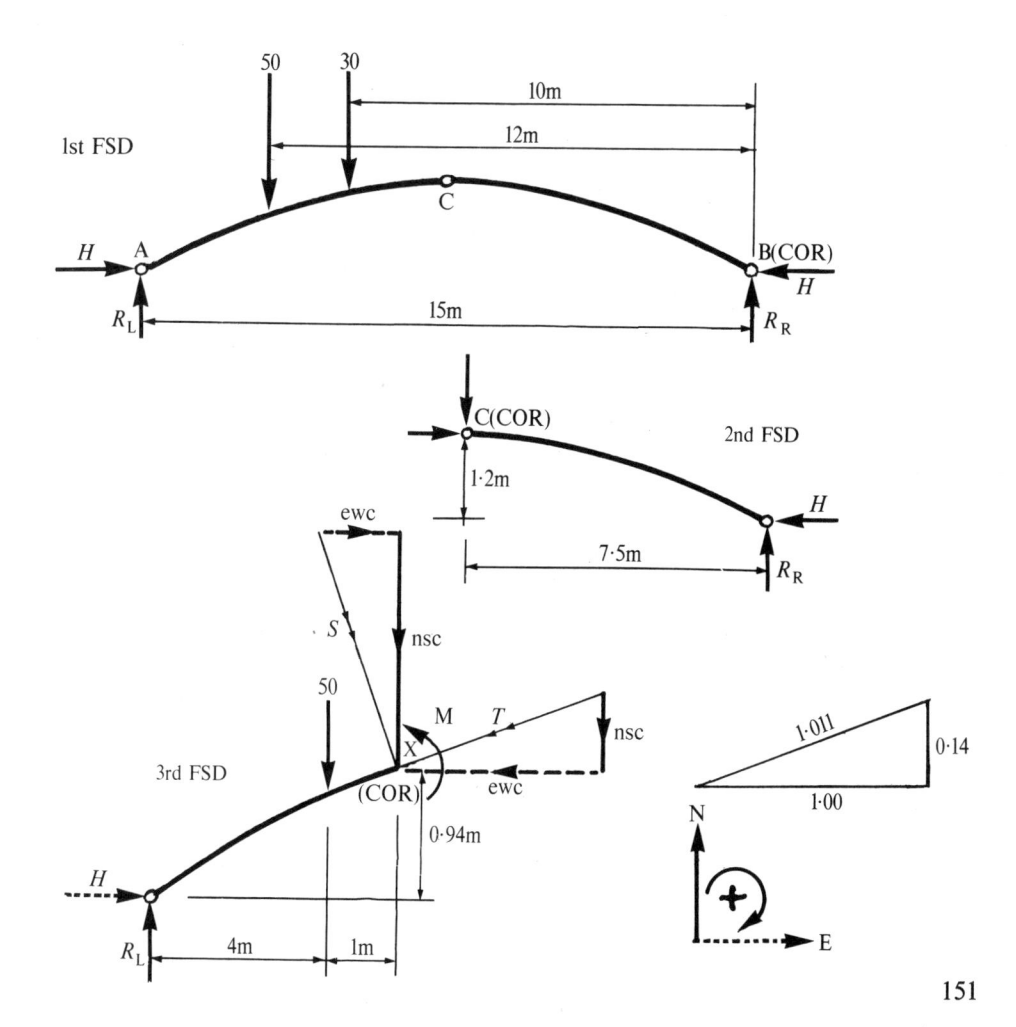

9/39

U *Unknowns and Units:*

There are three unknowns in addition to the reactions, and thus three equations will be required from the third FSD. Put in all unknown sense arrows at random. Use kN and m.

R *Rotation:*

COR for first FSD where two unknowns intersect (B).
COR for second FSD where two unknowns intersect (C).
COR for third FSD at cut section X.

E:T *Evaluate: Total is zero:*

FIRST FREE-STRUCTURE DIAGRAM

$$\Sigma M_B = 0$$
$$+R_L \times 15 - 30 \times 10 - 50 \times 12 = 0 \qquad +R_L = +60 \text{ kN}$$
$$+R_R = +20 \text{ kN}$$

SECOND FREE-STRUCTURE DIAGRAM

$$\Sigma M_C = 0$$
$$+H \times 1\cdot 2 - R_R \times 7\cdot 5 = 0 \qquad +H = +125 \text{ kN}$$

THIRD FREE-STRUCTURE DIAGRAM

Here, the cut section is a *held* support, and requires two forces (*T* which is a thrust tangential to the arch, and *S*, the shear, at right angles to the arch). Put in sense arrows at random, showing the moment of the bending couple (*M*). The forces *S* and *T* should not, strictly, be shown together with their components, since *either* the one group *or* the other act, but not both. For the convenience of forming the equations the components are assumed to be acting and their senses are taken from the arrows on the FSD. Whether these are correct or not, will appear in the final result. Showing both *T* and *S* and their components assists the geometry, but draw another FSD if this is not clear.

Geometry for the third FSD

There are three steps: ($x = 4$ m)

Step 1: Calculate *y*, the height of the section above the base line, using the standard relationships obtained from the equation to a parabola:

$$y = \frac{Qx}{S^2}(2S - x) = \frac{1\cdot 2 \times 4}{7\cdot 5^2}(15 - 4) = 0\cdot 94 \text{ m}$$

Step 2: Find the slope of the rib to the horizontal at the point where it is cut ($x = 4$ m).

$$\frac{dy}{dx} = \frac{2Q}{S}\left(1 - \frac{x}{S}\right) = \frac{2 \times 1\cdot2}{7\cdot5}\left(1 - \frac{4}{7\cdot5}\right) = 0\cdot149$$

This result means that there is a slope of 0·149 to 1 at the cut section. The hypotenuse of the small triangle can then be found and the values of cos and sin determined. Never convert to angles and look up cosines and sines in trigonometrical tables. This is not accurate enough and is wasteful of time. Pythagoras is a good guide! The ratio of cosine/sine is useful in solving the simultaneous equations:

$$\text{Hypotenuse} = \sqrt{[1^2 + (0\cdot149)^2]} = 1\cdot011$$

$$\cos \alpha = \frac{1\cdot00}{1\cdot011} = 0\cdot9891$$

$$\sin \alpha = \frac{0\cdot149}{1\cdot011} = 0\cdot1474$$

$$\text{Ratio} \frac{0\cdot9891}{0\cdot1474} = 6\cdot71$$

Equations from the third FSD

$$\Sigma M_X = 0$$
$+R_L \times x - H \times y - 50 \times 1 - M = 0$
$+60 \times 4 - 125 \times 0\cdot94 - 50 - M = 0 \qquad +M = +12\cdot5 \text{ kN m}$
$$\Sigma F_{NS} = 0$$
$+R_L - 50 - \text{nsc } S - \text{nsc } T = 0$
$+60 - 50 - 0\cdot989 S - 0\cdot147 T = 0$
$$\Sigma F_{EW} = 0$$
$+H + \text{ewc } S - \text{ewc } T = 0$
$+125 + 0\cdot147 S - 0\cdot989 T = 0$

Multiply this by ratio 6·71 and solve the NS and EW equations.

$$+T = +125\cdot1 \text{ kN}$$
$$+S = -8\cdot5 \text{ kN}$$

Yes or No?

Y The shear has been judged to be in the opposite sense from its correct one. Draw a correct FSD.

In problems of this complexity, a book of tables giving squares and square roots (Barlow's), or a desk calculator should be used.

9/40

9/40 THREE-HINGED ARCH: Semi-circular: Symmetrical

Find the thrust, shear and bending moment at a point on the arch rib at 10 m horizontally from A.

S *Structure:*

Three FSD will be required; for the horizontal components of the abutment reactions, for the vertical components of the abutment reactions, and for the values of the quantities at the cut section.

U *Unknowns and Units:*

Apart from the reactions there are three unknowns: S, T, and M. Put in sense arrows at random. Use kN and m.

R *Rotation:*

COR for first FSD is where two reactions intersect (B).
COR for second FSD is where two unknowns intersect (C).
COR for third FSD is where the arch rib is cut.

E:T *Evaluate: Total is zero:*

FIRST FREE-STRUCTURE DIAGRAM

$$\sum M_B = 0$$
$+R_L \times 32 - (5 \cdot 18 \times 32)16 = 0;$ $R_L = R_R = \mathbf{82 \cdot 88}$ **kN**
which comes also from the symmetry of the arch and load.

SECOND FREE-STRUCTURE DIAGRAM

$$\sum M_C = 0$$
$+R_L \times 16 - H \times 16 (5 \cdot 18 \times 16)8 = 0;$ $+H = +\mathbf{41 \cdot 44}$ **kN**

THIRD FREE-STRUCTURE DIAGRAM

Here, three steps must be taken before the equations can be written down. The height y above the baseline must be found. The slope of the arch at the cut point must also be known. The components of the appropriate forces must be determined, in the co-ordinate directions.
In this problem $R = S = Q$, and $h = y$.
If $x = 10$ m; $y = \sqrt{(R^2 - d^2)} = \sqrt{(16^2 - 6^2)} = \mathbf{14 \cdot 83}$ **m**

Equations for third FSD

$$\sum M_X = 0$$
$+R_L \times 10 - H \times 14 \cdot 83 - 51 \cdot 8 \times 5 = 0$
$+828 \cdot 8 - 614 \cdot 6 - 259 \cdot 0 = +M$ $+M = -\mathbf{44 \cdot 8}$ **kN m**

9/40

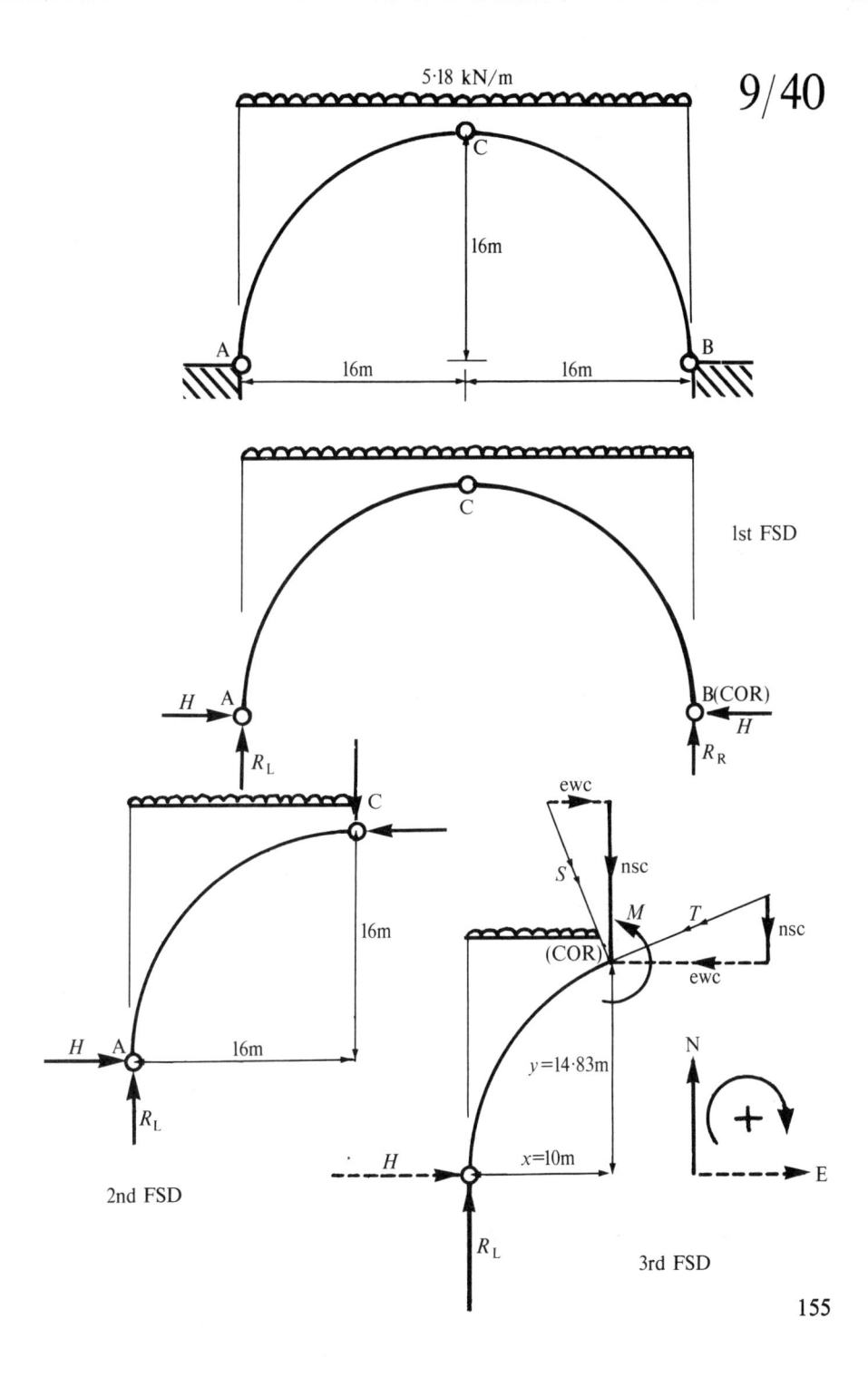

1st FSD

2nd FSD

3rd FSD

9/41

$$\Sigma F_{NS} = 0$$
$+R_L - \text{nsc } S - \text{nsc } T - 51.8 = 0$
$+31.08 - 0.93S - 0.38T = 0$
$$\Sigma F_{EW} = 0$$
$+H + \text{ewc } S - \text{ewc } T = 0$
$+41.44 + 0.38S - 0.93T = 0$

Multiply one of these equations by the ratio 2·47 and solve the two simultaneously:

$$+T = +50.0 \text{ kN}$$
$$+S = +13.3 \text{ kN}$$

Y *Yes or No?*

The sense of the bending couple must be changed.

9/41 **THREE-HINGED ARCH: Shear, thrust, bending couple: Parabolic**

Determine the shear, thrust, and bending moment at a point on the arch rib shown, 8 m horizontally from B. The arch is parabolic. It is one continuous parabola, broken by a hinge at the apex.

S *Structure:*

Three FSD will be required; the first to find the values of the vertical components of the reactions at the abutment hinges; the second to find the horizontal component of the abutment reactions and the third to determine the quantities asked for.

When the two abutments of a three-hinged arch are on the same horizontal line, the horizontal reaction disappears from the second equation. In this instance, *both* components of the abutment reaction appear in *both* second and third equations, and simultaneous solution is needed. This is the only difference made by one abutment hinge being lower than another. Otherwise, each side of the structure up to the central hinge is treated separately.

U *Unknowns and Units:*

There are three unknowns (apart from the reactions) and these are S, T, and M. Use kN and m. Sense arrows at random.

156

9/41

R *Rotation:*

COR for first FSD is at B where two unknowns intersect.
COR for second FSD is at C where two unknowns intersect.
COR for third FSD is at the cut section.

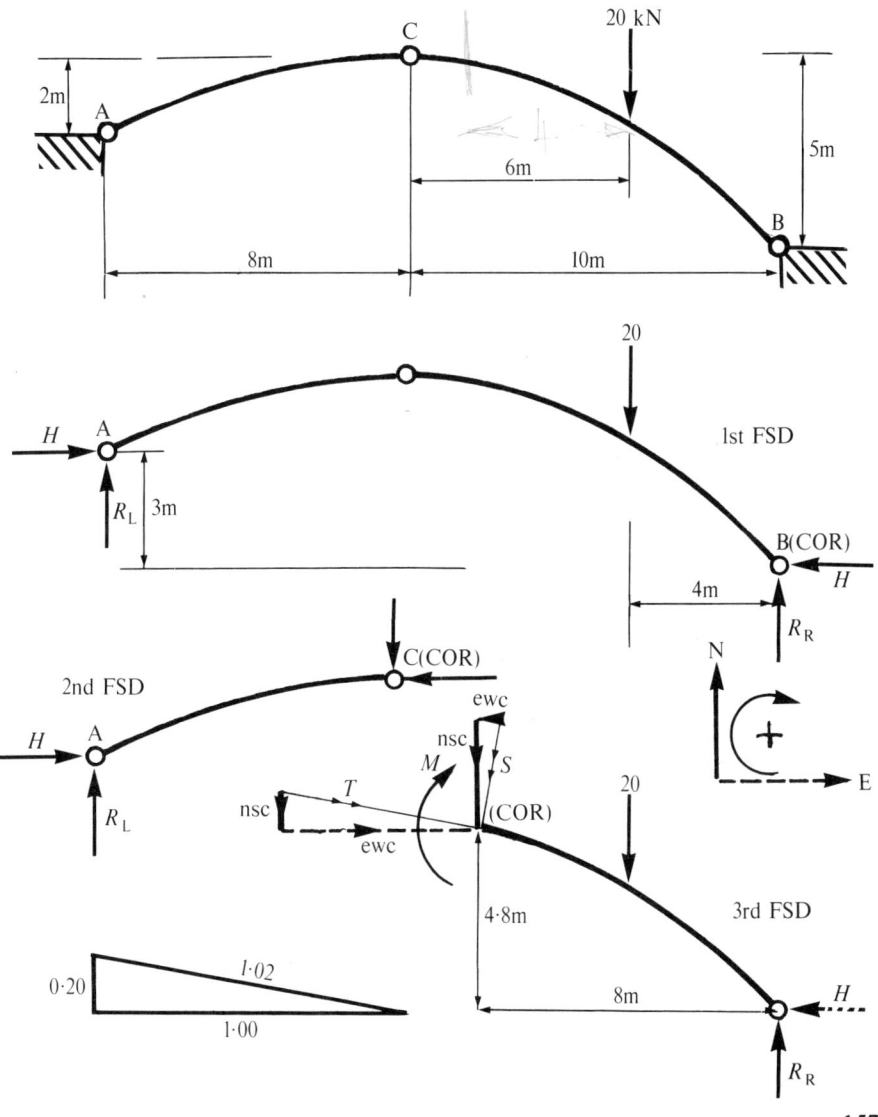

157

E: T

9/41

Evaluate: Total is zero:

FIRST FREE-STRUCTURE DIAGRAM

$$\sum M_B = 0$$
$$+R_L \times 18 + H \times 3 - 20 \times 4 = 0$$

SECOND FREE-STRUCTURE DIAGRAM

$$\sum M_C = 0$$
$$+R_L \times 8 - H \times 2 = 0$$

Solving these two simultaneously:

$$+R_L = +2 \cdot 67 : (R_R = +17 \cdot 33 \text{ kN}) : +H = +10 \cdot 67 \text{ kN}$$

THIRD FREE-STRUCTURE DIAGRAM

In this diagram, the cut section is a *held* reaction or support. There are two forces tangent to and at right angles to the section (thrust and shear) and a bending couple. The first step is to determine the slope of the arch rib at that point and transform the values of both T and S into two components in the co-ordinate directions which, in this instance are best vertical and horizontal. This is more an exercise in geometry than in the study of forces. Know the geometry of the parabola. Although S and T are shown to assist the geometry, only their components are acting as forces.

Geometry for the third FSD ($x = 8$ m)

There are three steps:

Step 1: Determine the value of y the height of section X above the base line, for $x = 8$ m:

$$y = \frac{Qx}{S^2}(2S - x) = \frac{5 \times 8}{100}(20 - 8) = 4 \cdot 8 \text{ m}$$

Step 2: Determine the slope of the arch rib at this point (X) to the horizontal:

$$\frac{dy}{dx} = \frac{2Q}{S}\left(1 - \frac{x}{S}\right) = \frac{10}{10}\left(1 - \frac{8}{10}\right) = 0 \cdot 20 \text{ m}$$

Step 3: Resolve the appropriate forces into the co-ordinate directions (in this instance, vertical and horizontal). It is best to sketch *another*

9/41

FSD showing these components until you are thoroughly familiar with the methods.

$$\text{Hypotenuse} = \sqrt{(1\cdot00)^2 + (0\cdot20)^2} = 1\cdot02$$

$$\cos \alpha = \frac{1\cdot00}{1\cdot02} = 0\cdot9804$$

$$\sin \alpha = \frac{0\cdot20}{1\cdot02} = 0\cdot1961$$

	T	S
nsc	$-0\cdot196T$	$-0\cdot980S$
ewc	$+0\cdot980T$	$-0\cdot196S$

Equations from the third FSD

$$\sum M_X = 0$$
$-R_R \times 8 - H \times 4\cdot8 + 20 \times 4 + M = 0$
$-17\cdot33 \times 8 - 10\cdot67 \times 4\cdot8 + 80 + M = 0 \qquad +M = +7\cdot47 \text{ kN m}$
$$\sum F_{NS} = 0$$
$+R_A - 20 - \text{nsc } S - \text{nsc } T = 0$
$+17\cdot33 - 20 - 0\cdot980S - 0\cdot196T = 0$
$$\sum F_{EW} = 0$$
$-H - \text{ewc } S + \text{ewc } T = 0$
$-10\cdot67 - 0\cdot196S + 0\cdot980T = 0$
Multiply by ratio $0\cdot980/0\cdot196 = 5$
$\left.\begin{array}{l} -53\cdot35 - 0\cdot980S + 4\cdot900T = \\ -2\cdot67 - 0\cdot980S - 0\cdot196T = \end{array}\right\} \qquad \begin{array}{l} +T = +9\cdot96 \text{ kN} \\ +S = -4\cdot64 \text{ kN} \end{array}$

Y   Yes or No?

The sense arrows were put on at random, and the result shows that S should be acting upwards. It is convenient when sketching components in this way, to place T and S so that the little triangles do not obstruct each other. The result is clearer even if you have to re-draw the FSD later when you find that S is not in the right sense. T is a thrust and there is very little doubt about its sense.

9/42

9/42 THREE-HINGED ARCH: Parabolic: Unsymmetrical loading

Determine the maximum bending moment on the arch rib shown, and the point at which it occurs. At that point, find the thrust and shear acting under the loading shown.

S *Structure:*

Two FSD's are required to determine the reactions. Thereafter, the bending moment at a cut section can be found, and its maximum value predicted, from a third FSD.

U *Unknowns and Units:*

There are three unknowns, maximum M and the values of S and T at that point. Mark sense arrows on the unknowns at random. Use kN and m.

R *Rotation:*

COR for the first FSD is at B where two unknowns intersect.
COR for the second FSD is at C where two unknowns intersect.
COR for the third FSD is at X where S, F, and max M act.

E:T *Evaluate: Total is zero:*

Follow arrows on FSD.

FIRST FREE-STRUCTURE DIAGRAM

$$\sum M_B = 0$$
$+R_L \times 90 - 650(45 + 22 \cdot 5) - 1250(22 \cdot 5) = 0 \qquad +R_L = +800 \text{ kN}$
$\qquad\qquad\qquad\qquad\qquad\qquad\qquad\qquad\qquad\qquad +R_R = +1100 \text{ kN}$

SECOND FREE-STRUCTURE DIAGRAM

$$\sum M_C = 0$$
$+R_L \times 45 - H \times 12 - 650 \times 22 \cdot 5 = 0 \qquad +H = +1780 \text{ kN}$

THIRD FREE-STRUCTURE DIAGRAM

It is probable that the maximum bending moment will be on the right-hand side, but a similar calculation can be made for the left-hand side. The maximum occurring on the left-hand side can then be compared with that on the right, so that the greater of these can be selected. For the purpose of this problem only one will be looked at.

9/42

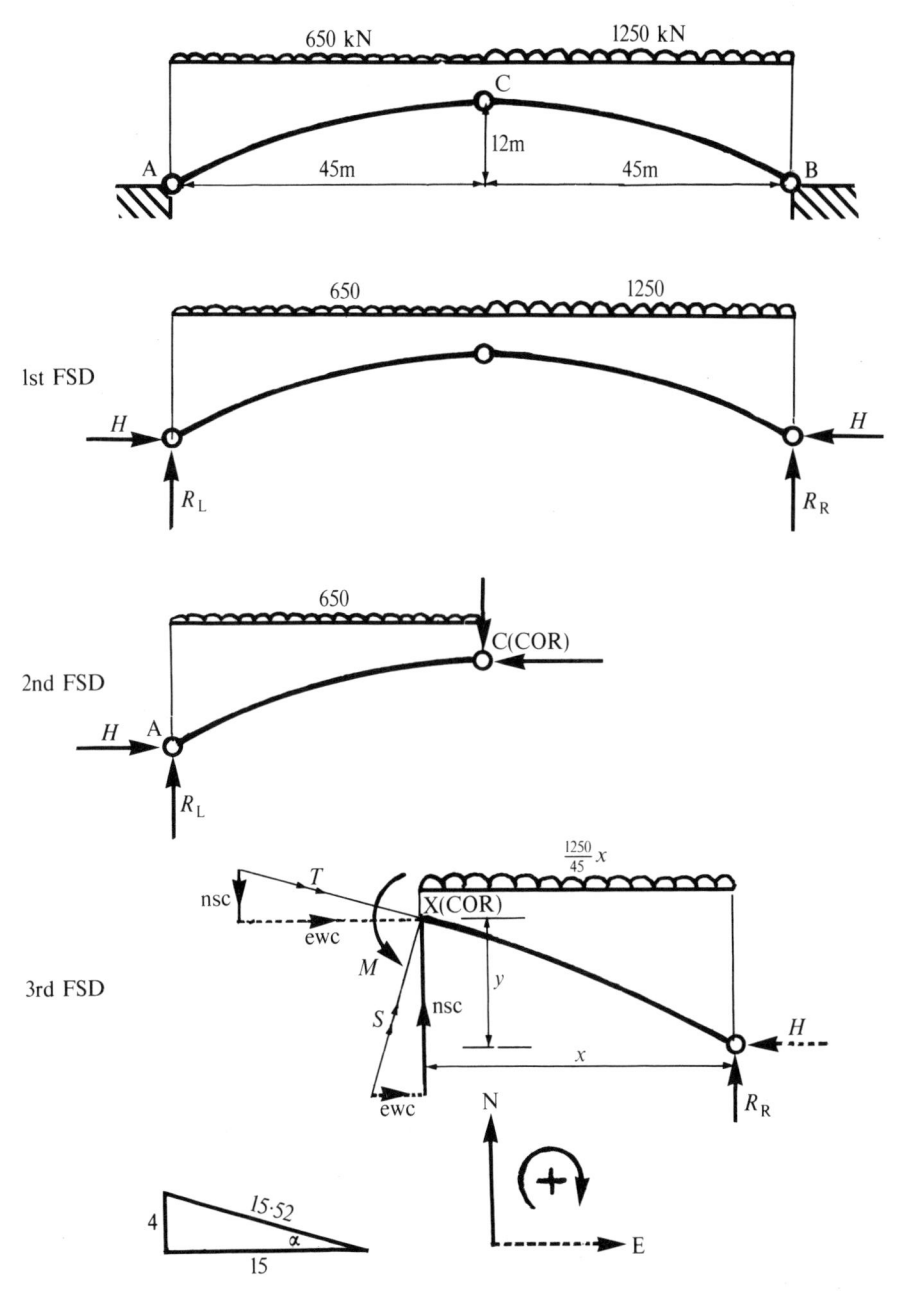

650 kN   1250 kN

C

45m 12m 45m

A   B

1st FSD

650   1250

H ← → H

R_L R_R

2nd FSD

650

C(COR)

H → A

R_L

3rd FSD

$\frac{1250}{45}x$

X(COR)

T, nsc, ewc, M, S, nsc, ewc

y

x

H

R_R

N ↑ → E

4, 15·52, α, 15

161

9/42

Geometry on the right-hand half

$$y = \frac{Qx}{S^2}(2S - x) = \frac{12x}{45^2}(90 - x)$$

Moment equation for the right-hand half

$$\sum M_x = 0$$

$$+ Hy - R_R x + \frac{1250}{45}\frac{x^2}{2} = M = 0$$

$$+ 1780 y - 1100 x + \frac{1250}{90} x^2 - M = 0$$

$$+ \frac{1780 \times 12x}{45^2}(90 - x) - 1100 x + \frac{1250}{90} x^2 - M = 0$$

$$\frac{dM}{dx} = 150 - 6 \cdot 67\, x = 0$$

$$x = 22 \cdot 5 \text{ m}$$

The maximum moment on the right-hand half occurs at the quarter point. Similarly for the left-hand half its maximum moment will occur at its quarter point. Check its value which is a little smaller than that on the right-hand half.

Geometry of the arch rib at quarter point ($x = 22 \cdot 5$ m)

$$y = \frac{12x}{45^2}(90 - x) = 9 \text{ m at } x = 22 \cdot 5 \text{ m}$$

$$\frac{dy}{dx} = \frac{2Q}{S}\left(1 - \frac{x}{S}\right) = \frac{4}{15} \text{ at } x = 22 \cdot 5 \text{ m}$$

$$\text{Hypotenuse} = \sqrt{(16 + 225)} = 15 \cdot 524$$

$$\left. \begin{array}{l} \cos \alpha = \dfrac{15 \cdot 0}{15 \cdot 524} = 0 \cdot 9898 \\[6pt] \sin \alpha = \dfrac{4 \cdot 0}{15 \cdot 524} = 0 \cdot 2577 \end{array} \right\} \text{ratio} = 3 \cdot 841$$

Equations at quarter point on right-hand half

$$\sum M_x = 0$$
$$+Hy - R_R x + \frac{1250 \times 22\cdot5^2}{90} - M = 0$$
$$+1780 \times 9 - 1100 \times 22\cdot5 + 7031 - M = 0 \qquad +M = -1700 \text{ kN m}$$
$$\sum F_{NS} = 0$$
$$+R_R + \text{nsc } S - \text{nsc } T = \frac{1250}{45} \times 22\cdot5 = 0$$
$$+1100 + 0\cdot99S - 0\cdot26T - 625 = 0$$
$$\sum F_{EW} = 0$$
$$-H + \text{ewc } S + \text{ewc } T = 0$$
$$-1780 + 0\cdot26S + 0\cdot99T = 0$$
Multiply by 3·841 and solve with NS equation.
$$-6837 + 0\cdot99S + 3\cdot80T = 0 \qquad\qquad +T = +1800 \text{ kN}$$
$$+S = -11 \text{ kN}$$

Y *Yes or No?*

The moment M should be acting clockwise instead of anti-clockwise as it is shown in the FSD. It will produce tension on the inside of the arch. The shearing force S should be acting downward on the cut section.

9/43   THREE-HINGED ARCH: Unsymmetrical loading: Circular arc

Find the thrust, shear, and bending moment on the arch at a section cut through at quarter span.

S *Structure:*

Three FSD will be required; one for vertical and one for horizontal components of the reactions, and one for the required values at the section cut through at one-quarter span.

U *Unknowns and Units:*

There are three unknowns in addition to the values of the reactions. Put in sense arrows for unknowns at random. Use kN and m.

9/43

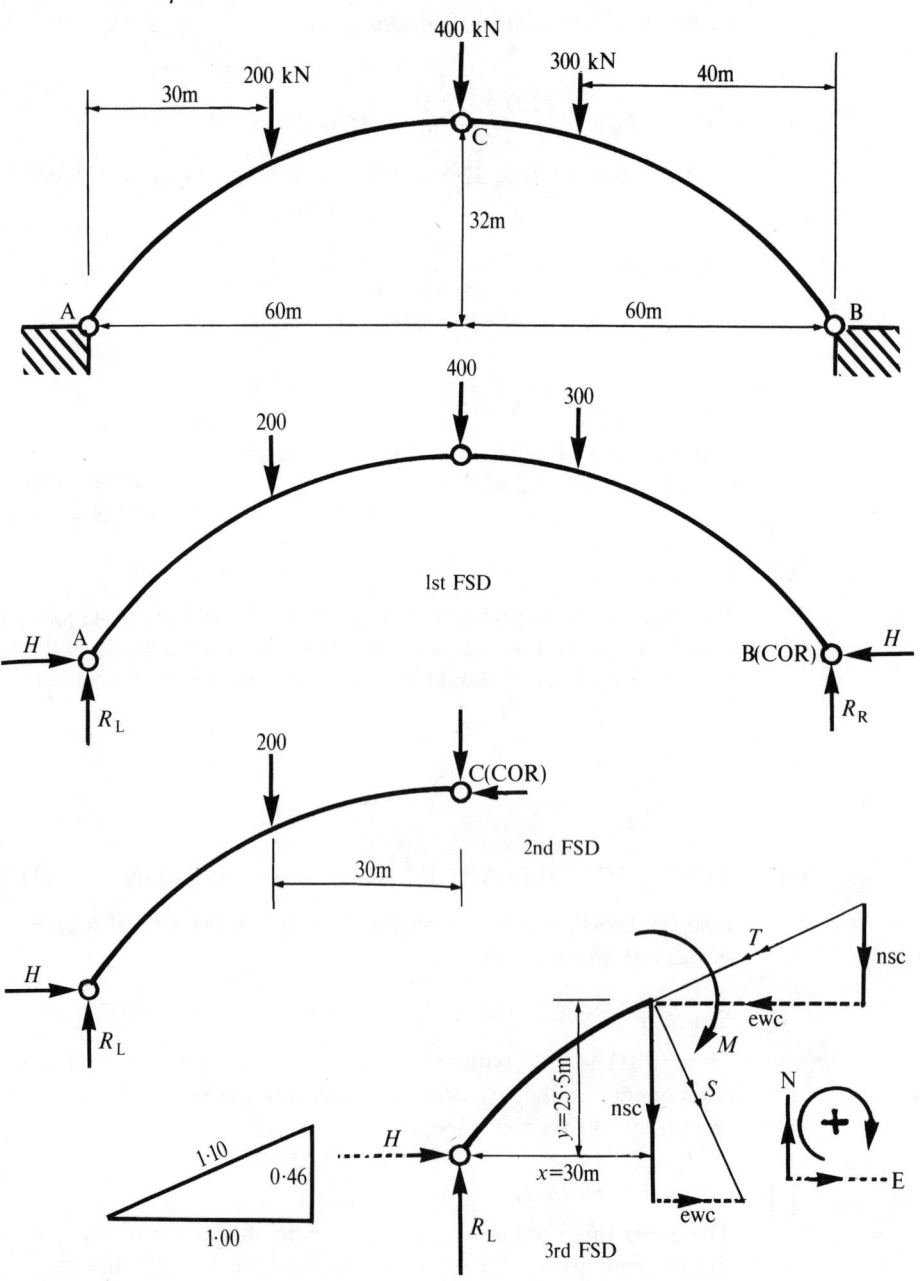

400 kN
200 kN
300 kN
30m
40m
C
32m
A
60m
60m
B

400
200
300

1st FSD

$H \rightarrow$ A B(COR) $\leftarrow H$
$R_L \uparrow$ $\uparrow R_R$

200
C(COR)

2nd FSD

30m

$H \rightarrow$
$\uparrow R_L$

1·10
0·46
1·00

$H \rightarrow$
$\uparrow R_L$
$x = 30\text{m}$
$y = 25\cdot5\text{m}$

T
nsc
ewc
M
S
nsc
ewc

N
$\oplus$
E

3rd FSD

164

R *Rotation:*
COR for first FSD is at B where two unknowns intersect.
COR for second FSD is at C where two unknowns intersect.
COR for third FSD is at the cut section X.

E:T *Evaluate: Total is zero:*

FIRST FREE-STRUCTURE DIAGRAM
$$\sum M_B = 0$$
H does not appear in this equation because its moment about B is zero.
$+R_L \times 120 - 200 \times 90 - 400 \times 60 - 300 \times 40 = 0$
$$+R_L = +450 \text{ kN}$$

SECOND FREE-STRUCTURE DIAGRAM
$$\sum M_C = 0$$
$+R_L \times 60 - H \times 32 - 200 \times 30 + 400 \times \text{zero} = 0$
$$+H = +656 \text{ kN}$$

THIRD FREE-STRUCTURE DIAGRAM

It is first necessary to determine the various dimensions of the circular arc. Only span and rise have been given. You particularly wish to know the value of y and the slope of the rib at a distance of 30 m measured horizontally from A (quarter span).

$$R = \frac{S^2 + Q^2}{2Q} = \frac{60^2 + 32^2}{64} = 72 \cdot 25 \text{ m}$$
$d = S - x = 60 - 30 = 30 \text{ m}$
$h = \sqrt{(R^2 - d^2)} = \sqrt{(72 \cdot 25^2 - 30^2)} = 65 \cdot 7 \text{ m}$
$y = +h - R + Q = +65 \cdot 7 - 72 \cdot 25 + 30 = 25 \cdot 5 \text{ m}$

$$\frac{dy}{dx} = \frac{d}{h} = \frac{30}{65 \cdot 7} = 0 \cdot 457$$

Hypotenuse $= \sqrt{[1^2 + (0 \cdot 457)^2]} = 1 \cdot 0995$

$\left. \begin{array}{l} \cos \alpha = 0 \cdot 9095 \\ \sin \alpha = 0 \cdot 4156 \end{array} \right\}$ ratio $= 2 \cdot 188$

Equations for the third FSD
$$\sum M_X = 0 \quad \text{at quarter span}$$
$+R_L \times 30 - H \times 25 \cdot 5 + M = 0$
$+450 \times 30 - 656 \times 25 \cdot 5 + M = 0 \qquad +M = +3228 \text{ kN m}$

9/43

$$\sum F_{NS} = 0 \quad \text{at quarter span}$$
$$+R_L - \text{nsc } S - \text{nsc } T = 0$$
$$+450 - 0{\cdot}91S - 0{\cdot}42T = 0$$
$$\sum F_{EW} = 0 \quad \text{at quarter span}$$
$$+H + \text{ewc } S - \text{ewc } T = 0$$
$$+656 + 0{\cdot}42S - 0{\cdot}91T = 0$$

Multiply this by 2·188 (see above) and subtract NS equation
$$+T = +782 \text{ kN}$$
$$+S = +134 \text{ kN}$$

The cut has been taken just to the left of the 200 kN load. If it had been taken just to the right of the load, both S and T would have been different, but M would have been the same. Try it!

Y *Yes or No?*

All quantities have been given the correct sense signs so the third FSD is a correct picture of the forces and couples which must be designed for at quarter point.

This problem is, again, one of geometrical investigation. Once the lengths and slopes have been determined, and you slavishly follow the arrows on the FSD whether they are correct or not, the final calculation for the three unknown quantities is relatively simple.

For a circular arch be sure you know, when given the span and the rise, how to find the values of R, d, h, y, and the slope, in that order.

Secondly, get $\cos \alpha$ and $\sin \alpha$ without looking up any tables except those of squares and square roots. And then obtain the ratio between the two (cotan) to use in the solution of the simultaneous equations.

Carry out one step at a time; in this way a problem is broken into small operations each easily accomplished.

9/44 THREE-HINGED ARCH: Inclined load: Abutments at different levels

Determine the thrust, shear, and bending moment at a point on the arch rib, 4·5 m measured horizontally from A, the left-hand abutments. The arch consists of a portion of a circle of radius 8 m, with the third hinge at a point to the right of the apex of the curve.

S *Structure:*

Here the abutments are at different levels, the third hinge is not at the highest point as it usually is, and there is an inclined load. This looks complex, but the system of attack on the problem is exactly as before. Three FSD's are required, one for the three quantities asked for, preceded by a study of two FSD's to determine the values of the components of the abutment reactions. Note that not only must the horizontal and vertical components of the reactions be found simultaneously (as always occurs when abutments are at different levels) but the two horizontal components are not equal as they are when loads are only vertical. The inclined load applies a horizontal component and must come into the E–W equations and the moment equations.

U *Unknowns and Units:*

After the reactions are determined, there are three unknowns at X. Use kN and m, and mark sense arrows at random on the unknown quantities.

R *Rotation:*

As for the simpler problems, the three COR's are:
COR for first FSD is at B where two unknowns intersect.
COR for second FSD is at C where two unknowns intersect.
COR for third FSD is at X where the cut has taken place.

E: T *Evaluate: Total is zero:*

Split inclined load into two components

FIRST FREE-STRUCTURE DIAGRAM

$$\sum M_B = 0$$
$$+ R_L \times 15\cdot2 + H_L \times 1\cdot5 + 100 \times 5\cdot2 - 173 \times 12\cdot2 - 500 \times 3\cdot3 = 0$$

9/44

SECOND FREE-STRUCTURE DIAGRAM

$$\Sigma M_\text{C} = 0$$
$$+R_\text{L} \times 9{\cdot}2 - H_\text{T.} \times 4{\cdot}8 - 100 \times 1{\cdot}1 - 173 \times 6{\cdot}2 = 0$$

Some dimensions in these equations have been determined in the geometry section, which has been printed in one piece so that references can be made to it in order to find out how a circular arch of this type can be investigated. Once the geometry is decided, the derivation goes on as it has in the earlier simpler problems.

Solving these two equations simultaneously, the reaction at the left abutment is seen to have as vertical and horizontal components:

Return to first FSD

$$\Sigma F_\text{EW} = 0$$
$$+H_\text{L} + 100 - H_\text{R} = 0$$

$$+R_\text{L} = +200 \text{ kN}$$
$$+H_\text{L} = +137 \text{ kN}$$
$$+H_\text{R} = +237 \text{ kN}$$

Geometry of the arch

Horizontal distances are given. Vertical distances from base lines (one left, one right) to arch are required at Y, X, and C. The slope is also required at X as in other problems, in order to obtain the components of the forces of Thrust and Shear at the cut section.

When several values must be found, it is best to use a tabular method, completing each *column* in turn.

Refer to unit 7, section 3:

Right or Left	Point on arch	x from A or B	$d = (S - x)$	$h = \sqrt{(R^2 - d^2)}$	$y = h - R + Q$
		(lengths are in metres)			
L (S 7·4)	Y	3·0	4·4	6·7	3·7
L (S 7·4)	X	4·5	2·9	7·5	4·5
R (S 7·8)	C	6·0	1·8	7·8	6·3

dy/dx required at x

$$\frac{dy}{dx} = \frac{d}{h} = \frac{2{\cdot}9}{7{\cdot}5} = 0{\cdot}39$$

9/44

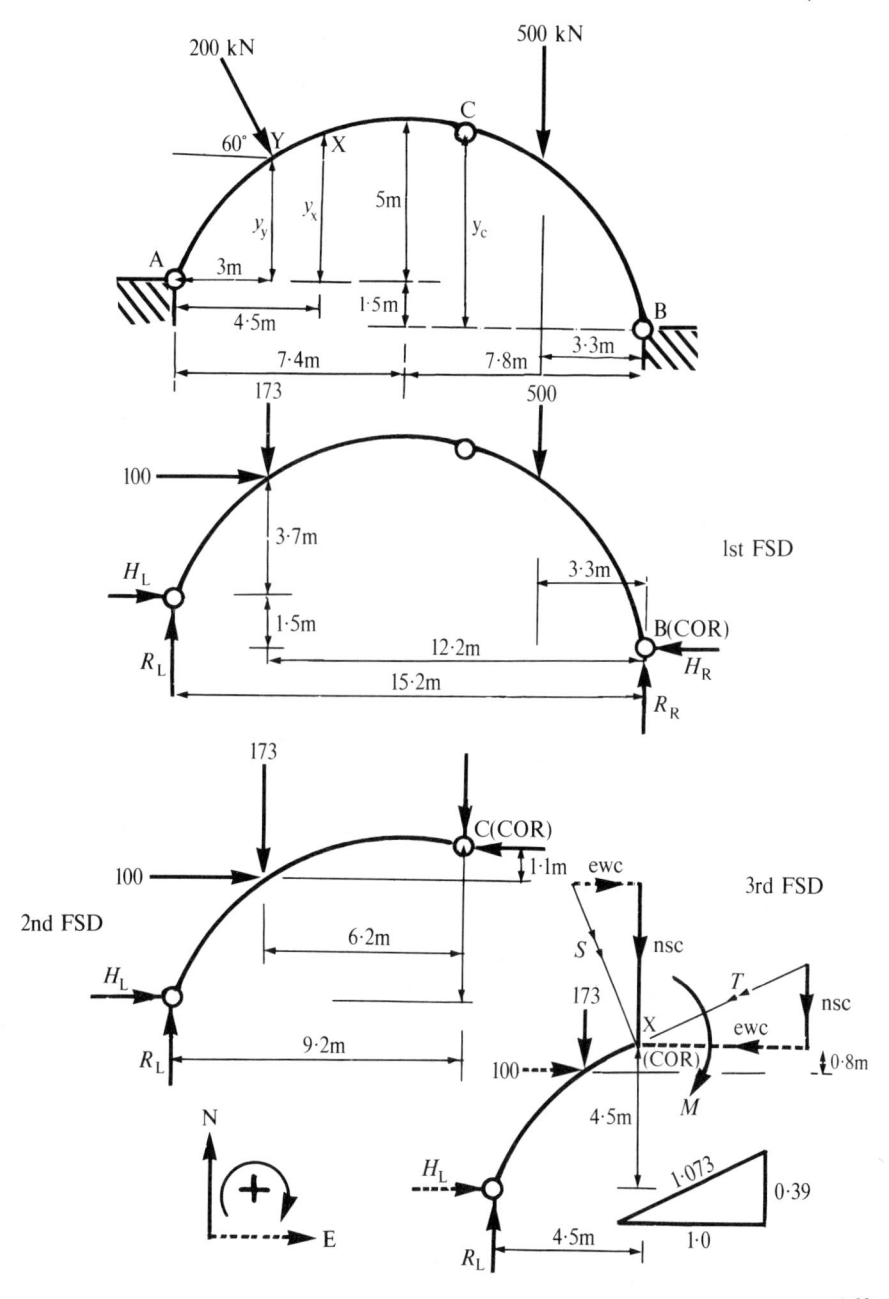

9/44

$$\text{Hypotenuse} = \sqrt{[1 \cdot 0^2 + (0 \cdot 39)^2]} = 1 \cdot 073$$

$$\left. \begin{array}{l} \cos \alpha = \dfrac{1 \cdot 00}{1 \cdot 073} = 0 \cdot 9320 \\[6pt] \sin \alpha = \dfrac{0 \cdot 39}{1 \cdot 073} = 0 \cdot 3625 \end{array} \right\} \text{ratio: } 2 \cdot 2571$$

From this table which should be prepared at the outset of any arch problem, especially if it is complex, it is possible to read off all the dimensions required in the construction of the three FSD.

THIRD (AND MOST IMPORTANT) FREE-STRUCTURE DIAGRAM

From this diagram you can determine the values of bending moment, thrust, and shear at the cut section X which is situated at 4·5 m measured horizontally from A.

$$\sum M_X = 0$$
$$+R_L \times 4 \cdot 5 - H_L \times 4 \cdot 5 - 100 \times 0 \cdot 8 - 173 \cdot 1 \cdot 5 + M = 0$$
$$+M = +57 \text{ kN m}$$

$$\sum F_{NS} = 0$$
$$+R_L - 173 - \text{nsc } S - \text{nsc } T = 0$$
$$+27 - 0 \cdot 93S - 0 \cdot 36T = 0$$

Multiply this by 2·571, the ratio obtained in the section on geometry, and use the resulting equation with the EW equation to find values of S and T.

$$+69 \cdot 4 \; -2 \cdot 4S - 0 \cdot 93T = 0$$

$$\sum F_{EW} = 0$$
$$+H_L + 100 + \text{ewc } S - \text{ewc } T = 0$$
$$+237 + 0 \cdot 36S - 0 \cdot 93T = 0$$

Solving NS and EW equations simultaneously, $\quad +S = -61 \text{ kN}$

Now, by substituting this value in the NS equation (or the EW) you can find the value of T. You must note, however, that, as the FSD is drawn, and you are working with the third FSD as it appears, the value of S has a negative sign.

$$+27 - 0 \cdot 93S - 0 \cdot 26T = 0$$
$$+27 - 0 \cdot 93(-61) - 0 \cdot 36T = 0 \qquad\qquad +T = +232 \text{ kN}$$

Y *Yes or No?*

In the corrected FSD, which you ought to draw, the shearing force S, should be acting upwards on the cut section, and not as shown here.

9/45

9/45 FIVE EQUATIONS: Three FSD'S: Hinged and held supports: Three-hinged arch

A bent beam is pin-jointed in three places. Find the bending moment, shearing force, and thrust at a point X as marked.

S *Structure and Supports:*

The supports are hinged, and the forces acting at the hinges can be divided into two components at the left-hand hinge and two at the right-hand hinge. The co-ordinate directions may be vertical and horizontal.

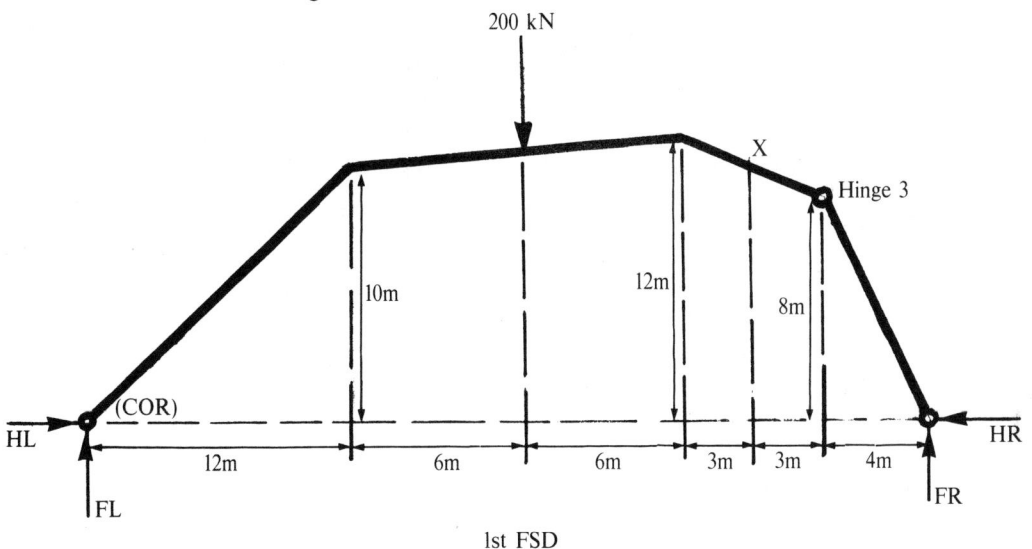

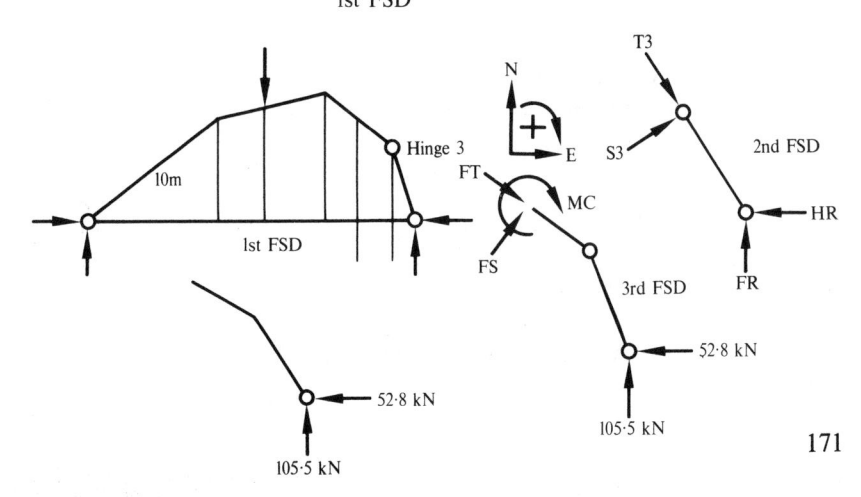

171

9/45

U *Unknowns:*

There are four unknowns on the original structure, and the three asked for in the question. For seven unknowns, more than one FSD will be required. Start with the original structure. Use kN and m.

FIRST FREE-STRUCTURE DIAGRAM

R *Rotation:*

Use the left-hand hinge, where the two unknowns H_1 and F_L intersect.

E: T *Evaluate and equate to zero:*

Forces acting: $\quad F_L \quad H_L \quad 200 \quad H_R \quad F_R \quad$ (kN)
Moment arm
about COR: $\quad\quad\quad 0 \quad 0 \quad 18 \quad 0 \quad 34 \quad$ (m)
Moments about COR: $+200 \times 18 - F_R \times 34 = 0$
$$+F_R = +105\cdot 5 \text{ kN}$$

SECOND FREE-STRUCTURE DIAGRAM

(Right-hand portion up to hinge)

R *Rotation:*

COR at the third hinge.

E: T *Evaluate and equate to zero:*

$$M_{H3} = 0$$

All forces acting: $\quad F_R \quad H_R \quad F_3 \quad T_3 \quad$ (kN)
Distance to
COR at hinge 3: $\quad 4 \quad 8 \quad 0 \quad 0 \quad$ (m)

Add applied moments at the cut section, and write moment equation. There is no moment applied at the cut section since this is at a hinge which can carry no moment in equilibrium.

Moment equation:
$-F_R \times 4 + H_R \times 8 = 0$
$-105\cdot 5 \times 4 + 8H_R = 0 \quad\quad\quad\quad\quad\quad +H_R = +52\cdot 8 \text{ kN}$

THIRD FREE-STRUCTURE DIAGRAM

(Cut made at X to make the internal quantities at that section, external)

R   *Rotation:*
COR at the cut section.

E: T   *Evaluate and equate to zero:*
$$\sum M_X = 0$$
Forces acting: F_R H_R F_S F_T
Moment arm to
COR at X: 7   10   0   0
Add applied moments (only the moment of the couple MC)
Moment equation:
$-7F_R + 10H_R + M_C = 0$
$-105 \cdot 5 \times 7 + 52 \cdot 8 \times 10 + M_C = 0$ $+M_C = +211 \text{ kN m}$
Continuing with the same FSD, use the remaining two equations and so determine
$$+F_S = -58 \cdot 5 \text{ kN} \quad \text{and} \quad +F_T = +102 \cdot 7 \text{ kN}$$

Y   *Yes or No?*

This a very common type of problem. Remember that when you cut through a section sustaining bending you are considering a *held* support, which is represented by two forces and the moment of a couple. The forces are always taken at right angles (*shearing force*) and in the line of the member ((*thrust*). The moment of the couple acting to maintain equilibrium in such a situation is called the *bending moment*.

The shearing force F_S in this problem is acting on the section in the opposite sense to that assumed (negative sign), so a new and correct third FSD should be drawn with all values marked on it.

Solve the PROBLEMS yourself. You will gain little from reading them over.

9/46

9/46 SUSPENSION CABLE: Uniformly-distributed load: Supports level

Find the tension in the suspension cable shown, both at the tops of the towers, and at the lowest point where the slope of the cable is horizontal. What is the slope of the cable at 50 m from B (horizontally) and what tension is developed at this point?

S *Structure:*

When a cable carries a uniformly distributed load, transferred, perhaps, through a stiff girder, hung on vertical hangers from the cable, the curve assumed is a parabola. To obtain the geometry of the structure, therefore, it is sufficient to re-draw the parabola of Unit 7 (sect. 3) upside down.

An advantage (so far as problems are concerned) of the suspension cable is that it cannot take shear or bending. This is shown if you hold a short length of string in both hands and then try to apply bending or shear. The string merely moves away and refuses to take the load. It can take the reverse of the thrust of arches, in the form of a tension, and this is the only load found in these problems.

U *Unknowns and Units:*

Three tensions have to be found, and three FSD's are required. Use kN and m and put arrows at random on the unknowns. However, it is evident from day-to-day experience the sense in which a pull acts in a suspended cable.

R *Rotation:*

In a stiff curved structure like a metal or reinforced concrete arch the moment equations are written for a COR at hinges where no moment can be developed; the work is thus much simplified. In a suspended cable every part of the length is effectively a hinge since it bends and moves whenever a moment is applied.

E: T *Evaluate: Total is zero:*

FIRST FREE-STRUCTURE DIAGRAM

$$\sum F_{NS} = 0$$
$$+R_L = +R_R = \text{half the load} = +375 \text{ kN}$$

174

9/46

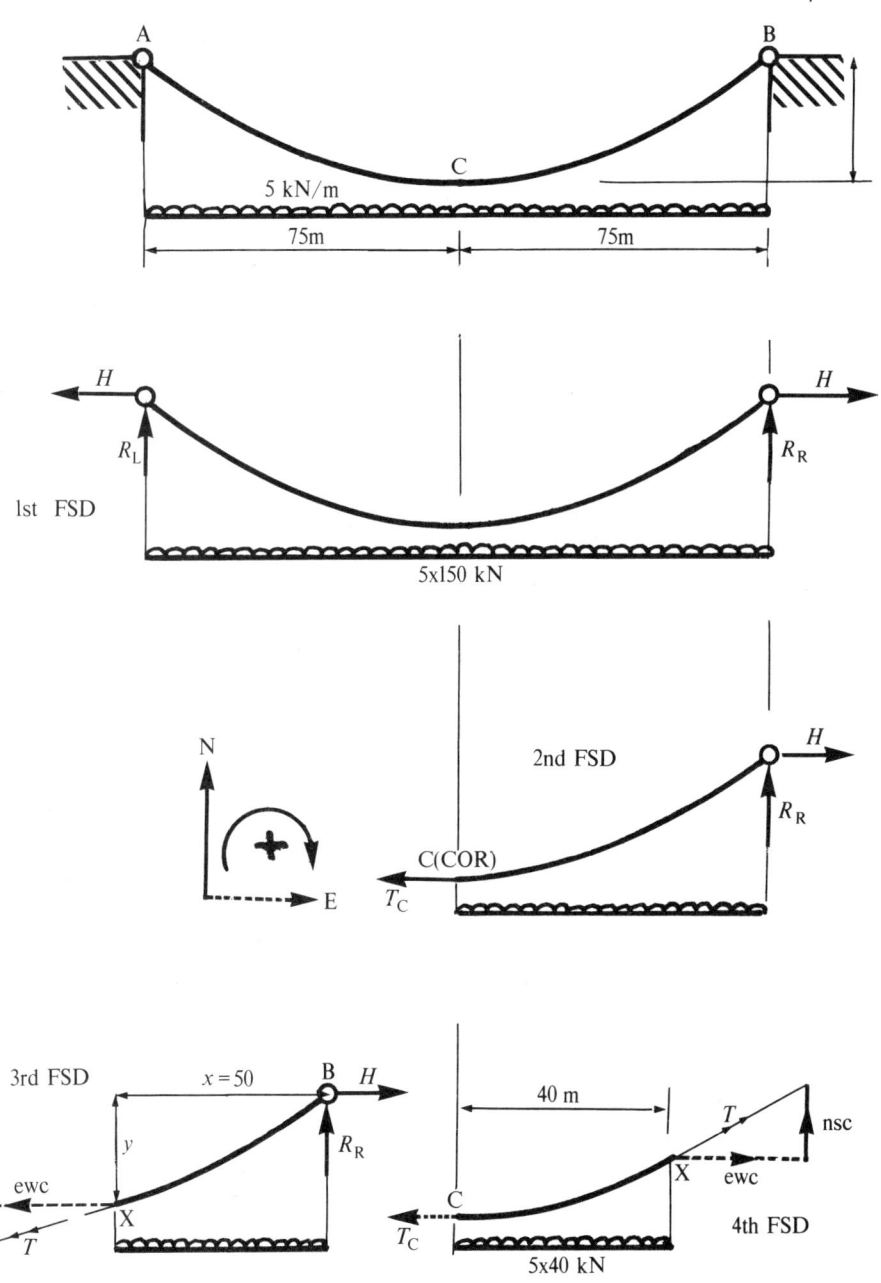

1st FSD

2nd FSD

3rd FSD

4th FSD

175

9/46

SECOND FREE-STRUCTURE DIAGRAM

Since any point may be taken as a hinge, the FSD cut-off at the lowest point of the dip can be used. COR is at C.
$$\Sigma M_C = 0$$
$+R_L \times 75 - H \times 20 - (5 \times 75)37 \cdot 5 = 0 \qquad\qquad +H = +703 \text{ kN}$

This value of H is not only the horizontal component of the reaction at B, but also the horizontal tension in the cable at the lowest point C.

The tension in the cable at B is found by combining the two components:
$$+F_B = \sqrt{(375^2 + 703^2)} = +797 \text{ kN}$$
tension in the cable at each abutment.

THIRD FREE-STRUCTURE DIAGRAM

Geometry of the cable:

Parabolic in shape under uniformly distributed load.
$$x = 50 \text{ m}$$
$$\frac{dy}{dx} = \frac{2Q}{S}\left(1 - \frac{x}{S}\right) = \frac{2 \times 20}{75}\left(1 - \frac{50}{75}\right) = 0 \cdot 178$$
$$\text{Hypotenuse} = \sqrt{[1 \cdot 0^2 + (0 \cdot 178)^2]} = 1 \cdot 016$$

$$\left.\begin{array}{l}\cos \alpha = \dfrac{1 \cdot 00}{1 \cdot 016} = 0 \cdot 9843 \\[2mm] \sin \alpha = \dfrac{0 \cdot 178}{1 \cdot 016} = 0 \cdot 1752\end{array}\right\} \text{ratio} = 5 \cdot 618$$

The answer to the first question (on the slope of the cable) is that it has, at 50 m from A or from B, a slope of 0·178 in 1 or 1 in 5·62.

Equation from the third FSD
$$\Sigma F_{EW} = 0$$
$+H - \text{ewc } T = 0$
$+703 - 0 \cdot 9843T = 0 \qquad\qquad +T = +714 \text{ kN at X}$

Y *Yes or No?*

All the evaluations show double plus signs, so all have been guessed in the right senses. This is not surprising, for it is easier to visualize the forces in a cable, than those in an arch. The forces acting are also much simpler, since there is no shear and no bending moment, merely a tension in the cable.

9/47

9/47 SUSPENSION CABLE: Three concentrated loads: Shape of cable?

The flexible suspension cable shown carries three concentrated loads. If the dip of the cable at the centre is 1·1 m, find the shape of the cable when in equilibrium.

S *Structure:*

This cable is not carrying a uniformly distributed load, so it does not hang in a parabolic curve. It is assumed to be very light in weight and thus stretches in straight lines between each of the heavy weights.

U *Unknowns and Units:*

The unknowns are the values of the dip of the cable at the three loads. Use kN and m.

R *Rotation:*

Any point on the cable can be taken as a COR, and at every point the bending moment must be zero, for a flexible cable cannot resist bending. It moves under any attempted bending until there remains only a tensile force in the cable. This fact can be used to determine forces and dimensions, for whatever moments are caused by the loading at points in the cable must cancel each other and become zero.

E: T *Evaluation: Total is zero:*

FIRST FREE-STRUCTURE DIAGRAM (calculate vertical reactions)

$$\sum M_B = 0$$
$$+6V_L + H \times 0 - 50 \times 4·9 - 40 \times 4 - 30 \times 2·5 \qquad +V_L = +81 \text{ kN}$$
$$+V_R = +39 \text{ kN}$$

SECOND FREE-STRUCTURE DIAGRAM (at centre dip is 1·1 m: find H)

$$\sum M_X = 0$$
$$-3V_R + 1·1H + 30 \times 0·5 = 0 \qquad +H = +9·3 \text{ kN}$$

THIRD FREE-STRUCTURE DIAGRAM (forces known: find dip)

$$\sum M_Y = 0$$
$$-2·5V_R + H \times \text{dip} = 0 \qquad \textbf{Dip} = \textbf{1·03 m}$$

9/47

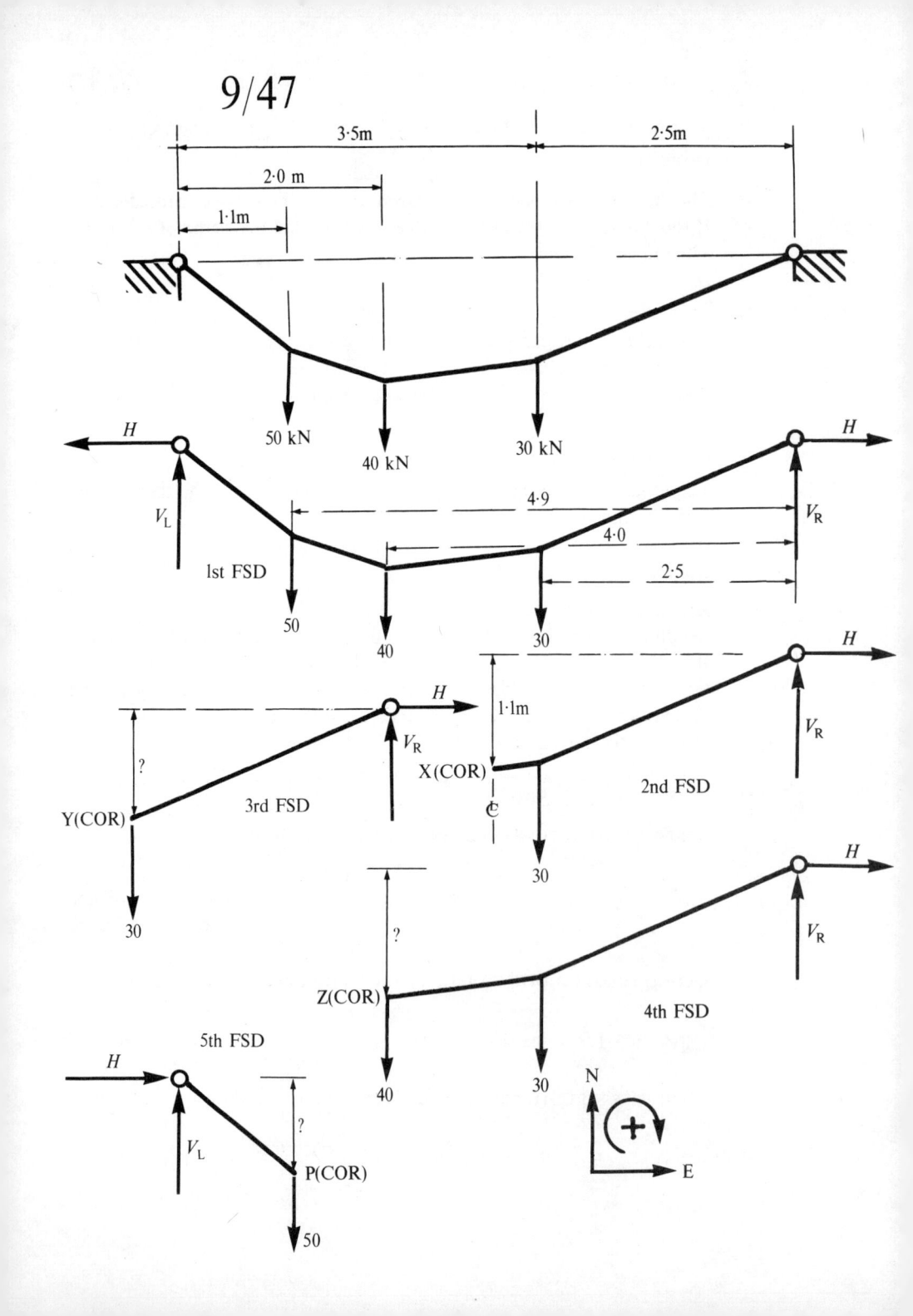

9/48

FOURTH FREE-STRUCTURE DIAGRAM (forces known: find dip)
$$\sum M_Z = 0$$
$-4V_R + H \times \text{dip} + 30 \times 1.5 = 0$ **Dip = 1·20 m**

FIFTH FREE-STRUCTURE DIAGRAM (forces known: find dip)
$$\sum M_P = 0$$
$+1·1 V_L - H \times \text{dip} = 0$ **Dip = 0·97 m**

9/48 SUSPENSION CABLE: Supports level: Temperature change

A cable is of 100 m horizontal span between supports at the same level. The central dip is 10 m. Loading is 0·5 kN/m, measured horizontally. What are the maximum and mimimum tensions in the cable? What alterations will occur in the tension H and in the central dip if the temperature rises by 30°F when the coefficient of expansion of the material is 6×10^{-6} strain per degree.

S *Structure:*

The maximum tension occurs at the supports and the mimimum at the lowest point of the dip where the tension is horizontal. When the temperature rises, the length of the cable increases, the dip increases, and the maximum tension decreases.

U *Unknowns and Units:*

There are five unknowns; the tension T at the supports before and after the rise in temperature, the tension at the lowest point before and after, and the dip after the rise in temperature. Use kN and m.

R *Rotation:*

Any point on the cable can be considered a point of zero bending moment.

E : T *Evaluate: Total is zero:*

The vertical components of the reactions T are clearly equal and equal to half the total load supported.

$$V_L = V_R = \tfrac{1}{2} \times 0·5 \times 100 \text{ kN} = 25 \text{ kN}$$

179

9/48

FIRST FREE-STRUCTURE DIAGRAM

$$\sum M_B = 0$$
$$-H \times 10 + (0.5 \times 50)25 = 0 \qquad\qquad +H = +62.5 \text{ kN}$$

SECOND FREE-STRUCTURE DIAGRAM

Remembering that T and H must intersect at the centre line of the load on the FSD, the gradient of T is at 10:25.

$$\text{Hypotenuse} = \sqrt{(10^2 + 25^2)} = 26.93$$

$$\cos \alpha = \frac{25.0}{26.93} = 0.93$$

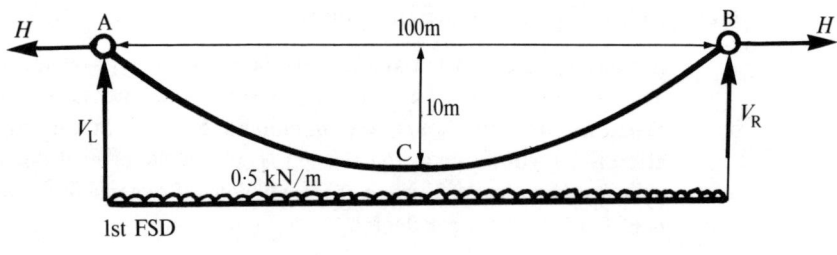

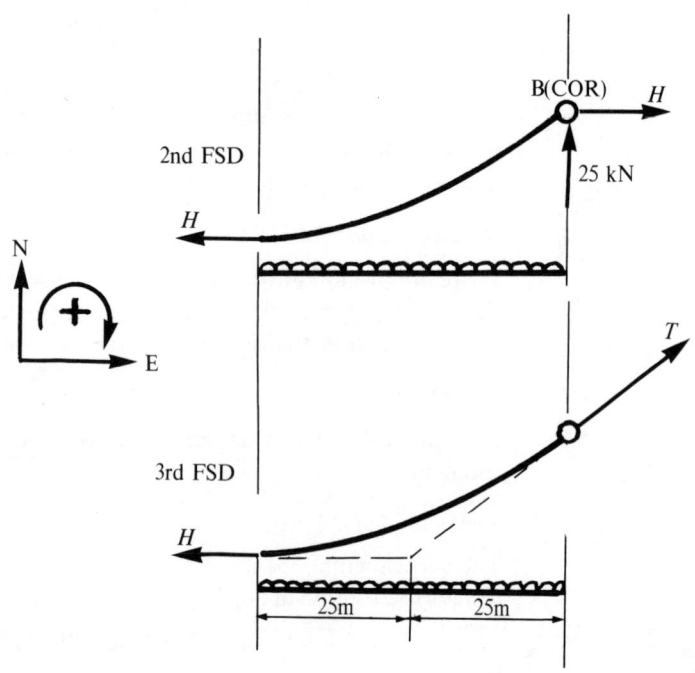

$$\sum F_{EW} = 0$$
$$-H + \text{ewc}\, T = 0 \qquad\qquad +T = +67\cdot3\,\text{kN}$$

The value of H is the minimum tension in the cable, and the value of T is the maximum tension.

Length and expansion of cable

The original length of the cable spanning 100 m with a 10 m dip can be calculated from the relationship given in section 7.3.

$$\text{Length} = 2\left(S + \frac{2Q^2}{3S}\right) = 2\left(50 + \frac{200}{150}\right) = 102\cdot66667\,\text{m}$$

After a rise in temperature of 30° the length increases to:

New length $= 102\cdot6667(1 + \alpha t) = 102\cdot6667(1 + 30 \times 6 \times 10^{-6})$
$$= 102\cdot6851\,\text{m}$$

This increased length produces dip which can be calculated from:

$$2\left(\frac{2Q^2}{3S}\right) = 2\cdot6851 : Q^2 = 100\cdot6875 \qquad Q = 10\cdot034\,\text{m}$$

Using second free-structure diagram

$$\sum M_B = 0$$
$$-H(10\cdot034) - (50 \times \tfrac{1}{2})25 = 0 \qquad\qquad +H = +62\cdot286\,\text{kN}$$

which is the least tension in the cable – at the lowest point.

9/49 SUSPENSION CABLE: Supports at different levels: Uniform loading

The cable is suspended from A and B and has its lowest point at C. A and B are 60 m apart horizontally, and C is 15 m below B and 5 m below A. A load of 600 kN is uniformly distributed horizontally between A and B. Find the position of C and the maximum tension in the cable.

S *Structure:*

There are only two horizontal forces acting – the horizontal components of the reactions. These must be equal to maintain equilibrium in the EW equation. The lowest point can be found from the equation to the parabolic curve to which the cable hangs.

9/49

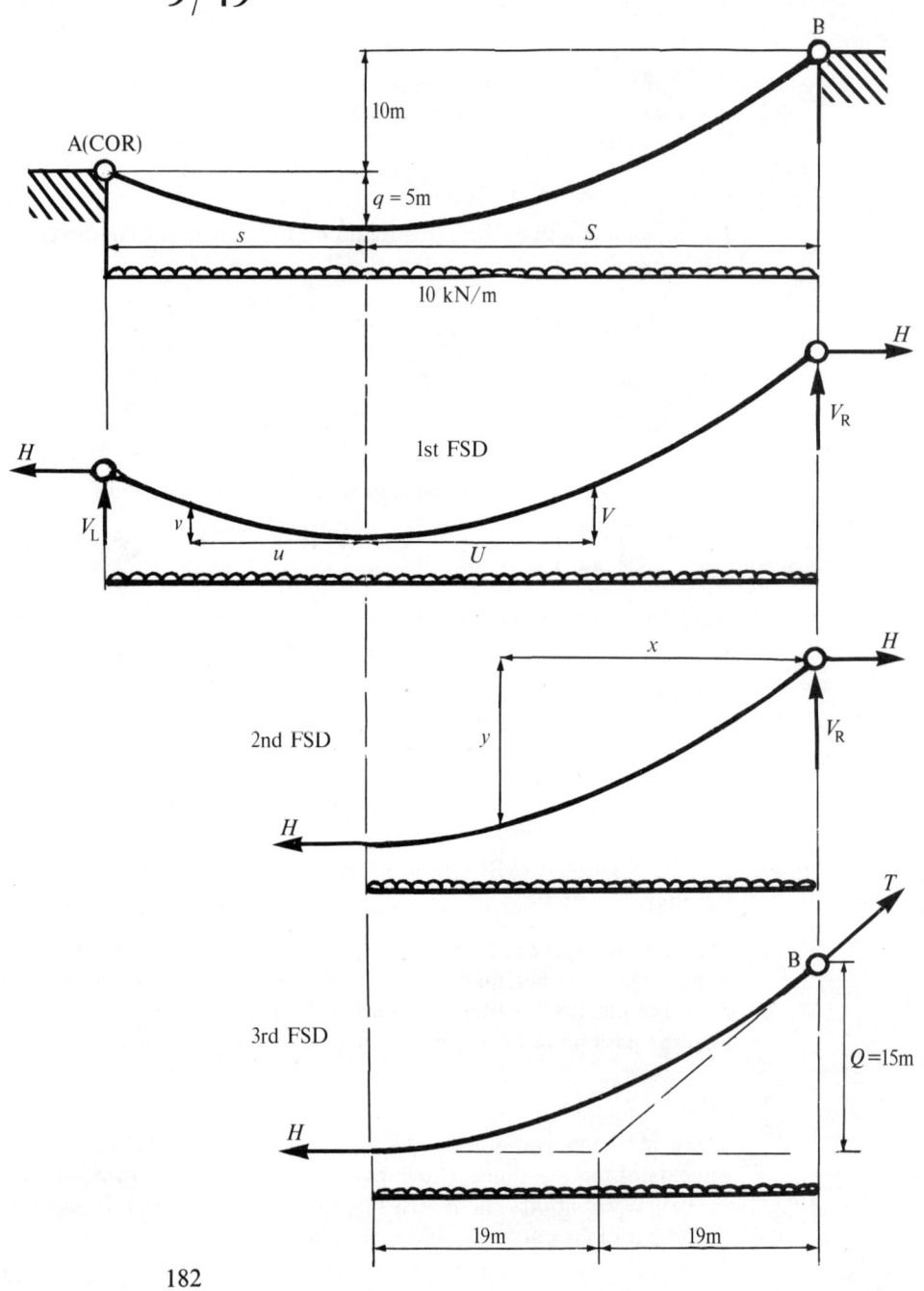

A(COR)   10m   B

$q = 5m$

s S

10 kN/m

1st FSD

H, V_L, v, u, U, V, V_R, H

2nd FSD

x, y, H, V_R, H

3rd FSD

T, B, $Q = 15m$, H, 19m, 19m

U *Unknowns and Units:*

Two unknowns: one geometrical and the other to be obtained from the equations of equilibrium. Use kN and m.

R *Rotation:*

Any point can be a rotational centre, but it is usual to use the supports and the apex (lowest point) for the FSD required.

E: T *Evaluate: Total is zero:*

FIRST FREE-STRUCTURE DIAGRAM

Geometry of the cable

On the left portion, the equation to the parabola is:

$$v = \text{constant} \times u^2$$

(for the reason for this see section 7.3). On the right-hand portion, the equation is

$$V = \text{constant} \times U^2$$

Since the same parabolic curve continues throughout, the constant is the same for both portions:

$$\frac{u^2}{v} = \frac{U^2}{V}$$
When u is s, then v is q.
When U is S then V is Q.

But $Q = q + 10$ and $S = 60 - s$. The basic relationship is

$$Vu^2 = vU^2$$

which becomes, for the largest values of v and V,

$$Qs^2 = qS^2$$
$$15s^2 = 5(60 - s)^2 \qquad +s = +22 \text{ m}$$
$$+S = +38 \text{ m}$$

Equation for the first FSD

$$\sum M_A = 0$$
$$+10H - 60V_R + 600 \times 30 = 0 \qquad\qquad +H = +482 \text{ kN}$$

SECOND FREE-STRUCTURE DIAGRAM

$$\sum M_B = 0$$
$$+15H - 380 \times 19 = 0$$

Substituting in the equation from the First FSD:

$$+482 \times 10 - 60V_R + 18\,000 = 0 \qquad\qquad +V_R = +381 \text{ kN}$$

9/49

Maximum tension in the cable (see third FSD)

This takes place at the supports, and is greater with a larger angle to the horizontal. The largest dy/dx must therefore, be found, and is clearly at B.

Method 1

Combine the two components of the tension at the support (H and V_R)

$$\text{Tension at } B = T = \sqrt{(482^2 + 381^2)} = \mathbf{614 \text{ kN}}$$

Method 2

Determine the slope of the cable at B from the equation to the parabola, and use the horizontal pull H to find T.

$$\text{When } x = 0 \quad \frac{dy}{dx} = \frac{2Q}{S} = \frac{2 \times 15}{38} = 0{\cdot}789$$

$$\text{Hypotenuse} = \sqrt{[1^2 + (0{\cdot}789)^2]} = 1{\cdot}275 \qquad T = 1{\cdot}275H$$

$$\mathbf{T = 614 \text{ kN}}$$

Method 3

Since only three forces are acting on the second FSD, they must meet at a point. The forces are T, H, and W, which is the weight of the uniformly distributed load on the length S (38 m). The junction of these three must be at half the length of S, and the proportions of the slope of the force T give a gradient of 15:19, which is the same figure as found in method 2.

$$\mathbf{T = 614 \text{ kN}}$$

Use the INDEX

Part I: Group C

10 Diagrams for design

There are two ways in which diagrams can be used in the subjects studied in this book. The first use is for calculation. Instead of carrying out the calculations as has been done so far in Part I, the problem may be set out graphically, to scale, and the results required measured off. This can be very accurate if the drawing is made with skill, but the technique requires drawing facilities, and the procedures form a complete branch of study, usually called *graphic statics*. This method of approach is not considered in this book.

The second use of diagrams is to provide a pictorial representation of the variation of important quantities within a given structure. A visual representation of this kind is much more valuable to the designer (all the studies here will eventually be directed to the design of structures) than are a number of figures set out to represent the results achieved. In this unit of study the sketching of these diagrams is described, and in unit 17, at least one of the methods of using sketched diagrams in finding important values, is the subject of exercises.

10.1 Representation of the effects of forces and couples

In the preceding units of study you have learnt how to obtain the values of *shearing forces*, *bending moments*, and *thrusts* in members of structures subjected to forces and couples. This is an essential step in the design of structures but individual values are not of maximum assistance to the designer. He must know the variation of the various quantities along the length of the member, and, in particular, the points where these quantities become large and reach a maximum value.

In this unit we shall study only straight beams, but they may be horizontal and may carry loads normal (at right angles) to the beam or inclined to it. The beams may also carry couples, externally applied. The aim is to draw diagrams which show how the three important quantities, *bending moment*, *thrust*, and *shearing force* vary from one end of the beam to the other. Whether the beam is horizontal or inclined,

10.2

the diagram is best plotted on a straight horizontal base, the vertical ordinates from that base giving the values of the quantities concerned, at that point in the beam.

Whatever forces are acting on the beam, and whatever the inclination of the beam to the horizontal, the force-components which are *normal (at right angles) to the beam* give rise to *shearing forces and bending moments*. The force-components which act *along the length of the beam* give rise to *thrusts*. The values of all of these are plotted vertically below the point on the beam to which they refer. Applied couples give bending moments only, and do not appear in the diagrams for shear and thrust.

The subject can be divided into three divisions:

(*a*) horizontal beams with vertical loads,
(*b*) horizontal beams with loads inclined to the vertical,
(*c*) beams inclined to the horizontal with loads both vertical and inclined to the vertical.

All these have the same treatment, but the first is simpler, for there is no thrust, the loads having no component in the direction of the length of the beam. This simpler case is considered first.

10.2 Diagrams for horizontal beams with vertical loads

Most beams in use are of this type, so you must be thoroughly familiar with the method of attack on their problems and with the results obtained. The two diagrams you must study are those showing *variation of shearing force* and those showing *variation of bending moment*. The third quantity, *thrust*, is zero when the loads are at right angles to the beam.

When the problem of drawing a diagram showing the variation in shearing force or bending moment is presented, you erect vertical lines under various points on the sketch of the beam. These vertical lines represent, to scale, the values of the quantities concerned. It is then only necessary to join the tops of these ordinates to have a pictorial representation of the variation of shear or bending along the beam. Clearly, it is necessary to know at what salient points the values should be obtained, and also by what type of line or curve the tops of the ordinates should be joined. If the type of curve appropriate to given conditions of loading is known, the sketch can be made rapidly. There is no need to use drawing instruments, as many students tend to do. Squared paper and a sketching pencil are sufficient, and ensure the rapid production of the diagrams.

The relationship between x, the distance along the beam, and v the vertical heights of the ordinates is the *equation* to the line. As a table of reference (which has more meaning later in the unit) the following should be noted:

Shearing force: The equations to the lines outlining the shearing force diagram for various types of loading are:

(i) Concentrated loads: $S = v$ Value of S is constant.

(ii) Uniformly distributed loads: $S = rx$ Sloping straight lines.

(iii) Loads distributed at a non-uniform rate: $S = qx^2 + px$ Quadratic curves.

Bending moment: The equations to the lines outlining the shearing force diagrams for various types of loading are:

(iv) Concentrated loads: $M = tx$ Sloping straight lines.

(v) Uniformly distributed loads: $M = lx^2 + mx$ Quadratic curves.

(vi) Loads distributed at a non-uniform rate: $M = ux^3 + nx^2 + kx$ Cubic curves.

The terms v, r, q, p, t, l, m, u, n, and k are numerical values which depend on the conditions of the problem. Other numerical values, not linked with the distance x along the beam, may also appear.

If you examine the $\sum F = 0$; $\sum M = 0$ equations in problems of Unit 9, from which the values of shearing force and bending moment were determined, you will see that, *numerically*, the values of these quantities are represented by the algebraic sum of the forces or moments on one side of the section concerned. The signs of the quantities are also important, and must be found separately.

The section concerned can be represented by the edge of a postcard which is used to cover one portion of a sketch of the beam, and leave the other portion exposed. The algebraic sum of the forces *which you can see* (positive up and negative down) gives the numerical value of the *shearing force* at the edge of the card. The sign to be given to this value

10.2

depends on the convention for shearing force, given in an earlier Unit (see index), but which is repeated below for convenience.

Similarly, the moments of the forces *you can see* taken about the edge of the card, when summed algebraically (clockwise, positive, anticlockwise, negative) give the values of the bending moments at different sections. The way in which a simple postcard can be used to determine the values of the ordinates of the shearing force and bending moment diagrams is as follows:

(a) Make a sketch of the beam and its loading reasonably to scale, but without any exact drawing techniques. Keeping the beam and the ordinates of the diagrams approximately to scale helps in the visualization.

(b) Cover the whole sketch with a postcard or other *opaque* sheet.

(c) Gradually slide the card towards the right or left, and as the forces appear, add and subtract their effects in terms of Newtons to give the shearing force at the edge of the card. (The ordinates representing these figures are drawn under the edge of the card.)

(d) As the card is once again laid over the whole diagram and slid towards the right or left multiply each force which appears by its distance to the edge of the card, and add and subtract the clockwise and anticlockwise effects to give the values of the bending moment at the edge of the card.

(e) If an applied couple appears from under the sliding card, its value is added to or subtracted from the bending moment at that point, but the moment of a couple is ignored when you draw the shearing force diagram.

(f) A positive shearing force or bending moment is represented by an ordinate rising above the base-line, and negative values are set off below the line. The conventions are:

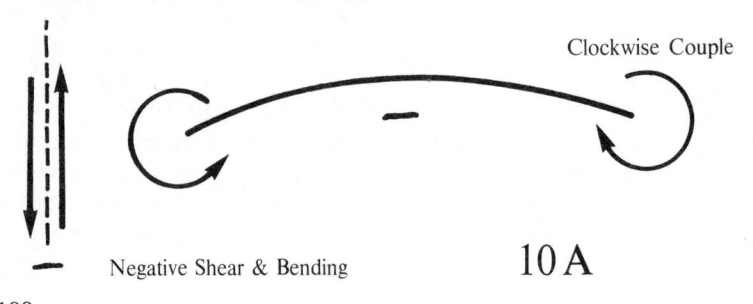

Clockwise Couple

Negative Shear & Bending

10 A

10.4

10A

Clockwise Couple

Positive Shear & Bending

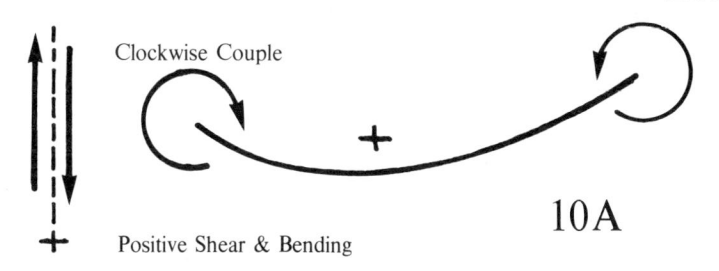

10.3 Deflected shapes

Although, strictly by the rules and limitations laid down in Unit 4, beams suffering bending and shear do not change shape, you have now reached a midway point between the theoretically limited RIGID structures, and the more realistic DEFORMABLE structures to be studied in PART II. Thus, it is quite permissible to think of the way in which beams do bend. The sketching of deflected shapes of the beams helps considerably in deciding on the shapes of bending moment diagrams, and also helps to link bending moment and shearing force diagrams.

Beams deform in three ways under bending: they are either concave-up (saucer), concave-down (umbrella) or straight (where there is no bending). The methods of sketching deflected shapes is studied in detail in sections 17.1 and 17.2, and you are strongly advised to turn to these sections now, and practise the methods presented. You will find that the drawing of the diagrams in the problem will be much easier if a deflected shape is known. They are used in the problems and labelled DS.

10.4 Couples and eccentric loads

It has been emphasized several times in this book that couples, which consist of *two* forces, equal in magnitude and opposite in sense, apply a pure moment. They do not appear, therefore, in any way, in the equations for shearing force.

However, if the moment applied to the beam is generated not by two balancing forces, as in a couple, but by only one force, conditions are different. A single force, not applied to the beam itself, but to a bracket fixed to the beam, not only provides a moment, which the beam must resist, but also applies a force.

There are two methods by which these *single* eccentric forces act on a beam to apply simultaneously a moment and a force, and cause their effects to appear in both the SFD and the BMD.

189

10.4

Type A: The force F_1 on its bracket acts at right angles to the beam. At the point where the bracket is attached to the beam, there is, clearly a moment of $F_1 \times a$ applied to the beam by the bracket. However, there is more to be considered. Nothing is disturbed if two equal and opposite forces are added to the arrangement. These are F_1 and F_3. The diagram 10/B(a) and the diagram 10/B(b) are identical for the two added forces cancel each other. If you imagine all the three forces to be equal, then the effect of F_1 on its bracket is:

> a couple of moment $F_1 \times a$, whose effect appears only in calculations for bending moment; and

> a force F_1 normal to the beam whose effect appears *both* in the calculations for shear and in those for bending moment.

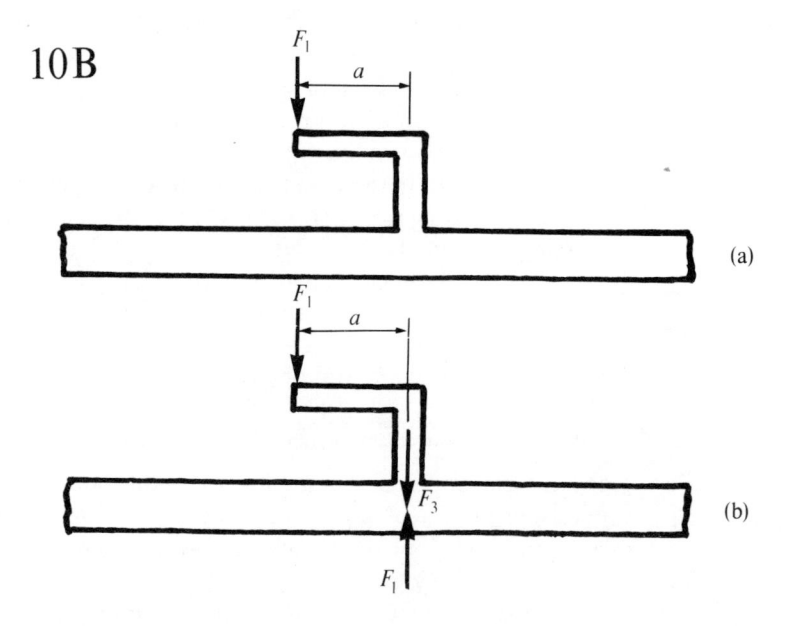

10B

(a)

(b)

Type B: The force F_4 acts parallel to the beam, and exerts a moment $F_4 \times b$ where the bracket joins the beam. Once again, there is no change made by the addition of two forces which cancel each other, so sketch 10/C(b) represents the same condition as sketch 10/C(a). Thus, once again, the single eccentric force on its bracket exerts a couple (F_4 and F_6 are equal) and a single force, this time along the beam. All three forces are equal to the original F_4. The couple has a moment which appears only in calculations for bending moment, while the force has an effect on the thrust (along the beam) but the bending moment, this time, is not affected. You cannot bend a beam (unstable buckling excepted) merely by pushing along its length.

There might be occasions when a compounding of A and B occurs, but this is unusual. If a problem of this kind arises you must resolve the inclined force into the two types A and B.

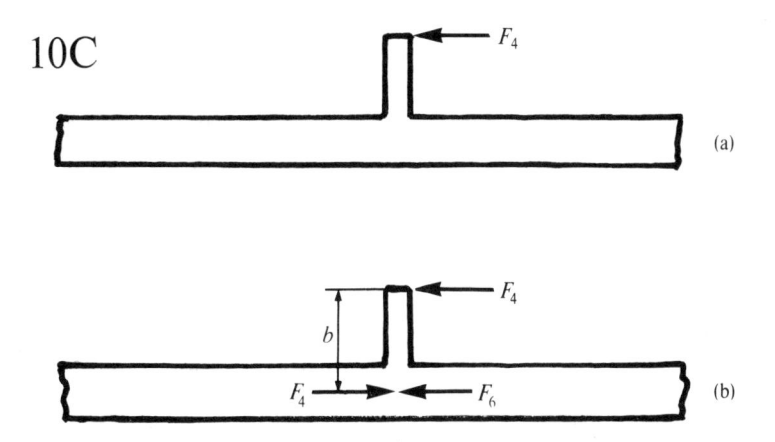

10C

(a)

(b)

10.5 Influence lines

When you have sketched some of the shearing force and bending moment diagrams in the following problems, you will be able to agree statement (*a*):

(*a*) A SFD or BMD, when only one load acts on the beam, shows the SF or BM at *all points* on the beam for *one position* of the load.

If, however, the load is moving over the beam, as loads often do, the SFD or BMD represents an instantaneous condition which changes immediately the load rolls forward. It would be an impossible task to draw an infinite number of SFD's and BMD's for every possible position of a rolling load, and it would be even more forbidding if there were more than one of these loads acting at the same time.

However, it is quite reasonable to suppose that the variation encountered at one selected salient point on the beam could be traced. If the variation were known for a few important sections on the beam, a

10.5

knowledge of the forces and moments applied to the beam by a rolling load could be ascertained. To show this variation in a diagram, you must invert statement (*a*) to statement (*b*).

(*b*) When only one load acts on the beam (unity for simplicity) the variation of SF, BM, or other quantities at *one point* on the beam, for *all positions* of the rolling load, is called an *influence line diagram* (ILD).

On each ILD there are two significant sections. The first is the section for which the diagram is drawn. All the values refer to this point. An ILD cannot, therefore merely be labelled ILD, as you have done with SFD and BMD. It must be labelled *ILD for* .... *at* ... ILD'S can be drawn for many quantities – deflection, moment, force, stress, and is always drawn for one point on the beam. Both the quantity and the section must be clearly defined.

The second point of importance and significance is the section at which the unit load is acting. The ordinates of the diagram all refer to some quantity (BM, SF, deflection, etc.) occurring at one point on the beam when the unit load is at a section immediately above the ordinate of the diagram. Thus in the sketch:

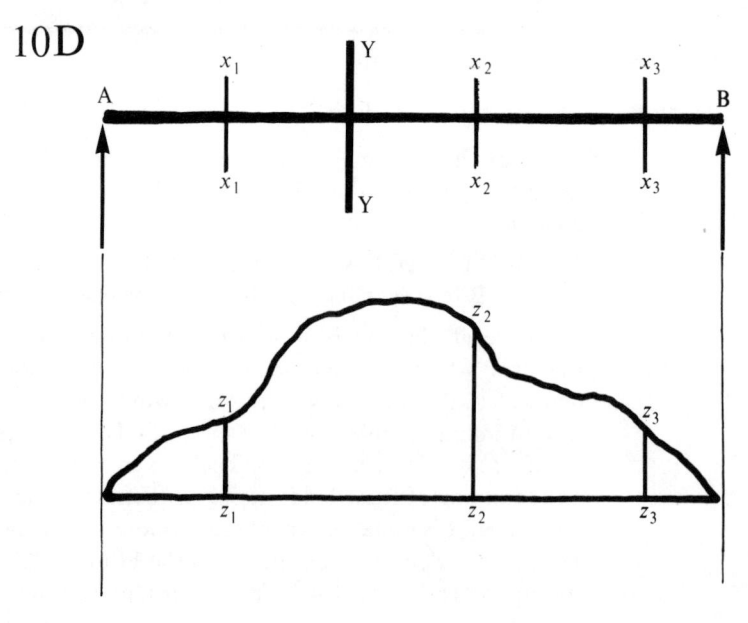

10D

10.5

Ordinate	Represents quantity concerned at	When unit load is at
$z_1 - z_1$	Y – Y	$x_1 - x_1$
$z_2 - z_2$	Y – Y	$x_2 - x_2$
$z_3 - z_3$	Y – Y	$x_3 - x_3$

The sketching of ILD's for various quantities, such as SF, BM, or Thrust in a beam, starts with the sketching of the ILD for the supporting reactions. This diagram shows the variation in one reaction or another as the unit load rolls over the whole length of the span. The height of the ordinate *under the position of the load* gives the value of the reaction for the load in that position. Since the load is unity, and there are no other loads on the beam, the sketching of the ILD's for reactions is not a subject for algebraic calculations, but rather for the application of common experience. If, for example, a unit load is placed just above one of the reactions, that reaction must be unity also. If the load rolls to the other reaction and is entirely supported there, the first reaction becomes zero. Join the two points and you have the ILD. Try the problems 10/27 *et seq*.

Only the forces you can see beyond the edge of the shielding card are used in calculations of conditions at the edge of the card. What is under the card is irrelevant.

In tabular calculations complete each column in turn, not each line.

10/1

10.6 THE PROBLEMS

10/1 SFD: U.D. Load: Cantilever

The cantilever shown carries a uniformly distributed load. Sketch the shearing force diagram.

SURE

Sketch beam approximately to scale. No reactions required for a cantilever.

Evaluation:

1. Cover the sketch of the cantilever with a card.
2. Slide the card towards the left.
3. Just as the card begins to show the end of the beam, there is as yet only a very small load visible, so the shear is zero at the end.
4. As you slide the card further, the shear increases at the rate of 1 kN for every metre of the beam uncovered. The sign is positive, for the forces are up-on-the-right of the edge of the card.
5. After the end of the UDL is reached there is no change in the shearing force, for no other force appears from under the card.

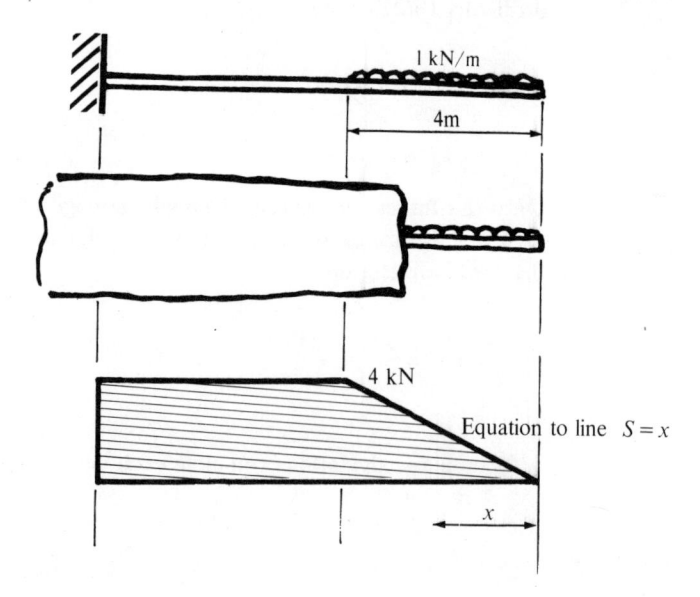

1 kN/m

4m

4 kN

Equation to line $S = x$

x

10/2

10/2 SFD: Concentrated and U.D. Loads: Cantilever with self-weight

The beam carrying the loads shown, itself weighs 10 kN. This weight is uniformly distributed over the whole length of the cantilever. Draw the SFD.

SURE

First steps scarcely required as this is a cantilever. Cover it with a card and withdraw so that the free end is exposed first.

Evaluation:
1. First draw the SFD for the applied loads. From A to B no force is uncovered by the moving card so there is no shearing force. From B to C the only force in sight is 6 kN, so the shear is 6 kN down on

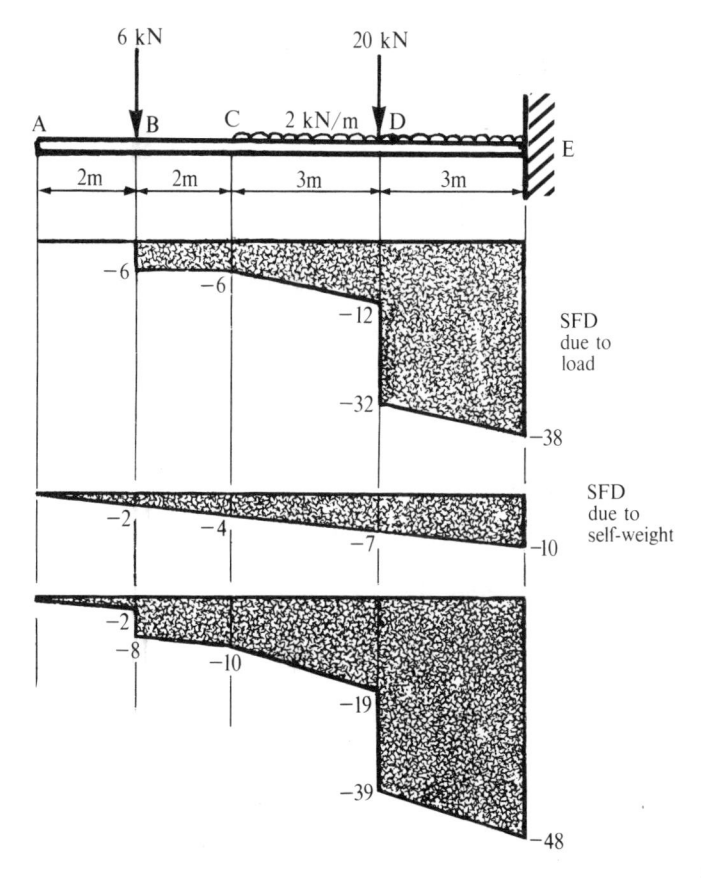

195

10/3

the left of the card and therefore negative. From C to D there is a gradual increase in shearing force at the rate of 2 kN for each metre, and at D there is a step of 20 kN before the sloping line resumes.

2. The weight of the beam is an additional UDL at the rate of 1 kN for each metre of length. A simple triangle gives the SFD for the self-weight of the cantilever.

3. Add the ordinates at critical points to obtain the combined and completed SFD.

10/3   SFD: Concentrated loads: Simple supports

The beam shown carries concentrated loads. Draw the shearing-force diagram.

S   *Structure:*
Sketch the beam roughly to scale, and show the loads.

U   *Unknowns and Units:*
The shearing-forces at the salient points – the concentrated loads are to be determined and joined by lines to equation (i), above. Note there are two values of shearing force for each load – one just to the left, and one to the right. Use kN. No distances required.

R   *Reactions:*
Unless the supporting reactions have been determined, you cannot draw the SFD. In this instance they are given.

E   *Evaluation:*
1. Cover the beam with a card.

2. Draw the card towards the left until 11 kN is uncovered. This is up-on-the-right of the edge of the card which is a conventional negative shear.

3. Draw the card further along. The shear does not change, for no further forces appear.

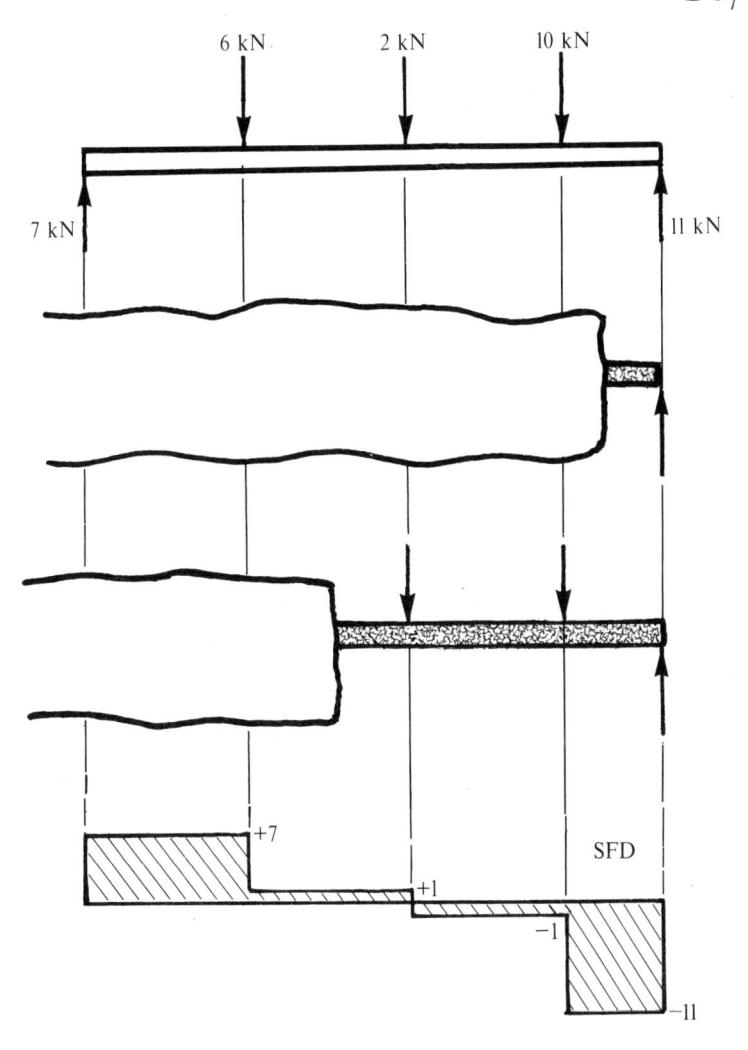

4. Draw the card along until 10 kN appears. Just before it appears, the shear is −11 kN. The downward acting 10 makes a step in the diagram of 10 kN, and the final shear just to the left of the 10 force is −1 kN.

5. When 2 kN appears at the edge of the card there is a further step up and the shear is then +1 kN.

197

10/4

6. It is quite convenient to continue in this way, but it is sometimes shorter to start at the other end once you have reached the middle. Uncover 7 and then 6 kN.

7. The values as obtained from either end should be identical in magnitude and in sign. If so you are SURE that your sketch of the SFD is correct.

10/4 SFD: U.D. Load and concentrated: Simple supports

The beam shown carries concentrated and uniformly distributed loads. Draw the shearing-force diagram.

SUR Sketch the beam approximately to scale. The reactions are known.

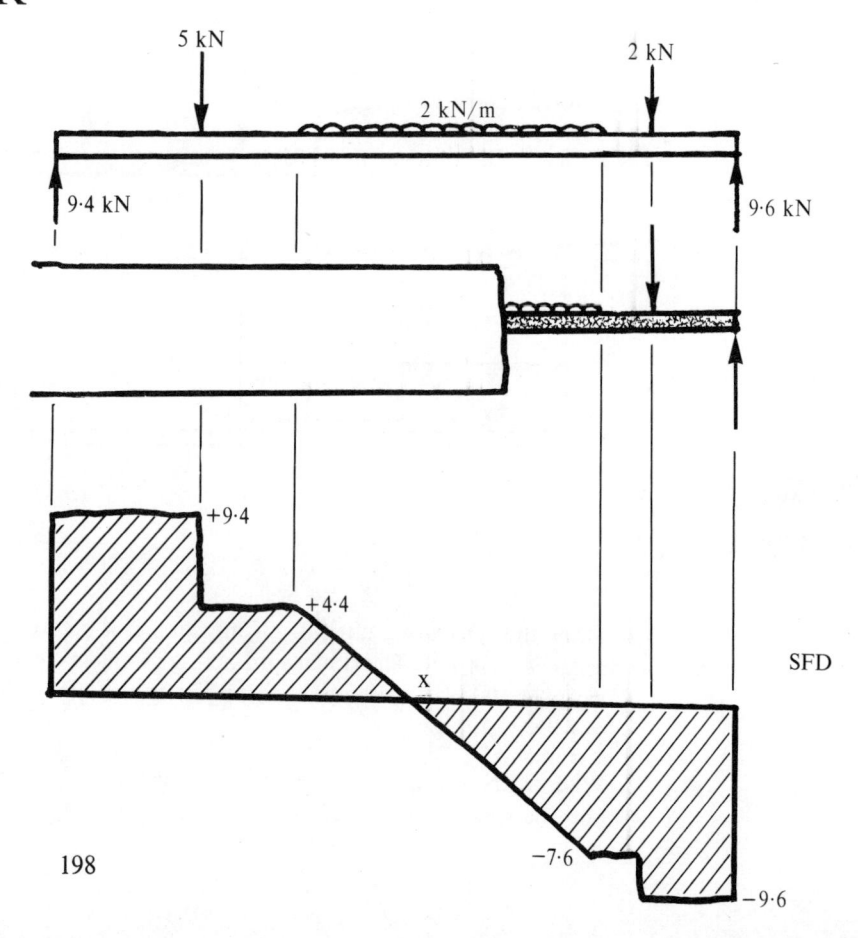

E *Evaluation:*

1. Cover beam with a card.

2. Draw the card towards the right uncovering 9·4 kN to give an up-on-the-left shear, which is positive.

3. Draw the card along to uncover the 5 kN. There is a 'step' and the values of shear are 9·4 kN on the left, and 4·4 kN on the right of the load.

4. There is no change in SF until the uniformly distributed load is uncovered, when the shearing force *gradually* reduces at a rate of 2 kN/m of length. This is the one part of the beam where the distance must be known. The final result is −7·6 kN at the end of the UDL (equation (ii)).

5. Again there is no change until the 2 kN is uncovered when there is another step, and the final figure of 9·6 kN agrees with the force represented by the reaction.

6. Note carefully, the point where the shearing force passes through zero and changes, *gradually* and without a step, from positive to negative. You must always be able to determine this point X when there is a distributed load. Now draw the card from right to left and re-draw the SFD.

10/5 SFD: Concentrated loads: Beam with cantilever

The beam carries concentrated loads, the values of the reactions having already been determined. Sketch the SFD.

S *Structure:*
Sketch the beam and cover it completely with a card. Gradually withdraw the card towards the right.

UR *Unknown Reactions:*
In this instance these have already been determined, but in any normal problem would first have to be found by the methods shown earlier in the book.

10/5

E *Evaluation:*

1. As the card moves to the right, the 3 kN reaction is first disclosed. This is acting on the left of the edge of the card and is thus, conventionally, positive shear.

2. No change takes place as the card moves along and the shearing force remains constant until changed by the appearance of the 5 kN force. This is down on the left of the edge of the card and thus changes the shearing force from $+3$ to -2 kN.

3. Continue in the same way to the end, and then cover the beam again and start at the right hand end, withdrawing the card towards the left. You ought to get identical results in value and sign.

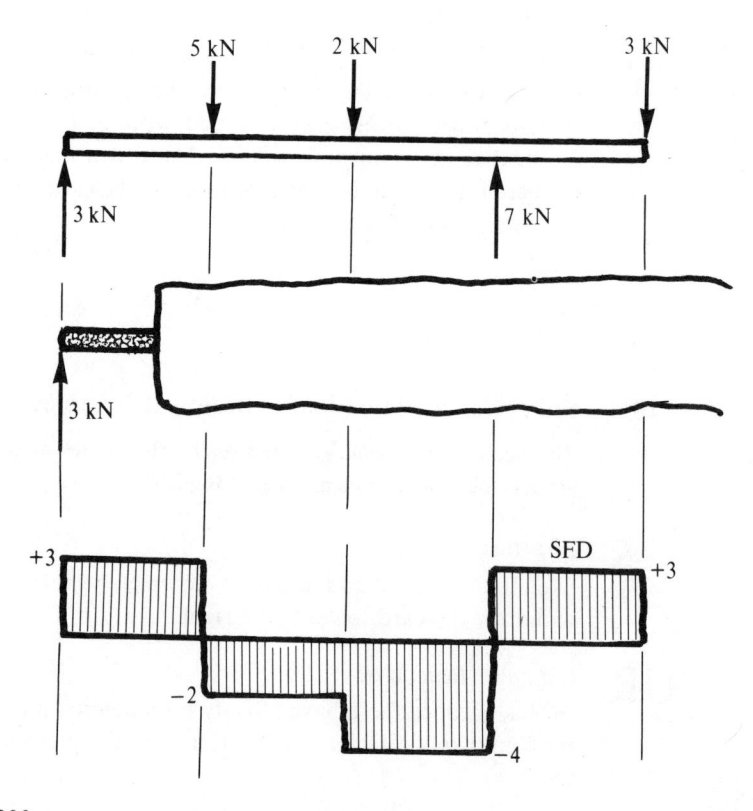

10/6 SFD: Concentrated and U.D. Loads: Beams and two cantilevers

This is an example where a concentrated load acts at the end of the uniformly distributed load. The reactions must also be determined. Draw the SFD.

S
UR

Sketch the beam approximately to scale and cover with a card.

Reactions:

Using the methods of previous units ($\sum M = 0$) you must find the values of the reactions before you can attempt to draw the SFD. The values turn out to be both upwards and of values 16 kN (left) and 8 kN (right). All lengths must be known if the reactions have to be determined.

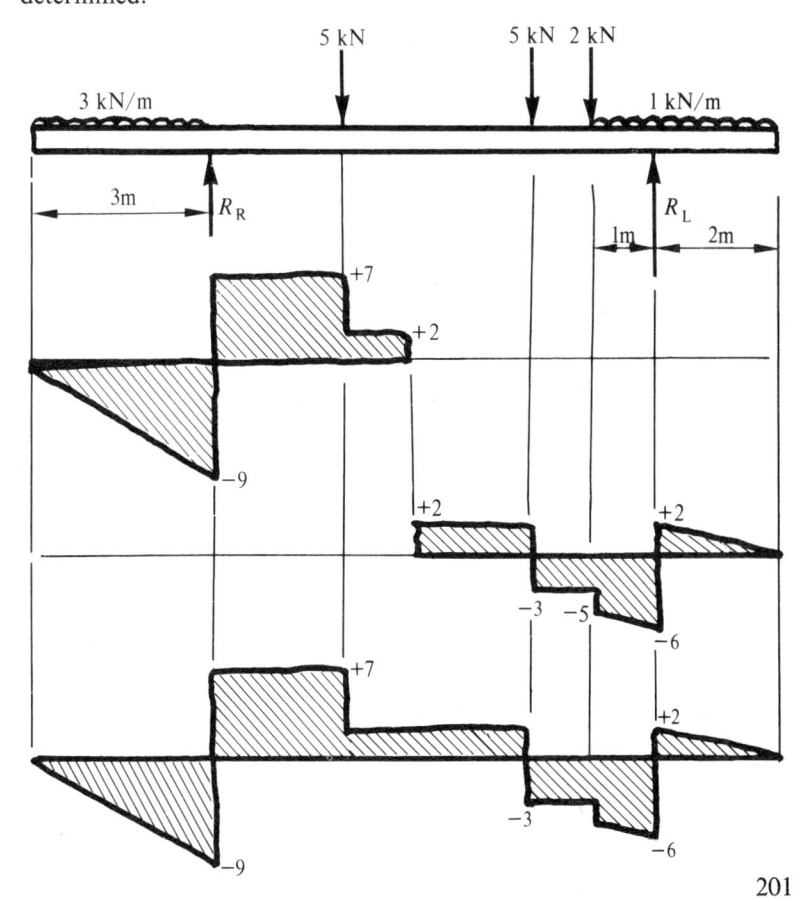

201

10/7

E *Evaluation:*

1. Use the methods already explained, and draw the card first from the left and then from the right.

2. The 16 kN and the 2 kN act at the end of a UDL. This means that there is a step *immediately* after the sloping line (representing the effect of the UDL on shear) has come to an end (16 and 8 are against and 2 with the effect of the UDL).

3. The 8 kN load acts within the UDL but in the opposite direction. It therefore, interrupts the smooth line, and makes it start again, parallel to its first direction but starting at −6 kN instead of at +2 kN.

4. The value +2 which is obtained in the centre stretch from *both* sides shows that the work has checked.

10/7 SFD AND TD: Inclined loads: Simple supports

A simply supported beam carries two inclined loads and one vertical load. Draw the SFD and TD. The beam is on rollers at the right-hand end and is hinged at the left-hand end.

S *Structures:*

Make two sketches – one as given and the other showing the components of the loads, along and normal to the beam. Those along the beam produce thrust (compression negative) and those normal to the beam produce shear.

UR *Unknown Reactions:*

Since the right-hand end of the beam is on rollers, the reaction there must be vertical. At the left-hand end there are both a vertical and a horizontal reaction to replace the hinge.

With L as the point of rotation:

Forces: 4 $5/\sqrt{2}$ 1 $5/\sqrt{2}$ $\sqrt{3}$ F_1 F_2 F_3
Moment arms
about L: 20   15   3   0   0   0   0   25
Moments with
signs: $+4 \times 20 + 5/\sqrt{2} \times 15 + 1 \times 3 - 25F_3 = 0$
$$+F_3 = +5\cdot 4 \text{ kN}$$

10/7

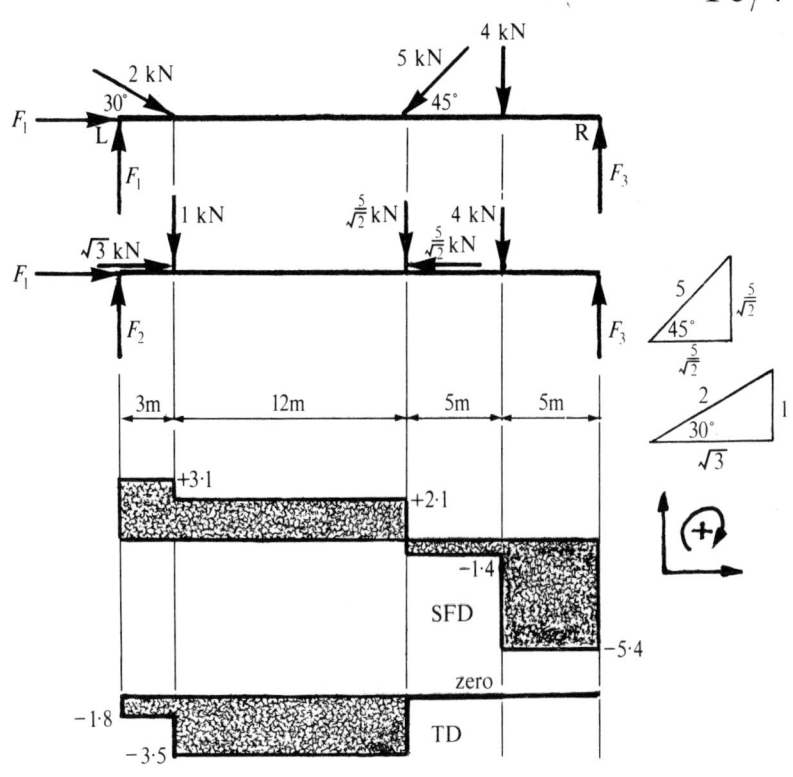

E *Evaluation:*

SFD: Uncover the beam from one end and draw the SFD from the visible forces which act normal to the line of the beam.

TD: Uncover the beam and observe the forces along the beam as they appear from under the card. Positive forces are those which act away from the edge of the card whether it moves from right to left or from left to right.

A hinge transmits shear but not moment.

10/8

10/8 SFD: Non-uniform load: Cantilever

The cantilever shown carries a load increasing uniformly from the free end. Draw the shearing force diagram.

SURE

Sketch the beam approximately to scale. No reaction required as this is a cantilever and can be studied from the free end.

Evaluation:

1. Cover the beam with a card and withdraw from the free end.

2. At a distance x from the free end, the edge of the card has uncovered a triangular portion of the load. The value of this load is the shearing force. You know that the horizontal length of beam exposed is x metres, and at 10 m the load is at the rate of 4 kN/m. At x metres, by proportion, the intensity of load must be $0·4x$ kN/m since the rate is increasing from zero at 0·4 kN/m for every metre from the free end.

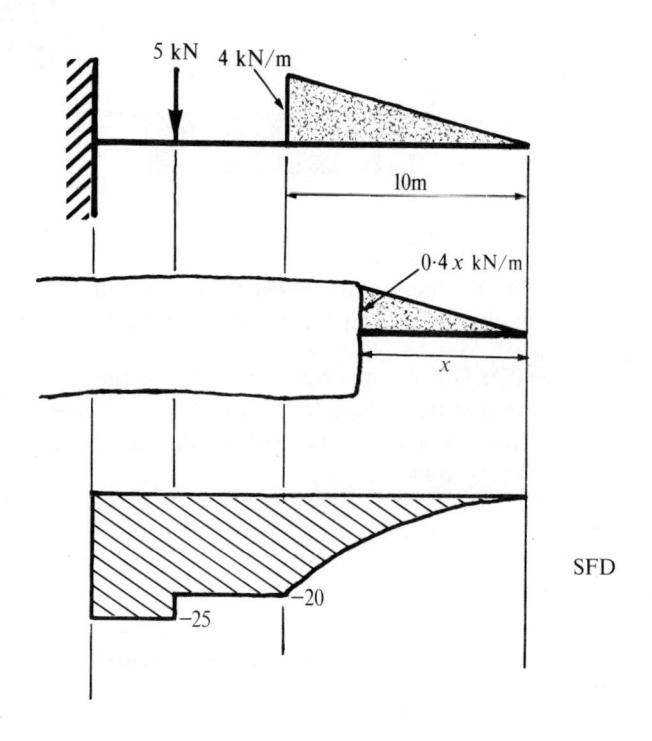

3. The area of the triangle now becomes

$$S = \tfrac{1}{2} \times 0\cdot 4x \times x = 0\cdot 2x^2 \text{ kN}$$

This is a quadratic (parabolic) line starting at zero at the free end, and becoming 20 kN at the end of the load. Note the shape of the parabola; It is tangential to the beam at the free end. You can check this by choosing x at various values between 0 and 10 m.

4. As the card uncovers more of the beam, no more loading is disclosed until the 5 kN appears and causes a step in the diagram.

10/9 SFD and TD: Inclined loads and couple: Cantilever

The cantilever shown carries an inclined load and an applied couple. Draw the shearing force diagram, and the thrust diagram.

S *Structure:*
Draw a sketch of the beam, and then draw a second sketch with the applied couple omitted. Couples do not appear in shearing force diagrams. In the *equations of equilibrium* earlier in the book, you will find that applied couples appear only in the moment equation.

U *Unknowns:*
The shearing force represents the sum of all the forces acting at right angles to the beam on one side of the section. One of the unknowns, therefore is the normal component of the inclined load. The other component, in the direction of the length of the beam, produced a thrust.

As the force 200 kN is at 45° to the horizontal, the horizontal and vertical components are both $200/\sqrt{2}$ or 141 kN.

R No reaction required; start from free end.

E *Evaluation:*
1. Cover the cantilever with a card and slide the card towards the left.
2. The forces which appear cause 'steps' and are in turn 141 kN and 50 kN.

10/9

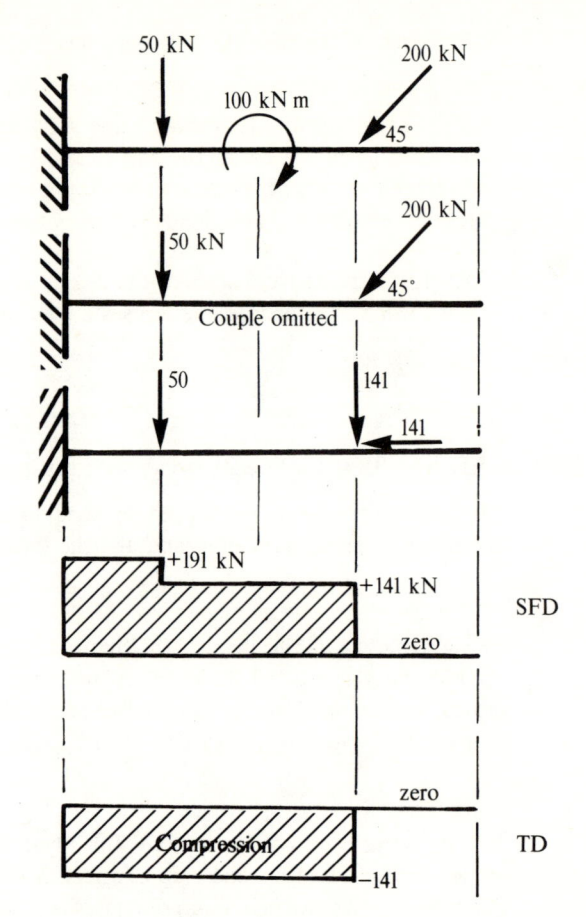

3. Cover the beam again and slide card to left.
4. No thrust force (parallel to beam) appears until the first load is reached, and after that thrust of 141 kN (towards the fixed end, and so compressive) has been passed no further thrust is added or subtracted.

Be sure you know thoroughly 10.5 (a) and (b).

10/10 SFD: U.D. Load: Beam with cantilever

A horizontal beam, 12 m long, is loaded with a vertical distributed force over its whole length, of 2 kN/m. It is held by a hinge at the left-hand end (L) and by a vertical simple support at some distance from the right-hand end. This distance (z) can be altered. The problem is to determine the value of z (where the support should be placed) to obtain the smallest possible maximum shearing force on the beam.

S *Structure:*

Sketch the beam with z of any value, and cover the sketch with a card.

UR *Unknown Reactions:*

Since there are only vertical forces on the beam, the reaction at A must be vertical. The horizontal component force at the hinge A is zero.
To find the value of one of the reactions take A as the point of rotation:

$$\sum M_A = 0$$

Forces	R_A	R_S	2×12
Moment arms about A:	0	$(12 - z)$	6
Moments with signs:	$-R_s(12 - z) + 24 \times 6 = 0$		$+R_S = +\dfrac{144}{(12 - z)}$

E *Evaluation:*

1. Draw a card over the beam and sketch the SFD.

2. There are three large values of shearing force – OR, OQ, and UT.

3. As the support moves towards the left, and z increases, OR, OQ, and UT change. If you plot their values it will be evident when the least values of shear develop.

4. The values of the three ordinates are:

$$OR = +2z \text{ kN}$$

which represents a straight line through the origin of co-ordinates.

$$OQ = -2z + R_s = -2z + 144/(12 - z) \text{ kN}$$
$$UT = +24 - R_s = +24 - 144/(12 - z) \text{ kN}$$

10/10

5. These graphs are best plotted from a tabular calculation:

z	$2z$	$12-z$	$144/(12-z)$	OQ	UT
0	0	12	12·0	12·0	+12·0
2	4	10	14·4	10·4	+9·6
4	8	8	18·0	10·0	+6·0
6	12	6	24·0	12·0	0·0
8	16	4	36·0	20·0	−12·0
10	20	2	72·0	52·0	−48·0
12	24	0	—	—	—

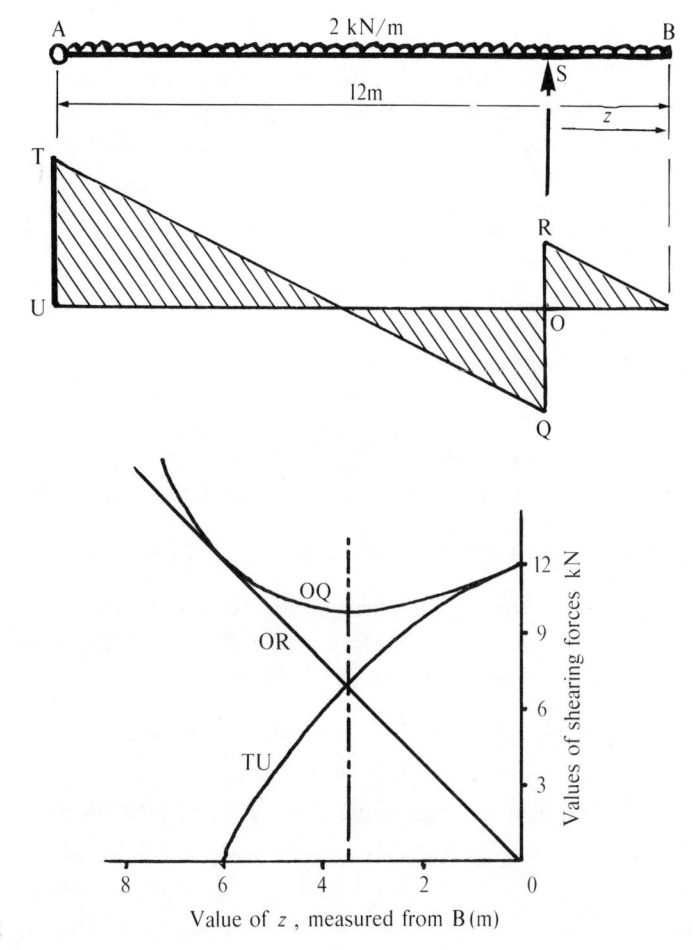

208

6. From the plot of these three, the value of z to give the least maximum shearing force is 3·5 m. If z varies above or below this point, the least of the three values will be greater than 7 kN, which is thus the shearing force for which the beam can be designed with z equal to 3·5 m.

The figures need be plotted for only the first 6 m. After that UT becomes negative and the other two are large.

10/11 SFD AND TD: Inclined load and couple: Inclined beam

A beam is inclined at a gradient of 1: $\sqrt{3}$ or 30°. It is held by a hinge at the lower end and at the top by a chain at 75° to the horizontal. Draw SFD and TD.

S *Structure:*
Sketch the beam and remember that a hinge can be portrayed in a FSD by two component forces. Place these along and at right angles to the beam.

U *Unknowns:*
The forces which must be known before such an inclined beam can be solved are those acting along the beam and normal to it. Break up the force in the chain and the 500 N into components along and normal to the beam. The numerical values of the force in the chain must still be found, but the proportions of the two components can be determined.

R *Reactions:*
By known methods, the reactions are found to be, approximately

$$+F_1 = +393 \text{ N} \qquad +F_2 = +28 \text{ N} \qquad +F_3 = 155 \text{ N}$$

and F_1 can now be resolved into 278 N along the beam and 278 N normal to the beam. Be careful to observe the correct senses.

E *Evaluation:*
The final sketch of the beam shows all loads either along or normal to the beam. By uncovering from one end and observing the conventions, those forces normal to the beam give the SFD and those along the beam give the TD.

10/11

210

Convention is:
Positive shear: up on the left of the card.
Positive 'thrust' – tension – away from edge of card.

It is convenient, for any inclined beam, to draw the diagrams on a horizontal base. Vertical thicknesses of the diagrams under a particular section indicate the shear or thrust. Remember that shear is always at right angles to the beam, and thrust along the beam, whatever its angle to the horizontal.

10/12 BMD: Concentrated loads: Simple supports

A beam, 20 m long, carries four loads as shown. Sketch the BMD.

S *Structure:*
Sketch the beam and its loading. Since the loads are all concentrated, the bending moment diagram consists of a series of straight lines joining up critical values of BM at the load points. See the introduction to the Unit if you have not yet read it.

UR *Unknown Reactions:*
All the forces acting, and this includes the supporting reactions, must be found before starting on any study of bending moment. By methods already practised, the reactions are found to be in the directions assumed in the sketch and are:

$$+R_L = +9.4 \text{ kN} \qquad +R_R = +6.6 \text{ kN}$$

E *Evaluation:*
Cover the beam entirely with a card and withdraw it first towards the right, uncovering the left end of the beam, and then towards the left, uncovering the right end of the beam. It is usually convenient to work in this way instead of entirely from one end. The forces *you can see* (and no others), multiplied by their respective distances to the edge of the card, form the bending moment. The most efficient way of recording these calculations is by a bending moment table.

10/12

Range	BM equation	When $x =$	When $z =$	$M =$
	Clockwise moment positive			
A to B	$+9.4x$	2 m (B)		$+18.8$ kN m
B to C	$+9.4 - 3(x - 2)$	5 m (C)		$+38.0$ kN m
F to E	$-6.6z$		3 m (E)	-19.8 kN m
E to D	$-6.6z + 1(z - 3)$		8 m (D)	-47.8 kN m
Check test:				
C to D	$+9.4x - 3(x - 2)$ $-5(x - 5)$	12 m (D)		$+47.8$ kN m

Clockwise moment on the left of the card, and counterclockwise moment on the right of the card, both indicate positive bending. There are no points of contraflexure.

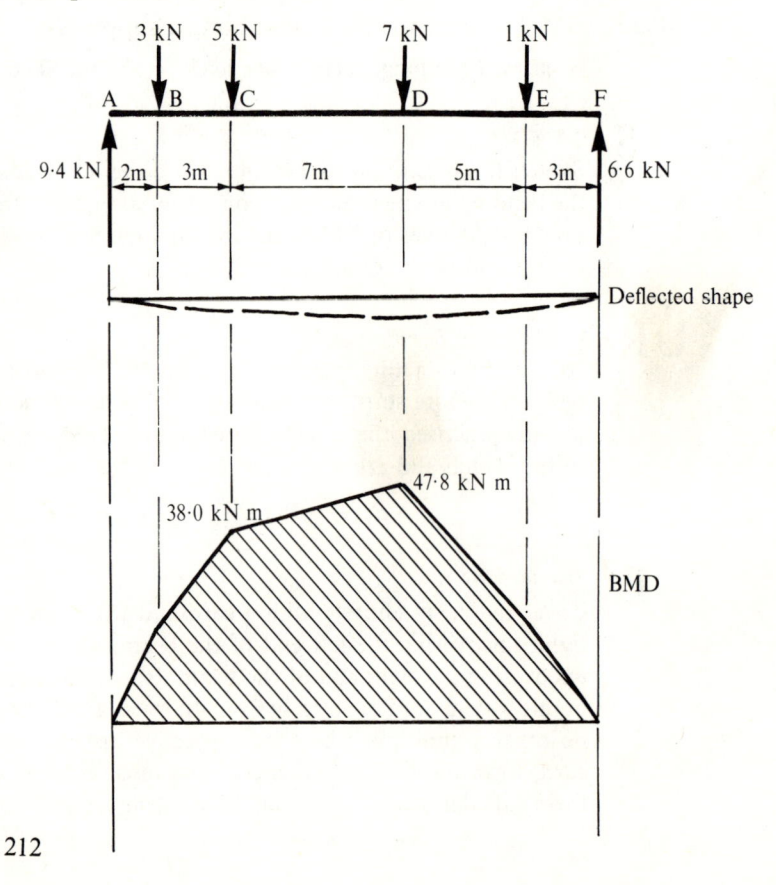

10/13 BMD: Concentrated loads: Simple supports with self-weight

This beam carries two concentrated loads of 40 kN and its self-weight is 0·2 kN/m. Sketch the BMD.

S *Structure:*

In this simple instance there is no real need to sketch the deflected shape, for the loading is symmetrical and the greatest bending must be in the centre.

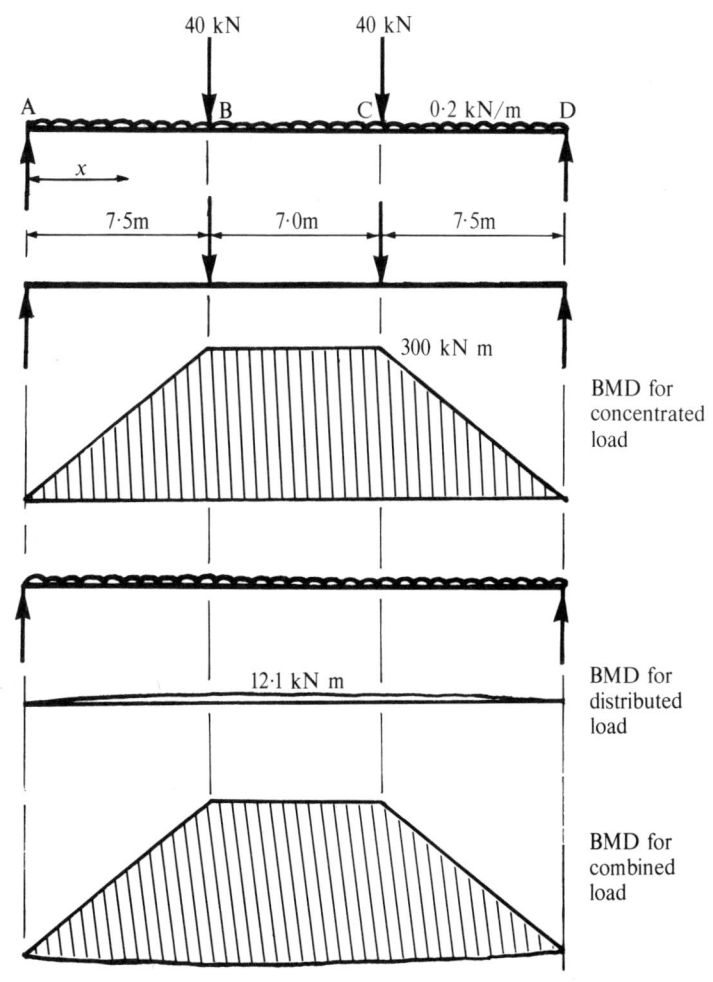

40 kN   40 kN

A   B   C   0·2 kN/m   D

x

7·5m   7·0m   7·5m

300 kN m

BMD for concentrated load

12·1 kN m

BMD for distributed load

BMD for combined load

10/13

UR *Reactions:*

The reactions from the concentrated loads must, from symmetry, be 40 kN at each end. The reactions from the distributed load must be each half of the total load of 0·2 × 22 or 2·2 kN each.

E *Evaluation:*

1. It is somewhat simpler, when seeking variation in bending moment along such a beam, to treat the two types of load separately. The results can then be added by placing one diagram against the other, so that the two depths are summed to give the total bending moment at each point.

2. The bending moment table then becomes:

Concentrated loads

Range	BM equation	When $x =$	$M =$
A to B	$M = 40x$ (straight line)	7·5 (B)	300 kN m
B to C	$M = +40x$ $-40(x - 7\cdot5)$	any value	300 kN m

Uniformly distributed load

A to D	$M = +2\cdot2x - 0\cdot2\dfrac{x^2}{2}$ (parabola)	7·5 (B) 11·0	10·9 kN m 12·1 kN m

3. The second diagram, although it gives positive bending, is, for convenience, inverted. The whole thickness or depth of the combined diagram gives the bending moment at any point.

Moments of couples affect the values of the reactions but do not appear in shear or thrust diagrams.

10/14

10/14 **BMD: U.D. Load and concentrated: Beam with cantilever**

Draw the bending moment diagram for the beam shown in the sketch, and indicate any points of contraflexure.

Structure:

S Sketch the structure with its assumed reactions. Sketch possible deflected shape.

UR *Reactions:*

Centre of Rotation: L $\sum M_L = 0$
Forces: R_R 2×50 R_L 110
Moments $-40R_R + 100 \times 15 - 110 \times 10 = 0$
with signs:

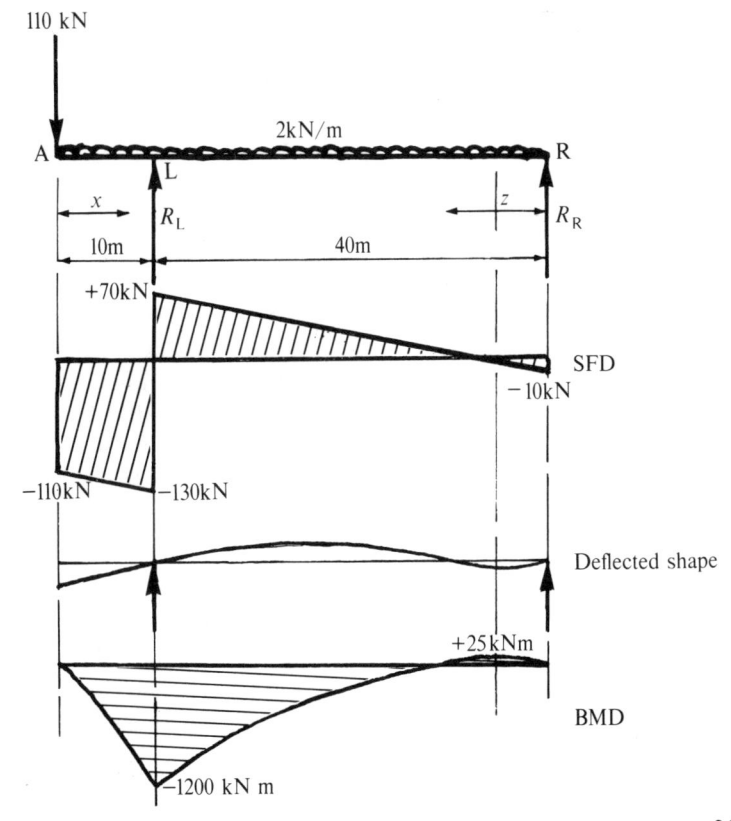

$+R_R = +10$ kN
$+R_L = +200$ kN

E *Evaluation:*

The DS diagram has shown a possible point of contraflexure where the bending moment passes through zero.

A quick sketch of the SFD shows that it also passes through zero near R. R_R which is 10 kN is cancelled out in shear in 5 m since the loading is 2 kN on each metre. Thus we have a summit in the BMD at 5 m from R, and a zero BM close by. These are useful hints in sketching.

The BM table is:

Range	BM equation (kN m)	When $x =$	When $z =$	$M =$
A to L	$M = -110x - (2x)\tfrac{1}{2}x$	5 m		-575 kN m
		10 m (L)		-1200 kN m
R to L	$M = -10z + (2z)\tfrac{1}{2}z$		2 m	-16
			5 m	-25
			10 m	zero
			20 m	$+200$
			40 m (L)	$+1200$

The line from A to L is slightly curved (concave-down) but note how the overpowering effect of the moment caused by the 110 kN overwhelms that of the UDL.

In the A to L range, the negative sign indicates counterclockwise moment to the left of the card (negative bending). The negative sign in R to L indicates counterclockwise moment to right of edge of card – positive bending. This changes to negative bending after 10 m.

This problem has been submitted to show that although the BM reaches a 'maximum' or summit when shear is zero, this is no guarantee that this figure is the maximum BM on the beam. The summit is 25 kN m, and the maximum BM is 1200 kN m.

When shear goes gradually through zero, the bending moment diagram reaches a summit, but not necessarily a maximum value.

10/15 BMD: Two U.D. Loads: Simple supports

A beam is simply supported at each end, and is 24 m long. It carries a uniformly distributed load of 6 kN over the left half of the beam, and 18 kN over the right half. Draw the bending moment diagram.

S *Structure:*

Sketch the beam and its loading. The deflected shape is a simple concave-up curve with a steeper slope at the right-hand end, and a deeper deflection under the heavier loading.

A sketch of the SFD shows that it passes through zero at 14 m from the left-hand end. This is the point at which the BMD rises to a summit.

UR *Unknown Reactions:*

These must first be found, and by the usual methods they are:

$$+R_L = +9 \text{ kN} \qquad +R_R = +15 \text{ kN}$$

E *Evaluation:*

The bending moment table:

Range	BM equation	When $x =$	When $z =$	$M =$
A to D	$+9x - (0.5x)\tfrac{1}{2}x$	4 m (B) 8 m (C) 12 m (D)		$+32$ kN m $+56$ kN m $+72$ kN m
G to D	$-15 + (1.5z)\tfrac{1}{2}z$		4 m (F) 8 m (E) 12 m (D) 10 m (H)	-48 kN m -72 kN m -72 kN m -75 kN m

Clockwise to the left of the edge of the card and counterclockwise to the right of the card both represent positive bending (concave-up). At 10 m from G there is a summit to the bending moment curve, which, in this example, represents a maximum value for the loading.

10/16

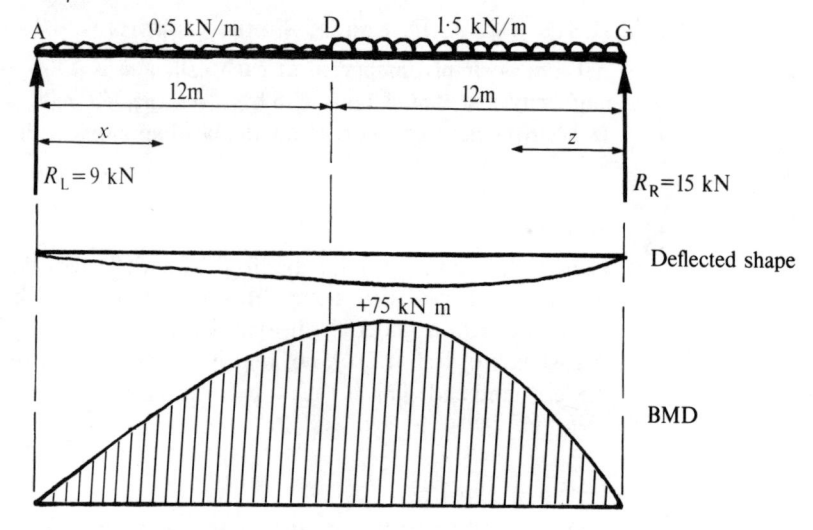

(Diagram: Beam AG, 24 m total. UDL 0.5 kN/m from A to D (12 m), UDL 1.5 kN/m from D to G (12 m). $R_L = 9$ kN, $R_R = 15$ kN. Distance x measured from A; distance z measured from G. Deflected shape shown. BMD shown with peak +75 kN m.)

10/16 BMD: Concentrated and U.D. Loads: Beam with one cantilever

A beam with one cantilever, and simply supported, carries both uniformly distributed loads and concentrated loads. Sketch the bending moment diagram.

S *Structure:*

Sketch the beam and its loading.

U *Unknowns:*

Two unknowns are (*a*) the deflected shape and (*b*) the shearing force diagram. The first gives some indication of how the bending moment outline will appear, and the second indicates if there is a 'maximum' point in the bending moment diagram. If the inclined line in the range of a distributed load passes through the baseline on a SFD, the bending moment rises to a high point and falls off again.

The shearing force diagram shows no such intersection of the base line by an inclined outline in the shearing force diagram, but the introduction of the 10 kN force breaks the line which would have gone through the baseline, and thus indicates a 'maximum' point. This, however, is not the highest bending moment on the beam.

218

The deflected form of the beam shows that there is certain to be one point of contraflexure, where the bending moment outline passes through the baseline. This is a great help to the development of the BMD.

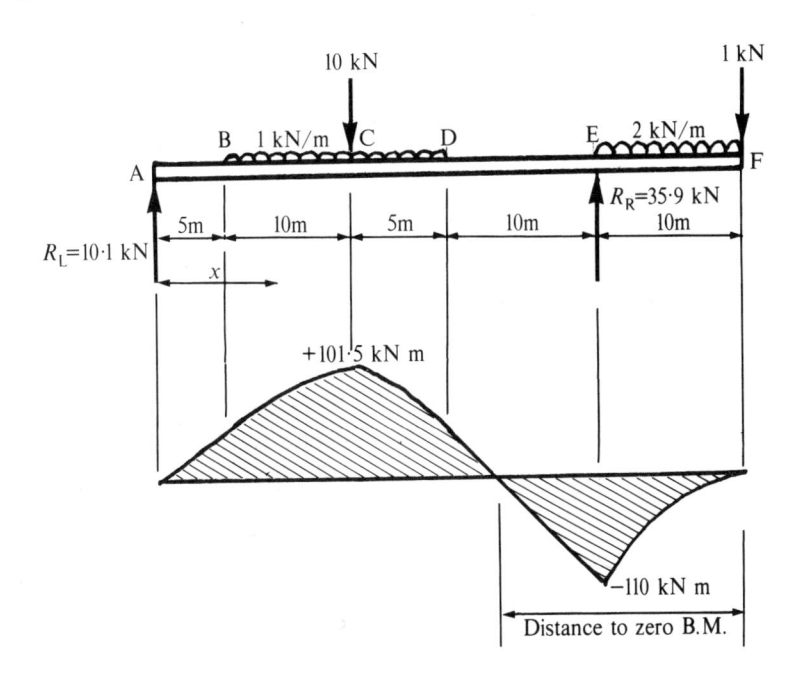

R   *Reactions:*

$R_L = 10.1$ kN and $R_R = 35.9$ kN in the senses shown.

E   *Evaluation:*

1. Draw the card over the sketch of the beam, disclosing first R_L. Here there can be no bending since the distance between R_L and the end of the beam is zero.

2. At B, the next salient point, the only force visible as the card reaches B is R_L. The bending moment in numerical terms is

$$+R_L \times 5 = 50.5 \text{ kN m} \qquad +R_L = +\mathbf{10.1} \text{ kN}$$

which is clockwise on the left of the edge of the card, and so gives what is called positive bending.

219

3. At C, 15 m from A, the forces visible are R and the UDL covering 10 m from B to C. The bending moment therefore is

$$+R_\text{L} \times 15 - (1 \times 10)10/2 = 101\cdot5 \text{ kN m}$$

4. According to the evidence of the shearing force diagram, this is an upper point on the BMD outline. Because of the presence of the concentrated load of 10 kN, the curve does not go smoothly through C, but has a break in curvature. Show that this is so by writing the equation to the moment in BC as follows, and substituting several values of x.

$$M_\text{BC} = +R_\text{L} \times x - 1(x-5)(x-5)\tfrac{1}{2}$$
$$= -x^2/2 + 15\cdot1x - 12\cdot5 \text{ kN m}$$

10/17 BMD: U.D. Load and one concentrated: Beam with cantilever

On this beam there is a heavy load at the end of the cantilever, and a UDL over the whole span. Draw the BMD.

S *Structure:*

Sketch the beam.

Sketch the deflected shape to determine whether there are any points of contraflexure (BM = 0). The DS must dip at the end of the cantilever, but the length BC is short and there is heavy loading on the span between A and B. There is likely to be positive bending in the span, and thus a point of contraflexure.

Another tip as to the shape of the bending moment outline is given by finding where the shearing force outlines pass through zero at a slope. This occurs when the reaction A is cancelled by the load on the beam, acting at 4 kN/m. The value of $R_\text{A} = 19\cdot3$ kN, which, divided by 4 kN/m, gives the number of metres from A to the point of zero shear. Sketch the whole of the SFD and prove this point. At 4·8 m from A, then, the bending moment diagram passes through a summit.

UR *Unknown Reactions:*

are already given: but must first be determined in other problems before the BMD is constructed.

E *Evaluation:*

The bending moment table:

Range	BM equation	When x =	When z =	M =
A to B	$+19 \cdot 3x - (4x)\frac{1}{2}x$	3 m		+40
		6 m		+44
		9 m		+12
		12 m (B)		−56
C to B	$+24z + (4z)\frac{1}{2}z$		1 m	+26
			2 m (B)	+56
A to B	A summit point at	4·8 m		+47
	A zero point when $+19 \cdot 3 - 2x^2 = 0$	9·65 m		0

Clockwise moment to the right of the card is negative bending.
Clockwise moment to the left of the card is positive bending.

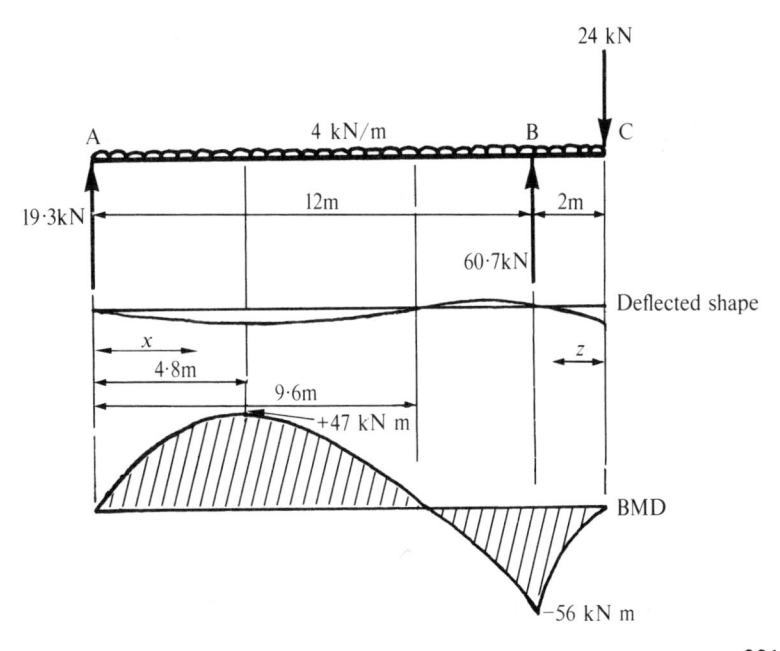

221

10/18

10/18 BMD: U.D. Loads: Foundation slab

The slab forming the foundation for a thick dock wall is 4·5 m wide, and the wall is 1·5 m thick. The total load being imposed on the foundation of the wall is 1200 kN/m of length. The base of the foundation is uniformly supported by the soil. Sketch the BMD.

S *Structure:*

Sketch the structure and then draw it again as a line diagram showing dimensions and loading. Think of a length of wall of 1 metre at right angles to the paper. The base is then 4·5 m² in area and the base of the wall is 1·5 m² for every metre of length.

UR *Unknown Reaction:*

The load is applied over an area of 1·5 m² for every metre of length of the wall, so the applied loading is at the rate of 1200/1·5 = 800 kN/m². Similarly the ground is pushing up with an intensity of 1200/4·5 kN/m². Since the wall is 1 m long perpendicular to the paper, these loadings can be taken as the same figures per metre of breadth.

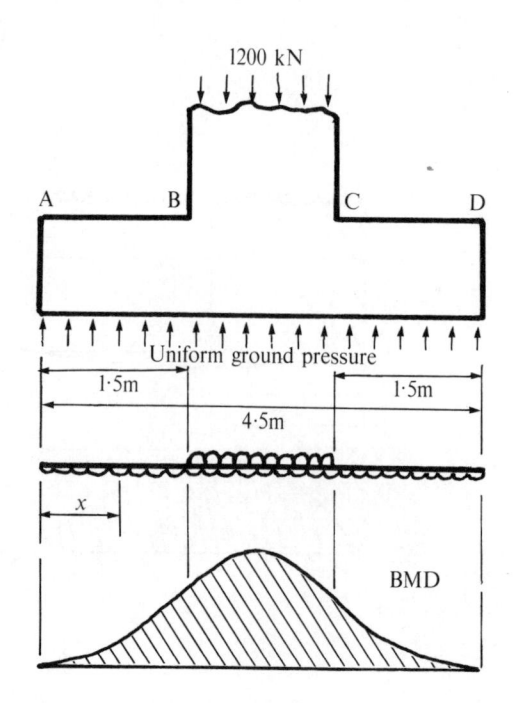

222

10/19

E *Evaluation:*

The bending moment table is:

Range	BM equation	When x =	M =
A to B	$+(267x)\tfrac{1}{2}x$	0·5 m	33 kN m
		1 m	134 kN m
		1·5 m	300 kN m
B to C	$+(267x)\tfrac{1}{2}x - 800(x - 1\cdot5)$ $\times \tfrac{1}{2}(x - 1\cdot5)$	2·0 m	434 kN m
		2·25 m	451 kN m

10/19 **BMD: U.D. Load: Two overhanging ends**

Draw the bending moment diagram for this beam, indicating any points of contraflexure.

S *Structure:*

Sketch the beam and its loading approximately to scale. Sketch also two diagrams which can help in giving a mental picture of how the bending moment diagram will develop. These two diagrams are the deflected shape and the SFD. The deflected shape shows any points of contraflexure by the change in curvature, and the SFD shows summit points on the BMD where the SF passes through zero on a sloping line.

Start the DS at either end, where it is clear that the two loaded cantilevers will show negative bending – they will droop down. Follow these two curves along, but you will find it impossible, while keeping a smooth line, to show any positive bending (concave-up). Thus it is unlikely that points of contraflexure will appear at all. This will be shown more definitely in the calculations.

The SFD shows a zero point at 6·4 m from A, and this must be a 'maximum' or summit point in the BMD.

UR *Unknown Reactions:*

These are found by the methods of earlier units to be: as shown, in the directions assumed on the sketch.

$$+R_B = +3\cdot2 \text{ kN}$$
$$+R_E = +4\cdot8 \text{ kN}$$

223

10/19

E **Evaluation:**

1. The bending moment table:

Range	BM equation	When $x =$	When $z =$	$M =$
A to B	$M = -(0·5x)\frac{1}{2}x$	4 m (B)		-4 kN m
B to C	$M = -(0·5x)\frac{1}{2}x + 3·2(x-4)$	6·4 m	(summit)	$-2·6$ kN m
		8 m (C)		$-3·2$ kN m
F to E	$M = +(0·5z)\frac{1}{2}z$		6 m (E)	$+9·0$ kN m
			8 m (D)	$+6·4$ kN m

2. The important aspect of this problem is to show that the facile statement, often heard, that 'when shear is zero, bending moment is a maximum' is misleading.

3. When shearing force outline passes through zero where there is a distributed load, the bending moment outline reaches a summit. This diagram shows that, far from being a maximum BM, this can be a minimum point.

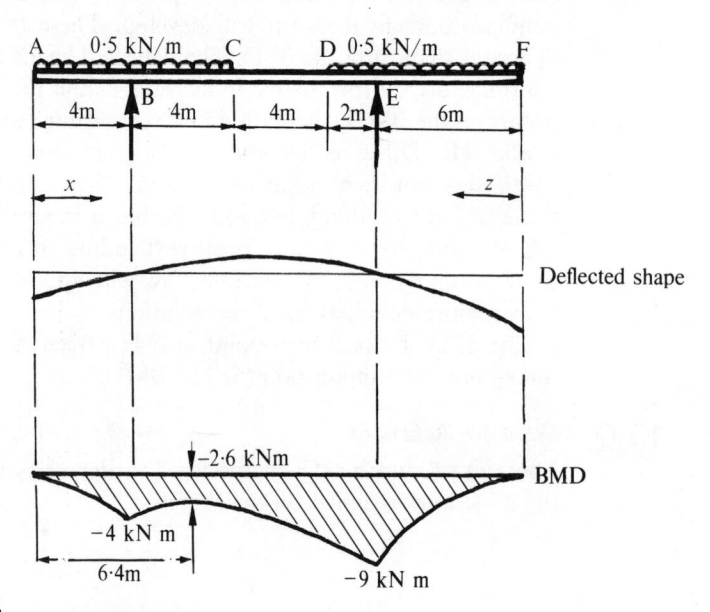

4. There are no points of contraflexure for the signs show counterclockwise moments to the left of the edge of the card in the first two groups, and clockwise moment on the right of the edge of the card when the moment is measured from the right hand end. These are both negative bending and the beam curves upwards over its whole length.

5. The line between C and D is straight, as you will find by writing the BM equation for that portion.

10/20 BMD: U.D. Loads: Canopy on column

A canopy is built on a column as shown in the sketch. Sketch the BMD for the whole structure.

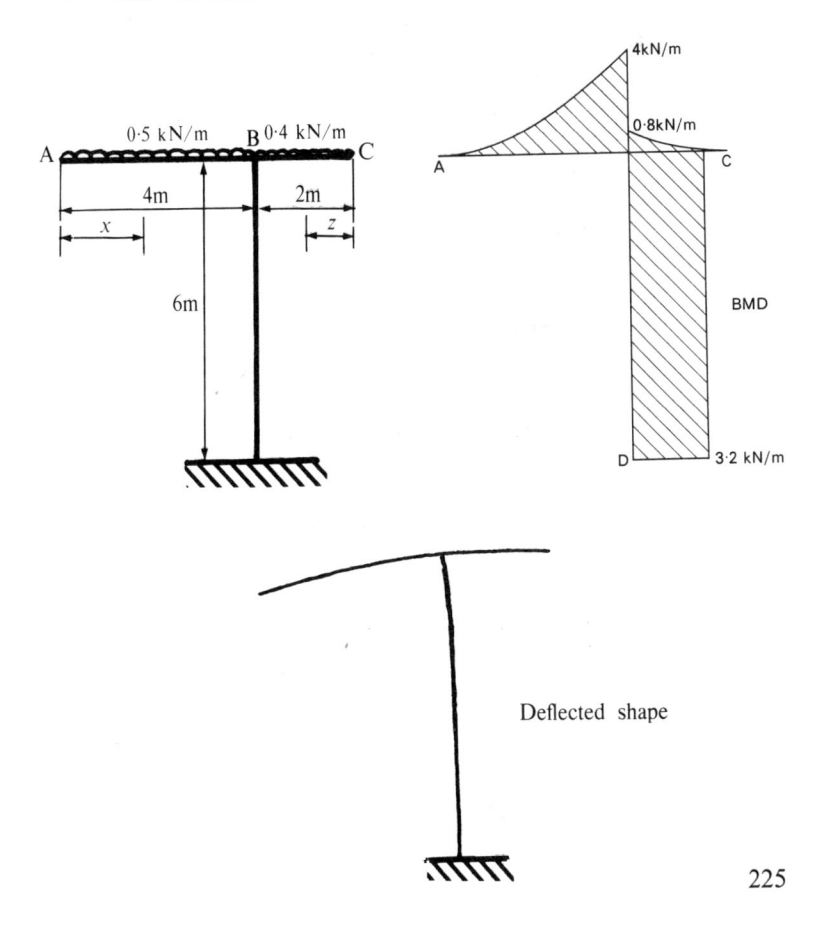

225

10/20

S *Structure:*
The whole structure is considered as a unit, the joint at B being rigidly built, and keeping its rectangular shape. The column is not held at the top and will deflect laterally towards the left. This deflection is small, and, like all structures in Part I of this book, the arrangement is considered to be rigid.

UR *Unknown Reactions:*
No further data required since the construction is cantilevered.

E *Evaluation:*
Place a card over the whole structure and withdraw it from C towards the left. Remember that the construction under the card – somewhat more complex than usual – is irrelevant. Only the forces *you can see* cause the bending moment sought.

The bending moment table:

Range	BM equation	When $x =$	When $z =$	$M =$
C to B	$(0 \cdot 4z)\frac{1}{2}z$		1 m	$+0 \cdot 2$ kN m
			2 m (B)	$+0 \cdot 8$ kN m
A to B	$(0 \cdot 5x)\frac{1}{2}x$	1 m		$-0 \cdot 25$ kN m
		2 m		$-1 \cdot 0$ kN m
		4 m (B)		$-4 \cdot 0$ kN m
B to D	There are two values of bending moment at B. For equilibrium, the difference must be carried by BD, and this value amounts to $3 \cdot 2$ kN m. There are no other forces and couples on BD to alter this, so the bending moment remains the same all the way to the base.			

Positive or clockwise moment to the right of the edge of the card and negative or counterclockwise moment to the left of the edge of the card both represent negative bending.

The convention for frames of this kind is that the BMD is drawn on the frame itself, and is sketched on the side of the member which is in tension – on the convex side. This is because there is difficulty in making distinction between positive and negative bending when the members of the frame are at different angles to the horizontal.

10/21 SFD, TD, BMD: Concentrated load: Inclined beam

An inclined beam is held by a hinge at the upper end and rests on rollers at the lower end. Sketch the SFD, TD, and BMD.

S *Structure:*

The rollers at the lower end cause the reaction there to be vertical. The hinge at the top can be represented by two component forces.

UR *Unknown Reactions:*

Since there is no other horizontal force on the beam, F_1 must be zero. Taking the upper end as the centre of rotation:

$+F_3 \times 13 - 10 \times 8 \cdot 5 = 0$ $\qquad +F_3 = +6 \cdot 55$ kN
$\qquad\qquad\qquad\qquad\qquad\qquad\qquad +F_2 = +3 \cdot 45$ kN

E *Evaluation:*

1. Using the values of sine and cosine for 35°, you must now break all the forces down into their components along and normal to the beam, and draw another sketch. Always make a number of sketches as calculations proceed; this avoids mistakes.

2. Slide your card across the beam, taking the forces normal to the beam as giving values of shear, and those along the beam as giving values of thrust (compression — ve).

3. The bending moment at the critical point under the 10 kN load can be obtained in four ways, from the two diagrams. Be sure you understand why you may use:

    $5 \cdot 37 \times 5 \cdot 49$, $2 \cdot 83 \times 10 \cdot 38$, $6 \cdot 55 \times 4 \cdot 50$, or $3 \cdot 45 \times 8 \cdot 50$.

    All these give the same result.

4. The diagrams are best drawn on horizontal baselines, the vertical thickness of the diagram giving the value of the quantity concerned at the equivalent point on the beam.

10/21

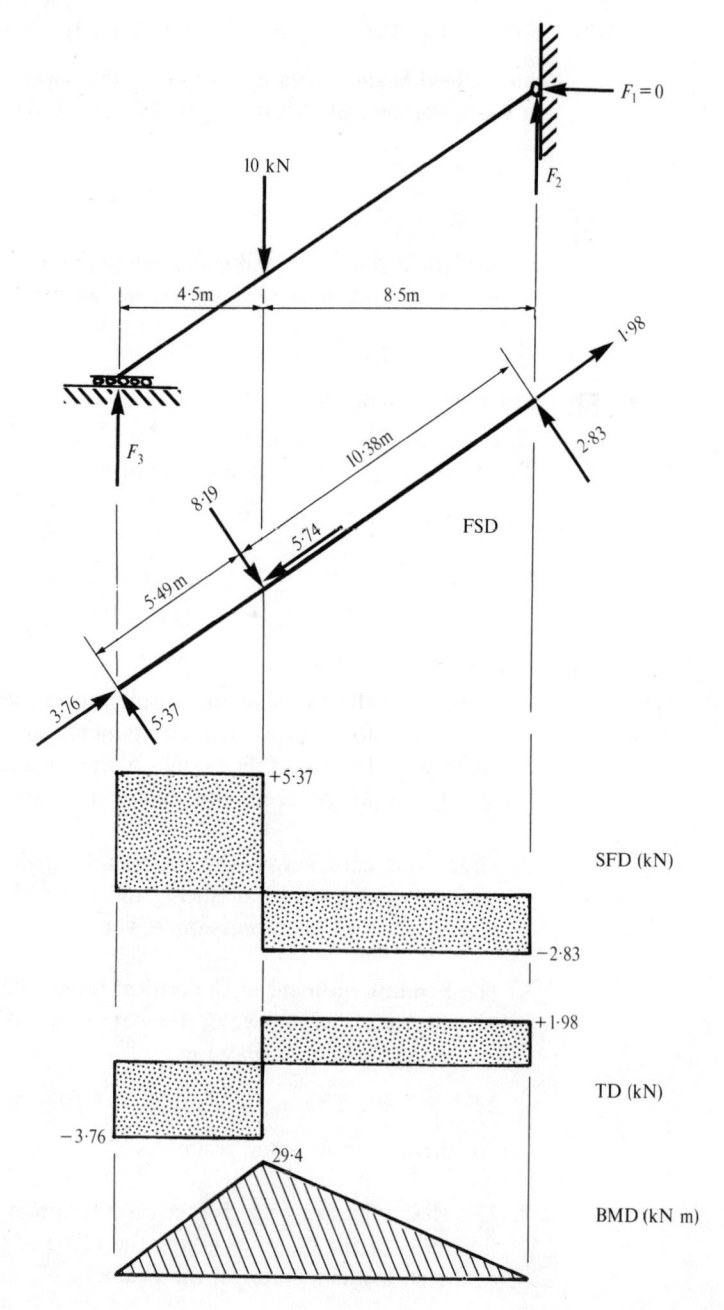

$F_1 = 0$

10 kN

4·5m   8·5m

F_2

F_3

1·98

2·83

10·38m

8·19

5·74

FSD

5·49 m

3·76   5·37

+5·37

SFD (kN)

−2·83

+1·98

TD (kN)

−3·76

29·4

BMD (kN m)

228

10/22 SFD, TD, BMD: Concentrated load: Inclined beam

An inclined beam is hinged at the bottom and rests, through rollers on a vertical wall. Sketch the SFD, TD, and BMD.

S *Structure:*
Draw the FSD and replace the hinge by two components – vertical and horizontal – and the rollers by a horizontal force.

UR *Unknown Reactions:*
Using L as the point of rotation:

$$\sum M_L = 0$$

$F_1 \times 8{\cdot}05 + 12 \times 11 = 0$ $\qquad +F_1 = +16{\cdot}4 \text{ kN}$

Since there are no other vertical forces, $\qquad +F_2 = +12{\cdot}0 \text{ kN}$
The only two horizontal forces are F_1 and F_3 $\quad +F_3 = +16{\cdot}4 \text{ kN}$
The senses shown in the sketch are correct.

E *Evaluation:*

1. Using the values of sine and cosine of 26° you now obtain the components of the forces in the directions normal to and along the length of the beam.

2. Also find, in the same way, the values of the lengths on the slope.

3. Uncover the beam from L; the forces normal to the length of the beam give the values of the shear, and those along the beam give the values of thrust.

4. Forces normal to the beam also provide the figures for bending moment. The diagram consists of two straight lines.

5. All these three diagrams should be drawn on a horizontal base, when the vertical ordinate of the diagram gives the value of the quantity concerned at the appropriate point on the beam. Remember that shear takes place at right angles to the beam, and thrust in the length of the beam.

230

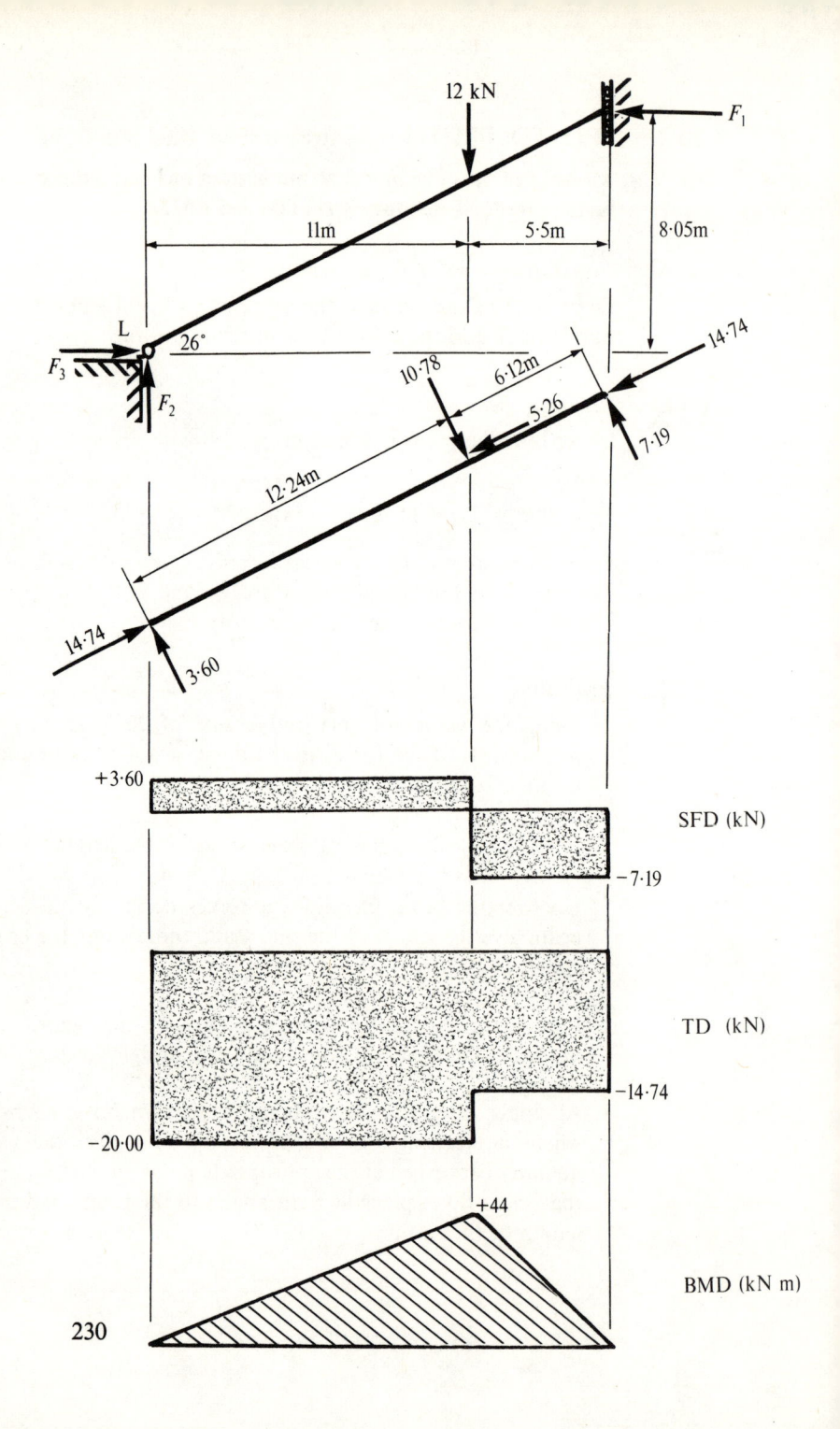

10/23 SFD, BMD: Applied couple: Simple supports

A freely supported beam of uniform cross section (hinged at each end) carries an applied couple at one end only. There is no other load on the beam which is 14 m long. The moment of the couple is 250 kN m. Draw the SFD and BMD for this condition, and show the differences which would occur if the couple were transferred to 4 m from one end and then to the centre of the beam.

S *Structure:*

Make a sketch of the beam with its applied couple.

UR *Reactions:*

An experiment with a rule will show you that you will have to hold the left end down, although this knowledge is not essential to the solution. First use the moment equation with the centre of rotation at R

$$\sum M_R = 0$$

$-R_L \times 14 + 250 = 0 \qquad\qquad +R_L = +17\cdot 86 \text{ kN}$

The double plus shows that the guessed sense of R_L was correct.
Second equation of equilibrium:

$$\sum F_{NS} = 0$$

$-R_L + R_R = 0$. There are no other forces on the beam, and couples do not come into the force equations. R_L and R_R are equal.

E *Evaluation:*

1. Slide a card over the beam and gradually expose the left end. The shear remains unaltered all the way along the beam, and is 17·86 kN down-on-the-left of the edge of the card which is, conventionally a negative shear.

2. The bending moment is also of a constant straight line all the way along the beam, the equation being

$$M = -R_L x \text{ (negative bending)}$$

3. If the couple is transferred to 4 m from the right hand end, or to the centre, the reactions remain equal and of the same value as before. The equation remains:

$$\sum M_R = 0$$

$-R_L \times 14 + 250 = 0 \quad$ wherever the couple is placed.

231

4. The shearing force diagram also remains unaltered for the couple does not appear in the shearing force equation.
5. The bending moment diagrams are different. Place the edge of the card at the point at which the couple acts. One value of BM at this point is $-R_1 x$, and the other is $+R_R(14-x)$. The difference is a step of 250 kN m, which is what one would expect.

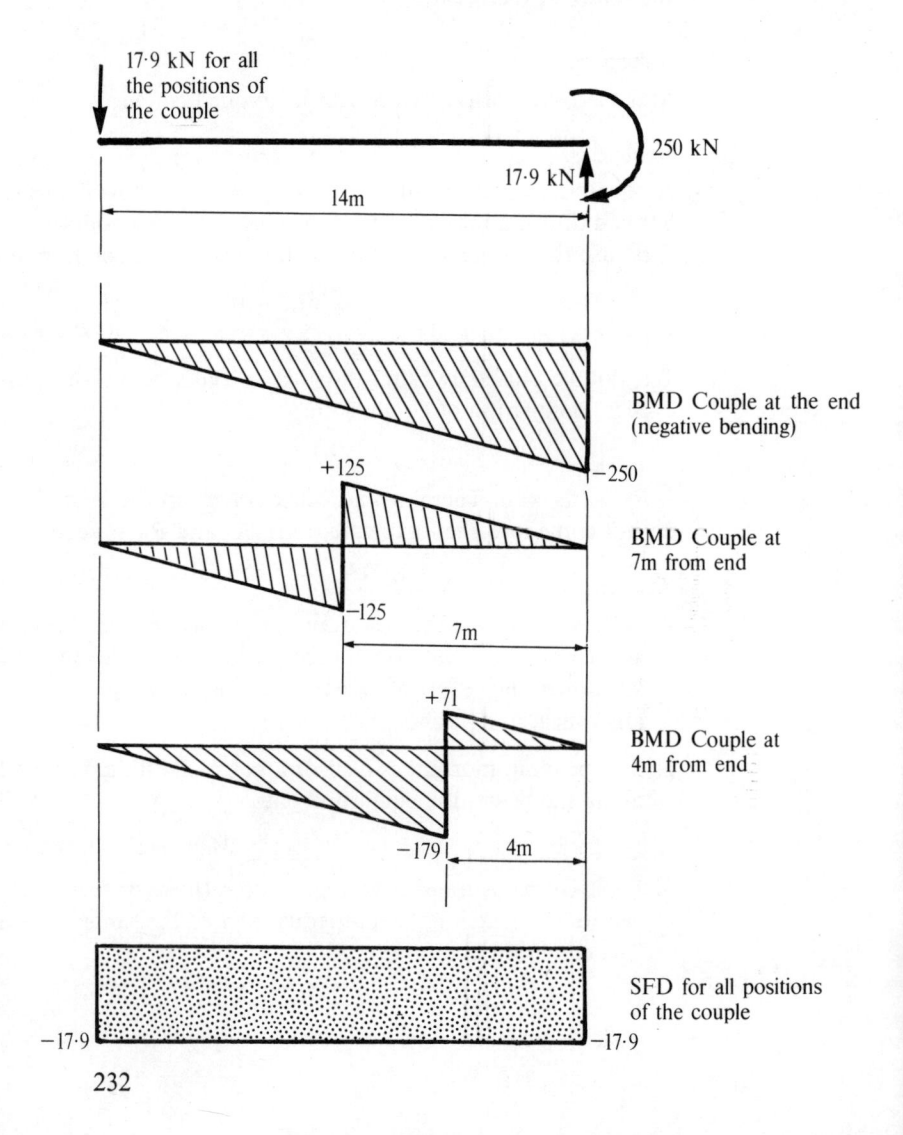

10/24 SFD, TD, BMD: Concentrated and eccentric loads: Column

A vertical column hinged top and bottom, carries a load of 450 kN. At half-height, a bracket 300 mm wide carries a vertical load of 90 kN. There is also a 20 kN load at this point acting horizontally. Sketch the SFD, TD, and BMD.

S *Structure:*
Sketch the column approximately to scale and show the two component forces at the hinges. Their sense must be guessed.

U *Unknowns:*
The 90 kN eccentric load must be transformed into a couple of $-90 \times 0.3 = -27$ kN m and a force of 90 kN along the beam. See the introduction to this Unit for the eccentric loading (B). Sketch a second column.

R *Reactions:*
Centre of rotation: $\sum M_L = 0$
Forces: F_1 450   90   20 F_2 F_3
Moment arms: 15   0   0   7.5   0   0
Moments with signs: $+15 F_1 + 150$
Add moments of couples: $+15 F_1 + 150 - 27 = 0$

$$+F_1 = -8.2 \text{ kN}$$

The negative sign indicates that F_1 was guessed in the wrong sense.
The other reactions can now be found in the usual way and are:

$$+F_2 = +11.8 \text{ kN}$$
$$+F_3 = +540 \text{ kN}$$

Now draw third stanchion, with forces and couple shown correctly.

E *Evaluation:*
SFD: Uncover the stanchion from the top. The forces which are normal to the line of the stanchion produce the SFD. If you turn the paper on its side, the stanchion is merely once more the familiar beam.
TD: Uncover the stanchion again from one end. The forces which act along the beam give the TD, compression being negative.

10/24

BMD: The best method is to complete a BM table for each of the lengths

Range	BM equation (kN m)	When $x =$	When $z =$	$M =$
L to T	$M = +11\cdot 8x$ (straight line)	7·5 (T)		$+88\cdot 5$ kN m
U to T	$M = +8\cdot 2z$ (straight line)		7·5 (T)	$+61\cdot 5$ kN m

There are two values of bending moment at the same point, the 'step' being equal to the moment of the couple (27 kN m) applied at that point.

10/25 SFD, TD, BMD: Inclined and eccentric loads: Overhang.

Draw the SFD, TD, and BMD for the beam shown.

S *Structure:*

It is unnecessary to point out that such a loading is highly improbable, but some of the items are frequently encountered by themselves – a simpler problem. The calculations below illustrate the effect of different types of eccentric forces, and also reiterates that of couples. In the introduction to this Unit it was shown that eccentric loads can be treated as a couple and a load. You ought to read introductions carefully. If you have not done so here, go back and study the (A), (B), and (C) arrangements of moment-producers.

The first step, therefore, is to change the two eccentric loads into a couple and a force. No further action is required for the couple already applied at A. The 10 kN force must also be split into two components, and a second representation of the beam drawn. It is always advisable, in problems of this kind, to re-draw the structure with the forces divided into components normal to and components parallel to the beam. Eccentric loads should be divided into couples and forces. Thereafter, the sketching of the diagrams is simple. The forces normal to the beam affect SFD and BMD; the forces parallel to the beam affect TD, and the couples affect only the BMD.

10/25

UR *Unknown Reactions:*

In complex problems it is well to obtain two of the reactions from the moment equation, which serves as a check on procedure.

Centre of rotation: R $\sum M_R = 0$
Forces: R_L 20   5 R_R
Moment arms: 12   6   2   0
Moments with signs: $+12R_L - 120 - 10$
Add moments of couples: $+12R_L - 120 - 10 + 70 + 4 - 8 = 0$
$$+R_L = +5\cdot 33 \text{ kN}$$

Centre of rotation: L $\sum M_L = 0$
Forces: R_L 20   5 R_R
Moment arms: 0   6   10   12
Moments with signs: $+120 + 50 - 12R_R$
Add moments of couples: $+120 + 50 - 12R_R + 70 + 4 - 8$
$$+R_R = +19\cdot 67 \text{ kN}$$

Check by: $\sum F_{NS} = 0$
$+R_L - 20 - 5 + R_R = +25 - 25 = 0$
Third equation of equilibrium:
$$\sum F_{EW} = 0$$
$+F_1 - 8 - 7 - 16\cdot 0 = 0$ $+F_1 = \mathbf{24\cdot 7 \text{ kN}}$

These results show the double plus sign which indicates that the senses of the forces have originally been guessed correctly. All forces, eccentric forces and couples on the beam have now been divided and are fully known.

E *Evaluation:*

SFD: Slide a card over the beam, disclosing the left end. The shearing force at this point is $+5\cdot 33$ kN. Thereafter add and subtract forces acting normal to the beam to obtain the SFD.

TD: Uncover the beam from the right hand end. The thrust of 16 kN, so disclosed, is acting inwards and is thus applying a compression ($-$). Add and subtract forces acting along the beam.

BMD: Using the variable x from one end and the variable z from the other end, write the equations to the bending moment and complete the BM table. The table below is fuller than it need be, chiefly to show

10/25

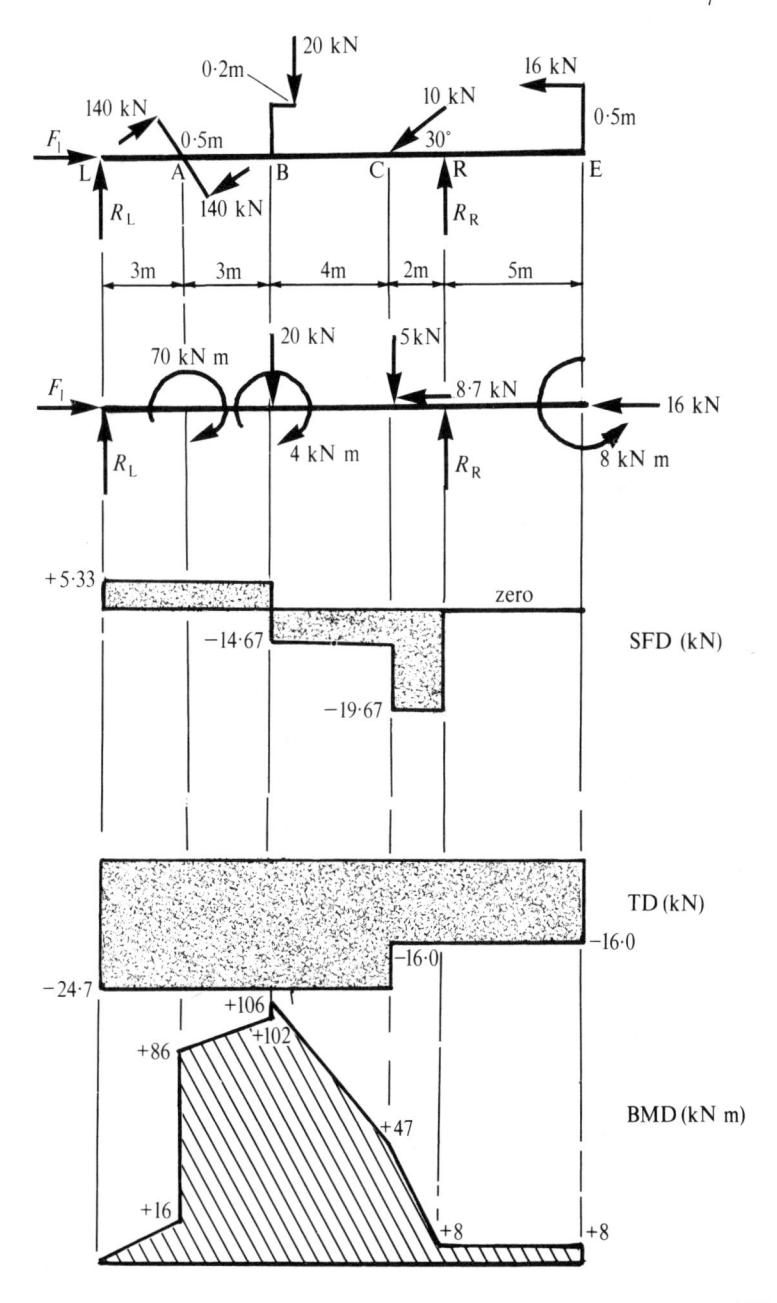

that a check can be obtained by working from both ends, and how 'steps' appear when couples are applied at a point.

Range	BM equation (kN m)	When $x =$	When $z =$	$M =$
L to A	$M = +5\cdot 33x$	3 m (A)		+16
A to B	$M = +5\cdot 33 + 70$	3 m (A)		+86
A to B		6 m (B)		+102
B to C	$M = +5\cdot 33 + 70$ $+4 - 20(x-6)$	6 m (B)		+106
B to C		10 m (C)		+47·33 Check
E to R	$M = -8$		0 m (E)	−8
E to R			5 m (R)	−8
R to C	$M = -8$ $-19\cdot 67(z-5)$		5 m (R)	−8
R to C			7 m (C)	−47·33 Check

The plus signs indicate clockwise moment applied to the beam. The first five show clockwise moments to the left of the edge of the card (do not disdain the use of the card). The four below show negative moments to the right of the edge of the card, which is also positive bending. So the whole beam, with the cantilever portion bends in positive bending. Sketch the deflected shape, which might have been difficult to predict at the start.

10/26 BMD: Non-uniform load: Simple supports

A beam, simply supported and 24 m long, carries a distributed load which varies from 50 N/m at one end to 150 N/m at the other end. The variation is linear. Draw the bending moment diagrams.

S *Structure:*

Draw a sketch of the structure with its varying load and cover it with a card.

U *Utilize:*

both the deflected shape and the shearing force diagram; these can give valuable hints on the shape of the bending moment diagram.

The deflected shape (DS) is a convex-up curve (positive bending) with a somewhat steeper slope towards the more heavily loaded end.

The shearing force diagram is a curved line which passes through the baseline at the point of maximum bending moment.

R *Reactions:*

The two reactions are obtained by standard methods. As this is an unusual loading the derivation of R_L and R_R may be helpful.

For any trapezoidal loading, it is best to divide the load into two triangles (whose centroids are known) or a rectangle and a triangle.

The value of the load per metre at any point, distant from L, is

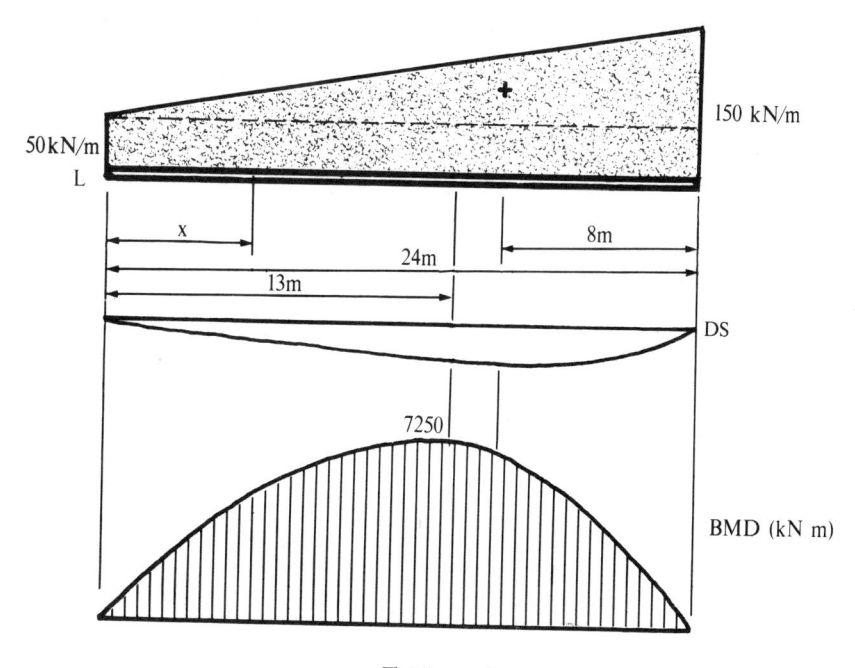

$$\sum M_R = 0$$
$$+ R_L \times 24 - (50 \times 24)12 + ((100/2) \times 24)8 = 0$$

$$R_L = 1000 \text{ kN}$$
$$\text{and } R_R = 1400 \text{ kN}$$

E *Evaluation:*

The first part of the evaluation is the sketching of the *shearing force diagram*. The equation to the line is:

Load to left edge of card:

$= +R_L - 50x - \frac{1}{2}((x/24) \times 100)x$ (positive shear)

If this is zero, the outline passes through the baseline, which gives the point of maximum bending moment.

$+1000 - 50x - x^2(1/0\cdot48) = 0$

Solving this quadratic equation: $\qquad x = \mathbf{13\ m}$

Which is the distance from L to the point of maximum bending moment.

As an exercise, sketch the SFD, using x with values of 6, 12, and 18, remembering the shear at 13 m is zero.

The *bending moment* at x from L is the moment of the loads *seen* to the left of the card, about the edge of the card. Take the load as a rectangle and a triangle.

The loads are:

R_L, $50x$ and $\frac{1}{2}x\left(\dfrac{x}{24} \times 100\right)$ kN

and the lever arms for the moments are:

x, $\frac{1}{2}x$ and $\frac{1}{3}x$ from the centroids to the edge of the card.

Moment is: $+1000x - 25x^{\frac{1}{2}} - x^3/1\cdot44$.

This cubic equation can be plotted by using trial values of x.

x	x^2	$-25x^2$	x^3	$-x^3/1\cdot44$	$+1000x$	M
6	36	-900	216	-150	$+6000$	$+4950$
12	144	-3610	1728	-1210	$+12\,000$	$+7180$
13	169	-4220	2197	-1530	$+13\,000$	$+7250$ (max)
18	324	-8100	5832	-4050	$+18\,000$	$+5850$

Use a book of tables for squares and cubes (e.g., Barlow's) and a slide rule for the rest. You can calculate to three significant figures on a slide rule, which is more than sufficient for plotting the BMD.

10/27 **ILD: Unit moving load: Simple supports and hinges**

Sketch the ILD's for the two reactions R_A and R_B for the three beams shown – simply supported over the whole length, simply supported but with a cantilever, and simply supported but with a hinge.

Beam 1 (upward forces positive):

(a) If the unit load is at A, R_A is unity, and R_B is zero.

(b) If the unit load is at B, R_B is unity and R_A is zero.

(c) Set the ordinates upward as positive and join the points to obtain the ILD's for both reactions.

Beam 2

(d) When the unit load rolls on to the cantilever, the beam being considered to have no weight, the end B must be held down, and the value of R_A must thus be increased above unity. Project the IL from Beam 1 to the end of the cantilever.

(e) Similarly for R_B the negative or downward value when the unit load is on the cantilever is shown by the line falling below the baseline to the negative side.

Beam 3

(f) Students are often puzzled by beams containing hinges. In earlier units it was noted that a hinge cannot transmit a moment. The hinge of a door moves when the handle is pulled so that no moment or bending can be applied. A hinge, however, can transmit forces. Since the only force is a vertical one, the portion of the beam between hinges acts as a suspended span, and can be imagined as laid on the main beam.

(g) Until the unit load rolls on to the suspended span, the ILD is as for beam 2. As the unit load rolls on the suspended span it sends back a reaction to point C, this reaction diminishing from unity to zero as the load rolls from C to D. So far as the beam ABC is concerned, the load at C has remained in position, but has gradually decreased. The values of R_A and R_B however, must still be plotted under the position of the unit load.

(h) A hinge, therefore is the start of a length over which the quantity represented in the ILD gradually decreases to zero.

10/27

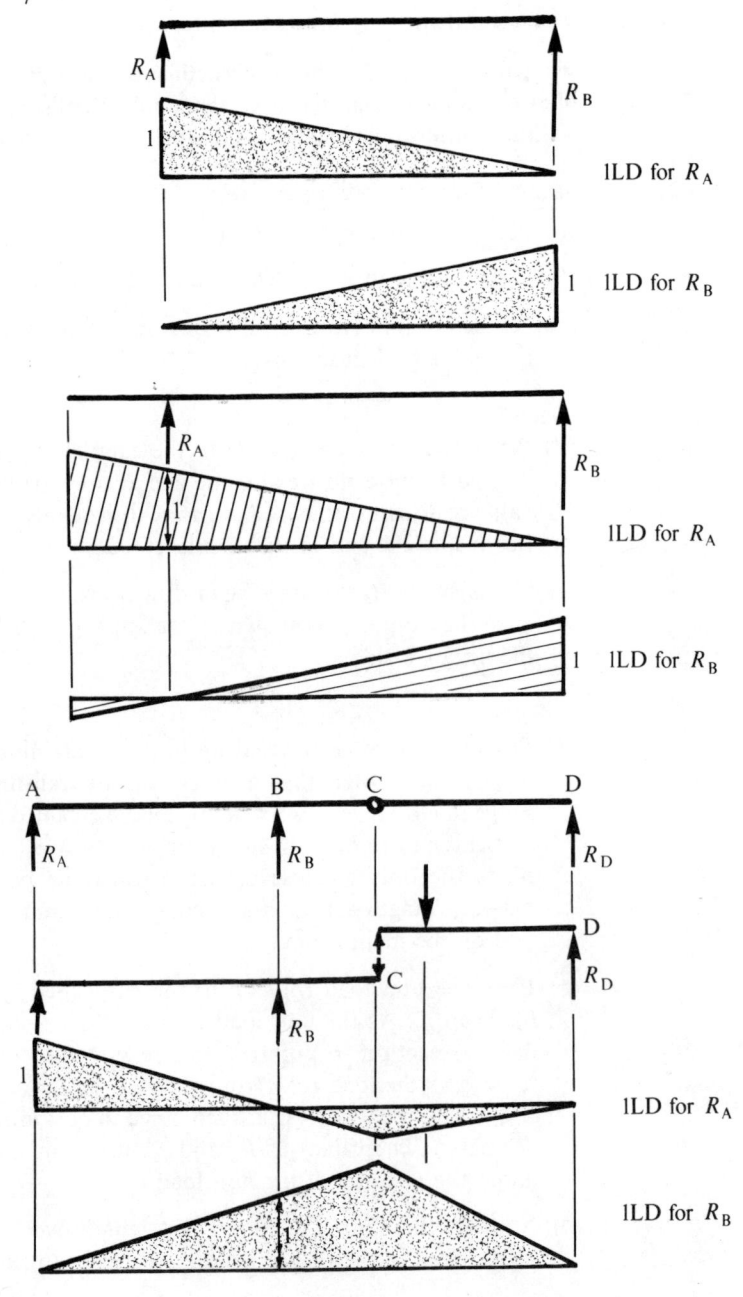

(i) The ILD for R_C is that for the left-hand reaction of a simply-supported beam CD. It takes the same shape as R_A's ILD in beam 1. There is no effect beyond the hinge, for, when the unit load is on ABC, the hinge merely moves and there is no effect on the value of R_C which remains zero.

10/28   ILD: Unit moving load: Beam with overhang

A simply supported beam ABC is being studied with a view to its carrying loads. For design purposes a critical section of the beam is at 1 m from A. If the beam eventually carries one rolling load of 4 kN, what is the worst shearing force at this critical section?

(a) Draw the ILD for the reaction at A (R_A). This is the same as for R_B in beam 2 in Problem 10/27.

(b) Cover the whole beam from X to C with a card.

(c) If the unit load is rolling out of sight under the card, the only force visible is R_A. Thus for positions of the load from X to C the shearing force at XX is the value of R_A.

(d) From X to C repeat the sketch of the ILD for R_A.

(e) When the unit load rolls out from under the card, there are two forces visible (R_A and unity). The shear at X–X is then $R_A - 1$.

(f) From a lightly dotted line representing the rest of the ILD for R_A drop ordinates of unity (or draw another line parallel and at a negative distance of unity below ILD for R_A). The ILD for shear at X–X is complete.

(g) Reply required is that at X–X the worst shearing force is 0·75 when a unit load acts, and it is thus 3·0 kN when 4 kN rolls over the span.

243

10/29

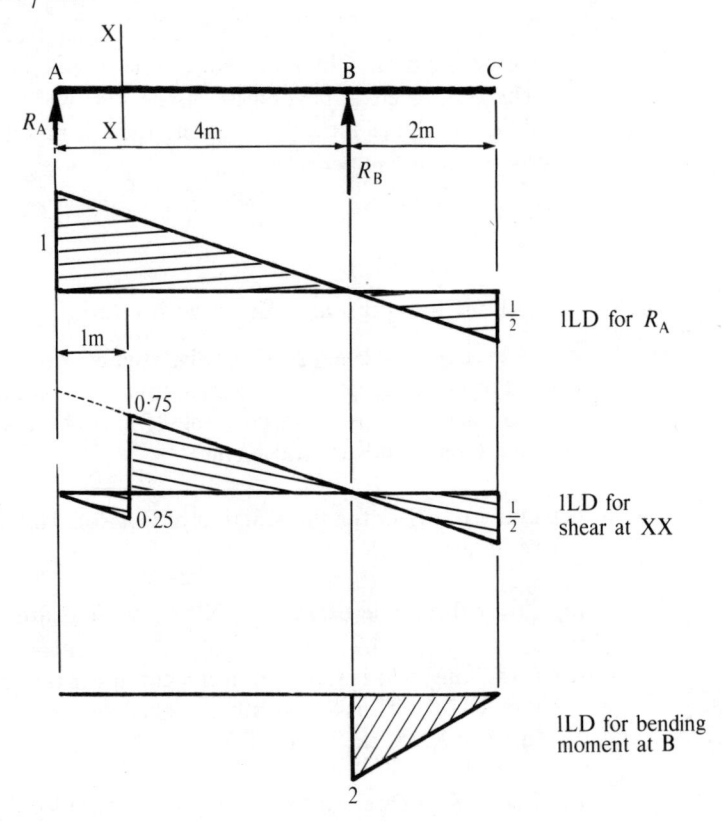

[Diagram: Beam ABC with R_A at A, R_B at B, section X-X at 1m from A; AB = 4m, BC = 2m. ILD for R_A showing ordinate 1 at A, ½ at B, 0 at C. ILD for shear at XX showing 0.25 below at 1m⁻, 0.75 above at 1m⁺, ½ at B. ILD for bending moment at B showing ordinate 2 at C (below line).]

10/29 ILD: Unit moving load: Several spans and hinges

In the beam shown, there is a simply supported beam with cantilever, a suspended span and a cantilever built in at one end. Sketch the ILD's for shearing force at the centre of AB and at the centre of BC. Sketch also the bending moment ILD's for the centre of AB, for the point B and for the point E.

(a) The first step is always to sketch the ILD's for the reactions – in this instance R_A and R_B. They may not both be required, but are the basis of many of the shear and bending moment ILD's.

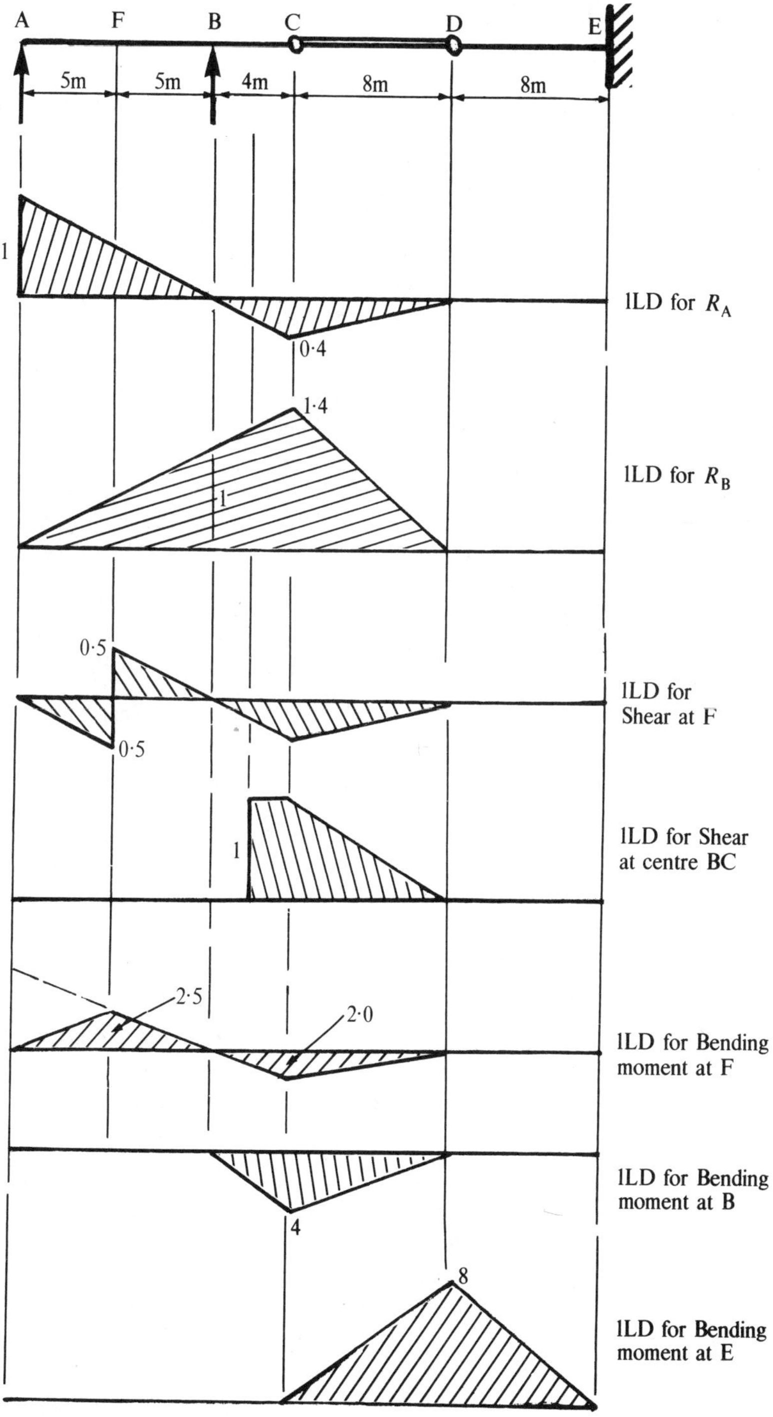
A
F
B
C
D
E
5m
5m
4m
8m
8m
1
0·4
ILD for R_A
1·4
1
ILD for R_B
0·5
0·5
ILD for Shear at F
1
ILD for Shear at centre BC
2·5
2·0
ILD for Bending moment at F
4
ILD for Bending moment at B
8
ILD for Bending moment at E

(b) In drawing these two diagrams remember how a suspended span brings the quantity down to zero.

(c) Place a card over the beam leaving visible only R_A and half of AB. When the unit load is rolling under the card, the shearing force at the edge of the card is represented by R_A. Thus from B to E the ILD for shear at the centre of AB is the same as for R_A.

(d) When the rolling unit load comes out from under the card on to the left half of AB, the shearing force is $R_A - 1$. Draw a line parallel to the R_A IL and at unity below it.

(e) The ILD for the shear at the centre of BC. Cover the beam from the centre of BC to A. When the load is under the card no force is visible to the right of the edge of the card. When the load rolls out, the shear is immediately unity and remains so until the load reaches C. Thereafter the hinged span reduces it gradually to zero. There is, therefore, no shear at the centre of BC until the load rolls over FD.

(f) The ILD for BM at F, the centre of AB. Cover the beam from F to E. The moment about the edge of the card is simply $5R_A$ kN m. The ILD from F to E, therefore, is that of R_A multiplied by 5 m.

(g) When the load rolls out along FA, the moment at the edge of the card is 5 R_A - unity × distance to the card, or $-x$. Subtract this from R_A by coming down from the dotted R_A IL.

(h) There is no moment at E until the rolling unit load reaches CD over which M_E gradually increases to a value of 8 kN m (unity × DE), and then reduces again as the distance between the load and E decreases.

10/30   ILD: Two rolling loads: Crane girder

A simply supported beam of span 36 m, carries a rolling load of two wheels 6 m apart and exerting 7·3 kN each. Find the maximum possible value of shearing force and bending moment at 14 m from the left-hand end.

(a) Draw the ILD for R_A (beam 1 in Problem 10/27).

(b) Cover the 22 m at the right of the section defined (CB). When unit rolling load is under the card, the shear is the same as R_A which is the only force visible to the left of the edge of the card. When the unit load is between A and C, shearing force is $R_A - 1$.

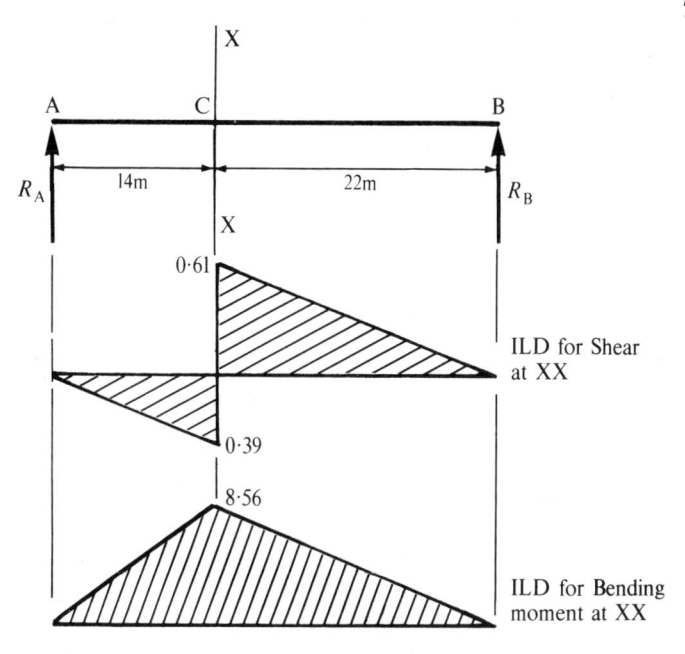

(c) Similarly, when unit load is under the card, the ILD for M_C is $R_A \times 14$ kN m. When the unit load is to the left of C the bending moment at C is $14R_A - 1 \times x$.

(d) You must now place the two rolling loads, held at 6 m apart, in such a way that they are over the largest possible ordinates. For shearing force, for example, both wheels should be on the right portion of the beam, when the shear will be:

One load at C: $+7.3 \times 0.61 = +4.45$ kN

One load at 6 m from C: $+7.3 \times \frac{16}{22} \times 0.61 = +3.24$ kN

Thus the maximum shear with two loads spaced 6 m is $+7.69$ kN.

(e) A similar placing for the loads on the BM. ILD is indicated by inspection:

Max BM: $+7.3 \times 8.56 + 7.3 \times 8.56 \times \frac{16}{22} = +107.9$ kN m.

In this way the effect of one unit load can be multiplied to show the effects of several moving loads on the beam.

RECAPITULATION

In Part II, which follows, you are expected to be able to carry out any of the operations in Part I. No further explanation is given as to the determination of forces and couples acting on a structure.

Success in doing this depends on various related factors; it would be well at this stage for you to assess your mental equipment before you attempt to solve Problems in Part II. Here is a selection of points on which you must be quite certain, and not depend on a vague idea. *Write* the answers down clearly and exactly: do not merely mutter them into your beard or your lipstick!

Do you know, and can you define exactly, the differences between the terms below, given in related pairs:

'Surface traction' and 'Body force'
'Mass' and 'Weight'
'Direction' and 'Sense'
'Couple' and 'Moment of couple'?

Can you identify and express the condition which makes equilibrium under more than three forces an entirely different conception from equilibrium under two or under three forces?

Can you write down, in symbols, the three Conditions of Equilibrium?

Do you know the limitations under which problems in Part I are solved?

Can you sketch, as forces and couples, the various categories of supporting systems for structures in equilibrium? How many are there?

How sound is your geometry? It has to be in good working order to find required distances in the problems of Part I.

Write down, without referring to Unit 8, the significance of each of the letters S-U-R-E-T-Y.

What is the significant difference between an influence line diagram for bending moment and a bending moment diagram?

Any 'fool' can solve the problems of Part I if given unlimited time—you have to deal with them FAST if you are to succeed.

PART II
DEFORMABLE STRUCTURES

GROUP D
The Working Conditions

Part II: Group D

The geometry of the cross-section

In Part I it is shown how important is a knowledge of the *geometry of the structure*. Indeed, many of the difficulties and failures which beset the student of numerical problems in these topics are due, not to a lack of knowledge of the subject, but to a lack of attention to the necessary geometry.

In part II also, in the study of *deformable structures*, a skill in geometry is a *sine qua non* for success in solving numerical problems. This time, you are faced, not with a free-structure diagram, but with the cross-section of a beam, tie-bar, column, or other member within which the effects of the applied loading are to be calculated.

The cross-section can be of a simple shape, or may be a combination of simple shapes. Commonly encountered cross-sections which are not merely triangles, rectangles, or circles, may be of many forms, such as are shown in the problems.

11.1 The geometrical calculations

The skills you must develop are:

(*a*) finding the surface areas of simple shapes (m^2);
(*b*) finding the first moment of any area about any given axis (m^3);
(*c*) finding the second moment of any area about any given axis (m^4).

It is assumed that you know how to calculate (*a*). This unit of study concentrates on (*b*) and (*c*), neither of which forms a difficult calculation, but both of which can be tedious and complex if not attacked systematically. The complexity might lead to error, so all calculations in the problems of *first* and *second moments of areas* must be carried out by tabular methods in which each operation is carried out in turn so that you can concentrate on one thing at a time, while the table records the sequence of operations.

11.2 The tabular method of recording operations

The tables used in the unit have horizontal lines and vertical columns. The horizontal lines each refer to one of the simpler areas of which the whole cross-section is built up. A final line refers to the whole cross-sectional area. Each vertical column refers to one property or dimension, either given in the question as an item of information, or derived from the given data.

In using the tables as frameworks in which the necessary calculation is to take place, always concentrate on one *column* at a time, writing in all the values of the one property to which the column refers, before going on to the next column. The lines and columns in the tables are similar to the jigs set up for a complex manufactured product. Within their framework, the operations cannot go out of truth, provided each step is carried out effectively and correctly. It is quicker and more efficient to carry out the same step a number of times in succession, than to carry out a number of different operations successively. So, deal with one column at a time.

Suppose you have to deal with the highly unlikely area shown in sketch 11/A and a table to complete. Nothing could look much more complex, but each of the operations is equally simple whatever the complexity of the cross-section. The *first step* is to break the area into simple geometrical shapes. The *second step* is to give them numbers, prefacing these with R, T, or C (or other appropriate letters) for rectangle, triangle, or circle (or hexagon, etc.). The *third step* is to mark the centroids of these simple areas. You know how to find these centroids, and they can be located easily. The *fourth step* is to mark the axis about which the first and second moments are to be taken. This can

Table of quantities

	1 b	2 d	3 r	4 bd	5 d^2	6 bd^3	7 A	8 y	9 Ay
R1	7	17	—	119	189	225×10^2	119	8·5	1012
R2	20	5	—	100	25	25×10^2	100	2·5	250
R3	8	16	—	128	256	328×10^2	128	8·0	1024
T4	20	9	—	180	64	115×10^2	90	8·0	720
C5	—	—	6	—	—	—	36π	20·0	2261
Total							550		5267

11.2

be chosen at random, but in each problem it is usually obvious what axis is appropriate.

The table shown as an example is made up almost at random with some of the quantities you will require in your calculations. There is no need to use them at present, but they must be obtained accurately and recorded in the table for later use. The columns should bear a number and the units represented. In this one the units are m, m^2, m^3, and m^4, but mm units are just as likely to appear.

This table illustrates the type of tabular calculation used in this unit. The quantities obtained are some of those needed in obtaining the first and second moments of areas. The following procedures should be observed.

1 In this unit and in others, it is highly advantageous to use a book of tables giving squares, cubes, fourth powers, and other similar quantities. Such calculations cannot be done accurately enough on the slide rule. *Barlow's Tables* give the quantities required.

2 Put R (rectangle), T (triangle), C (circle), etc. in front of the number defining the area. These letters remind you of how to obtain the areas (column 7, above).

3 The breadth of each simple area is measured parallel to the chosen axis, and depth, *d* at right angles to it. If the area is circular, use the radius, *r*.

4 The height *y* (column 8) is the distance from the chosen axis XX to the centroid of the area concerned.

5 The figures in each column are either dimensions read off from the drawing or some product of these.

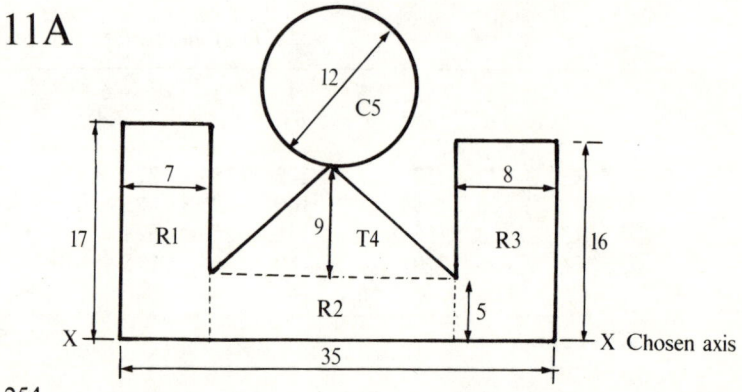

11A

254

11.3

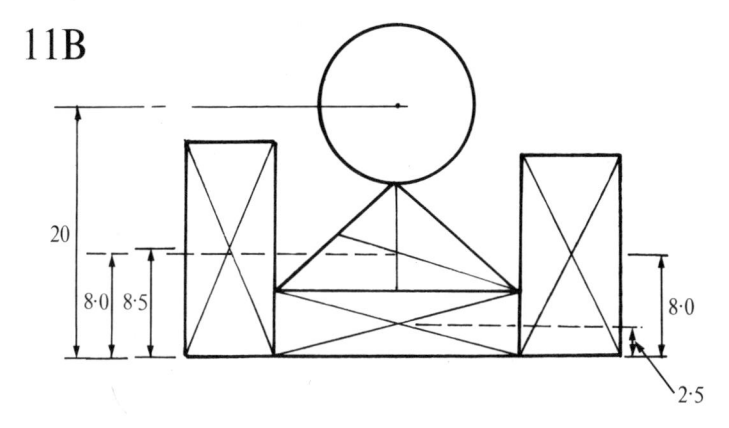

11B

11.3 First moment of Area

The first moment of an area about an axis is obtained by multiplying the area (in m² or mm²) by its distance y from the axis about which the moment is being calculated (m). This gives a product in m³ (Ay). This is simple if the area is small and y large, for all parts of the area are then at about the same distance from the axis. If, however, the axis XX about which the moment is to be taken, is close to or even within the area concerned, a large number of moments of elementary areas must be

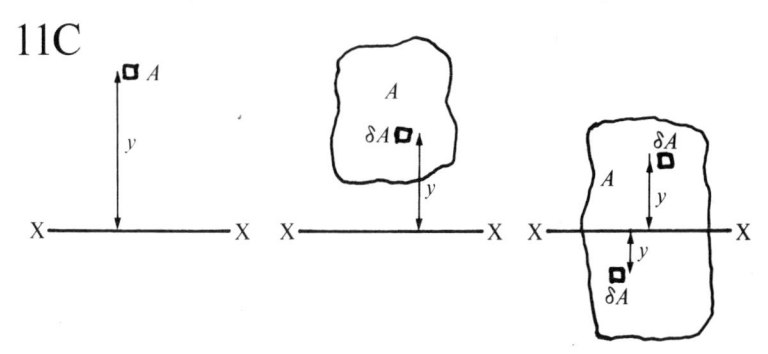

11C

summed. The result of this summation shows that the whole value of the area may be considered to be concentrated at its centroid. The whole area, multiplied by the distance of the centroid from XX is the *first moment of the area*. The distance y is from XX to the centroidal axis (CA).

255

11.3

If the axis XX passes through the centroid, the distance between the centroidal axis and XX is zero. The first moment of the area is, therefore, zero also. This is another way of saying that the area is balanced on its centroid.

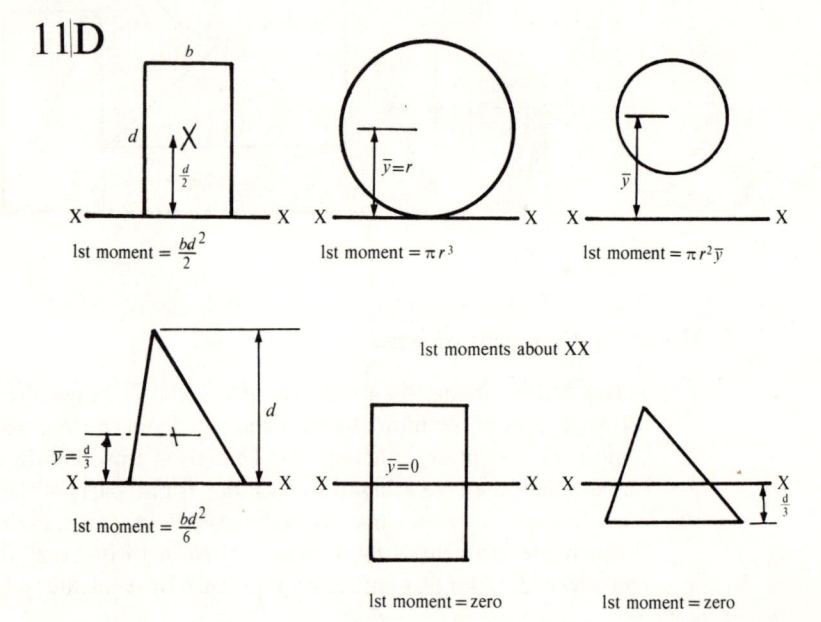

11D

1st moment = $\frac{bd^2}{2}$ 1st moment = πr^3 1st moment = $\pi r^2 \bar{y}$

1st moments about XX

1st moment = $\frac{bd^2}{6}$ 1st moment = zero   1st moment = zero

The *first moment of simple areas* is used to find the centroid of complex areas. The sum of the moments of all the parts of the complex area about XX must be equal to the moment of the whole area about the same axis. We know the value of the whole area and can thus determine $\bar{y}$, the distance from the centroid to XX.

$$\sum Ay = \bar{y} \sum A$$
$$\bar{y} = \sum Ay / \sum A$$

In the table drawn up earlier $\sum A\bar{y}$ and $\sum A$ can be found by adding their respective columns, so

$$\bar{y} = 5267/550 \text{ m}$$

This the distance between XX and CA. Thus, since XX is known CA can be located.

256

11.4 Second moment of Area

As the name implies, the *second moment of area* (symbol I) is merely the first moment multiplied again by the length y from the centroid to the axis concerned.

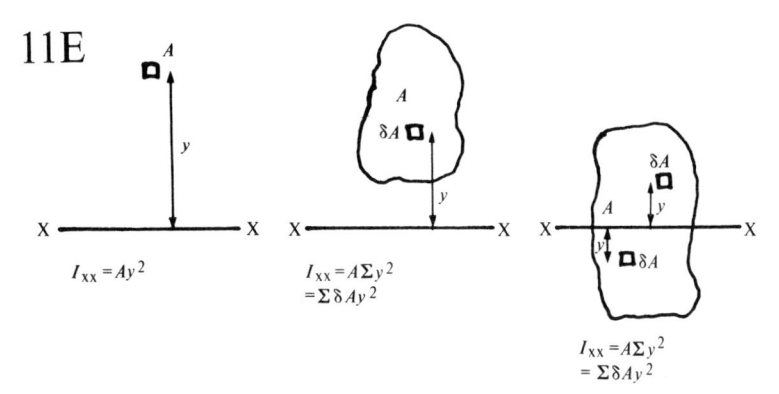

11E

$I_{xx} = Ay^2$

$I_{xx} = A\Sigma y^2$
$= \Sigma \delta A y^2$

$I_{xx} = A\Sigma y^2$
$= \Sigma \delta A y^2$

Again, this is simple if the area is small and y large. If, however, the axis XX is near to or passes through the area, there are many elementary areas which must be multiplied by y twice, or by y^2.

The significance of the second moment in mathematical and logically argued terms is explained in any text-book on strength of materials. In more direct terms, the *second moment of area* is a measure of the *stiffness* of the section, or of its resistance to being distorted by a bending couple. A plank is much more easily bent by a man walking across it, if the wide dimension is horizontal, and the thickness vertical. If the wide dimension of the plank is vertical and the thickness horizontal, the man, walking 'tight-rope' on the edge of the plank, scarcely bends it at all. This is because the second moment of the area on edge is very much greater than the second moment of area when the plank is laid flat.

The first moment of area about an axis passing through the centroid is zero, because the area balances at the centroid. A piece of card will balance on the point of a pin if the pin makes contact at the centroid. The second moment of area about a centroidal axis is not zero, for the second moment of area indicates stiffness against bending moments, and any area certainly shows some resistance to being bent about a centroidal axis. Since the second moment of area of simple geometrical shapes about selected centroidal axes are required in the calculations of numerical examples, it is convenient to have a set of them to refer to.

11.5

These are listed in this unit for use in the next group of units. For example:

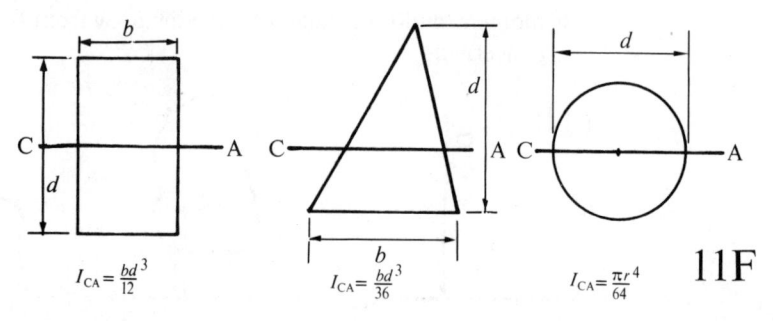

$I_{CA} = \frac{bd^3}{12}$ $I_{CA} = \frac{bd^3}{36}$ $I_{CA} = \frac{\pi r^4}{64}$

11F

When the resistance to bending couples is required about an axis parallel to the centroidal axis, but at some distance from it, the more direct second moment comes into the reckoning. The second moment of area of a geometrical shape about an axis, is the sum of its I about a parallel centroidal axis, and the value $A y_C{}^2$.

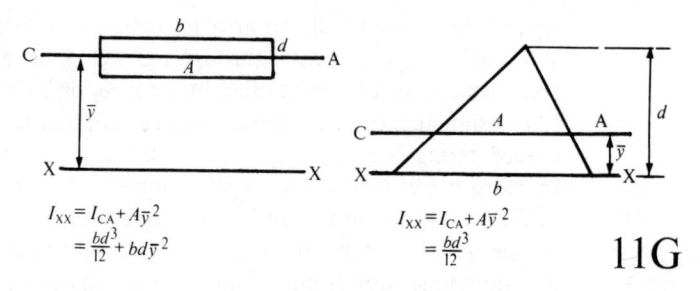

$I_{XX} = I_{CA} + A\bar{y}^2$
$= \frac{bd^3}{12} + bd\bar{y}^2$

$I_{XX} = I_{CA} + A\bar{y}^2$
$= \frac{bd^3}{12}$

11G

The second moment of area about an axis through the centroid is the smallest value of I which the area can show. Thus, if the value of a second moment about XX is found (an axis not passing through the centroid), the value of I_{CA} must be less than this by the value of $Ay_C{}^2$ (CA indicates centroidal axis):

$$I_{CA} = I_{YY} - Ay_C{}^2$$

or

$$I_{YY} = I_{CA} + Ay_C{}^2$$

11.5 Polar Second moment of Area

So far, you have looked at the problems of finding the second moment of area about axes in the plane of the section concerned. If, however, the section is not bent about such an axis, but twisted, the incipient rotation is about an axis at right angles to the plane of the section. The resistance to torsion or twisting is also measured (as was resistance to bending) by a second moment of area. To indicate that this particular value of the second moment refers to torsion it is called the *polar second moment of area*. Again, the units in which it is measured are m^4 or mm^4, taken about the axis, which is seen 'end on' as a point at the centroid of the section. In this volume you will deal with only one type of section where torsion is concerned, so there is only one value of polar moment of area to be considered – that of a circular section about an axis through the centre and at right angles to the plane of the section. The symbol for the *polar moment* is J, as I was for the *second moment of area*.

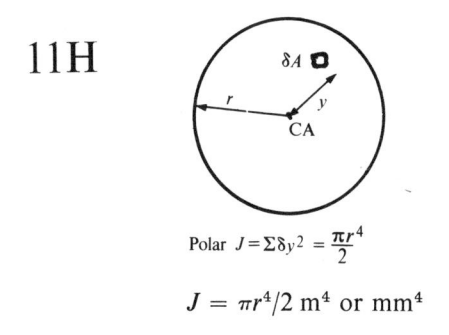

11H

Polar $J = \Sigma \delta y^2 = \dfrac{\pi r^4}{2}$

$$J = \pi r^4/2 \text{ m}^4 \text{ or mm}^4$$

11.6 The three categories

The following problems fall into three categories. They concern the finding of the *first moments of area* (with the object of locating the centroid), the *second moment of area*, and the *polar second moment of area*. The table required for the *first moments* and the *centroids* is a preliminary to the table of calculations of r the *second moments*. The whole table, therefore can be looked upon as one. The quantities required are:

b d bd r A y Ay d^2 bd^3 r^4 I_c y_c y_c^2 Ay_c^2

The first seven columns refer to the first moments and to finding the centroid, and the other seven are used in finding the second moments. Sometimes it is advantageous to repeat d and bd from the first portion, when the second moment is in question.

11.6

The first seven quantities have already been explained in section 11.2. The other quantities are obtained as follows:

bd^3 which is useful for rectangles and triangles is obtained by multiplying bd and d^2.

I_C is the second moment of area of that particular simple area, and is obtained from bd^3 or from πr^4.

y_C is the distance between the centroid of a simple area and the centroid of the whole section as found from $\sum Ay/A$.

Ay_C^2 is to be added to I_C to obtain the second moment of that particular area about the centroidal axis of the whole section.

Always complete the table column by column. Do not work along a line.

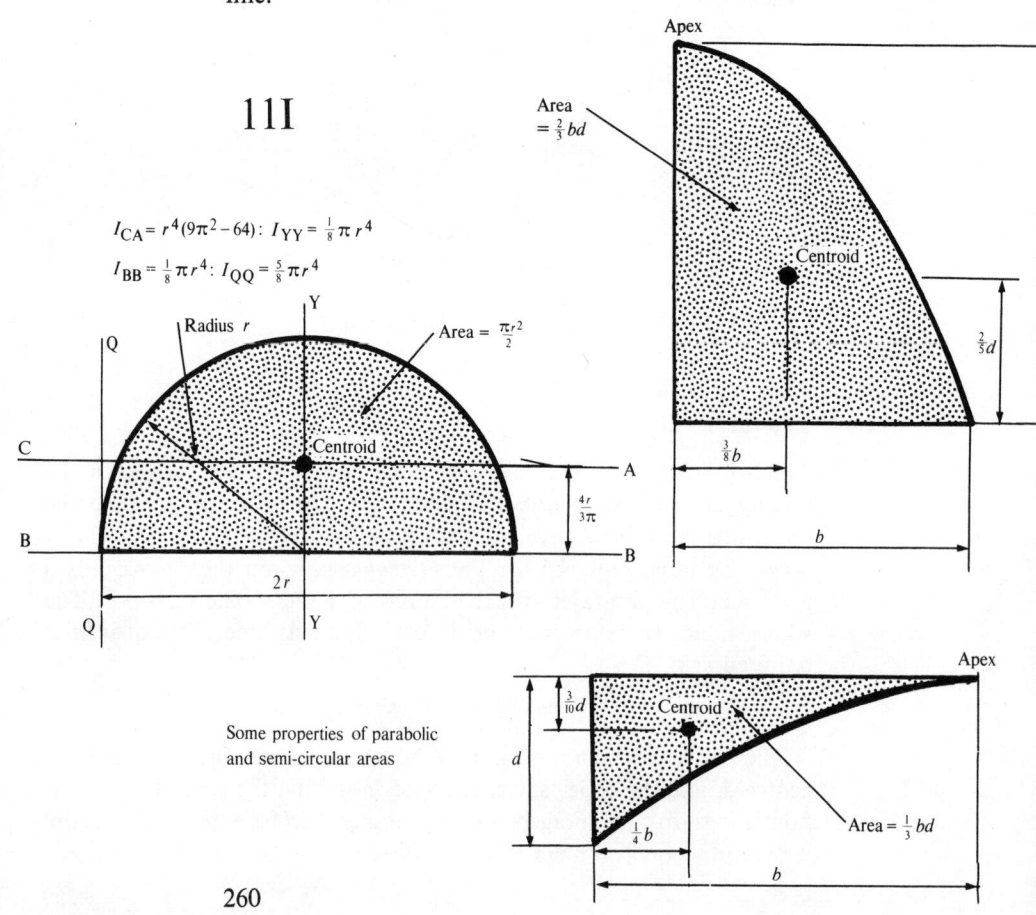

11I

$I_{CA} = r^4(9\pi^2 - 64)$: $I_{YY} = \frac{1}{8}\pi r^4$

$I_{BB} = \frac{1}{8}\pi r^4$: $I_{QQ} = \frac{5}{8}\pi r^4$

Some properties of parabolic and semi-circular areas

260

11.6

Common Cross Sections: Dimensions, Centroids and Areas

Shape	Dimensions and axes	Height of centroid above base	Area
Triangle	Base (length b) is one of the sides (horizontal) Height (h) is vertical distance from base to apex	$\frac{1}{3}d$	$\frac{1}{2}bd$
Rectangle	One side forms horizontal base. Vertical height (h) is from horizontal base to parallel side	$\frac{1}{2}d$	bd
Parabola	Height (h) is vertical from base to apex Base is formed by horizontal line cutting through the symmetrical parabola (b)	$\frac{2}{5}d$	$\frac{2}{3}bd$
Circle	r is the radius Base is the horizontal tangent to the circle	r	πr^2
Semi-circle	r is radius Base is a diameter (b)	$\dfrac{4r}{3\pi}$	$\frac{1}{2}\pi r^2$

Horizontal direction is AXIS XX, BB or CA
Vertical direction is AXIS YY or QQ

Common Cross Sections: Second Moments of Area

	I_{xx} through centroid	I_{xx} through base	I_{yy} through centroid
Triangle	$\frac{1}{36}bd^3$	$\frac{1}{12}bd^3$	Depends on shape of triangle
Rectangle	$\frac{1}{12}bd^3$	$\frac{1}{3}bd^3$	$\frac{1}{12}b^3d$
Parabola	$\frac{24}{525}bd^3$	$\frac{80}{525}bd^3$	$\frac{1}{30}b^3d$
Circle	$\frac{1}{4}\pi r^4$	$\frac{5}{4}\pi r^4$	$\frac{1}{4}\pi r^4$
Semi-circle	$(9\pi^2 - 64)r^4$	$\frac{1}{8}\pi r^4$	$\frac{1}{8}\pi r^4$

11/1

11.7 THE PROBLEMS

11/1 CENTROID: Symmetrical area

A symmetrical area (symmetrical about a vertical line) has a centroid which must lie on this vertical line. Find the position of the centroid.

1. Break the area into triangle T1 and rectangle R2.

2. Locate and mark the centroids of these areas.

3. Choose the axis for first moments at the lower edge of R2. This is not the best choice, but allows the method to work out clearly.

4. Set up the *table of first moments*.

5. The columns are always b, d, bd, A, y, Ay unless a circle is included when there is a column for r.

	b	d	bd	A	y	Ay
T1	20	$10\sqrt{3}$	$200\sqrt{3}$	$100\sqrt{3}$	$10 + \dfrac{10}{\sqrt{3}}$	$1000[\sqrt{3} + 1]$
R2	20	10	200	200	5	1000
Total				$100[\sqrt{3} + 2]$		$1000[\sqrt{3} + 2]$

$$\bar{y} = 1000/100 = 10 \text{ mm above } XX_1$$

or on the dotted line dividing the areas.

If the axis is taken at the level of the centroid of R2, the table becomes somewhat simpler:

T1	b	d	bd	A	y	Ay
T1		as above		$100\sqrt{3}$	$5 + \dfrac{10}{\sqrt{3}}$	$500(2 + \sqrt{3})$
R2		as above		200	0	0
Total				$100[\sqrt{3} + 2]$		$500[\sqrt{3} + 2]$

$$\bar{y} = 500/100 = 5 \text{ mm above } XX_2$$

which is also on the dotted line dividing the areas.

11/2

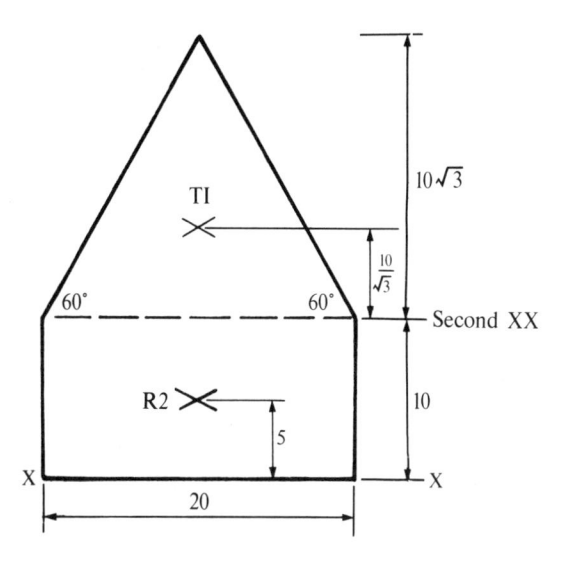

11/2 CENTROID: Symmetrical area

The Tee section is symmetrical about a central vertical line. The centroid is on this line. Determine its position.

1. Break the area into two rectangles R1 and R2.

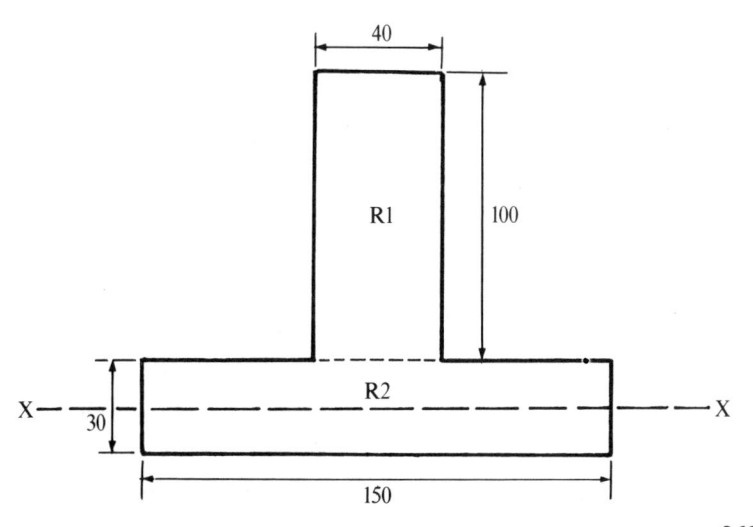

263

11/3

2. Locate and mark the centroids of these areas.

3. Choose an axis for first moments through the centroid of R2.

4. Set up table of first moments:

	b	d	bd	A	y	Ay
R1	40	100	4000	4000	65	260×10^3
R2	150	30	4500	4500	0	0
Total				8500		260×10^3

$\bar{y} = 260/8{\cdot}5 = 30{\cdot}6$ mm above XX

11/3 CENTROID: Unsymmetrical area

When an area is not symmetrical about any axis, two derivations must be made. Determine the centroid of the area shown.

1. Break the area into T1 and R2.

2. Locate and mark the centroids of these areas.

3. Choose the axes for the first moment through the centroid of R2, first in the horizontal direction and then in the vertical direction.

4. Complete two tables of first moments:

	b	d	bd	A	y	Ay
T1	15	18	270	135	16	2160
R2	15	20	300	300	0	0
				435		2160

$\bar{y} = 2160/435 = 5{\cdot}0$ mm above XX_1

	b	d	bd	A	y	Ay
T1	18	15	270	135	−2·5	337·5
R2	20	15	300	300	0	0
				435		

$\bar{y} = 337{\cdot}5/435 = 0{\cdot}8$ mm below XX_2

Cut out a card of this shape. It should balance on the point shown as C.

11/4

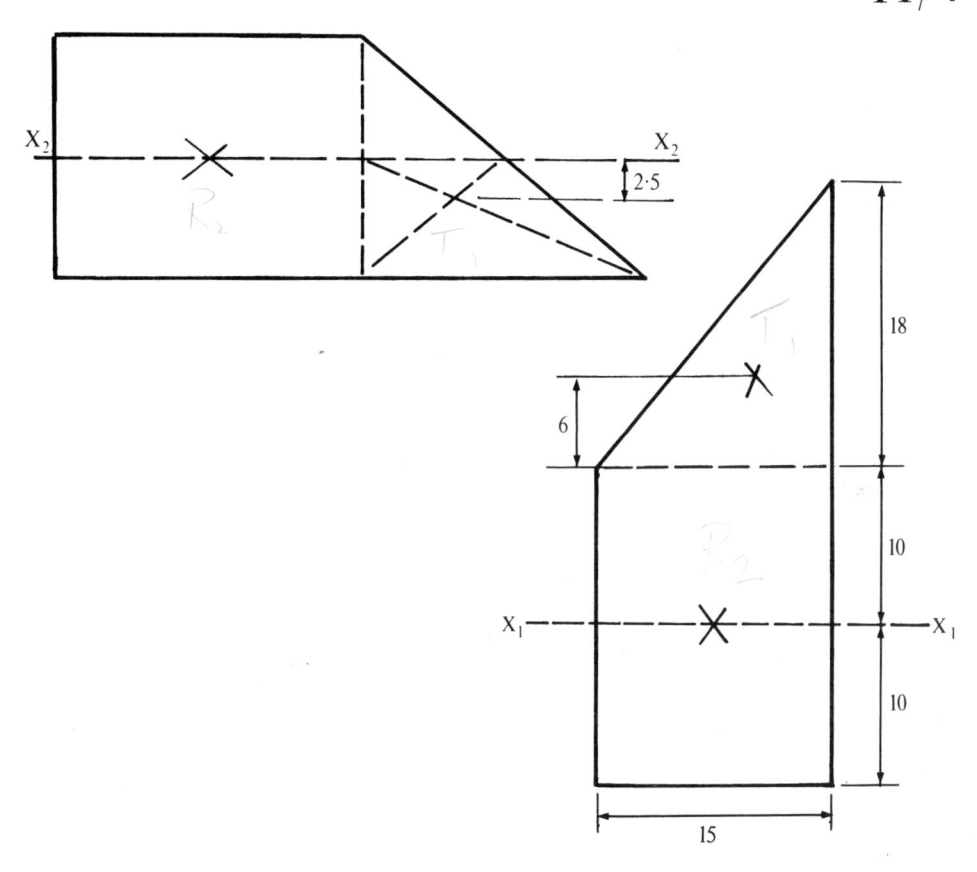

11/4 CENTROID: Complex unsymmetrical area

Any complex area can be investigated for the position of the centroid in the way already described. Since the portions of the table concerned with area do not change, the one table can contain all the data for first moments about two axes. Find the position of the centroid of the area shown.

1. Divide the area into four rectangles.

2. Locate and mark the centroids of these areas.

3. Mark the axes XX and YY.

265

4. Draw up the table of first moments.

	b	d	bd	A	y	Ay	x	Ax
R1	20	20	200	200	10	2000	5	1000
R2	20	10	200	200	5	1000	20	4000
R3	20	30	600	600	15	9000	40	24 000
R4	10	20	200	200	0	0	55	11 000
Total				1200		12 000		40 000

$\bar{y} = 12\,000/1200 = 10$ mm above XX
$\bar{x} = 40\,000/1200 = 33\cdot3$ mm to right of YY

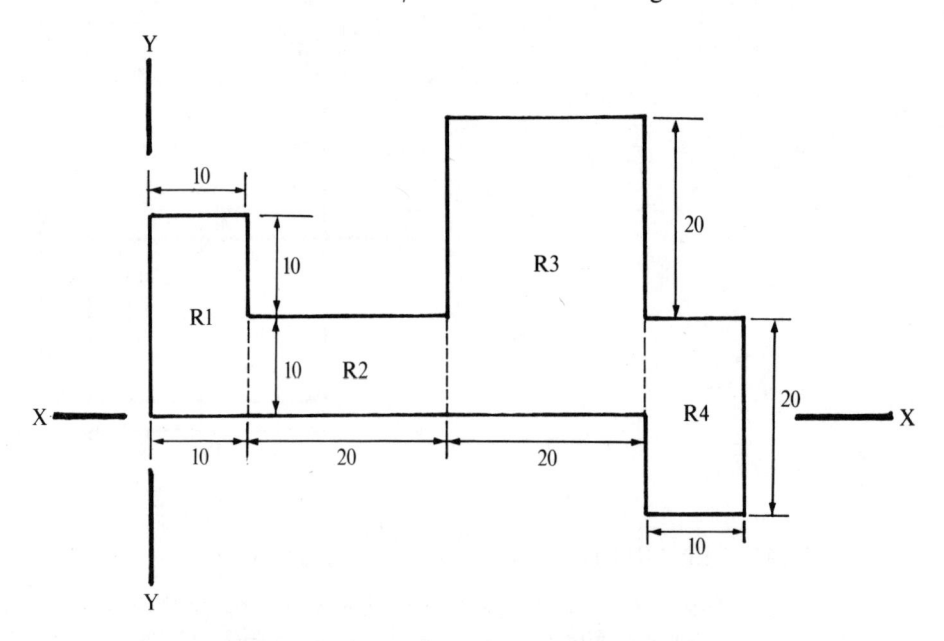

11/5 CENTROID: Punched plate: Negative area

When an area is punched by a hole forming a void, this void can be considered as a negative area. The first area, therefore, is the whole, and the second, negative area is the aperture or void. The area shown has a hole of 11 mm square punched through it. Find the position of the centroid.

1. The area R1 is 42 × 23. The area R2 is −(11 × 11).

2. Choose the axis XX_1 through the centroid of R1, and the axis YY also through the same point.

3. Set up the tables of first moments.

	b	d	bd	A	y	Ay
R1	42	23	966	966	0	0
R2	11	11	−121	−121	2·5	−302·5
Total				845		−302·5

$\bar{y} = 302\cdot 5/845 = 0\cdot 36$ mm above XX

		A	x	Ax
R1	as before	966	0	0
R2	as before	−121	9	−1089
Total		845		−1089

$\bar{x} = 1089/845 = 1\cdot 29$ mm to left of YY

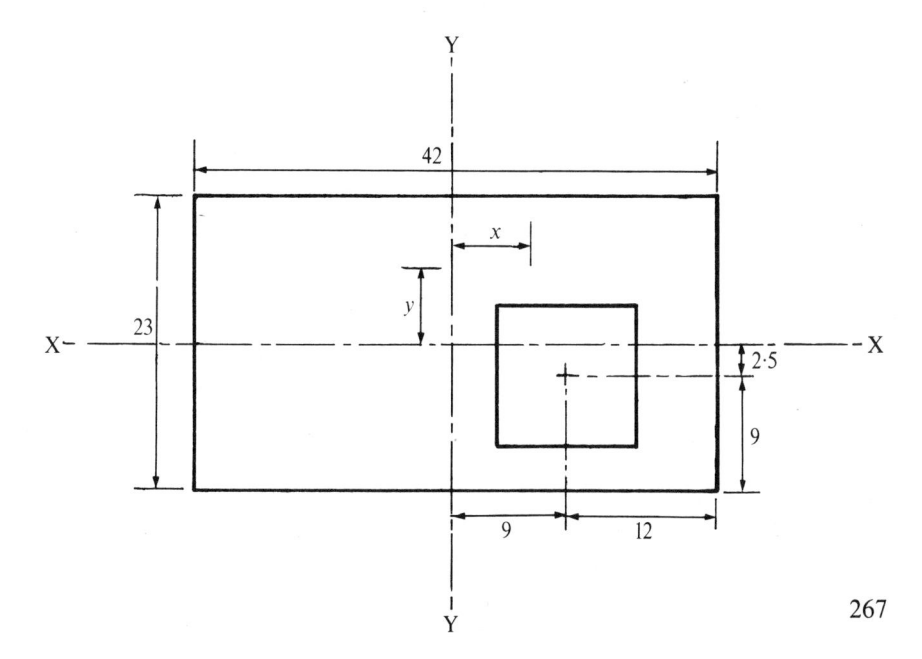

11/5

267

11/6

11/6 CENTROID: Punched circular plate: Unsymmetrical

Find the centroid of the punched plate shown in the sketch. The area of the whole plate is 44·0 $\#^2$, that of the larger hole, 4·4 $\#^2$ and that of the smaller hole, 2·2 $\#^2$.

Note:

To find the centroid, rather than one of the centroidal axes, it is necessary to find the intersection of two centroidal axes, usually taken at right angles to each other. The most convenient axes to select for the calculation of first moments are XX and YY, both acting through the centre of the circular plate.

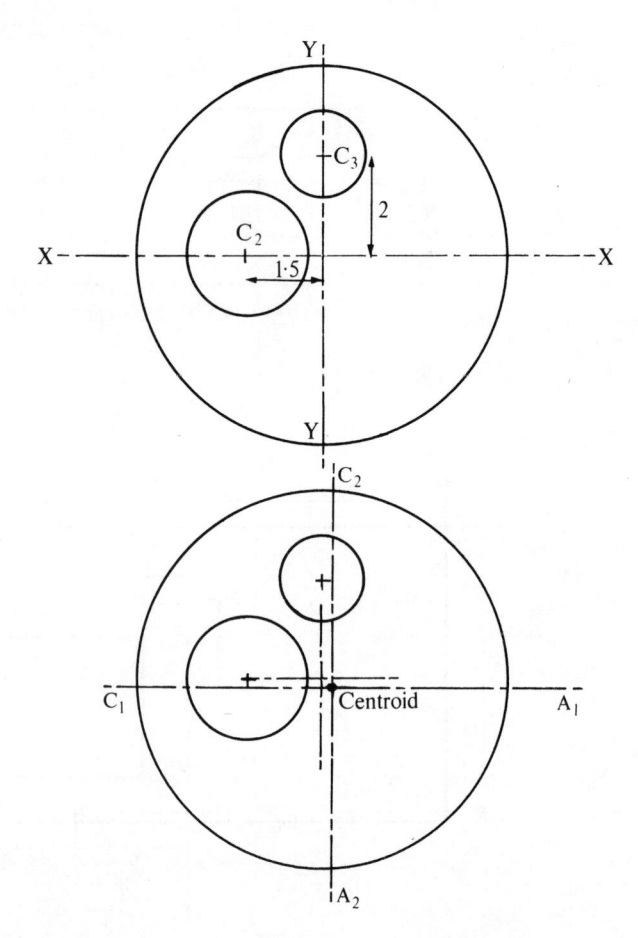

Table of first moments (about XX)

	A	y	Ay
C1	+44·0	0	0
C2	−4·4	0	0
C3	−2·2	2·0	−4·4
Total	+37·4		−4·4

$$\frac{\sum Ay}{\sum y} = \frac{-4\cdot 4}{37\cdot 4}$$

$\quad\quad = -0\cdot 123 \,\#$ (on the other side of XX from the smaller hole)

Table of first moments (about YY)

	A	y	Ay
C1	+44·0	0	0
C2	−4·4	1·5	−6·6
C3	−2·2	0	0
Total	+37·4		−6·6

$$\frac{\sum Ay}{\sum y} = \frac{-6\cdot 6}{37\cdot 4}$$

$\quad\quad = -0\cdot 176 \,\#$ (on the other side of YY from the larger circle)

11/7 SECOND MOMENT OF AREA: Square with hole: Unsymmetrical

Find the second moment of area of the punched square about the base XX. Since the centroid of the whole area is not required, a table of first moments can be omitted.

Table of second moments (area and moments of R2 are negative)

	b	d	bd	d^2	bd^3	A	y	Ay^2
R1 (whole)	2	2	4	4	16	—	—	—
R2 (aperture)	0·8	0·8	0·64	0·64	0·41	0·64	1·2	0·92

11/8

$$I_{xx} = \frac{bd^3}{3} \text{ (for all R1) less} \left(\frac{bd^3}{12} + Ay^2\right) \text{ (for R2)}$$

$$= \frac{16}{3} - \frac{0{\cdot}41}{12} - 0{\cdot}92 \qquad I_{xx} = \mathbf{4{\cdot}38} \;\#^4$$

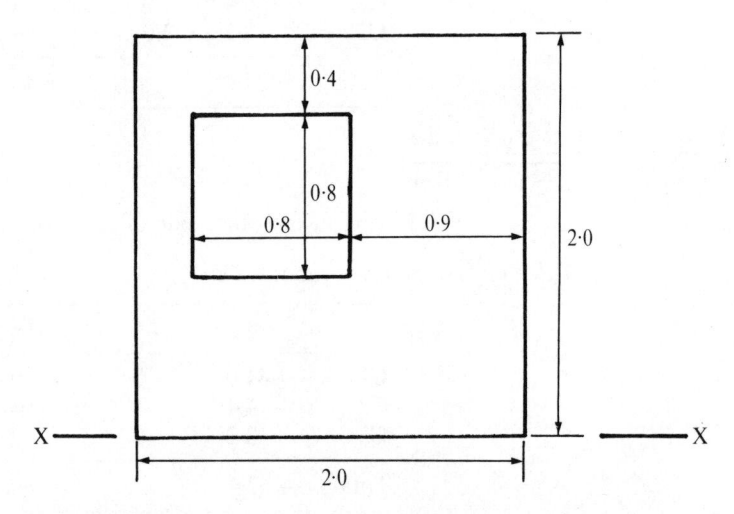

11/8 SECOND MOMENT OF AREA: Angle section: Unsymmetrical

Find the second moment of area of the angle section shown in the sketch, about a centroidal axis parallel to the base XX.

1. This requires both a table of first moments and a table of second moments.
2. Break the shape into two rectangles.
3. Mark the dimensions including the centroids: it is best to sketch approximately to scale.

Table of first moments

	b	d	bd	A	y	Ay
R1	2	10	20	20	5	100
R2	8	2	16	16	1	16
Divide totals ...				36		116

270

11/9

Table of second moments

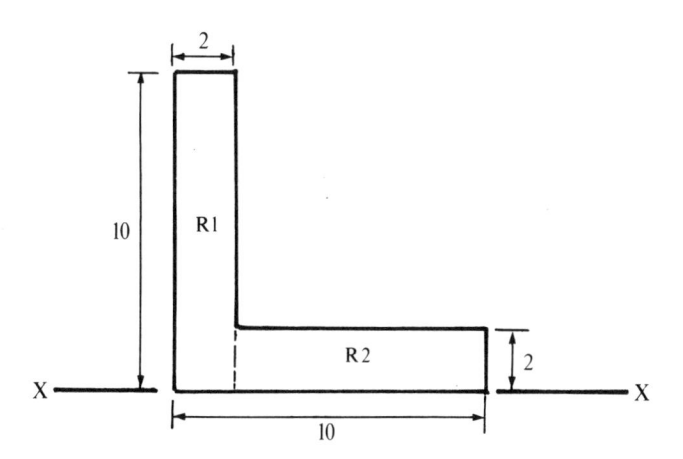

	bd	d^2	bd^3	$bd^3/12$	y_c	y_c^2	Ay_c^2
R1	20	100	2000	167	1·77	3·13	63
R2	16	4	64	5	2·23	4·97	80
Add totals...				172			143

11/9 SECOND MOMENT OF AREA: Combination of rectangles: Unsymmetrical

In this Tee-section, the centroidal axis is unknown, and must first be found by a table of moments. Then find the second moment of area about the centroidal axis by a table of second moments.

Remember:
Breadth b is measured parallel to the chosen axis XX. The values of y_c are the distances between the centroid of the rectangles selected, and the centroidal axis of the whole section. Units used are general and shown as #.

271

11/9

Table of first moments (about XX)

	b	d	bd	A	y	Ay
R1	60	5	300	300	30	9000
R2	5	55	275	275	0	0
Divide...				575		9000

Centroidal axis lies 15·65 # above XX.

Table of second moments (about CA)

	bd	d^2	bd^3	$bd^3/12$	y_C	y_C^2	Ay_C^2
R1	300	25	7 500	625	14·35	205·9	61 777
R2	275	3025	831 800	69 317	15·65	244·9	67 354
Total							129 131

Total $I_{CA} = 625 + 69\ 317 + 61\ 777 + 67\ 354 = \mathbf{199} \times \mathbf{10^3}\ \#^4$

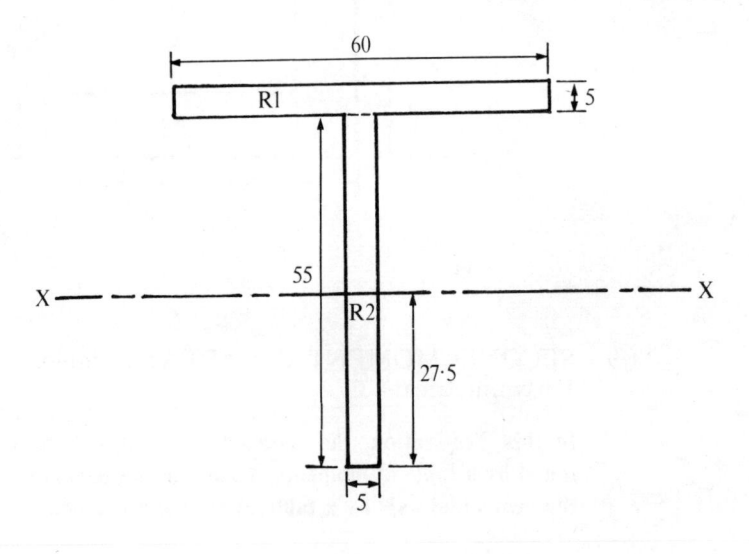

11/10

11/10 SECOND MOMENT OF AREA: Triangles and rectangle: Unsymmetrical

Find the second moment of area of the shape shown in the sketch, about its centroidal axis parallel to the base.

Notes:

The derivation here is taken using XX as the base line of the area. It would be quicker to use XX as passing through the centroid of the triangles or of the rectangle. Try both in the table of first moments. The results obtained should be the same.

Table of first moments (about XX)

	b	d	bd	A	y	Ay
T1	10	10	100	50	3·33	167
T2	10	10	100	50	3·33	167
R3	20	10	200	200	5·00	1000
Total				300		1333 (to nearest digit)

Centroidal axis is 1333/300 above the line of XX = 4·44 #

Table of second moments (about CA)

	y_C	y_C^2	Ay_C^2	bd^3	$bd^3/36$ (about TT)	$bd^3/12$ (about RR)
T1	1·11	1·23	61·5	10 000	278	—
T2	1·11	1·23	61·5	10 000	278	—
R3	0·56	0·31	62·0	20 000	—	1667

Total I_{CA} = Twice $(I + Ay_C^2)$ (for T1 and T2) + $(I + Ay_C^2)$ (for R3)
      = 2(+278 + 61·5) + (+1667 + 62)
$\mathbf{I_{CA}} = +\mathbf{2408}$ #⁴

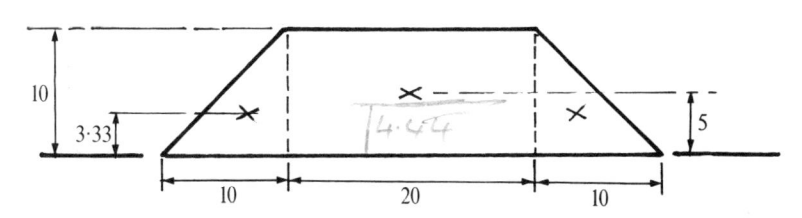

11/11

11/11 SECOND MOMENT OF AREA: Hexagon: Symmetrical

A hexagon has a side of 8 # in length. Find the second moment of area about a centroidal axis passing through two of the apex points.

Notes:

Two possible methods are shown below. The first of these is used here. Try the second for yourself, using T3 and T4 (identical) and R5. The axis XX is also the centroidal axis CA, so there is no need for a table of first moments.

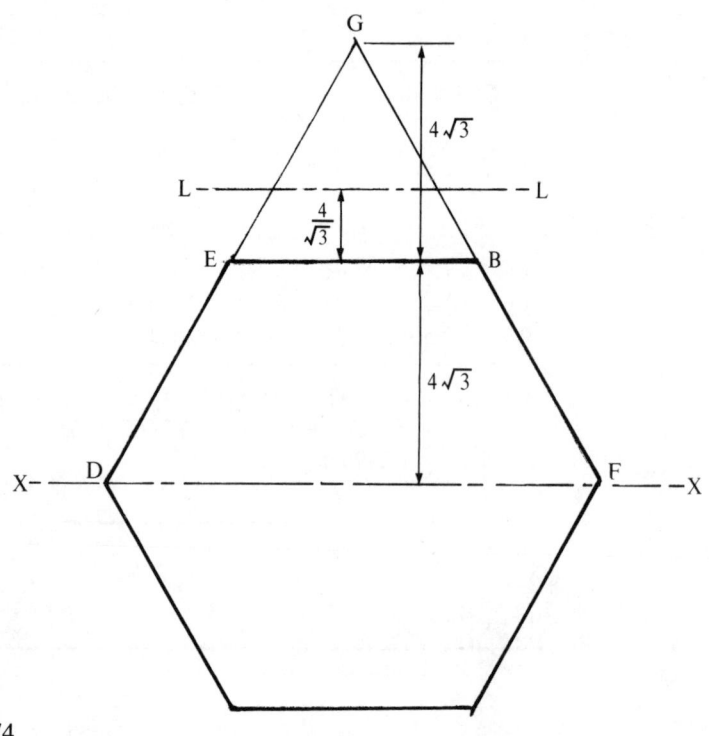

Table of second moments (about CA, or XX)

	b	d	bd	d^2	bd^3	A
T1 (GDF)	16 (DF)	$8\sqrt{3}$ (GO)	$128\sqrt{3}$	192	42 566	$64\sqrt{3}$
T2 (GEB)	8 (EB)	$4\sqrt{3}$ (GM)	$32\sqrt{3}$	48	2660	$16\sqrt{3}$

274

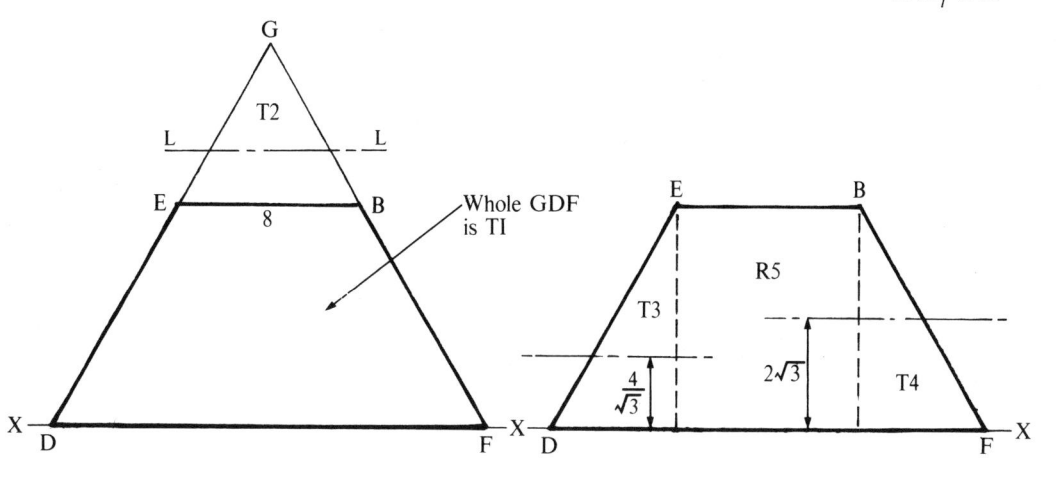

Table of second moments (continued)

	$bd^3/12$ (about CA)	$bd^3/36$ (about LL)	y_C	y_C^2	A	Ay_C^2
T1 (GDF)	3547	....	....	....	....	....
T2 (GEB)	—	74	$16/\sqrt{3}$ (PO)	256/3	$16\sqrt{3}$	2365

$$\text{Total } I_{CA} = I_{T1} - I_{T2} = \frac{bd^3}{12} \text{ (for TI)} - \left(\frac{bd^3}{36} + Ay_C^2\right) \text{ (for T2)}$$

$$= 3547 - 74 - 2365 = 1108 \; \#^4$$

For the **whole hexagon** which is twice this value: $\mathbf{I_{CA} = 2216} \; \#^4$

Know your geometry; enlarge table in Section 11·6 with values for other shapes.

11/12

11/12 SECOND MOMENT OF AREA: Triangle and semi-circle: Unsymmetrical

Find the second moment of area about a centroidal axis parallel to the line joining the triangle and the circle.
Second moment of area of triangle about its CA = $bd^3/36$
Second moment of area of semi-circle about CA = $0\cdot 11r^4$.

Table of first moments (about XX)

Take XX through centroid of triangle

	b	d	bd	r^2	A	y	Ay
T1	10	12	120		60	0	0
C2	—	—	—	25	39	6·1	238
Totals					99		238

Centroid of whole section lies 2·4 # below XX.

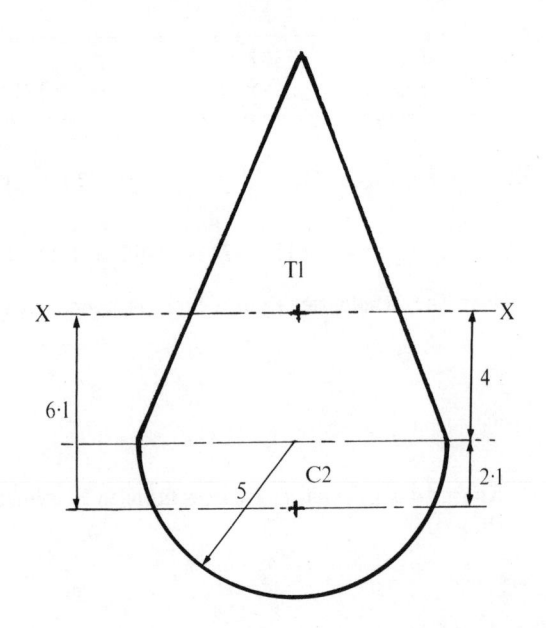

11/13

Table of second moments (about CA)

	d^2	bd	r^4	bd^3	I	y_C	y_C^2	Ay_C^2
T1	144	120		$17{\cdot}28 \times 10^3$	480	2·4	5·8	348
C2	—	—	625	—	$0{\cdot}11r^4 = 69$	3·7	13·7	533
Add totals:					549			881

$$I_{CA} = I + Ay_C^2 = 549 + 881 = 1430 \ \#^4$$

11/13 SECOND MOMENT OF AREA: Perforated rectangle: Symmetrical

Find the second moment of area of all three sections shown. The rectangle is punched by four holes. The second moment is required about CA.

	b	d	bd	d^2	bd^3	$bd^3/12$ or $bd^3/36$	r	$\pi r^4/4$
R1	40	60	2400	3600	8640×10^3	720×10^3		
R2	20	40	800	1600	1280×10^3	107×10^3		
C3							15	$39{\cdot}8 \times 10^3$
T4	15	18	270	324	$87{\cdot}5 \times 10^3$	$2{\cdot}4 \times 10^3$		

So long as the areas are symmetrical about CA, or have their centroids on CA (triangle) the second moment of area of the perforated section is the difference of that for the full section and that for the hole, as taken about their own centroidal axes. If the centroidal axes of the whole rectangle and of the hole do not coincide, then a further calculation must be made for Ay_C^2. This is shown in subsequent problems.

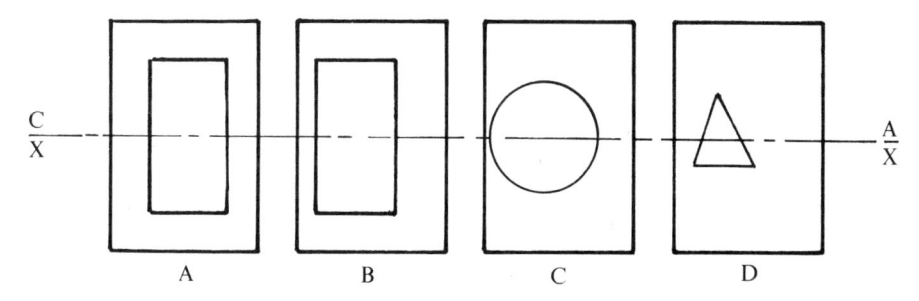

A B C D

11/14

Second moment of area of each of the four sections

Section	Second moment of R1	Second moment of hole	I_{CA}
A	720×10^3	107×10^3	613×10^3
B	720×10^3	107×10^3	613×10^3
C	720×10^3	39.8×10^3	680×10^3
D	720×10^3	2.4×10^3	718×10^3

All the *I* values are in linear units to the fourth power – $\#^4$.

11/14 **SECOND MOMENT OF AREA:** Notched rectangle: Unsymmetrical

Find the second moment of area of the section about its centroidal axis, and also about XX.

Note:

In an instance like this it shortens the work considerably to take the second moment about the base XX, since both the areas rest on this axis. The second moment of a rectangle about such an axis is $bd^3/3$, and the centroidal axis for the whole section is obtained by *subtracting* the value of Ay_C^2 since the *I* about the centroidal axis is less than about any other axis. Values of the area of the notch and its moments are negative.

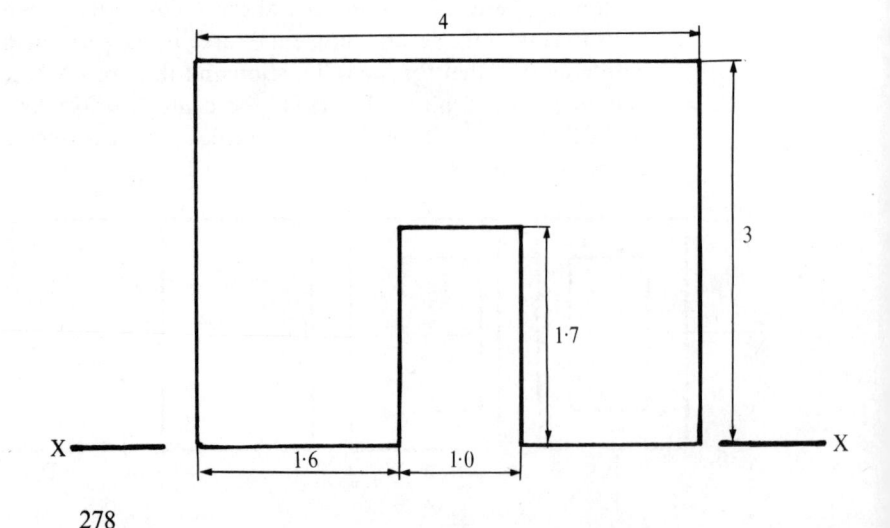

Table of first moments (about XX)

	b	d	A = bd	y	Ay	Ay²
R1(whole)	4·00	3·00	12·00	1·50	18·00	27·00
R2(notch)	1·00	1·70	1·70	0·85	1·45	1·23
Total			10·30		16·55	25·77

$$\frac{\sum Ay}{\sum A} = \frac{16·55}{10·30} = 1·61 \,\#$$

Table of second moments (about XX)

	bd^3	\multicolumn{3}{c}{For whole area}
		A	y_C	y_C^2
R1(whole)	108·00			
		10·3	1·61	2·58
R2(notch)	4·92			
Total	103·08			

$$I_{CA} = \frac{bd^3}{3} - Ay_C^2 \text{ (for whole area)} = \frac{103·1}{3} = 10·3 \times 2·58$$

$$= 34·4 - 26·6$$

$$I_{CA} = \mathbf{7·8} \,\#^4$$

A missing area is a negative area.

11/15

11/15 SECOND MOMENT OF AREA: Punched circular plate: Unsymmetrical

The centre of a circular plate of radius 13 mm is at O. The plate is perforated by a circular hole of radius r. If the centre of the hole is at 2·5 mm to the left of O and the centroid of the punched plate is at 2·5 mm to the right of O, find the radius of the hole and the second moment of area on two axes at right angles, both passing through the centroid.

Table of first moments (use CA as XX for moments.)

	Radius	A	y	Ay
Unperforated plate	13	169	2·5	422·5
Hole (negative)	r	$-\pi r^2$	5·0	$-5\pi r^2$
Total		$(169 - r^2)$		$(422·5 - 5r^2)$

$$\frac{Ay}{y} = \frac{422·5 - 5r^2}{169 - r^2} = \bar{y}$$

Since this value of $\bar{y}$ gives the distance from the chosen axis for moments (XX) to the centroidal axis (CA), in this instance, the value must be zero for XX and CA are one and the same.
Thus:

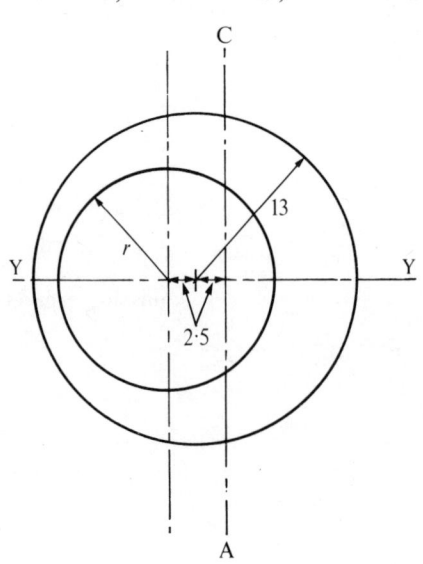

$$422·5 - 5r^2 = 0; \quad r^2 = 84·5; \quad r = 9·2 \text{ mm}$$

Area of hole:
$$\pi r^2 = 265 \text{ mm}^2$$

Table of second moments

	y_C	y_C^2	Ay_C^2	r^4	$\pi r^4/4$
Plate	2·5	6·25	1056	28 561	22 400
Hole	5·0	25·00	6625	7164	5620

$$I_{CA} = (1056 + 22\,400) - (6625 + 5620) = \mathbf{11\,211 \text{ mm}^4}$$

About YY

$$I_{YY} = \frac{\pi}{4}(R^4 - r^4) = \frac{\pi}{4}(28\,561 - 7164) = \mathbf{13\,300 \text{ mm}^4}$$

11/16 SECOND MOMENT OF AREA: Oval link: Symmetrical

Find the second moments of area about CA and NA. Since this link is symmetrical about both axes, no table of first moments is required.

Note:

There are two semi-circles at both ends. About CA, these may be combined into circles. About NA, only one of them need be considered.

Tables of second moments

About CA

	b	d	bd	d^2	bd^3	$bd^3/12$	r	$\pi r^4/4$
C1	—	—	—	—	—	—	26	$0·36 \times 10^6$
C2	—	—	—	—	—	—	19	$0·10 \times 10^6$
R3	40	52	2080	2704	$5·6 \times 10^6$	$0·47 \times 10^6$	—	—
R4	40	38	1520	1444	$2·2 \times 10^6$	$0·18 \times 10^6$	—	—

$$I_{CA} = (0·47 - 0·18 \quad 0·36 - 0·10) \times 10^6 = \mathbf{0·55 \times 10^6 \text{ mm}^4}$$

11/16

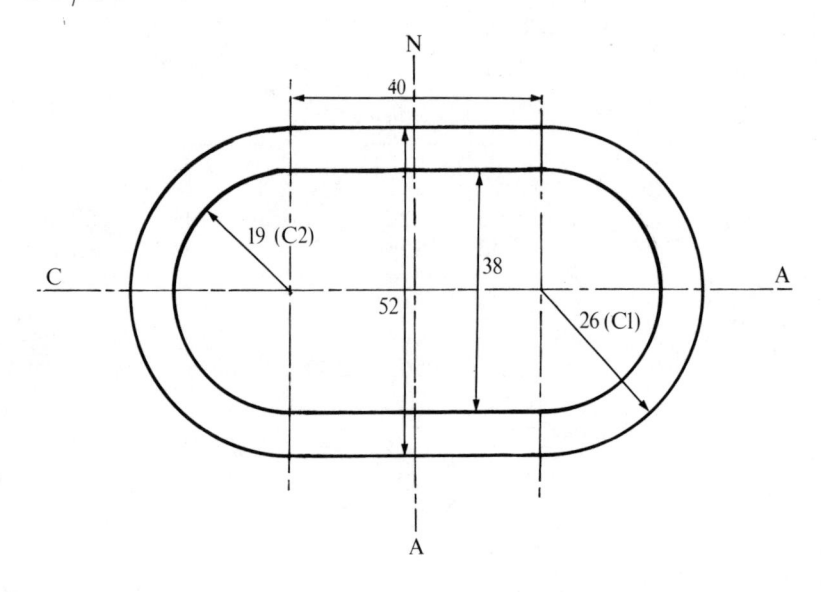

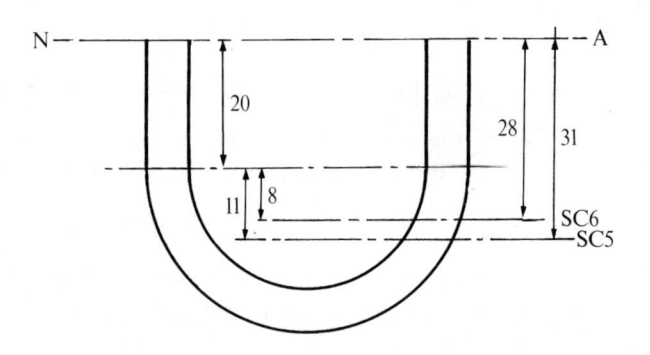

About NA $0.425R = 0.425 \times 26 = 11 \#$; $0.425r = 0.425 \times 19 = 8 \#$

	b	d	bd	d^2	bd^3	$bd^3/12$	r	$0.11r^4$
SC5	—	—	—	—	—	—	26	0.014×10^6
SC6	—	—	—	—	—	—	19	0.14×10^6
R7	52	20	1040	400	0.42×10^6	0.03×10^6	—	—
R8	38	20	760	400	0.30×10^6	0.025×10^6	—	—

	A	y_C	y_C^2	Ay_C^2
SC5	1062	31	961	1.021×10^6
SC6	567	28	784	0.445×10^6
R7	1040	10	100	0.104×10^6
R8	760	10	100	0.076×10^6

I_{NA} for half section = $(0.046 \ 0.104 - 0.076 \ 1.021 - 0.445)10^6$

I_{NA} for whole section is twice this: $\mathbf{1.3 \times 10^6} \; \#^4$

11/17 SECTION MODULUS: Truncated square: Symmetrical

Find the extent to which a square section on its diagonal (as shown in the sketch) must be truncated in order to obtain the maximum section modulus.

Notes:

The section modulus is the second moment of area divided by the distance from the CA to the outer fibres (distance t, in the sketch). The height t is unknown, and the problem is best solved by a general approach. ($l/2 = (H - x)$ since triangle has 45° slopes).

The area of the section is:

$$2 \int_0^h l\,dx = 2 \int_0^h 2(H - x)\,dx$$

The *first moment* of the section is:

$$2 \int_0^h lx\,dx = 2 \int_0^h 2(H - x)x\,dx$$

The *second moment* of the section is:

$$4 \int_0^h (H - x)x^2\,dx$$

The *second moment* divided by the half depth is:

$$z = 4 \int_0^h \frac{(H - x)x^2}{h}\,dx$$

283

If this, the *section modulus* is to be a maximum, then:

$$\frac{dz}{dh} = 0; \qquad h = \frac{8}{9}H$$

Thus to achieve a maximum value of the section modulus, the square on its diagonal must not be left complete, but reduced in depth by one-ninth.

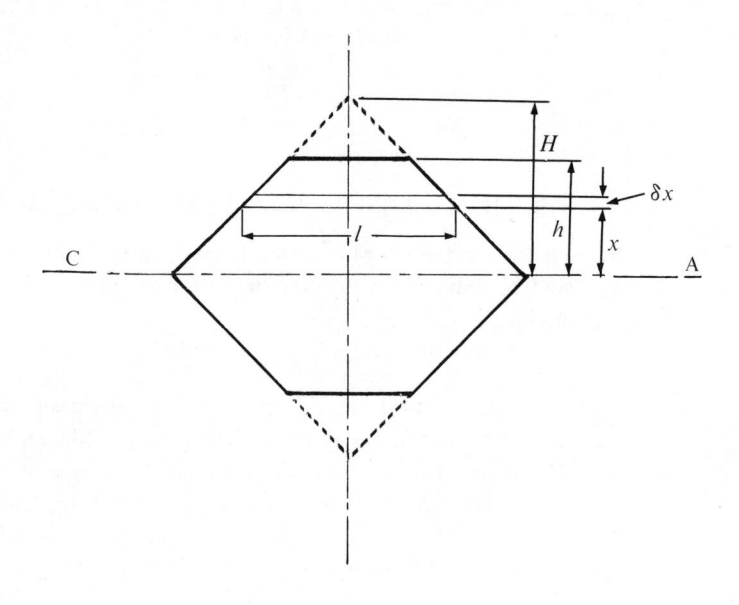

Don't irritate an examiner by confused expression of your ideas.

Part II: Group D

12 Stress, deformation and energy

The first part of this book teaches the evaluation of the effects of forces and couples on *rigid structures* in equilibrium. The fiction of rigidity was convenient for the time being, and introduced no significant error into the calculation for planar structures under the actions of forces and couples. The convention of thinking of each structure as being paper-thin, and contained in one plane along with its loading was also useful for a first attack on these problems. Part I sets the scene for the more detailed work of Part II.

In Part II the bodies or structures are not considered to be rigid, but to be deformable; two dimensions are no longer enough, and forces and couples may not be co-planar with the structure if the third dimension enters into the problem. There is a distinct break, therefore between the conceptions of *co-planar statics* or *elementary theory of structures* (much the same subject with different names) and the second topic of *strength of materials*. These titles, although hallowed by long use, are not correctly descriptive, nor do they underline the fundamental difference between the subjects. Be sure to remember therefore, that Part I deals with *rigid structures*, and Part II with *deformable structures*. Part II deals with the real effects of the forces and couples defined in Part I.

12.1 The three effects

What are the effects of forces and couples acting on a deformable body? They had no effect at all on the fictitious rigid body, but if the body is allowed to take up its natural shape under the loading, three things happen:

1. *Deformation:* By the title *deformable body* it is clear that the body changes its shape. If you apply couples to the end of a rule it bends. It is not a rigid body because it is made of real materials and has a real cross-section which is impressionable to forces and moments applied.

2. *Stress:* This is more difficult to quantify. It is the distress experienced by the material as the applied forces and couples cause it to deform. Stress is what your arm experiences when you carry a heavy suitcase. Stress is visible under the microscope, or, when the deformation is large, the distress can be observed directly. Because of the difficulty of definition, stress is usually defined by what causes it, rather than by what it is. You can guess that it is not force alone which causes stress, but rather the intensity of force. A suitcase which causes distress in a weakling's arm, would cause none in that of a weight-lifter. His muscles have a larger cross-sectional area and the force is distributed more widely so that the muscle fibres do not experience much stress. Stress, therefore, is defined by the *intensity of loading*, or the force per unit of area (N/m^2).

3. *Storage of energy:* This third effect is the result of the other two. We assume that the material, under the applied loading, remains elastic – it returns to its original size when the load is released. Energy must be applied to deform and stress the structure, and energy is given out when the deformation and stress are released. The best way to illustrate this is to think of a clock spring. This is really a very long, very flexible cantilever, clamped at one end and having a couple continuously applied to the other end (when you wind up the clock). This couple produces a great deal of deformation. If the couple were suddenly withdrawn, you are quite well aware that there would be a great release of energy with the spring unwinding instantly. The mechanism of the clock is to ensure that the energy supplied when the deformation is cancelled is released gradually.

12.2 The elastic condition

In Part I, there were a number of conditions within which you had to work. In order to obtain the results required, you were obliged to impose artificial conditions and conceptions. Now, in Part II, these artificially-imposed restrictions are not necessary; you have released the structure so that it behaves in a natural manner. There is, however, one condition which must be imposed. When a material is loaded it resists the load easily for a time, and when the load is removed, the material returns to its original state. If the load is increased too much, the material gives up the struggle to resist and becomes plastic like a piece of putty. You are not concerned with this second state, and all you do in this Part II must be within the first state. The condition therefore, and the only one you need bother with, is as follows:

> In all problems in deformable structures, the materials of which the structure is composed, remain elastic under the loading. When the load (externally applied forces and couples) is released, the structure reverts to its original unloaded shape, and there is no permanent deformation or damage.

12.3 Three origins

The studies of Part I have shown you that there are three distinct ways in which applied forces and couples act on a structure. These are usually combined in any one structure under load, but the results of their action are best studied separately in the first instance. The three phenomena are the application of

1 *Direct force* (tension of compression)

2 *Shearing force*

3 *The opposing couples which produce bending*

Each of these three (and a fourth which is a subdivision of shearing) produce *deformation, stress* and the *storage of energy*. As the last of these phenomena is more easily discussed in general terms, it is studied in a Unit of its own. We are left, therefore, in the next five Units of Study with the need to evaluate stress and deformation by the three types of loading listed.

UNIT 13 considers the *stress and deformation* caused by *direct force*. This is relatively simple and both can be taken together.

UNIT 14 considers the *stress* caused by bending only. The *deformation* caused by bending only, is a wide subject and is taken separately.

UNIT 15 considers the *stress* caused by transverse shear – this is the shearing force S determined so often in Part I. It acts, as all shear does, across the section cut at right angles to the tangent of the member of the structure.

These three provide values of the stresses caused by *direct force, bending*, and *transverse shear*, together with the deformation caused by *direct force*.

12.4

Torsion: There is, however, another type of shearing stress which must be brought into the study. It has not appeared in Part I for its occurrence depends on the third dimension, and the stipulation of Part I that all structures and loading must be co-planar, eliminated any possibility of considering torsion or twisting. Hold a circular pencil at each end and apply couples in opposite directions to apply a twist. Various stresses are produced within the rod, but you are particularly interested in the shearing stress across the circular section of the pencil. This shearing stress depends on the couples, which are acting round an axis up the centre of the rod (three dimensions), and on the size of the circular cross-section. When members whose cross-sections are not circular, are twisted or submitted to torsion, the problem becomes difficult and outside the scope of this book. You will consider, therefore, only members of circular cross-section under torsion.

UNIT 16 considers the *stress and deformation* caused by *torsion*. This is a variation of stress produced by transverse shear, and the stress is thus a shearing stress.

UNIT 17 considers the rather complex problem of how a structure *deforms under bending*.

These five Units cover the simpler conditions of loading. You have probably noticed that nothing has been said of the deformation produced by transverse shear. This *can* be calculated but is so small as to be negligible in comparison with the great deformations produced by bending couples, usually acting at the same time as transverse shear. Accordingly, you may, for the time of reading this book, forget about deformation due to transverse shearing force.

12.4 Fundamental relationships

Although all these four types of loading – *direct force, bending couples, transverse shear* and *torsional couples* – produce different types of stress and deformation, they have effects which are relatively of greater or less importance in design; they show common fundamental relationships, making it easier to understand the methods of solving the numerical problems which are the subject of this book. There are three groups of quantities which, for any type of loading, are mutually equal. They are:

12.4

$$\begin{pmatrix} \text{A term} \\ \text{concerned} \\ \text{with the} \\ \text{loading} \\ \text{applied} \end{pmatrix} = \begin{pmatrix} \text{A term} \\ \text{relating} \\ \text{to the} \\ \text{stress} \\ \text{produced} \end{pmatrix} = \begin{pmatrix} \text{A term} \\ \text{relating} \\ \text{to the} \\ \text{deformation} \\ \text{produced} \end{pmatrix}$$

These relationships can be further specified as:

$$\begin{pmatrix} \text{The loading} \\ \text{divided by a} \\ \text{property of the} \\ \text{cross-sectional} \\ \text{area} \end{pmatrix} = \begin{pmatrix} \text{The stress} \\ \text{divided by a} \\ \text{dimension} \\ \text{of the} \\ \text{cross-section} \end{pmatrix} = \begin{pmatrix} \text{A modulus} \\ \text{of the material} \\ \text{multiplied by a} \\ \text{ratio representing} \\ \text{the deformation} \\ \text{produced} \end{pmatrix}$$

As a record of these relationships for all the four conditions considered, the mathematical expressions are given below together with an explanation of the symbols. You will not understand these at present, but will continually refer to them in solving problems. What is important at this first reading is to realize that all the problems of stress and deformation can be solved by similar routines, the blocks of terms being similar.

Direct Force: $\quad \dfrac{F}{A} = \sigma = \dfrac{Ed}{l}$

Bending: $\quad \dfrac{M}{I} = \dfrac{\sigma}{h} = \dfrac{E}{R}$

Transverse Shear: $\quad \dfrac{S}{I} = \dfrac{\tau t}{m}$

Torsion: $\quad \dfrac{T}{J} = \dfrac{\tau}{r} = \dfrac{C\theta}{l}$

There is no third term in transverse shear because the deformation is negligible.

Explanation of the terms:

Direct force

F is the direct force, or compression, applied through the

12.4

    centroid of the cross-section or uniformly over its surface (N, kN, MN);

A is the cross-sectional area of the cross-section (m^2);

σ is the *uniform* direct stress produced over the cross-section, *sigma* (N/m^2);

E is the *modulus of elasticity* – a property of the material of which the member is made (N/m^2);

d/l is the ratio of deformation to total length. This is called the *strain*. This ratio is dimensionless.

Bending

M is the bending moment applied at the point in the beam under consideration (N m);

I is the second moment of area of the cross-section about a centroidal axis (m^4);

σ is stress produced in the beam. As this stress varies across the cross-section, and is *not uniform* as for direct stress, *sigma* is usually calculated as the maximum stress (which occurs on the upper and lower faces of the beam, one in compression and the other in tension). The stress produced is similar to a direct stress and is given the same symbol (N/m^2);

h is the distance outward from the centroidal axis. This is usually taken to the upper or lower face of the beam. If the cross-section is unsymmetrical about the CA there are two values of this extreme value of h (m);

E is the *modulus of elasticity*, a property of the material (N/m^2);

l/R is the inverse of the radius to which the beam is bent by the bending couples. This radius is not necessarily constant along the beam, and the figure obtained refers only to the point at which the measurement of bending moment has been made. The reciprocal of the radius can be referred to as the curvature (m^{-1}).

Transverse shear:

S is the transverse shearing force acting at the section considered;

12.4

- *I* is the *second moment of area* of the cross-section about a centroidal axis at right angles to the direction of shearing force;

- τ is the shearing stress produced; the letter is the Greek *tau* (pronounced to rhyme with 'cow') (N m^{-2});

- m/t is in terms of an area, but is a complex conception not worth describing at this stage. It is fully worked out in Unit 15 (m^2)

Torque:

- *T* is the torque of a pair of couples applied in the plane of the cross-section, at right angles to the axis of the bar. They produce a twist in the bar (N m);

- *J* is the *polar second moment of area* of the circular cross-section. The relationships in Unit 16 refer only to circular cross-sections (m^4);

- τ is the same shearing stress as in *transverse shear*, above. It occurs at a radius *r* from the axis of the bar (N/m^2);

- *C* is the *modulus of rigidity*, a property of the material of which the member is made, which denotes its resistance to shearing stress (N/m^2);

- θ/l is the rate of twist or the intensity of twist. It is the angle through which one unit of length of the rod distorts under the torque applied. The term *l* is the length of the rod, and θ is the angle through which the whole rod turns. θ is usually very small, and can be measured in radians or as a tangent value.

Notes on the symbols:

1. In giving dimensions, the values of *newtons* and *metres* have been used, but other variations are, of course, possible: mm, kN/m^2, N/mm^2, etc.
2. The values of stress (which varies across a cross-section) can be obtained at any distance from the neutral axis but you are not usually interested in any but the maximum values. Thus, unless something is said to the contrary, problems all concentrate on the maximum values of stress.

12.5

12.5 Mnemonics or G-U-I-D-E-S

In Part I you used the word S-U-R-E-T-Y as a means of reminding you of the procedure, and the steps to solution. In Part II the problems are relatively simpler, for most of them require only the substitution of values in the expressions discussed earlier in this unit. The substitution, however, must be done with knowledge and carried out accurately. Six steps again bring you to the final answer. The mnemonic is G-U-I-D-E-S. The steps referring to each of these letters are:

G Decide on the *Group* of expressions, from the four discussed earlier and write them down. Often, the whole three terms of the group will not be required, but write down all three each time. Such repetition ensures recall when you need the group of terms next time.

U Problems are often presented in such a tangled way, that it pays dividends to extract, from the jungle of data and questions, what is asked for, and to write down the *Unknowns* clearly. The less time you take to disentangle information and unknowns from each other, the more efficient you will become.

This letter can also be used for *Units*. Consistency in units is very important, and a short statement noting which units you are to use in the question makes sure that you will not make a mistake.

I An *Inventory of Information* of the data you have been explicitly given comes next. In this block of work, always change the data as it appears in the examination questions into the units you have selected. Never go on unless all information has been transformed to the correct units.

D *Derived Data* refers to those parts of the group of terms which have not been given as explicit pieces of information, but which must be calculated from other data not required for substitution in the terms of the expression being used. This is usually where the flavour or twist to the problem occurs. You may have to argue conditions in order to appreciate the direction the problem takes. *Second moment* and *polar second moment of area* are usually calculated at this stage also, so be sure you have absorbed the techniques of Unit 11.

E Only now can you go ahead and *Evaluate* the terms of the expressions you wrote down under G. This is usually a short piece of work, for you

12.5

merely pick up from I and from D all the numerical values of the necessary terms in the expression. They have already been transformed to consistent units, so you take them as they stand and substitute in the group of terms. The Unknowns show up clearly and can be evaluated. NEVER try to carry out this stage until all terms are in the correct units. It can be done, and often is shorter, but the danger of making a mistake is great, and the derivation of the Unknowns is better understood if all conversion to consistent units has already been accomplished.

S Now write down the *Solutions* to the specific questions asked. Never go beyond what you have been asked; volunteered information scores no marks, but a clear statement at the end helps the examiner and puts him in a favourable frame of mind!

Only repeated solution of Problems will give the facility required.

Complete tabular work column by column.

Use the INDEX.

RECAPITULATION

There are two Units of Study in Group D. Unit 11 deals with the geometry, not of the outline of a structure as in Unit 7, but with the geometry of the cross sections of parts of the structure – the parts which are stressed and deformed. Be sure that you can carry out the work in Unit 11 without stumbling. The tabular method of writing down results allows you to do one step at a time (one column) so that you do not require to hold operations in your mind. If you are interrupted, you can come back and see exactly at what point in the Table you left.

The First, Second and Polar Second moments of areas are indicative of the resistance to types of manipulation to which the cross section may be submitted. Juggle with as many problems as possible, from this and from other textbooks until obtaining a Second Moment is second nature!

Define the three ways in which forces and couples can act on a structural member. What three effects do they produce? If you do not know the answers to these questions exactly and without doubt, then do not proceed further until you do.

What condition is implicit in all that is done in Part II?

The Fundamental Relationships given in 12.4, have common forms. Study these forms and know them in detail. Be able to change them into various expressions depending on which terms are known and which are unknown.

In Part I the mnemonic was SURETY. In Part II it is GUIDES. Be certain of the significance of these letters and do not depart from the routines they bring. These routines, in Part I and Part II take the drudgery out of the solutions, and make it more likely that you will get a correct result.

GROUP E
Stresses and Deformations

Part II: Group E

13 Stress and deformation caused by direct force

In this Unit you must refer back to Unit 12 and be sure that you understand the meaning of the symbols concerned with direct force:

F, A, σ, E, d, l

These are all defined in Unit 12 and their meanings must be memorized.

13.1 Direct stress and deformation

The relationships are:
Term referring to applied loading = Term referring to stress
= Term referring to deformation

$$\frac{F}{A} = \sigma = \frac{Ed}{l}$$

The direct force F which is applied to the end of a rod in tension or compression – as in the members of a pin-jointed frame – is assumed to be acting through the centroid of the section, so that the stress produced (σ) is uniform and is the same at any part of the cross-section (F/A). You may say that the tensile or compressive force applied may well be at some distance from the centroid and apply more stress to one part of the rod than to the others. This is perfectly true, and this arrangement is considered in Unit of Study 18. In this present Unit 13, the stress produced in the rod or column or tie-bar is uniform across the section. To increase stress, therefore, either F must be increased or A decreased, since stress is force divided by area.

The determination of the stress produced by direct force is rather simpler than the determination of the deformation. Deformation under a known direct force depends on two quantities – the *modulus of elasticity* (E) and the length (l) of the rod. The modulus is a measure of the deformability of the material of which the rod is composed; the higher the modulus, the more difficult it is to deform the material. A high modulus shows a stiffer material. Some typical values of E are given in the problems. You ought to extract these and note them for future work. E is measured in the same terms as stress (N/mm^2, kN/m^2, etc.).

All of the three terms of the relationship given above are equal, so if F and A are fixed then the stress is fixed, and the term Ed/l must also be equal to the other two terms. The deformation of the material, measured by d, can be altered only by changing the material (E) or by changing the length of the rod or tie-bar. Suppose you wish the deformation d to be doubled. The whole term Ed/l must not alter, but must remain equal to F/A (assumed fixed in value). Thus doubling of the value of d means that the E/l portion of the term must be halved in order to keep Ed/l constant. This can be done by reducing the value of E (choosing a more easily deformed material) and increasing the length of the rod until E/l is half of what it was before. Of course this can be done by altering only one of the two (E or l). Note that all this 'juggling' with quantities, in order to control the deformation, can go on without having any effect on the stress produced by the loading. The stress remains the same, so long as F and A are unaltered, regardless of the type of material used, or of the length of the loaded member.

Try similar jugglings until you really feel what is going on in a bar pulled or pushed by a direct force. What must take place, for example, if the deformation (d) is to remain constant, but the force (F) must be increased? What changes would you make in the material, area or length of bar? Or, suppose that the bar has to be changed to a more deformable material, the length remaining unaltered, but that the cross-sectional area must be reduced (A). What happens to the deformation? Does it increase or decrease?

13.2 Longitudinal strain

In common parlance you may speak of 'the stress and strain of modern life' using the terms 'stress' and 'strain' to mean very nearly the same thing. In the study of material under loading, however, *strain* and *stress* are quite different. They must be clearly defined in your mind, and the terms must never be used loosely, but only with specific intent. You have already defined *stress* as the *intensity of force*. Similarly, *strain* might be defined as the *intensity of deformation*. Strain is the proportion of the original size which represents the deformation. If a bar 100 mm long shortens by 1 mm, the strain is the ratio of the deformation to the original length and is 0·01. No units are shown, since strain is a ratio. When the original size is a length, then the strain is a longitudinal or a

13.3

lateral strain. Longitudinal strain occurs in the direction of the applied direct force, and lateral strain at right angles to this direction.

Longitudinal strain is a useful quantity, for it allows you to discuss the intensity or scale of the deformation without discussing its real value in mm, or knowing the length of the rod which is under the load. Strain, since it is useful, is given a special symbol: e.

Longitudinal strain is given the symbol e_l
So,
$$e_l = d/l$$
and thus
$$\sigma = Ee_l$$
or, if you like, the characteristic of the material which defines its natural extensibility, is equal to the ratio (stress/strain).

13.3 Lateral strain

If you pull a piece of rubber or thick elastic you will see that the longitudinal extension produced, is accompanied by a decrease in the thickness of the material – there is a lateral deformation and, therefore, a lateral strain. This type of decrease also occurs in the elastic material discussed in Part II of the book – steel, copper, aluminium, etc. It is more convenient to talk in terms of strain rather than of deformation. The ratio between the lateral and the longitudinal strain is known for the materials generally in use. It is a ratio and is called *Poisson's ratio*.

For most of the materials we deal with, PR varies from one-third to one-quarter. This means, for example, that if the longitudinal deformation of a rod 12 m long were to be 1·44 mm (a strain of 0·000 12), the lateral strain with a PR of one-quarter, would be 0·000 03. So, if the bar is of square cross-section of, say 60-mm square, the lateral deformation in both directions would be 0·000 03 × 60 mm, or 0·018 mm. If the original force (F) is tension, the lateral dimension would decrease, and become 59·9982 mm in each direction. If the loading is compressive the lateral dimension would increase and become 60·0018 mm in each direction.

Note that the lateral deformation is always of the opposite sense to the longitudinal, and the lateral strain of opposite sense to the longitudinal strain. If the bar lengthens under tension, its lateral dimensions decrease.

$$\text{Poisson's ratio} = v = e_t/e_l$$

where e_t is the strain on the thickness or breadth of the cross-section. Sometimes both e_t and e_b are used to denote the lateral strains in the two lateral directions at right angles to each other. (ν is a Greek letter and is pronounced 'new'.)

13.4 Change in volume

Since, under stress, the longitudinal and lateral dimensions all alter, the volume of the original rod may also alter. If, for example, we think of the stressed bar as being of rectangular cross-section, its volume is represented by the product of length, breadth and thickness ($V_0 = lbt$). When the bar is stressed longitudinally, its length alters to $(l + e_l l)$ if the force is a tensile one, and $(l - e_l l)$ if the force is compressive. Let us assume a tensile force. The following arguments apply in both instances. The breadth decreases if the length increases and the ratio of strains depends on PR. The new breadth, therefore, is $(b - be_l \nu)$ and the new thickness is $(t - te_l \nu)$. Let $e_l \nu = e_b$ and also e_t. The new volume, therefore is:

$$V_N = lbt(1 + e_l)(1 - e_b)(1 - e_t)$$

Multiplying out and discarding second order products such as $e_b \times e_t$,

$$V_N = lbt + lbt(e_l - e_b - e_t) = V_0 - V_0(e_l - e_b - e_t)$$

The change in volume is, therefore

$$\delta V_0 = lbt(e_l - e_b - e_t)$$

and the *volumetric strain* or

$$\delta V_0/V_0 = e_l - e_b - e_t$$

The above refers to the change in volume when there is one longitudinal force. There may also be forces and pressures applied laterally and simultaneously with longitudinal force, as, for example, in the triaxial testing of materials. These lateral pressures, then, have similar effects in the two other co-ordinate directions as those described above. All the effects can be superimposed to obtain the final deformations in the three directions and the volumetric strain, or the change in volume divided by the original volume.

If the pressure on all faces of a block of material is the same – as under hydrostatic pressure – a special case arises, for the pressure on each of the six faces produces longitudinal and lateral strains in the three directions. There are, therefore, nine linear strains to be considered. It is much more convenient to use the conception of a *bulk modulus* which relates the *bulk strain* or *volumetric strain* ($\delta V/V$) to the bulk or hydrostatic, all-round pressure or stress σ_B.

13.5

As before:
    Stress   = Modulus × Strain
    Longitudinal stress = $E \times e_l$
    Hydrostatic stress = Bulk modulus × Bulk strain
                                = $K \times (\delta V/V)$

There is, fortunately, a relationship between the *modulus of elasticity* of a material and its *bulk modulus,* so the value of K can be found if E and ν are known:

$$K = \frac{E}{3(1 - 2\nu)}$$

13.5 Controlled or restrained strain

When F is applied and produces a stress, there is often no opposition to the deformation which develops in accordance with the terms of the expressions for load, stress, and deformation. Sometimes, however, as with lateral pressures combined with longitudinal force, one movement or change in shape may be cancelled to a greater or less degree by a deformation imposed by a second force or stress in another plane. There are also situations of loading when one of the deformations which would normally take place is physically prevented or controlled wholly or partially. These circumstances often form the subject of examination questions, and are also frequently found in constructional work. They can usually be divided into three categories, which, of course, may also overlap:

(*a*) Shared loads on members of the same material with some imposed physical condition (Problems 13/6, 13/7).

(*b*) Shared loads on members of different materials with some imposed physical condition (Problems 13/8, 13/9, 13/10).

(*c*) The prevention or partial prevention of a deformation which would have taken place under unrestrained conditions (e.g. under change of temperature). Such prevention gives rise to stress (Problems 13/11, 13/12, 13/13).

You have, therefore, in the study of stress and deformation caused by direct force, the following types of problem:
    Longitudinal force: no restraint.
    Longitudinal force taking account of PR: no restraint.

13.5

Longitudinal and lateral forces, taking account of PR.

Shared loads on compound members, perhaps of different materials.

Loads induced by change in temperature, with imposed conditions.

The first two of these are unrestrained, but the other three have some condition of restraint of strain.

NOW refer to Unit 12 to refresh your memory on the G-U-I-D-E-S.

Some Problems, as printed are not entirely complete: fill in the gaps.

Memorize the Groups of Terms (Sect. 12.4) and know their significance.

Make sketches for every Problem, whether given in the text or not.

301

13/1

13.6 THE PROBLEMS

13/1 COMPRESSION: No restraints

A circular cast-iron column, 125 mm diameter and 3·03 m long is loaded with an axial compressive force of 996 kN. What is the stress in the column and how much does it shorten if the modulus of elasticity for cast-iron is 120×10^3 N/mm²?

G *Group of Terms:*

$$\frac{F}{A} = \sigma = \frac{Ed}{l}$$

U *Unknowns and Units:*
1. Deformation of the column.
2. Stress in the column.
Use N and mm.

I *Inventory of Information:*
$F = 996$ kN $= 996 \times 10^3$ N $A = (62·5)^2 \pi$ mm²
$l = 3·03$ m $= 3030$ mm $E = 120 \times 10^3$ N/mm²

D *Derived data:*
None; the column is free to contract under load and there are no restricting conditions.

E *Evaluation:*

$$\frac{F}{A} = \sigma = \frac{Ed}{l}$$

$$\frac{9996 \times 10^3}{(62·5)^2 \pi} = \sigma = \frac{120 \times 10^3 d}{3030}$$

S *Solution:*
From a solution of the above, the answers required are:
1. Deformation of the column: **2·05 mm**
2. Stress in the column: **81·2 N/mm²**

13/2

13/2 TENSION: No restraints

A long steel wire (E: 207×10^3 N/mm^2) operates a signal. It is actuated by a pull at one end which causes a movement at the other end. If the signal wire is 610 m long and 5 mm diameter, find the stress in the wire and the movement required at the end remote from the signal, if the signal requires a movement of 180 mm to operate it. The force applied in the signal box is 1780 N.

G *Groups of Terms:*
$$\frac{F}{A} = \sigma = \frac{Ed}{l}$$

U *Unknowns and Units:*
1. Movement required at end remote from signal.
2. Stress in the wire.
Use N and m.

I *Inventory of Information:*
$F = 1780$ N $\qquad A = (2\cdot5)^2\pi$ mm$^2 = (2\cdot5)^2\pi \times 10^{-6}$ m^2
$E = 207 \times 10^3$ N/mm$^2 = 207 \times 10^9$ N/m^2 $\qquad l = 610$ m

D *Derived Data:*
If the wire did not extend under the tension of 1780 N, the movement at the signal box would be the same as that at the signal – 180 mm or 0·18 m. But the wire does extend, so the movement at the signal box is 0·18 m plus the extension of 610 m of wire.

E *Evaluation:*
$$\frac{F}{A} = \sigma = \frac{Ed}{l}$$

$$\frac{F}{A} = \sigma \quad \frac{1780}{6\cdot25\pi \times 10^{-6}} \quad 94\cdot5 \times 10^6 \text{ N/m}^2$$

$$d = \frac{\sigma l}{E} \quad \frac{94\cdot5 \times 10^6 \times 610}{207 \times 10^9} \quad 0\cdot278 \text{ m}$$

S *Solution:* Answers required are:
1. Movement required is $+0\cdot180 + 0\cdot278$ $\qquad = +0\cdot458$ m.
2. Stress in the wire is $\qquad\qquad\qquad\qquad\qquad\qquad$ 94·5 N/mm^2.

13/3

13/3 CYLINDRICAL BOILER: Internal pressure: No restraints

A cylindrical boiler is 1852 mm external diameter and 1820 mm internal diameter. The safe working stress in the boiler plate is 69 MN/m² and the efficiency of the riveted joints is 75%. What is the safe working pressure in the boiler.

G *Group of Terms:*

$$\frac{F}{A} = \sigma$$

U *Unknowns and Units:*
1. The safe working pressure in the boiler. Use N and mm.

I *Inventory of Information:*
External diameter = 1852 mm
Thickness of plate = 16 mm   Total thickness = 32 mm
Safe working stress in plate = 69 N/mm²
Efficiency of riveted joints 75%.

D *Derived Data:*
Since the joints develop only 75% of the strength of the plate, the safe working stress must be reduced to 75% of 69 N/mm² = 51·8 N/mm².

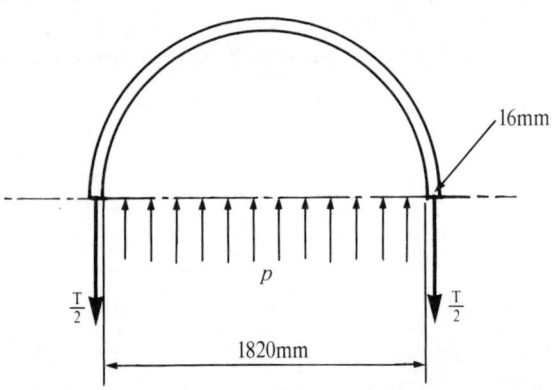

E *Evaluation:*
At a cross-section of the boiler across a diameter, the force trying to tear the boiler apart over one millimetre of its length is
  = pressure within the boiler × 1820 mm (N)

13/4

The resistance to this is the safe working stress in the plates, multiplied by the total area of plate cut by the cross-section:
$$= 51{\cdot}8 \times 32 \text{ N}$$
Thus unknown p is found by:
$$+p \times 1820 - 32 \times 51{\cdot}8 = 0$$

S *Solution:*
The safe working pressure within the boiler $\quad +\mathbf{p} = +\mathbf{0{\cdot}91}$ **N/mm²**.

13/4 TENSION: Varying stress and load: No restraints

A wire rope is of uniform circular section and 304 m long. It hangs freely down a pit shaft, supported on a rigid beam by its upper end. The cross-sectional area of the rope is 100 mm². What is the extension of the rope if the density of its material is 7700 kg/m³? Find the stress at the top of the rope, and the extension of the upper half. E: 207×10^3 N/mm².

G *Groups of Terms:*
$$\frac{F}{A} = \sigma = \frac{Ed}{l}$$

U *Unknowns and Units:*
1. Extension of whole rope.
2. Extension of half rope at upper end.
3. Stress at top of the rope.

Use N and m.

I *Inventory of Information:*
$l = 304$ m $\qquad A = 100$ mm² $= 100 \times 10^{-6}$ m²
$E = 207 \times 10^3$ N/mm² $= 207 \times 10^9$ N/m²

D *Derived Data:*
Volume of rope per metre of length: 100×10^{-6} m³
Mass of this rope/m of length: $7700 \times 100 \times 10^{-6}$ kg
Weight of one metre of length:
$$= 9{\cdot}81 \times 7700 \times 100 \times 10^{-6} \text{ N}$$
$$= 7{\cdot}56 \text{ N}$$
Weight of whole rope of 304 m $= 2296$ N
Weight of half rope of 152 m $= 1148$ N

13/4

E *Evaluation:*
Stress at the top

$$\frac{F}{A} = \sigma = \frac{2296}{100} = 22 \cdot 96 \text{ N/mm}^2$$

Extension of whole rope:
The force exerted on the rope at the top is 2296 N, and at the bottom, zero. Thus the mean force exerted is 1148 N (N and m).

$$\frac{F}{100 \times 10^{-6}} = \frac{207 \times 10^9 \, d}{304}$$

$$d = \frac{1148 \times 304}{100 \times 207 \times 10^3} = 0 \cdot 0169 \text{ m} = 16 \cdot 9 \text{ mm}$$

Extension of the top half of the rope:
The extension of the upper half of the rope is represented by the downward movement of the middle point of the rope. This is caused by two forces; the weight of the lower half (1148 N) and the force causing extension of the upper half due to its own weight. This second force is the mean of 1148 N and zero. Thus the effective force causing extension of the upper half is the weight below the middle plus half the weight above the middle:

$$= 1148 + 574 = 1722 \text{ N}$$

$$d = \frac{1722 \times 152}{100 \times 207 \times 10^3} = 0 \cdot 0126 \text{ m} = 12 \cdot 6 \text{ mm}$$

S *Solution:*
The three answers required are:
1. Extension of the whole rope: **17 mm.**
2. Extension of the upper half of the rope: **13 mm.**
3. Stress in the metal of the rope at the top: **23 N/mm².**

13/5 REVERSAL OF STRESS: No restraints

The piston of a double-acting hydraulic cylinder moves under a pressure on the upper face of the piston, and a back pressure on the lower face. On the return stroke the pressures are reversed. The diameter of the piston rod is 75 mm and its length 3 m. The diameter of the piston is 250 mm. If the main pressure is 6.9 N/mm^2 and the back pressure, 0.3 N/mm^2, what is the maximum stress in the piston rod in compression and in tension? What is the difference in length of the rod between the two strokes? E is 207×10^3 N/mm^2.

G *Group of Terms:*

$$\frac{F}{A} = \sigma = \frac{Ed}{l}$$

U *Unknowns and Units:*
1. Maximum stress in rod in tension.
2. Maximum stress in rod in compression.
3. Difference in length of the rod between strokes.

Use N and mm.

I *Inventory of Information:*

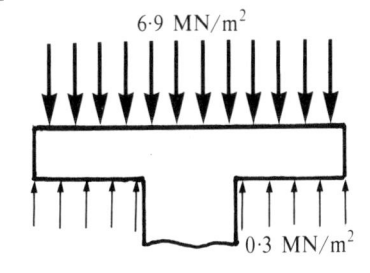

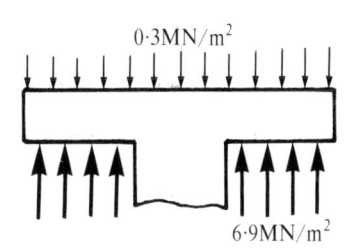

Diameter of piston rod = 75 mm: $A = 4420$ mm^2
Length of piston rod = 300 mm
Main pressure = 6.9 and back pressure = 0.3 N/mm^2
Diameter of piston, 250 mm $E = 207 \times 10^3$ N/mm^2

D *Derived Data:*
Upper face of piston has area $(125)^2\pi$ mm^2.

Lower face has area $(125)^2\pi - (37.5)^2\pi$ mm^2, for the presence of the piston rod reduces the area on which pressure can act.

13/5

The two cases are: pressure on upper face with back pressure on lower face, and pressure on lower face with back pressure on upper face.
Nett forces developed in these two cases are:

$F_1 = -6.9 \times$ large area $+ 0.3 \times$ smaller area
$ = 324.7$ kN compression on rod
$F_2 = -0.3 \times$ larger area $+ 6.9 \times$ smaller area
$ = 292.9$ kN tension on rod

E *Evaluation:*
Stresses:

Case 1: $\dfrac{F_1}{A} = \sigma_1 = \dfrac{324.7 \times 10^3}{4420} = 73.5$ N/mm² (compression)

Case 2: $\dfrac{F_2}{A} = \sigma_2 = \dfrac{292.9 \times 10^3}{4420} = 66.2$ N/mm² (tension)

Deformations:

Case 1: $\sigma_1 = \dfrac{Ed_1}{l}$: $d_1 = \dfrac{3 \times 10^3 \times 73.5}{207 \times 10^3} = 1.02$ mm (shortening)

Case 2: $\sigma_2 = \dfrac{Ed_2}{l}$: $d_2 = \dfrac{3 \times 10^3 \times 66.2}{207 \times 10^3} = 0.96$ (lengthening)

S *Solution:*
The answers required are:
1. Maximum stress in rod in tension: **66.2 N/mm².**
2. Maximum stress in rod in compression: **73.5 N/mm².**
3. Maximum difference in length of rod between strokes: **1.98 mm.**

Solve problems from other textbooks by the methods given here.

13/6

13/6 SHARED LOAD: One material: Lack of fit

A weight is lowered on to three short vertical steel cylinders standing on a rigid horizontal base. The cylinders are in a straight line and equally spaced. The centre of gravity of the weight is directly over the central cylinder. The diameter of all the cylinders is 30 mm, and the height of two of them 50 mm. The central cylinder, however, is 50·015 mm high. If the modulus of elasticity is 206×10^3 N/mm² and the total weight of 150 kN is applied to all cylinders simultaneously through a rigid, horizontal slab, what is the stress on the central cylinder, and how is the weight shared between the three cylinders?

G *Group of Terms:*

$$\frac{F}{A} = \sigma = \frac{Ed}{l}$$

U *Unknowns and Units:*
1. Stress on the central cylinder.
2. Loads on all three cylinders.
Use kN and mm.

I *Inventory of Information*
For central cylinder, until all three are 50 mm high:
$A = \pi \times 15^2$ mm² d (deformation) $= 0·015$ mm
l(height) $= 50·015$ mm $E = 207 \times 10^3$ N/mm²

D *Derived Data:*
Load required to bring the central cylinder down to 50 mm height of the other two:

$$\frac{Ed}{l} = \sigma$$

$$\frac{207 \times 10^3 \times 0·015}{50·015} = \sigma$$

$$\sigma = 62 \text{ N/mm}^2$$

Stress to bring central cylinder to 50 mm height $= 62$ N/mm².
Load to bring the central cylinder to 50 mm height (changing to kN):

$$62 \times 10^{-3} \times \pi \times 15^2 = 43·8 \text{ kN}$$

309

E *Evaluation:*
When load comes on all three cylinders:
Part of the total load (43·8 kN) has been 'used up' in bringing the central cylinder to a height of 50 mm. The rest, 150 − 43·8 kN is now distributed equally between the three cylinders:

Load on outer cylinders: 106·2/3 kN
Load on central cylinder: 43·8 + 106·2/3 kN

S *Solution:*
2. Loads on the three cylinders: **35·4 kN; 79·2 kN; 35·4 kN**
1. Stress on the central cylinder: **$F/A = 112$ N/mm².**

13/7 SHARED LOAD: One material: Differential extension

Two parallel rods of equal length, and of the same material and cross-sectional area, hang vertically, carrying a horizontal bar. The rods are 6 m in length and of 160 mm² cross-sectional area. The distance between the rods is 65 mm, and a load of 8·9 kN is attached to one of the rods at the level of the horizontal bar. The effect of this is to cause the horizontal bar to tilt through 1·1°. What is the modulus of elasticity of the rods and what is the strain in the loaded wire? The second rod does not extend, as it is unloaded.

G *Group of Terms:*

$$\frac{F}{A} = \sigma = \frac{Ed}{l}$$

U *Unknowns and Units:*
1. Modulus of elasticity of the material.
2. Strain in the loaded rod.
Use MN and m.

I *Inventory of Information:*
$F = 8.9$ kN $= 0.0089$ MN
$A = 160$ mm² $= 160 \times 10^{-6}$ m²
$l = 6$ m;   Distance between rods $= 75$ mm
Tilt of cross-bar $= 1.1°$

D *Derived Data:*
The second rod is unloaded, and does not extend. The extension of the loaded rod can thus be obtained from the tilt of the cross-bar:
Extension, $d = 75 \tan 1\cdot 1° = 75 \times 0\cdot 0192$ mm $= 1\cdot 44 \times 10^{-3}$ m

E *Evaluation:*

$$\sigma = \frac{F}{A} = \frac{0\cdot 0089}{160 \times 10^{-6}} = 55\cdot 6 \text{ MN/m}^2$$

$$\text{Strain } e = \frac{d}{l} = \frac{1\cdot 44 \times 10^{-3}}{6}$$

$$E = \frac{\sigma l}{d} = \frac{55\cdot 6}{0\cdot 24 \times 10^{-3}}$$

S *Solution:*
2. Strain in the loaded rod: $\quad\quad\quad\quad\quad\quad\quad\quad 0\cdot 24 \times 10^{-3}$
1. Modulus of elasticity: $\quad\quad\quad\quad\quad\quad\quad\quad 234 \times 10^3 \text{ MN/m}^2$

13/8 SHARED LOAD: Two materials: Equal extension

Two vertical rods, equal in length, one of copper and one of steel, hang from a horizontal rigid support. They are connected at the lower end by a rigid cross-bar. This cross-bar is loaded but the loads are so arranged that the cross-bar remains horizontal. The total load carried on the cross-bar in various places is 8500 N. If the diameter of the copper rod is 25 mm and the diameter of the steel rod, 18 mm, find the load, stress and strain in each rod. The modulus of elasticity for copper is 108×10^3 MN/m², and that for steel is 207×10^3 MN/m².

G *Group of Terms:*

$$\frac{F_s}{A_s} = \sigma_s = \frac{E_s d_s}{l_s} : \quad \frac{F_c}{A_c} = \sigma_c = \frac{E_c d_c}{l_c}$$

U *Unknowns and Units:*
1. Load in each rod.
2. Stress in each rod.
3. Strain in each rod.
Use MN and m.

13/8

I *Inventory of Information:*

Steel
$A_s = 81\pi \times 10^{-6}$ m²
$E_s = 207 \times 10^3$ MN/m²

Copper
$A_c = (12 \cdot 5)^2 \pi \times 10^{-6}$ m²
$E_c = 108 \times 10^3$ MN/m²

D *Derived Data:*

In any problem dealing with two materials, look for the links between the two as given in the conditions of the particular problem. In this instance the cross-bar remains horizontal, so the two rods remain of equal length. Thus both extend equally, and since both were originally the same length, they both have the same value of strain, which is the ratio deformation/length.

$$d_s = d_c = d \qquad d/l = e \text{ (strain)}$$

Secondly here, the two loads in the rods must be equivalent to the loading, for

$$+F_s + F_c - 8500 \text{ MN} = 0$$

Using these two relationships, the evaluation is made easy.

E *Evaluation:*

Stress in steel: $\sigma_s = \dfrac{207 \times 10^3 d}{l}$

Stress in copper: $\sigma_c = \dfrac{108 + 10^3 d}{l}$

or $\dfrac{\sigma_s}{\sigma_c} = 1 \cdot 92$

Thus: $\dfrac{\text{load in steel}}{\text{Load in copper}} = \dfrac{F_s}{F_c} = \dfrac{A_s \sigma_s}{A_c \sigma_c} = \dfrac{81 \times 1 \cdot 92 \, \sigma_c}{(12 \cdot 5)^2 \, \sigma_c}$

But
$$+F_s + F_c = +8500 \text{ kN}$$

which makes
$$+F_c = +0 \cdot 004 \, 26 \text{ MN} \quad \text{and} \quad +F_s = +0 \cdot 004 \, 24 \text{ MN}$$

S *Solution:*
1. Load in steel — 4·24 kN
   Load in copper — 4·26 kN
2. Stress in steel — 16·7 N/mm²(MN/m²)
   Stress in copper — 8·7 N/mm²(MN/m²)
3. Both strains are equal: $0 \cdot 0805 \times 10^{-3}$ (dimensionless)

13/9 SHARED LOAD: Combined section: Two materials

A bar of steel is riveted, over its whole length, to a bar of brass. The combined bar is subject to a tensile force of 200 kN. The width of both bars is the same (100 mm) but the thickness of the steel bar is 25 mm and the thickness of the brass bar is 10 mm. If the modulus of elasticity of brass is 93×10^3 N/mm², and that of steel is 208×10^3 N/mm² find the forces and stress in each bar.

G *Groups of Terms:*

$$\frac{F_s}{A_s} = \sigma_s = \frac{E_s d_s}{l_s}$$

$$\frac{F_b}{A_b} = \sigma_b = \frac{E_b d_b}{l_b}$$

U *Unknowns and Units:*
1. Forces in brass and steel rods.
2. Stresses in brass and steel rods.

Use kN and mm.

I *Inventory of Information:*

$A_s = 2500$ mm² $A_b = 1000$ mm²
$E_s = 208$ kN/mm² $E_b = 93$ kN/mm²

Force to be divided between the two: 200 kN.

D *Derived Data:*

Since the bars are of the same length and held firmly together, their elongations and strains must also be the same for both:

$$d/l = e \text{ for both steel and brass}$$

Further, $+F_s + F_b = +200$ kN as the total force carried by the two bars together.

E *Evaluation:*

$$\sigma_s = \frac{208d}{l}: \quad \sigma_b = \frac{93d}{l} \quad \frac{\sigma_s}{\sigma_b} = 2\cdot 24$$

$$\frac{F_s}{F_b} = \frac{\sigma_s A_s}{\sigma_6 A_b} \quad \frac{2\cdot 24 \times 2500}{1000} = 5\cdot 59$$

313

13/10

but $+F_s + F_b = +200$.
Replacing F_s by $5\cdot 59\ F_b$, the results are obtained.

S *Solution:*

The answers required are:
1. Force in steel rod   169·7 kN
   Force in brass   30·3 kN
2. Stress in steel   67·9 N/mm²
   Stress in brass   30·3 N/mm²

13/10 SHARED LOAD: Two materials: Combined section

A composite conductor consists of six copper wires tightly arranged round a steel core wire. The whole conductor is subjected to a tensile force of 2·9 kN. If the diameter of all the seven wires is 2·6 mm, and the length of the conductor 30 m, find the stresses in the two materials and the elongation of the whole conductor. The modulus of elasticity of steel is 207×10^3 N/mm² and that of copper can be taken as 108×10^3 N/mm².

G *Group of Terms:*

$$\frac{F_s}{A_s} = \sigma_s = \frac{E_s d_s}{l_s} : \quad \frac{F_c}{A_c} = \sigma_c = \frac{E_c d_c}{l_c}$$

U *Unknowns and Units:*

1. Stresses in both materials.
2. The elongation of the conductor.

Use kN and mm.

I *Inventory of Information:*

$A_s = (1\cdot 3)^2 \pi$ mm² A_c is the same as A_s for one wire but $\times 6$
$F = 2\cdot 9$ kN   for the whole cable.
$E_s = 207$ kN/mm² $E_c = 108$ kN/mm²
l = Length of both, 30×10^3 mm.

D *Derived Data:*

Strains are equal: $e_s = e_c = e$
The sum of the forces equals the load, thus $+F_s + F_c = +2\cdot 9$ kN

E *Evaluation:*

$$\sigma_s = \frac{207d}{30 \times 10^3}; \quad \sigma_c = \frac{108d}{30 \times 10^3}; \quad \frac{\sigma_s}{\sigma_c} = 1\cdot92$$

$$F_s = (1\cdot3)^2\pi\,\sigma_s; \quad F_c = 6(1\cdot3)^2\pi\,\sigma_c; \quad \frac{F_s}{F_c} = 0\cdot32$$

$$+F_s + F_c = +2\cdot9; \quad F_s = 0\cdot7\text{ kN}; \quad F_c = 2\cdot2\text{ kN}$$

$$\sigma_s = \frac{F_s}{(1\cdot3)^2\pi} = 133\text{ N/mm}^2; \quad \sigma_c = 69\text{ N/mm}^2$$

$$d = \frac{\sigma_s \times 30 \times 10^3}{207 \times 10^3} = 19\text{ mm}$$

S *Solution:*

The answers required are:
1. Stress in steel   **133 N/mm²**
   Stress in copper   **69 N/mm²**
2. Elongation of the conductor   **19 mm**

13/11   RESTRAINED STRAIN: Change in temperature

Both ends of a steel beam, (6 m long and of an area of 0·01 m²) are built into brick walls. On a hot day the temperature rises 10°C in the steel of the beam. If the coefficient of expansion of steel is 10·7 × 10⁻⁶ per degree centigrade, and the modulus of elasticity of steel is 207 × 10³ N/mm², what force is exerted on the walls? What is the additional stress in the beam due to the rise in temperature?

G *Group of Terms:*

$$\frac{F}{A} = \sigma = \frac{Ed}{l} = Ee$$

U *Unknowns and Units:*

1. Outward force exerted on the walls.
2. Stress induced in beam due to rise in temperature.
Use kN and m.

I *Inventory of Information*

$A = 0\cdot01$ m² $\alpha_s = 10\cdot7 \times 10^{-6}$ per °C
$E = 207 \times 10^3$ N/mm² (MN/m²) $= 207 \times 10^6$ kN/m²

13/12

D *Derived Data:*

The coefficient of expansion is the proportional change in length for each degree. It is, in fact a strain (d/l). Strain usually has the symbol e. Thus the strain induced in this beam due to the rise in temperature would be

$$e = 10.7 \times 10^{-6} \times 10 \text{ (dimensionless)}$$

This strain is resisted and thus gives rise to stress. Normally, stress gives rise to strain but if strain is prevented, stress occurs without the application of load.

E *Evaluation:*

$$\frac{F}{A} = \sigma = Ee$$

$$\frac{F}{0.01} = \sigma = 207 \times 10^6 \times {\cdot}107 \times 10^{-6}$$

S *Solution:*

Evaluation of F and σ in the above expression gives:
1. Outward force exerted due to temperature: **221·5 kN**
2. Stress in the steel of the beam: **221 50kN/m² = 22·1 MN/m²**

13/12 RESTRAINED STRAIN: Change in temperature

In order to strengthen a cottage, a steel bar was passed through both outer walls and fitted with bearing plates on the outside faces. The bar, 25 mm in diameter and 4·6 m long, was then raised in temperature through 84°C. The nuts were tightened and the bar allowed to cool. During the cooling process the walls were observed to move towards each other by 2·5 mm. If the modulus of elasticity of steel is 207×10^3 MN/m² and the coefficient of expansion of steel is 10.7×10^{-6} per °C, calculate the load applied to the walls by the tie-bar, and the stress in the bar.

G *Group of Terms:*

$$\frac{F}{A} = \sigma = \frac{Ed}{l}$$

316

13/12

U *Unknowns and Units:*
1. Load applied to the walls by the tie-bar.
2. Stress in the tie-bar.
Use MN and m.

I *Inventory of Information:*
$A = (12 \cdot 5)^2 \pi$ mm² $= (12 \cdot 5)^2 \pi \times 10^{-6}$ m²
$E = 207 \times 10^3$ MN/m² $l = 4 \cdot 6$ m

D *Derived Data:*
Expansion of bar on increase in temperature:
  = strain × length
  = (rise coefficient) × length
  = $10 \cdot 7 \times 10^{-6} \times 84 \times 4 \cdot 6 = 4135 \times 10^{-6}$ m
Extension retained in bar after walls moved towards each other by 2·5 mm,
  = $(4135 - 2500)10^{-6}$ m $= 1635 \times 10^{-6}$ m
Strain (e) is deformation remaining divided by length.
  Strain $= (1635 \times 10^{-6})/4 \cdot 6 = 363 \cdot 4 \times 10^{-6}$ (dimensionless)

E *Evaluation:*
$\sigma = Ee = 207 \times 10^3 \times 363 \cdot 4 \times 10^{-6} = 75 \cdot 2$ MN/m²
$F = \sigma A = 75 \cdot 2 \times (12 \cdot 5)^2 \pi \times 10^{-6}$ MN

S *Solution:*
The two answers required are:
  1. $F = \mathbf{36 \cdot 9}$ **kN**;   2. **75·2 MN/m² (N/mm²)**

Ease in solution comes from familiarity in the method; much practice is essential.

317

13/13

13/13 RESTRAINED STRAIN: Two materials: Change in temperature

A steel rod is fitted inside a copper tube and nuts are tightened so that the two are the same length but there is no stress on either. The temperature is then raised by 60°V. The diameter of the steel rod is 19 mm and the internal and external diameters of the tube 50 and 56 mm. After the arrangement has adjusted itself to the higher temperature, what is the stress in the rod, and the stress on the cross-section of the tube?
$\alpha_3 = 6.6 \times 10^{-6}$; $\alpha_0 = 10.0 \times 10^{-6}$ (dimensionless);
$E_P = 207 \times 10^3$ N/mm^2; $E_0 = 108 \times 10^3$ N/mm^2.

G *Groups of Terms:*

$$\frac{F}{A} = \sigma = \frac{Ed}{l}$$

U *Unknowns and Units:*
1. Stress in the rod after heating.
2. Stress in the tube after heating.
Use kN and mm.

I *Inventory of Information:*

$A_s = (9.5)^2 \pi$ mm^2 = 283 mm^2
$A_c = (28^2 - 25^2)\pi$ mm^2 = 500 mm^2
$E_s = 207 \times 10^3$ N/mm^2 = 207 kN/mm^2
$E_c = 108 \times 10^3$ N/mm^2 = 108 kN/mm^2
$\alpha_s = 6.6 \times 10^{-6}$ $\alpha_s = 10.0 \times 10^{-6}$ (dimensionless)

D *Derived Data:*
Expansion of copper due to rise in temperature:
$$60(10 \times 10^{-6}) \times \text{length} = 600 \times 10^{-6} \times \text{length}$$
Expansion of steel due to rise in temperature:
$$60(6.6 \times 10^{-6}) \times \text{length} = 396 \times 10^{-6} \times \text{length}$$
Stress in copper due to F_h, the force induced by heat:
$$\frac{F_h}{500} \text{ kN/mm}^2$$

318

13/13

Strain in copper due to F_h:

$$\frac{F_h}{500 \times 108} = 18{\cdot}5 \times 10^{-6} \text{ (dimensionless)}$$

Strain in steel due to rise in temperature and F_h:

$$\frac{F_h}{283 \times 207} = 17{\cdot}0 \times 10^{-6} \text{ (dimensionless)}$$

E *Evaluation:*

These four changes in length, two due to rise in temperature and two to the force developed between the rod and the tube, have been calculated as if they were free to take effect without restraint. However, they are acting together on two materials of different coefficients of expansion, the larger being that of copper. The materials, starting with equal lengths, try to expand, find a force developing between them due to the rise in temperature, and remain still of equal (but different) lengths after the temperature has had its effect. The higher coefficient of expansion of copper puts a tension in the steel and a compression in the copper, the force F_h being the same in both materials. The situation develops as (in terms of strain).

Expansion strain +in copper due to 60°F rise. $= +600 \times 10^{-6}$	−	Compressive strain in copper due to F_h (shortening) $(18{\cdot}5 \times 10^{-6})F_h$
Expansion strain +in steel due to 60°F rise $= +396 \times 10^{-6}$	+ +	Tensile strain in steel due to F_h (expansion) $(17{\cdot}0 \times 10^{-6})F_h$

These two combined strains must be equal, since equal lengths become other lengths, but still equal. Thus:

$$+600 - 18{\cdot}5F_h = +396 + 17{\cdot}0F_h \qquad +F_h = +5{\cdot}75 \text{ kN}$$

S *Solution:*

1. Stress in steel rod after heating:

$$\frac{F_h}{283} \text{ kN/mm}^2 = 0{\cdot}0203 \text{ kN/mm}^2 = \mathbf{20{\cdot}3 \text{ N/mm}^2}$$

319

13/14

2. Stress in copper tube after heating:

$$\frac{F_h}{500} \text{ kN/mm}^2 = 0\cdot 0114 \text{ kN/mm}^2 = \mathbf{11\cdot 5 \text{ N/mm}^2}$$

13/14 LATERAL STRAIN: One applied stress: No restraint

A bar, 100 mm long, is compressed longitudinally by a force of 40 kN. If the cross-section of the bar is 50 mm × 25 mm, what are the changes in dimensions of the bar? Poisson's ratio is 0·25, and the modulus of elasticity is 93×10^3 N/mm².

G *Group of Terms:*

$$\frac{F}{A} = \sigma = \frac{Ed}{l}$$

$$e_b = e_t = 0\cdot 25 e$$

U *Unknowns and Units:*
1. Change in length of the bar.
2. Change in the lateral dimensions of the bar.
3. Change in volume of the bar.

Use kN and mm.

I *Inventory of Information:*

$A = 50 \times 25 = 1250$ mm²
$V = 50 \times 25 \times 100 = 125\,000$ mm² $= 125 \times 10^3$ mm³
$F = 40$ kN;   P.R. $= 0\cdot 25$
$E = 93 \times 10^3$ N/mm² $= 93$ kN/mm²

D *Derived Data:*

Lateral strain $= 0\cdot 25 \times$ longitudinal strain.
Longitudinal strain:

$$\frac{d}{l} = \frac{F}{AE} = \frac{40}{1250 \times 93} = 0\cdot 344 \times 10^{-3}$$

Volumetric strain $= (+e - e_b - e_t)$

320

E *Evaluation:*
$$d = \text{length} \times \text{longitudinal strain}$$
$$= 100 \times 0.344 \times 10^{-3} = 0.0344 \text{ mm}$$
Deformation of breadth $= 50 \times 0.25 \times 0.344 \times 10^{-3}$ mm
Deformation of thickness $= 25 \times 0.25 \times 0.344 \times 10^{-3}$ mm
Change in volume $= 125 \times 10^3 (+0.344 - 0.086 - 0.086)10^{-3}$
$$= 21.5 \text{ mm}^3$$

S *Solution:*
1. Bar is shortened by   **0.0344 mm**
2. Breadth changes, from 50 mm, by   **0.004 mm**
3. Thickness changes from 25 mm, by   **0.002 mm**
   Both breadth and thickness increase.

13/15 VOLUMETRIC CHANGE: One applied stress: No restraint

A steel rod is 0.23 m long and 24 mm square in cross-section. It sustains an axial tension of 86.4 kN. Poisson's ratio is 0.3 and the modulus of elasticity is 207×10^3 N/mm². What is the change in volume?

G *Group of Terms:*
$$\frac{F}{A} = \sigma = \frac{Ed}{l}$$

U *Unknowns and Units:*
1. Change in volume.
Use N and mm.

I *Inventory of Information:*
$F = 86.4$ kN   or 86.4×10^3 N
$A = 24^2 = 576$ mm² $E = 207 \times 10^3$ N/mm²
$l = 230$ mm $\nu = 0.3$

D *Derived Data:*
σ: Stress (tension) $= \dfrac{86.4 \times 10^3}{576} = 150$ N/mm²

e_l: Strain $\dfrac{d}{l} = \dfrac{150}{207 \times 10^3} = 0.725 \times 10^{-3}$

$e_b = e_t$: Lateral strain $= 0.3 e_l = 0.218 \times 10^{-3}$

E *Evaluation:*

Change in volume (see 13.4) = +Volume $(+e_l - e_b - e_t)$
$$= +23 \times 576 \times 0.289 \times 10^{-3}$$

S *Solution:*

The change in volume is an increase of $\qquad$ +38·3 mm³

13/16 VOLUMETRIC CHANGE: One applied stress: No restraint

A steel bar, of length 250 mm, is rectangular in section (75 × 50 mm). A stress is applied to the cross-section in the direction of the length. If Poisson's ratio is 0·25, what is the percentage change in volume? The stress applied is 200 MN/m². E: 207 × 10³ N/mm². If the stress is tension is the change an increase or decrease?

G *Groups of Terms:*

$$\frac{F}{A} = \sigma = \frac{Ed}{l}$$

Volumetric strain $= \dfrac{\delta V}{V} = (+e_l + e_b + e_t)$

U *Unknowns and Units:*

1. Change in volume.
2. Percentage change in volume.

Use kN and mm.

I *Inventory of Information:*

$l = 250$ mm $\qquad A = 75 \times 50 = 3750$ mm²
$\sigma = 200$ MN/m² or N/mm² = 0·2 kN/mm²
$\nu = 0.25$ $\qquad E = 207 \times 10^3$ N/mm² = 207 kN/mm²

D *Derived Data:*

Longitudinal strain: $e_l + \dfrac{\sigma}{E} = \dfrac{+0\cdot 2}{207} = +0\cdot 97 \times 10^{-3}$

Lateral strain: $e_b = e_t = -\dfrac{0\cdot 2}{207 \times 4} = -0\cdot 243 = 10^{-3}$

322

E *Evaluation:*

Change in volume: $\delta V = (+e_l - e_b - e_t)lbt$
$= (+0 \cdot 970 - 0 \cdot 486)(250 \times 75 \times 50) \times 10^{-3}$

Percentage change in volume: $\dfrac{lbt}{lbt}(+0 \cdot 97 - 0 \cdot 486) \times 100 \times 10^{-3}$

S *Solution:*
1. Change in volume **453 mm³**
2. Percentage change in volume **0·048%**

13/17 **RESTRAINED LATERAL STRAIN: Three applied stresses**

A prism of square cross-section (20 mm × 20 mm) is subjected to an axial pressure of 50 kN. If the length of the prism is 100 mm, what lateral pressure on all four faces would prevent lateral strain completely? Poisson's ratio is 0·5 and E is 93×10^3 N/mm². What would be the change in length of the prism in these circumstances?

G *Groups of Terms:*

$$\frac{F}{A} = \sigma = \frac{Ed}{l}; \qquad e_l = \frac{\sigma}{E} = \frac{F}{AE}$$

$$e_b = e_t = 0 \cdot 25 e_l = \frac{\sigma}{4E}$$

U *Unknowns and Units:*
1. What lateral pressure prevents lateral strain?
2. What is the change in length in these conditions?
Use kN and mm.

I *Inventory of Information:*

$F = 50$ kN $A = 20 \times 20 = 400$ mm²
$\nu = 0 \cdot 25$ $E = 93 \times 10^3$ N/mm² = 93 kN/mm²
$l = 100$ mm

323

13/17

D *Derived Data:*

You are interested in the lateral strain. On the two lateral dimensions the change in dimension depends on the following strains:
+ Outward strain caused by longitudinal compression
− inward strain caused by applied pressure p,
+ outward strain caused by applied pressure p on the other lateral face

$$= \frac{e}{4} - \frac{p}{E} + \frac{p}{4E} = \frac{F}{4AE} - \frac{3p}{4E} = \frac{1}{4E}\left(\frac{F}{A} - 3p\right)$$

E *Evaluation:*

1. If the lateral strain is to be eliminated,

$$\frac{1}{4E}\left(\frac{F}{A} - 3p\right) = 0; \qquad \frac{50}{400} - 3p = 0$$

$$p = \frac{1}{24} \text{ kN/mm}^2$$

2. The change in the longitudinal length is dependent on the strains:
− Longitudinal compressive strain (shortening) due to 50 kN load
+ longitudinal strain (lengthening) due to compression p on one pair of lateral faces
+ longitudinal strain (lengthening) due to compressive strees p on the other pair of faces.

$$= -\frac{F}{AE} + \frac{p}{4E} + \frac{p}{4E}$$

$$= -\frac{1}{93}\left(\frac{50}{400} - \frac{3}{24}\right) \text{ (dimensions)}$$

= zero strain; so zero change in length

S *Solution:*

The answers required are:
1. The lateral pressure which would prevent lateral strain if applied to all four sides of the prism is **41·7 N/mm²**
2. The change in axial length of the prism if this lateral pressure were applied, would be **zero**, since the lateral pressure eliminates lateral expansion and the rod stays its original shape.

13/18 RESTRAINED LATERAL STRAIN: Three applied stresses

A rectangular prism (E: 200 kN/mm^2) is subjected to compressive stresses on all six faces. The dimensions of the prism are 120 × 60 × 30 mm. If Poisson's ratio is 0·25, what are the strains and deformations in each direction and what is the change in volume?

G *Groups of Terms:*

$$\frac{F}{A} = \sigma = \frac{Ed}{l} = Ee$$

$$\delta V = (e_l + e_b + e_t)$$

U *Unknowns and Units:*
1. Strains in each of the three directions.
2. Deformations in each of the three directions.
3. Change in volume.
Use N and mm.

I *Inventory of Information:*

$\sigma_l = 20$ N/mm^2 acting on the 60 × 30 mm faces
$\sigma_b = 40$ N/mm^2 acting on the 120 × 30 mm faces
$\sigma_t = 80$ N/mm^2 acting on the 120 × 60 mm faces
$E = 200 \times$ kN/mm$^2 = 200 \times 10^3$ N/mm^2
$\nu = 0.25$

D *Derived Data:*

The strain in any direction is the algebraic sum of the strain in that direction caused by stress in the same direction, together with the strains in the other two directions multiplied by the Poisson's ratio.

Strain is obtained from stress by dividing by the modulus of elasticity.

$$e_l = + \frac{\sigma_l}{E} - 0.25 \frac{\sigma_b}{E} - 0.25 \frac{\sigma_t}{E}$$

is one of the three equations; write the others.

The dimension in any of the three directions is the unstressed dimension plus or minus the unstressed dimension multiplied by the strain caused by the stress. Each of the three dimensions, after stressing is thus:

$$l - l\left(+ \frac{\sigma_l}{E} - 0.25 \frac{\sigma_b}{E} - 0.25 \frac{\sigma_t}{E}\right), \text{ etc.}$$

13/18

The change in volume is:
$$(e_l + e_b + e_t)lbt$$

the volumetric strain being:
$$e_l + e_b + e_t$$

E Evaluation:

The various terms required in the expressions for the combined strains in each direction are best found by a tabular calculation. Draw up the table and put a dot in each of the spaces where a figure is required. Mark out the others and then fill in the figures. This prevents mistakes in signs:

	σ_l	$\sigma_l/4$	σ_b	$\sigma_b/4$	σ_t	$\sigma_t/4$	Strain $\times$ E
Expression for $e \times E$	-20	—	—	$+10$	—	$+20$	$+10$
Expression for $e_b \times E$	—	$+5$	-40	—	—	$+20$	-15
Expression for $e_t \times E$	—	$+5$	—	$+10$	-80	—	-65

Strains in all three directions:

$$e_l = +\frac{10}{200 \times 10^3} = +0.050 \times 10^{-3} \text{ (extension)}$$

$$e_b = -\frac{15}{200 \times 10^3} = -0.075 \times 10^{-3} \text{ (contraction)}$$

$$e_t = -\frac{65}{200 \times 10^3} = -0.325 \times 10^{-3} \text{ (contraction)}$$

Changes in length in all three directions:

$d_l = +120 \times 0.050 \times 10^{-3} = +6 \times 10^{-3}$ mm (extension)
$d_b = -60 \times 0.075 \times 10^{-3} = -4.5 \times 10^{-3}$ mm (contraction)
$d_t = -30 \times 0.325 \times 10^{-3} = -9.75 \times 10^{-3}$ mm (contraction)

Change in volume
$$\delta V = 120 \times 60 \times 30(+0.05 - 0.075 - 0.325)10^{-3}$$

S *Solution:*

The three answers required are:
1. $+0.050 \times 10^{-3}$: -0.075×10^{-3}: -0.325×10^{-3}
2. $+6.00 \times 10^{-3}$: -4.50×10^{-3}: -9.75×10^{-3} mm
3. -75.5 mm^3 decrease in volume.

13/19 HYDROSTATIC PRESSURE: Bulk modulus

A cube of metal is lowered into deep sea water, and sustains hydraulic pressure on all of its sides, which are each 150 × 150 mm. If the depth to which the cube is lowered is 3250 m, and the density of sea water is 1·05 Mg/m³, what is the pressure on the faces of the cube and what is the change in volume due to this pressure? $E = 102 \times 10^3$ N/mm², and Poisson's ratio $= 0.25$.

G *Groups of Terms:*

$$\frac{F}{A} = \sigma = \frac{Ed}{l}$$

$$\sigma_B = \frac{K\delta V}{V}$$

$$K = \frac{E}{3(1 - 2\mu)}$$

U *Unknowns and Units:*
1. Pressure at this depth.
2. Volumetric change due to pressure.
Use kN and m.

I *Inventory of Information:*
Side of cube $= 150$ mm $= 0.15$ m
Depth $= 3250$ m   Density of water $= 1.05$ Mg/m³
$E = 102 \times 10^3$ N/mm² $= 102 \times 10^6$ kN/m²

327

13/19

D *Derived Data:*

Pressure per metre of depth caused by density of water: $1 \cdot 05 \times 9 \cdot 81$ kN/m²

p = Pressure at 3250 m: $1 \cdot 05 \times 9 \cdot 81 \times 3250$ kN/m²
$$= 33\,480 \text{ kN/m}^2$$

K = Bulk modulus: $\dfrac{E}{3(1 - 2\mu)} = \dfrac{102 \times 10^6}{3(1 - 2 \times \frac{1}{4})} = 68 \times 10^6$ kN/m²

V = Volume of cube: 150^3 mm³ = $150^3 \times 10^{-9}$ m³

E *Evaluation:*

$$\sigma_B = \frac{K \delta V}{V} \qquad 33\,480 = \frac{68 \times 10^6 \, \delta V}{150^3 \times 10^{-9}}$$

$$\delta V = \frac{33\,480 \times 150^3 = 10^{-9}}{68 \times 10^6} = 1 \cdot 665 \times 10^{-6} \text{ m}^3$$

$$= 16665 \text{ mm}^3$$

S *Solution:*

1. Pressure on cube **33·48 MN/m²**
2. Reduction in volume **1665 mm³**

State explicitly the *Unknowns* before starting solution.

Part II: Group E

14 Stress caused by bending

Bending is caused by a couple applied against the resistance of another couple. This is studied in detail in Part I. In this present Unit of Study, you are assumed to know how to determine the couples which cause bending, and to calculate the value of the moment they produce – the *bending moment*. If you are not absolutely certain of how such calculations are made, you are recommended to turn to Part I and revise.

When a bar or a beam is bent, the couple applied causes tension on one side and compression on the other. This is common knowledge. You apply couples in bending by breaking a stick over your knee. You know it will break and be torn in tension on the top face. You are reminded that bending couples which cause tension on the lower surface of the bent bar are conventionally said to be in positive bending, and tension on the upper face represents negative bending.

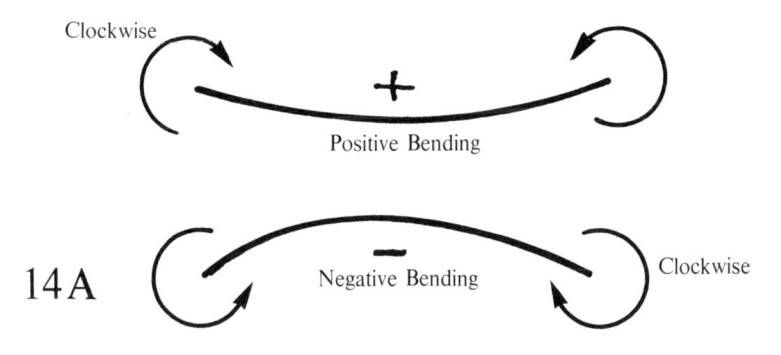

14A

Positive Bending — Clockwise

Negative Bending — Clockwise

The couples applied at the ends of the bar have equal and opposite moments. This value is the *bending moment*.

14.1 Translation of bending moment into stress

The stress produced in a cross-section by a bending couple depends, not only on the magnitude of the moment of the couple – the bending

329

14.2

moment – but also on properties of the cross-section. For the determination of bending stress and the deformation due to bending – which occur together, the six quantities concerned are:

$$\frac{\text{Moment of applied couple}}{\text{Property of the cross-section}} = \frac{\text{Bending stress}}{\text{Dimension of the cross-section}} = \frac{\text{Modulus of elasticity}}{\text{Term representing deformation}}$$

In symbols, this is

$$\frac{M}{I} = \frac{\sigma}{h} = \frac{E}{R} \text{ N/m}^3$$

which compares with the group of quantities for stress and deformation due to direct stress:

$$\frac{F}{A} = \sigma = \frac{Ed}{l} \text{ N/m}^2$$

The so-called *bending stress* has the same symbol (σ) *sigma* as the tensile and compressive stress due to direct force. Bending stress is also tensile or compressive. The difference between the values of *direct stress* and those of *bending stress* lies in the fact that direct stress is uniform over the whole of the cross-section, and bending stress varies from tension to compression across the cross-section. Direct stress is tensile *or* compressive over any cross-section, and bending stress is both tensile *and* compressive.

14.2 The significance of the six quantities

$$\frac{M}{I} = \frac{\sigma}{h} = \frac{E}{R}$$

M   *moment* of the couple causing bending – *bending moment* (N m).
I   the *second moment of area* about the centroidal axis (m^4).
σ the *stress* set up in the material of the bar, either tension or compression (N/m^2).

14.2

h   the *distance* measured outwards from the centroidal axis to the 'horizon' at which the stress is being calculated (m). (The full line in the sketch is the 'horizon'.)

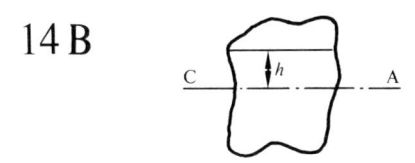

14 B

E   the *modulus of elasticity* of the material of which the bent beam or bar is made (N/m^2).

R   the *radius* to which the beam is bent at the section considered. This value of radius may vary from one point on the beam to another, but in the calculation R refers only to the radius of curvature at the section under consideration (m).

It is of value, in understanding the significance of these six quantities for you to consider how a variation in the value of one of the quantities affects the values of the others. All of the three terms which are equated must remain equal to each other in all circumstances of bending, so any change in one term must be accompanied by the same change in the other two. For example:

An increase in the second moment of area results in a decrease in the value of all the three terms. For a given distance h from the centroidal axis, the value of the stress (σ) must decrease in order to show a decrease in the second term. Similarly, the third term must decrease in value, and therefore, for the same material (E), the radius of bending must increase. An increased radius means a flatter curve, and thus less bending. These considerations show that a higher second moment of area provides greater stiffness, and lower stresses.

A decrease in h, for example, causes an increase in the second term, if the stress remains the same, and therefore an increase in the value of the other two terms. But this is not possible for both M and I are fixed quantities and do not depend on a measurement from the centroidal axis (h). Thus, if h decreases, the stress must also decrease in order to maintain constant the ratio in the second term. Thus, at a small distance from the centroidal axis the stress is low, and at a larger distance, the stress is higher, the ratio of σ to h remaining unchanged.

Since you are always interested in the greatest stress which can be

14.3

developed, this figure is clearly found at the extreme values of h – at the upper and lower surfaces of the cross-section. The variation of stress across the cross-section is shown in the sketch. At the centroidal axis there must be a zero or neutral stress. For this reason the *centroidal axis* is often called the *neutral axis*. If the neutral axis is at the centre of the height of the section, the compressive and tensile stresses at the outer faces are equal. If the centroidal or neutral axis is not symmetrically placed, one of the maximum stresses is greater than the other.

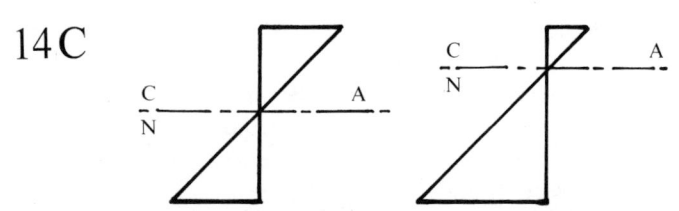

14 C

A change in the material to a more compressible or extensible material – a smaller E – means that if the other terms remain unaltered, the value of R must decrease. This indicates a greater curvature. Alternatively, if, with a smaller E, the curvature must remain unaltered, then the other two terms must decrease in order to retain parity. This means a smaller maximum stress and a larger I or a smaller M.

Write down, in this way, what happens when some of the other terms alter. In carrying out hypothetical changes of this kind you will gradually find yourself understanding, almost visually, what happens when a beam is bent. The relationship of the quantities to each other will not only be recognized intellectually, but intuitively appreciated.

14.3 Sections unsymmetrical about the neutral or centroidal axis

In the first few problems which follow, the cross-sections are all symmetrical about the centroidal axis (or neutral axis). The distances from the CA to the upper and lower faces of the beam (h) are, therefore equal, and the stresses caused by bending at these two faces are also equal, the one being tension and the other compression.

If, however, the section is unsymmetrical about its centroidal (or neutral) axis, the maximum values of h above and below are not equal. It is important, therefore, when the section is unsymmetrical about the NA (or CA) to know whether the bending is positive or negative. In the first, tension occurs on the lower face, and in the second, compression occurs on the lower face. This is a very important distinction, since many materials have different resistances to tension and compression. It is not sufficient to give the answer to the problem in terms of stress only. The kind of stress must be specified in words – 'tension' or 'compression'. When a vertical or inclined beam is loaded a similar careful decision must be made as to which side of the beam is in tension.

14.4 Deformation due to bending

Under *direct force* (Unit 13) it was convenient to determine both stresses and deformations together. The deformation caused by bending cannot so easily be determined along with the values of the stresses produced. It is true that the radius of the bent beam at a particular point can be found, but this is not of great value. The designer looks for the linear movement of the beam from its original position, and the translation of E/R into terms of linear deflection is complex. It requires a Unit of Study to itself. There are, in fact, two relationships which are studied separately when bending is in question:

$$\frac{M}{I} = \frac{\sigma}{h}$$

studied here in Unit 14, and

$$\frac{M}{I} = \frac{E}{R}$$

studied later in Unit 17.

14.5 Solving the relationships

The determination of stress due to bending is such a direct calculation that examination questions in this topic are deliberately made more complex by requiring the analysis to be started at an earlier stage than is represented by the simple relationship

$$\frac{M}{I} = \frac{\sigma}{h}$$

Usually, the values of one or more of the four quantities must be calculated from raw data before the *Evaluation* can begin. Such quantities are M, the bending moment, or I the second moment of area. The value of h may also be concealed in some way or left to your choice. Thus, if

14.5

you are in the least doubt about how to find M or I or the section modulus, do a little revision.

In the problems presented in this unit, the data is directly related to the solution of the bending relationship, but be prepared for a more complex type of problem in which all the data are not given.

As for the accurate solution of all problems, the secret of success is to maintain a strict routine and a proved technique. Use the same attack each time, writing down each step, even if you think it looks too simple for *your* attention. In the basic problems of part II there is one line of attack as there was in part I. This time, you are reminded that the mnemonic is

G-U-I-D-E-S

and that the meaning of each of these letters is given in unit 12 (section 12.5). Do not proceed until you have revised this section.

Define the units to be used in calculation.

Keep Section 12.4 in mind: it is very important.

14.6 THE PROBLEMS

14/1 SYMMETRICAL STEEL JOIST: Safe bending moment?

A beam consists of a single standard steel joist, whose section is symmetrical above and below the neutral or centroidal axis. From a published list of the properties of standard steel beams, the second moment of area of this section is found to be 33×10^6 mm^4. Find the bending moment which can be safely supported if the maximum stress permitted is 160 N/mm^2. The beam is 250 mm deep. What is the section modulus?

G *Group of Terms concerned:*

$$\frac{M}{I} = \frac{\sigma}{h} = \frac{E}{R}$$

U *Unknowns and Units:*
1. Section modulus (z).
2. Safe bending moment (M).
Use N and mm.

I *Inventory of Information supplied:*

$\sigma = 160$ N/mm^2
$I = 33 \times 10^6$ mm^4
$h = 250/2 = 125$ mm

D *Derived Data:*
None required.

E *Evaluation:*

$$\frac{M}{33 \times 10^6} = \frac{160}{125}$$

$$\frac{I}{h} = z = \frac{33 \times 10^6}{125}$$

S *Solution:*
1. The section modulus of the beam: .... 264×10^3 mm^3
2. The safe bending moment: .... $42 \cdot 2 \times 10^3$ kN m.

Note: The safe bending moment has been transformed into kN m in order to make the figure smaller, but that this is done at the very end.

14/2

The calculation in the *Evaluation* gives $42 \cdot 2 \times 10^6$ N m. If you are not sure of the significance of section modulus, revise Unit 11.

14/2 SYMMETRICAL SECTION: Solid circle: Maximum stress?

A solid circular shaft, 400 mm in diameter, carries its own weight between bearings, and sustains a bending moment of 100×10^3 N m. Calculate the maximum stress developed in the steel. What would be the difference in the stress if the metal were bronze?

G *Group of Terms:*

$$\frac{M}{I} = \frac{\sigma}{h} = \frac{E}{R}$$

U *Unknowns and Units:*
1. Maximum stress (σ).
2. Effect of change of material.
Use N and mm.

I *Inventory of Information:*
$M = 110 \times 10^3$ N m $= 110 = 10^6$ N m
$h =$ half diameter $= 200$ mm

D *Derived Data:*
(a) Second moment of area
I (for circle 200 mm radius) $= \pi r^4/4 = 1260 \times 10^6$ mm^4

E *Evaluation:*

$$\frac{110 \times 10^6}{1260 \times 10^6} = \frac{\sigma}{200}$$

S *Solution:*

1. Maximum stress developed: .... **17·5 N/m²**.
2. Change in material can make no difference to the stress, since it alters only the final and third term of the relationship. If this term is to remain unaltered with a more extensible material (lower E), then R must be less. There is thus a greater curvature for a softer material, but the stress remains the same if the material continues to be elastic.

14/3

14/3 **SYMMETRICAL RECTANGULAR JOIST**: Determine the dimensions

A wooden floor is supported by timber joists of rectangular cross-section ($b \times d$). The maximum bending moment in one of the joists is found to be 7.6×10^3 N m. The working stress allowed is 7 N/mm². Find the necessary dimensions of the cross-section of the joists, if the ratio of breadth to depth is 1:4. Find the modulus of section.

G *Group of Terms:*

$$\frac{M}{I} = \frac{\sigma}{h}$$

U *Unknowns and Units:*
1. Depth of joist (d).
3. Breadth of joist (b).
3. Modulus of section (z).
Use N and mm.

I *Inventory of Information:*
$M = 7.6 \times 10^3$ N m $= 7.6 \times 10^6$ N mm
$\sigma = 7.0$ N/mm²
$b:d = 1:4$

D *Derived Data:*
(a) I for a rectangle, when $b = d/4$, is $d^4/48$
(b) $h = d/2$

E *Evaluation:*

$$\frac{(7.6 \times 10^6)48}{d^4} = \frac{7 \times 2}{d} \quad d^3 = 26.1 \times 10^6 \text{ mm}^3$$

$$z = \frac{I}{h} = \frac{d^4 \times 2}{48 \times d} = \frac{d^3}{24} = \frac{26.1 \times 10^6}{24} \text{ mm}^3$$

S *Solution:*
1. Depth of section: .... **296 mm.**
2. Breadth of section: .... **74 mm.**
3. Modulus of section: .... **1080 × 10³ mm³.**

337

14/4

14/4 SYMMETRICAL COMPOUND GIRDER: Safe bending moment?

A compound girder, consisting of a steel I beam with a plate (225 × 25 mm) on each flange, is 300 mm deep over the beam, and 350 mm deep over the whole girder. The second moment of area of the beam alone is 80×10^6 mm^4. What safe bending moment can the beam carry, if the working stress permitted in the steel is 155 MN/m^2?

G *Group of Terms:*

$$\frac{M}{I} = \frac{\sigma}{h}$$

U *Unknowns and Units:*
1. Safe bending moment.
Use N and mm.

I *Inventory of Information:*
$\sigma = 155$ MN/m$^2 = 155$ N/mm^2
$h = 350/2 = 175$ mm
I of beam alone $= 80 \times 10^6$ mm^4

D *Derived Data:*
(a) Second moment of area
I for the whole girder $= I$ of beam $+ Ay^2$. This is calculated in one of the problems of Unit 12, and amounts to 377×10^6 mm^4.

E *Evaluation:*

$$\frac{M}{377 \times 10^6} = \frac{155}{175}$$

S *Solution:*
1. Safe bending moment: .... 334×10^3 N m.
    Again, although the moment is found in terms of N mm it is transformed into the more usable N m.

14/5 SYMMETRICAL DOUBLE CHANNELS WITH PLATES: Well designed?

A box girder consists of two steel channels placed back to back. The channels are each 300 mm deep and each has a second moment of area of 83×10^6 mm^4. The bending moment applied is 0.22×10^6 N m. The channels are strengthened by the welding of a 200×12 mm steel flange plate to each flange, and the backs of the channels are 100 mm apart. Find the stress developed and decide whether the beam is under- or over-designed.

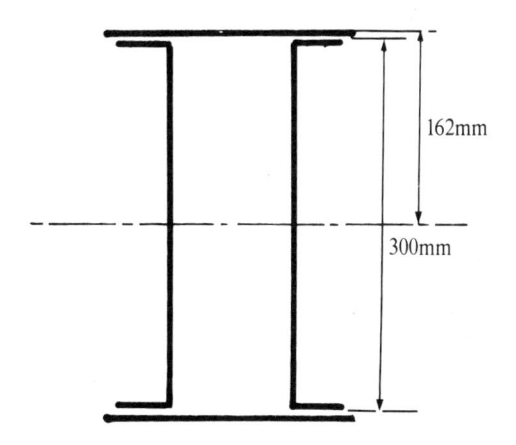

G *Group of Terms:*

$$\frac{M}{I} = \frac{\sigma}{h}$$

U *Unknowns and Units:*
1. Bending stress developed
2. Decision whether this is acceptable
Use N and mm.

I *Inventory of Information:*

$M = 0.22 \times 10^6$ N m $= 220 \times 10^6$ N mm
Plates are 300×12 mm
I of one channel only $= 83 \times 10^6$ mm^4
$h = 300/2 + 12 = 162$ mm

339

14/5

D *Derived Data:*

(a) The *centroid* is at half the depth, since the girder is symmetrically built.

(b) The derivation of the *second moment of area* for this girder is the longest piece of calculation in the problem. It has been shown in one of the problems of Unit 12.

$I = 341 \times 10^6$ mm^4

E *Evaluation:*

$$\frac{220 \times 10^6}{341 \times 10^6} = \frac{\sigma}{162}$$

S *Solution:*

1. Stress in outer fibres – the maximum: .... 105 N/mm^2.
2. Over-designed since this is low for steel.

Work out each Problem; do NOT read them over.

Read Section 12.5 if you find difficulties in solution.

14/6 SYMMETRICAL SECTION: Effect of strengthening while partially loaded

A standard steel beam was 250 mm deep, had a second moment of area of 126×10^6 mm^4 and was subjected to a maximum bending moment of 180 kN m. The beam was to be strengthened after it was in place (in order to reduce the maximum stress, which had been incorrectly calculated) by welding a flange plate (150 × 12 mm) to each flange. While this operation was being performed, it was found impossible to reduce the maximum bending moment, temporarily, to 40 kN m. Find the maximum stress due to bending on the original beam, on the strengthened girder, the distribution of stress through the depth of the section when the full original moment is re-applied, and the percentage reduction in maximum stress achieved by the strengthening.

G *Group of Terms:*

Original beam: $\dfrac{M_0}{I_0} = \dfrac{\sigma_0}{h_0}$

Beam relieved of part of B.M.: $\dfrac{M_L}{I_0} = \dfrac{\sigma_L}{h_0}$

Strengthened beam: $\dfrac{M_0}{I_s} = \dfrac{\sigma_s}{h_s}$

U *Unknowns and Units:*
1. Maximum stress in original beam.
2. Maximum stress in beam when plates being welded on.
3. Maximum stress in strengthened beam.
4. Distribution of stress through the strengthened beam.
5. Percentage reduction in stress after strengthening.

Use N and mm.

I *Inventory of Information:*

I of beam only $= 126 \times 10^6$ mm^4 (I_0)
Flange plates $= 150 \times 12$ mm
h for original beam $= 250/2 = 125$ mm (h_0)
h for strengthened beam $= 125 + 12 = 137$ mm (h_s)
Bending moment to be supported $= 180$ kN m $= 180 \times 10^6$ N mm (M_0)
Reduced bending moment during welding $= 40 \times 10^6$ N mm (M_L)

14/6

D *Derived Data:*

Second moment of area

I for the whole girder after strengthening is calculated by the tabular method of Unit 12 where this particular beam is studied. The second moment of area amounts to 362×10^6 mm^4.

E *Evaluation:*

Original beam: $\qquad \dfrac{M_0}{I_0} = \dfrac{\sigma_0}{h_0}$

At the outer faces of the beam

$$\frac{180 \times 10^6}{126 \times 10^6} = \frac{\sigma_0}{125}; \qquad \sigma_0 = 179 \text{ N/mm}^2$$

Beam before welding of plates but with decreased loading:
At the outer faces of the beam

$$\frac{M_L}{I_0} = \frac{\sigma_L}{h_0}; \qquad \frac{40 \times 10^6}{126 \times 10^6} = \frac{\sigma_L}{125}; \qquad \sigma_L = 40 \text{ N/mm}^2$$

Beam with plates welded on:
At the outer faces of the flange plates

$$\frac{180 \times 10^6}{362 \times 10^6} = \frac{\sigma_S}{137}; \qquad \sigma_S = 68 \text{ N/mm}^2$$

At the inner faces of the flange plates

$$\frac{180 \times 10^6}{362 \times 10^6} = \frac{\sigma_S}{125}; \qquad \sigma_S = 62 \text{ N/mm}^2$$

Stress in outer faces of the beam only, consists of the stress applied by the bending moment of 40×10^6 N mm (M_L) together with the extra stress imposed by the addition of $(180 - 40)$ kN m to bring the moment back to its working value. This extra stress is calculated at an h of 125 mm.

$$\sigma = 40 + \frac{140 \times 10^6}{362 \times 10} \times 125$$

$$= 88 \text{ N/mm}^2$$

S *Solution:*

Answers required are
1. **179 N/mm²**
2. **40 N/mm²**
3. **68 N/mm²**
4. **68, 62, 88 N/mm²**
5. **91/179 = 51%**

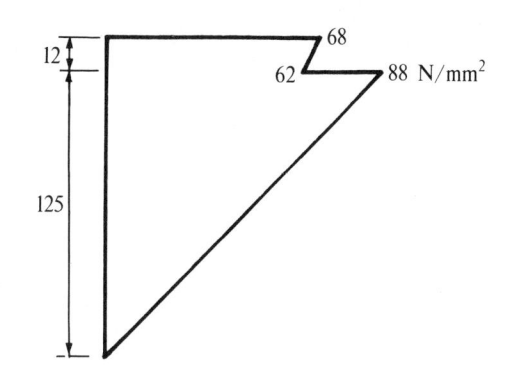

14/7 CIRCULAR CROSS-SECTION: Tapering beam: Varying bending moment

A beam of circular cross-section is subjected to a bending moment which varies along the beam according to the expression

$$560 \times 10^3 x - 35 \times 10^3 x^2 \text{ kN m}$$

where x is the distance in metres from the end of smaller diameter. The beam is 16 m long and its diameters are 0·5 m at the left-hand end, and 1·0 m at the right-hand end. Find the value of the maximum stress in the beam and the distance along the beam at which it occurs.

G *Group of Terms:*

$$\frac{M}{I} = \frac{\sigma}{h} \quad \text{or} \quad \sigma = \frac{M}{z}$$

U *Unknowns and Units:*

1. Point at which maximum stress occurs.
2. Value of the maximum stress.
Use kN and m.

I *Inventory of Information:*

$$M = (560 \times 10^3 x - 35 \times 10^3 x^2) \text{ kN m}$$

Beam tapers according to sketch.

343

14/7

D *Derived Data:*

(a) *Diameter* of beam at x metres from the left-hand end

$$D = (x + 16)\frac{0.5}{16} = \frac{(x + 16)}{32} \text{ m}$$

(b) *Second moment of area* of a circle:

$$I = \frac{\pi D^4}{64} = \frac{\pi}{64}\left(\frac{x + 16}{32}\right)^4 \text{ m}^4$$

(c) *Radius* (h) at a distance x from narrow end.

$$h = \frac{x + 16}{64} \text{ m}$$

(d) *Modulus of section:*

$$z = \frac{I}{h} = \frac{\pi(x + 16)^4}{64 \times 32^4} \times \frac{64}{(x + 16)} = \frac{\pi(x + 16)^3}{32^4} \text{ m}^3$$

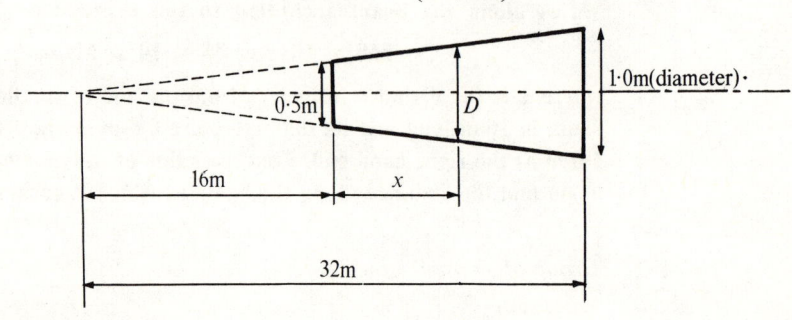

E *Evaluation:*

$$\sigma = M/z$$

which can be written as the expression given for bending moment, divided by the value of z calculated as a derived quantity.

A tabular form of evaluation together with a graph is sufficiently accurate for engineering purposes, and for examinations.

$$\sigma = \frac{M}{z} = (560 \times 10^3 x - 35 \times 10^3 x^2)\frac{32^4}{\pi(x + 16)^3}$$

$$= (A - B)\frac{C}{D} \text{ kN/m}^2$$

344

x (m)	(A) $560 \times 10^3 x$	(B) $35 \times 10^3 x^2$	$(D$ $\pi(x+16)^3$	$\dfrac{(A-B)}{(D)}(C)$
3	1680	315	21·5	66·6 × 10⁶
4	2240	560	25·0	70·5 × 10⁶
5	2800	875	29·1	69·7 = 10⁶
6	3360	1260	33·4	66·0 × 10⁶
8	4480	2240	43·4	54·2 × 10⁶
				kN/m²

(A), (B), and (D) values must be $\times 10^3$.
$(C) = 32^4 = 1 \cdot 05 \times 10^6$

S *Solution:*
1. A graph shows that the maximum stress in the beam occurs at a section 4·3 m from the end with the smaller diameter......... **4·3 m.**
2. Substituting 4·3 m for x in the expression above the table, the value of the maximum bending stress is found to be .... **70·7 kN/m².**

The procedure in dealing with varying dimensions and/or loading is the same as for problems with purely numerical data. The labour is heavier and it takes longer to reach a solution, but the same steps must be followed.

14/8 UNSYMMETRICAL TEE SECTION: Design the dimensions

A cantilever of 'Tee' section can carry safely a bending stress of 47 N/mm² in tension, and 117 N/mm² in compression. What should be the breadth of the section at the top, and what is the safe bending moment? The bending is, conventionally, negative (tension on the top).

G *Group of Terms:*

$$\frac{M}{I} = \frac{\sigma}{h}$$

U *Unknowns and Units:*
1. Breadth B of the 'table' of the 'Tee' beam.
2. Safe bending moment.
Use N and mm.

14/8

I *Inventory of Information:*

$$\sigma_t \not> 47 \text{ N/mm}^2; \quad \sigma_c \not> 117 \text{ N/mm}^2$$

D *Derived Data:*

(a) *Centroid:* The centroid of the section must be determined by the methods of unit 11, since the neutral axis, from which values of h are measured, passes through the centroid. The value of B is not known, so the position of the centroid is

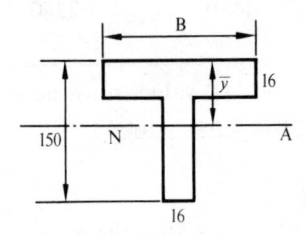

$$\bar{y} = \frac{\Sigma Ay}{\Sigma A} = \frac{128B + 0.18 \times 10^6}{16B + 2144} \text{ m from XX}_1 \text{ the upper surface}$$

(b) *Relationship between tensile and compressive stresses*

If both 47 N/mm² and 117 N/mm² are to act simultaneously (the former at the upper end of the stem) the values of h measured up and down from the N.A. must be proportional to these values, since stress varies directly with h.

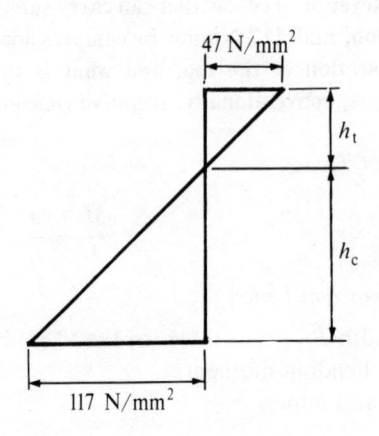

346

You thus, have two statements about the values of h

$$h_c + h_t = 150 \text{ mm}$$

$$\frac{h_t}{h_c} = \frac{47}{117}$$

From these it is not difficult to calculate that

$$h_t = 43 \text{ mm} \quad \text{and} \quad h_c = 107 \text{ mm}$$

E *Evaluate:*

The fact that stress is proportional to h is important here. If you did not remember this fact the problem could become much more complex. Read the earlier part of this unit where the effect of changes in the quantities used in calculation are assessed. Now, with the value of h from the N.A. to the upper face being established as 43 mm, the value of B can be found:

$$43 = \frac{128B + 0 \cdot 18 \times 10^6}{16B + 2144}$$

and

$$\mathbf{B = 153 \text{ mm}}$$

Second moment of area for the 'Tee' must now be determined by the methods of Unit 11. This particular section is studied in one of the problems of that unit. The value of I is $23 \cdot 5 \times 10^6$ mm .

Safe bending moment:

$$\frac{M}{I} = \frac{\sigma}{h}$$

$$\frac{M}{23 \cdot 5 \times 10^6} = \frac{47}{43}; \quad M = 26 \times 10^6 \text{ N mm}$$

This should be the same value whether it is calculated from the safe tensile stress or the safe compressive stress. From the compressive side, where $h = 107$ and $\sigma = 117$:

$$\frac{M}{23 \cdot 5 \times 10^6} = \frac{117}{107}; \quad M = 26 \times 10^6 \text{ N mm}$$

347

14/9

S *Solution:*

1. Breadth of the table: ....153 mm
2. Safe bending moment: .... 26 kN m

14/9 **UNSYMMETRICAL I SECTION: Check stresses: Acceptable design?**

A beam, whose section is shown in the sketch, carries a maximum bending moment of 132 kN m. Decide whether the section is over- or under-designed, if the working stresses are 110 N/mm² in tension, and 90 N/mm² in compression. The bending is negative.

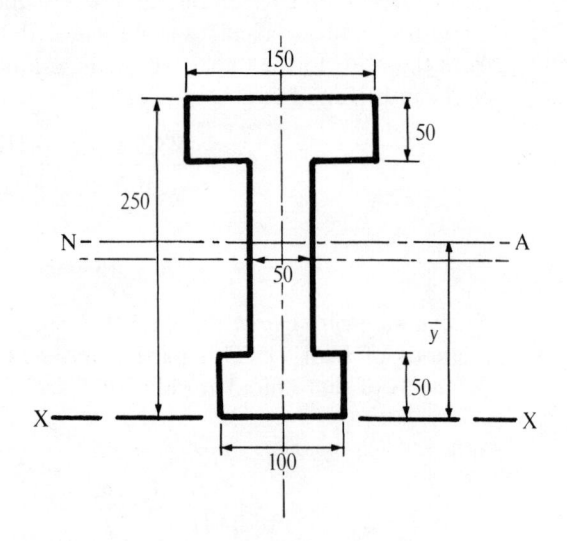

G *Group of Terms:*

$$\frac{M}{I} = \frac{\sigma}{h}$$

U *Unknowns and Units:*

1. Working stresses in tension and compression.
2. Decide whether these are acceptable.

14/9

I *Inventory of Information:*
$$M = 312 \text{ kN m} = 132 \times 10^6 \text{ N mm}$$
$$\sigma_t = 110 \text{ N/mm}^2$$
$$\sigma_c = 90 \text{ N/mm}^2$$

D *Derived Data:*
(a) *Centroid* must be found by the methods of Unit 11, in order to locate the neutral axis.
$$\bar{y} = 138 \text{ mm from XX}$$

(b) *Second moment of area* about the N.A. This cross-section is studied in Unit 11, and the value of
$$I_{NA} = 141 \times 10^6 \text{ mm}^4$$

E *Evaluate:*
$$\frac{M}{I} = \frac{\sigma}{h}$$

(1) Tensile stress (upper face) $\dfrac{132 \times 10^6}{141 \times 10^6} = \dfrac{\sigma_t}{112}$; $\sigma_t = \mathbf{106 \text{ N/mm}^2}$

(1) Compressive stress (lower face) $\dfrac{132 \times 10^6}{141 \times 10^6} = \dfrac{\sigma_c}{138}$; $\sigma_c = \mathbf{128 \text{ N/mm}^2}$

This section is not well designed, for, although the tensile stress is within the allowable figure, the compressive stress is greater than the allowable. The bending moment applied must be reduced, or, better still, the section changed in shape to give a greater area towards the lower edge. This would not only increase the second moment of area of the section, but would produce a shorter *h* on the compressive side (because of the downward movement of the N.A.) and so reduce the stress developed. Repeat this with a larger bottom flange.

S *Solution:*
The answers specifically asked for are:
1. Working stresses: tensile, **106 N/mm²**; compressive, **128 N/mm²**.
2. Not an acceptable design.

14/10

14/10 UNSYMMETRICAL STEEL-PILE SECTION: Safe bending moment?

The approximate section of a patent steel sheet pile is shown in the sketch. What maximum safe bending moment can this section withstand if the working stress is 160 MN/m²? These piles are driven vertically and loaded on the side of what is the lower edge in the diagram. The 'lower' edge carries the tension in such a vertical cantilever.

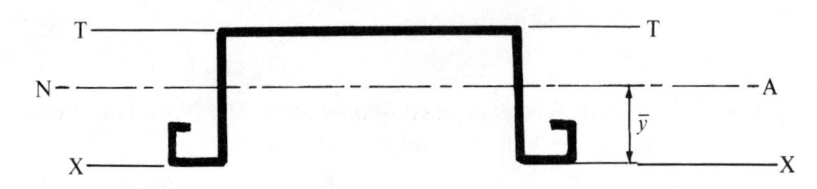

G *Group of Terms:*

$$\frac{M}{I} = \frac{\sigma}{h}$$

U *Unknowns and Units:*

1. Safe bending moment which can be resisted.
Use N and mm.

I *Inventory of Information:*

σ: the working stress = 160 MN/m² = 160 N/mm²

D *Derived Data:*

(a) *Centroid*, for the location of the N.A. is found, by the methods of Unit 11 to be

$\bar{y}$ = 107 mm above XX, and 93 mm below TT

these are the values of h_t and h_c respectively.

(b) *Second moment of area* about the N.A. is found to be

I_{NA} = 137 × 10⁶ mm⁴ (see Unit 11 for this)

E *Evaluate:*

Safe bending moment on the tension (lower) side:

$$\frac{M_t}{137 \times 10^6} = \frac{160}{107}$$

Safe bending moment on the compression (upper) side:

$$\frac{M_c}{137 \times 10^6} = \frac{160}{93}$$

The smaller of these must be taken.

S *Solution:*

1. The two possible safe bending moments are **205 kN m** and **236 kN m**. The smaller must not be exceeded. The safe bending moment is **205 kN m**.

Figures in Problems are often rounded off; use a calculator and see what effect is produced by a greater number of significant figures. Is it worth the extra effort?

Part II: Group E

15 Stress caused by transverse shear

Direct force produces direct stress, and shearing force produces shearing stress. However, whereas direct stress was uniform across the section, you cannot divide the transverse shearing force by the area on which it acts in order to find the shearing stress. This would give the mean shearing stress which is of little significance, since the shearing stress at certain parts of the cross-section of a beam is very much greater than the mean shearing stress, and it is the maximum value in which you are interested.

In this unit of study, therefore, the position on the cross-section at which the stress is measured is of major importance if the maximum shearing stress is to be determined.

The relationship between the various terms follow the same pattern as was outlined in unit 12, which you should again study. The three parts of the relationship are:

$$\begin{pmatrix} \text{A term} \\ \text{concerned} \\ \text{with the loading} \\ \text{applied} \end{pmatrix} = \begin{pmatrix} \text{A term} \\ \text{relating to} \\ \text{the stress} \\ \text{produced} \end{pmatrix} = \begin{pmatrix} \text{A term} \\ \text{relating to} \\ \text{the deformation} \\ \text{produced} \end{pmatrix}$$

This relationship can be further specified by noting that, for transverse shear, the deformation is very small and can be neglected. You have, then only two terms:

$$\begin{pmatrix} \text{The loading divided} \\ \text{by a property of} \\ \text{the cross-section} \end{pmatrix} = \begin{pmatrix} \text{The stress divided} \\ \text{by a dimension} \\ \text{of the cross-section} \end{pmatrix}$$

15.1

15.1 Evaluation of the relationship

For transverse shear (the shearing force calculated as S in part I) the terms can be written as:

$$\frac{S}{I} = \frac{\tau t}{m}$$

S is the shearing force acting across the cross-section in question (N);

I *second* moment of area of the whole cross-section about a centroidal axis at right angles to the direction of the shearing force (m^4);

τ is the shearing stress produced (N/m^2). The Greek letter is *tau*, pronounced to rhyme with 'cow';

m is the *first* moment of the area lying beyond the horizon selected for discussion of shearing stress. The moment of this area (shaded in the diagram) is taken about the centroidal axis;

t is the thickness of the cross-section at the horizon selected – the thick lines in the diagrams.

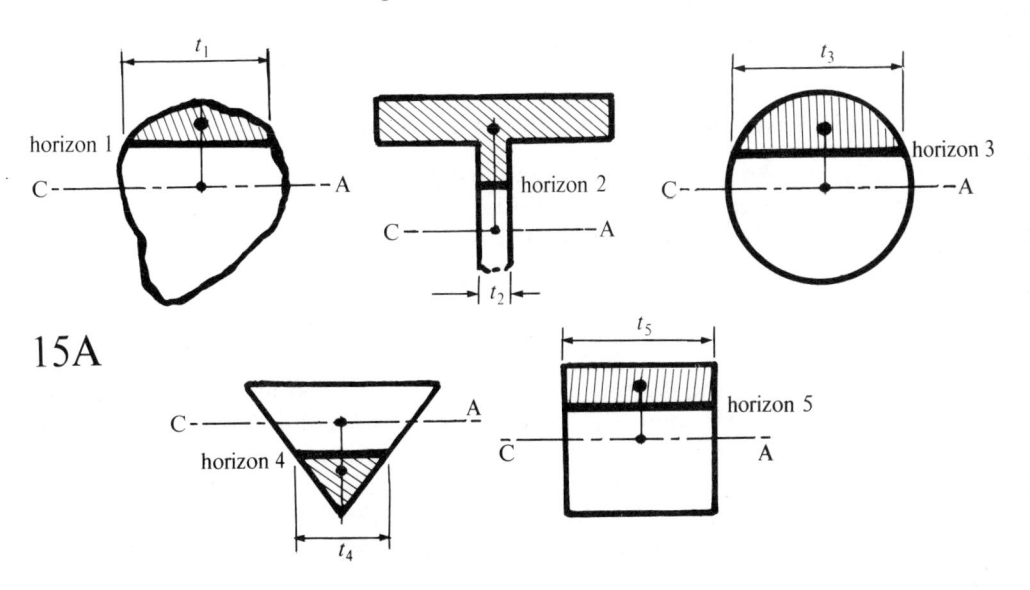

15A

To evaluate the terms of the relationship, you must first decide at what point in the cross-section you wish to know the value of the shearing stress. At this point, measure t the thickness of the section

353

15.2

parallel to the centroidal axis which is at right angles to the direction of shearing force. You then determine the area beyond the horizon (dark line), and also the distance of the centroid of that area (cross-hatched) from the centroidal axis of the whole cross-section. The product of this distance and the cross-hatched area gives the first moment of the area about CA. The second moment of the *whole* area of the cross-section about CA must also be found.

The solution of problems in shearing stress depends, as so many solutions do, on your being able to handle the geometry of areas. Before any result can be obtained you must be able to find:

the centroidal axis of the whole section,

the second moment of area of the whole section about this,

the length t representing the thickness of the section at the selected horizon,

the area of the cross-hatched portion of the cross-section beyond the chosen horizon,

the centroid of this area and,

the distance from this centroid to the centroidal axis of the whole section.

15.2 The influence of variations in individual quantities

An understanding of what happens in the distribution of shearing force over the section as shearing stress, can be obtained by imagining variations in the terms used. Take two examples:

$$\frac{S}{I} = \frac{\tau t}{m}$$

If the shearing force and the second moment of area of the section remain constant, the shearing stress varies directly with m/t. The value of the shearing stress is thus equal to constant $\times$ m/t. Thus if the thickness t decreases, the shearing stress increases; if the moment of the shaded area about the CA decreases, the shearing stress also decreases. The maximum shearing stress on the section is thus obtained at a horizon where the thickness of the section is as small as possible and the moment of the cross-hatched area as large as possible.

In a rectangular beam, for example, t is constant so all you have to do to obtain the maximum value of the shearing stress is to make the moment of the shaded area as large as possible. This means bringing the horizon down to the centroidal axis. For an I-section, as used in steel beams, the thickness t of the web is small, and the shearing stress, therefore, large. In the flange of such a beam, the width of the flange represents t – much larger than that of the web – and the area above this horizon is small. The shearing stress in the flange is, therefore, much smaller than it is in the web.

For the second example, determine what occurs if the shearing stress must not exceed a certain figure – a common requirement in design. If the stress is found to be too high, a reduction in S produces a linear reduction in stress. An increase in I also produces a reduction in stress, but as I is measured in the fourth power of the linear dimension, a small linear change has a considerable effect. Such a change in I will also effect changes in t and in m.

15.3 Suggestions and the mnemonic

The following problems cover a range of types of cross-section. A simple rule helps to decide how to tackle the calculation:

If the thickness t does not vary as the distance from the CA increases, the calculation can be carried out numerically.

If the thickness t varies as the distance of the chosen horizon from the CA increases, it is usually quicker to carry out a general analysis before substituting numerical data.

The mnemonic is again G-U-I-D-E-S. See the explanation in Unit 12 (section 12.5).

15/1

15.4 THE PROBLEMS

15/1 **SYMMETRICAL SECTION: Rectangle: t Constant**
A rectangular beam has a cross-section in which the depth is twice the breadth. What must the breadth be if the maximum shearing stress is not to exceed 50 N/mm²? The loading is a shearing force of 240 kN.

G *Group of Terms:*
$$\frac{S}{I} = \frac{\tau t}{m}$$

U *Unknowns and Units:*
1. Breadth b for $\tau = 50$ N/mm²

I *Inventory of Information:*
$S = 240$ kN $= 240 \times 10^3$ N
$t = b \quad d = 2b$

D *Derived Data:*
$$I_{CA} = \frac{bd^3}{12} = \frac{b(2b)^3}{12} = \frac{2b^4}{3} \text{ mm}^4$$

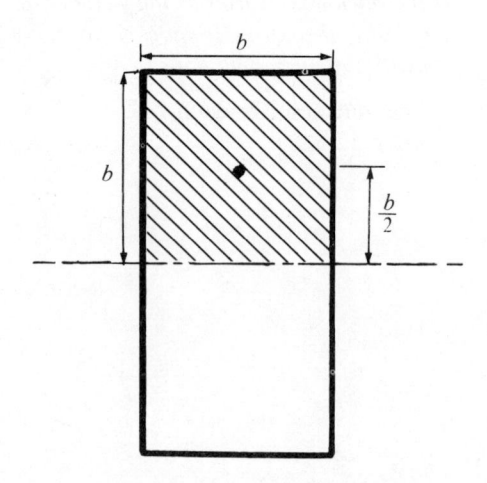

Maximum shearing stress occurs at the centroidal axis, where the cross-hatched area is greatest.

356

The moment of this area (half area of cross-section) about the CA:

$$m = b^2 \times \frac{b}{2} = \frac{b^3}{2} \text{ mm}^3$$

E Evaluation:

$$\frac{S}{I} = \frac{\tau t}{m} : \quad \frac{240 \times 10^3}{2b^4/3} = \frac{\tau b}{b^3/2} = \frac{50b}{b^3/2}$$

Thus $\quad \dfrac{240 \times 10^3}{2b^2/3} = 100 \quad b^2 = 3600 \text{ mm}^2$

S Solution:
1. The breadth required to reduce the maximum shearing stress to 50 N/mm² is **b = 60 mm** and **d = 120 mm**.

15/2 SYMMETRICAL SECTION: Rectangle: *t* Constant

In a beam of rectangular section a shearing force is applied. Show how the shearing stress varies, at any given horizon, with
  (*a*) **variation in the breadth of section,**
  (*b*) **variation in the depth of section,**
  (*c*) **variation in the position of the horizon at which the shearing stress is to be found.**

G Group of Terms:

$$\frac{S}{I} = \frac{\tau t}{m}$$

U Unknowns and Units:
1. General expression for shearing stress at any horizon distant *h* from the upper face.
2. Interpretations of (*a*), (*b*), and (*c*).
Use general units.

15/2

I *Inventory of Information:*
Shearing force: S
Other data in diagram.

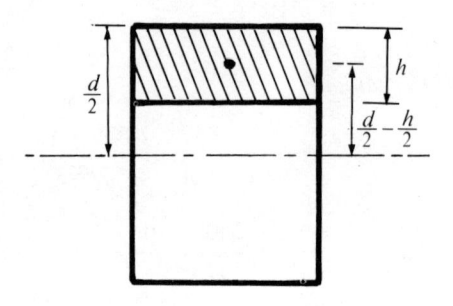

D *Derived Data:*
$$t = b;$$
$$m = \frac{bh}{2}(d - h) \qquad I = \frac{bd^3}{12}$$

E *Evaluation:*
$$\frac{S}{I} = \frac{\tau t}{m}; \qquad \frac{12S}{bd^3} = \frac{\tau b}{\frac{bh}{2}(d-h)}$$

$$\tau = \frac{6hS(d-h)}{bd^3} = \frac{6S}{bd^3}(hd - h^2)$$

$$\frac{d\tau}{dh} = d - 2h. \qquad \text{For maximum } \tau: d - 2h = 0$$

S *Solution:*
(a) The shearing stress varies directly with breadth.
(b) The shearing stress decreases rapidly with increase in depth of section.
(c) The shearing stress is zero when $h = 0$ (outer face) and maximum when $h = \frac{1}{2}d$ (on the CA).

15/3 SYMMETRICAL SECTION: I Beam: Two values of t

A steel joist has the cross-section shown. The shearing force is 390 kN. Find the variation of shearing stress over the depth of the section.

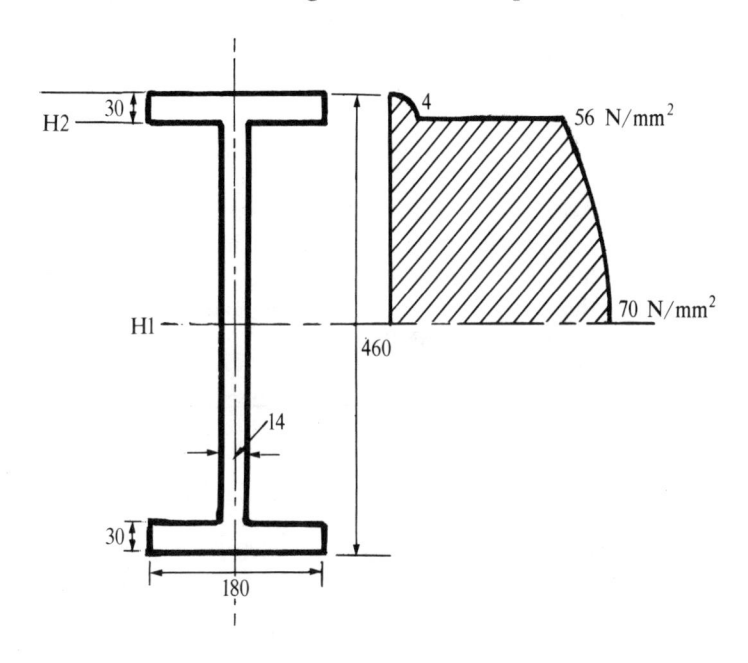

G *Group of Terms:*

$$\frac{S}{I} = \frac{\tau t}{m}$$

U *Unknowns and Units:*
Since t has only two values (14 mm and 180 mm), it is best to obtain the shearing stress at selected critical horizons:
1. at the centroidal axis (H1).   3. at the bottom of the flange (H3).
2. at the top of the web (H2).   4. at the top of the flange (H4).
Use N and mm.

I *Inventory of Information:*
$S = 390 \text{ kN} = 390 \times 10^3 \text{ N}$
$t = 14$ mm in web; 180 mm in flange

15/3

D *Derived Data:*

$$I_{CA} \text{ for flanges} = 2\left\{\frac{bd^3}{12} + Ay^2\right\} = 2\left\{\frac{180 \times 30^3}{12} + (180 \times 30)215^2\right\}$$

$$I_{CA} \text{ for web} = \frac{bd^3}{12} = \frac{14 \times 400^3}{12}$$

I_{CA} for whole section is sum of these $= 575 \times 10^6$ mm^4
Area of flange $= 180 \times 30 = 5400$ mm^2
Area of web $= 14 \times 400 = 5600$ mm^2

In calculating the stresses at the four horizons, there are two cross-hatched areas used. One is half the area of the section, and the other is the area of one flange. The calculations for the four horizons are best carried out in tabular form.

	H1		H2	H3	H4
t	14		14	180	180
Area above selected horizon	Flange 5400	Web 2800	5400	5400	0
Moment arm about CA	215	100	215	215	—
(M) First moment of area	$1{\cdot}44 \times 10^6$		$1{\cdot}16 \times 10^6$	$1{\cdot}16 \times 10^6$	—
m/t	$102{\cdot}9 \times 10^3$		$82{\cdot}9 \times 10^3$	$6{\cdot}4 \times 10^3$	—

E *Evaluation:*

$$\frac{S}{I} = \frac{\tau t}{m} \qquad \tau = \frac{390 \times 10^3}{575 \times 10^6}\frac{m}{t} = 0{\cdot}678 \times 10^{-3}\frac{m}{t}$$

Multiply the values of m/t from the table by $0{\cdot}678 \times 10^{-3}$.

S *Solution:*

1. At horizon 1: $\tau = 70 \text{ N/mm}^2$
2. At horizon 2, top of web: $\tau = 56 \text{ N/mm}^2$
3. At horizons 3, just inside flange: $\tau = 4 \text{ N/mm}^2$
4. At horizon 4: τ is zero.

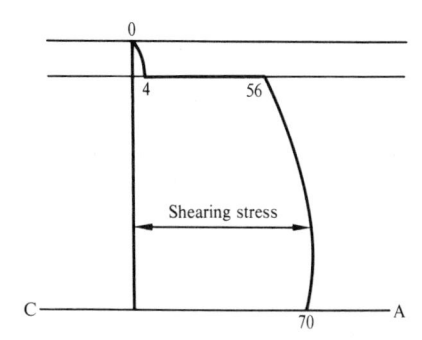

This diagram for I beams is sometimes referred to as the 'bowler hat' diagram.

15/4 SYMMETRICAL SECTION: Solid circle: Varying t

What should be the radius of a solid shaft of circular cross-section which carries a shearing force of 500 kN and whose working stress is 150 N/mm²?

G *Group of Terms:*

$$\frac{S}{I} = \frac{\tau t}{m}$$

U *Unknowns and Units:*

1. Minimum safe radius of solid shaft.
Use N and mm.

I *Inventory of Information:*

$S = 500 \text{ kN} \quad 500 \times 10^3 \text{ N}$
$\tau = 150 \text{ N/mm}^2$

15/4

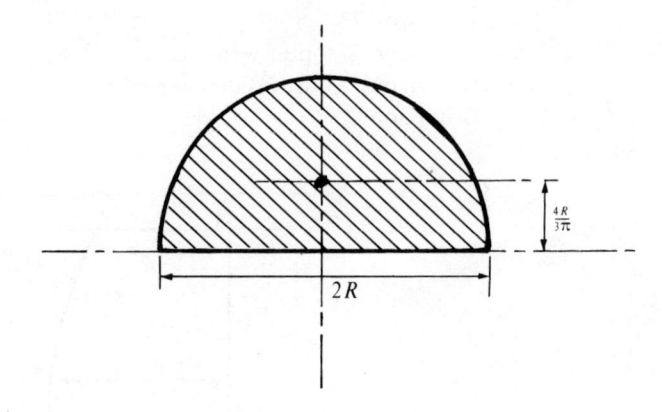

D *Derived Data:*

Maximum shearing stress is on a diameter (the centroidal axis).

Cross-hatched area (CHA) and its moment (m) about CA

CHA	Moment arm about CA	Moment (m) about CA
Semi-circle $\pi r^2/2$	$4R/3\pi$	$2R^3/3$ mm^3

Second moment of area of whole shaft

$$I_{CA} = \frac{\pi R^4}{4} \text{ mm}^4$$

Thickness (t) at horizon CA = 2R mm

E *Evaluation:*

$$\frac{4 \times 500 \times 10^3}{\pi R^4} = \frac{50 \times 2R \times 3}{2R^3}; \quad R^2 = 1414 \text{ mm}^2$$

S *Solution:*

1. Minimum safe radius of the shaft to carry 500 kN at maximum shearing stress of 150 N/mm² is **38 mm**.

15/5 SYMMETRICAL SECTION: Solid circle: Varying m/t

Trace the variations of m/t in the relationship $\tau =$ Constant $\times\ m/t$ for a solid circular shaft. Draw a diagram showing this variation.

G *Group of Terms:*

$$\frac{S}{I} = \frac{\tau t}{m} \qquad \tau = \frac{S}{I} \cdot \frac{m}{t}$$

U *Unknowns and Units:*
1. A study of m/t.
2. A diagram of the variation found.

I *Inventory of Information:*

The section is circular. The value t is the breadth of the circle at the horizon selected. The value m is the moment of the cross-hatched area (CHA) beyond the selected horizon, about the CA.

D *Derived Data:*

$$t = R\cos\theta_t \qquad m = \int_h^R wy\,dy$$

where w is the width of an elementary area and y is its moment arm about CA.

Transform to polar co-ordinates by making
$w = 2R\cos\theta$
$y = R\sin\theta$
$dy = R\cos\theta\,d\theta$
At h, θ is θ_t; at R, θ is $\pi/2$.

$$m = \int (2R\cos\theta)(R\sin\theta)(R\cos\theta)\,d\theta = 2R^3 \int (\cos^2\theta \sin\theta)d\theta$$

$$m = \frac{2}{3} R^3 \cos^3\theta_t \qquad \frac{m}{t} = \frac{R^2}{3}\cos^2\theta_t$$

15/5

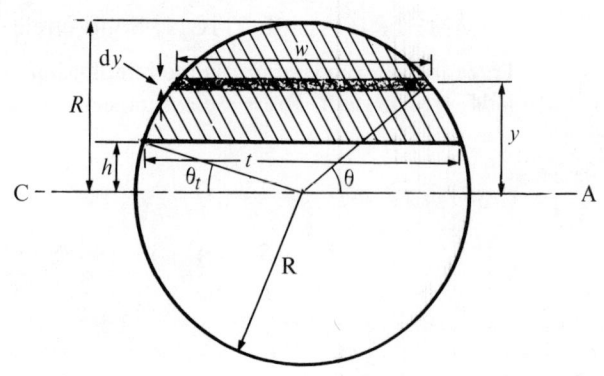

E *Evaluation:*

Since $R^2/3$ is a constant, the shearing stress at the horizon defined by θ_t is equal to a constant multiplied by $\cos^2 \theta_t$. A tabular calculation gives the diagram of variation asked for.

θ_t	$h = R \sin \theta_t$	$\cos \theta_t$	$\cos^2 \theta_t$	$R^3/3 \cos^2 \theta_t$
0	0	1	1	$0 \cdot 33 R^2$
30°	$0 \cdot 5R$	3/2	0·75	$0 \cdot 25 R^2$
60°	$0 \cdot 87R$	0·5	0·25	$0 \cdot 08 R^2$
90°	R	0	0	0

S *Solution:*

This table can be calculated more exactly and with more values of h or θ_t. When plotted, the diagram allows of the definition of m/t in terms of R^2 when h (the height to the selected horizon) is known as a proportion of R.

For example, in the first line of the table h is zero, which defines the diameter, and m/t is then $0 \cdot 33R^2$. The maximum shearing force, therefore, which occurs on the diameter, the centroidal axis, is

$$\tau = \frac{S}{I} \times \frac{m}{t} = \frac{4S}{\pi R^4} \cdot \frac{R^2}{3} = \frac{4S}{3\pi R^2}$$

This is four-thirds of the mean shearing stress which is merely the value S divided by the area of the circle. In a circular section, therefore, the

364

maximum shearing stress due to a transverse shearing force is four-thirds of the mean shearing stress.

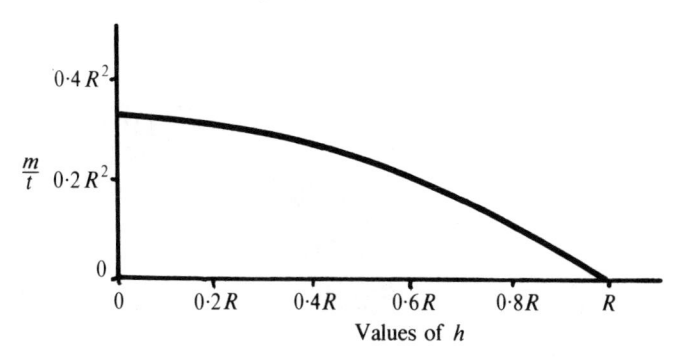

[Graph: vertical axis m/t from 0 to $0.4R^2$ (with $0.2R^2$ marked); horizontal axis "Values of h" from 0 to R (with $0.2R$, $0.4R$, $0.6R$, $0.8R$ marked); curve starts near $0.4R^2$ and decreases to 0 at R.]

15/6 SYMMETRICAL SECTION: Annular area: Varying m/t

Determine the maximum intensity of shearing stress for a hollow shaft, 228 mm external diameter and 150 mm internal diameter, if it is subjected to a shearing force of 400 kN. What is the ratio of maximum to mean shearing stress?

G *Group of Terms:*

$$\frac{S}{I} = \frac{\tau t}{m}$$

U *Unknowns and Units:*
1. Maximum intensity of shearing stress.
2. Ratio of maximum to mean shearing stress.
Use N and mm.

I *Inventory of Information:*
$S = 400 \text{ kN} = 400 \times 10^3 \text{ N}$ $R = 114 \text{ mm}$
$r = 75 \text{ mm}$ $t(\text{ at diameter}) = 2(R - r) = 78 \text{ mm}$
CA is on diameter.

D *Derived Data:*
Maximum shearing stress occurs at centroidal axis.

$$I_{CA} = \frac{\pi}{4}(114^4 - 75^4) = 112 \cdot 7 \times 10^6 \text{ mm}^4$$

15/6

$$\text{Mean shearing stress} = \frac{400 \times 10^3}{\pi (R^2 - r^2)}$$

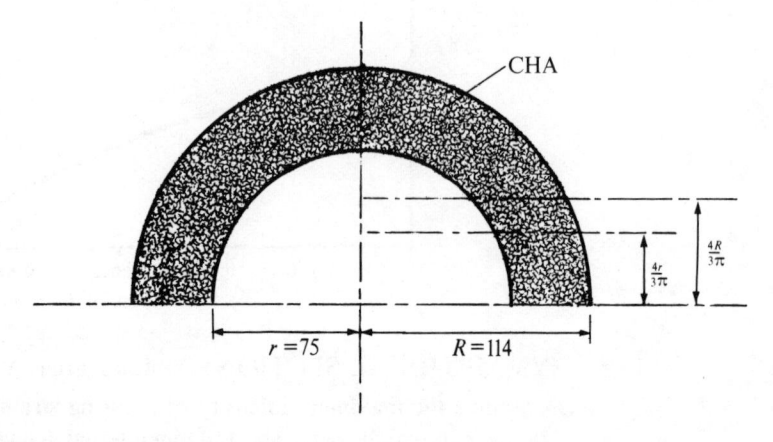

[Diagram: semi-circular cross-section (CHA) with $r = 75$, $R = 114$, moment arms $4r/3\pi$ and $4R/3\pi$]

Table of first moments of CHA (semi-circle)

	CHA	Moment arm about CA	Moment (m) of CHA about CA
Large semi-circle	$\pi R^2/2$	$4R/3\pi$	$\tfrac{2}{3}R^3 - \tfrac{2}{3}r^3$
Small semi-circle	$\pi r^2/2$	$4r/3\pi$	$= \tfrac{2}{3}(114^3 - 75^3)$
			$= 0.705 \times 10^6 \text{ mm}^3$

E *Evaluation:*

$$\frac{S}{I} = \frac{\tau t}{m}$$

$$\tau = \frac{Sm}{It} = \frac{400 \times 10^3 \times 0.705 \times 10^6}{112.7 \times 78}$$

S *Solution:*

1. Maximum shearing stress is on the diameter and has a value of **32 N/mm²**.
2. Mean shearing stress is **17·3 N/mm²**, and the ratio of maximum to mean is **1·86**.

15/7

15/7 UNSYMMETRICAL SECTION: Tee beam: Two values of t

A 'Tee' section has dimensions 150 × 150 × 12 mm. It is used as a beam with the web vertical and above the flange, as shown in the sketch. The shearing force applied is 60 kN. Find the maximum shearing stress and the horizon at which it occurs.

G *Group of Terms:*
$$\frac{S}{I} = \frac{\tau t}{m}$$

U *Unknowns and Units:*
1. Horizon at which maximum shearing stress occurs.
2. Value of maximum shearing stress.
Use N and mm.

I *Inventory of Information:*
$S = 60 \text{ kN} = 60 \times 10^3 \text{ N}$ $\qquad t = 12 \text{ mm}$

D *Derived Data:*

Centroid of area: Moments about base:

	b	d	bd	A	y	Ay
R1	12	150	1800	1800	75	135×10^3
R2	$(150 - 12)$	12	1656	1656	6	9.9×10^3
Total				3456		144.9×10^3

Height of CA above base is 42 mm.

Second moment of whole area of section about CA

	d^2	bd^3	I	y	y^2	Ay^2	I_{CA}
R1	22.5×10^3	40.5×10^6	3.4×10^6	33	1089	2.0×10^6	5.4×10^6
R2		Negligible		36	1296	2.1×10^6	2.1×10^6

$$I_{CA} = 7.5 \times 10^6$$

15/7

In problems of this kind the second moments of areas which are narrow at right angles to the CA are usually so small relative to the values of Ay^2 that they can usually be neglected.

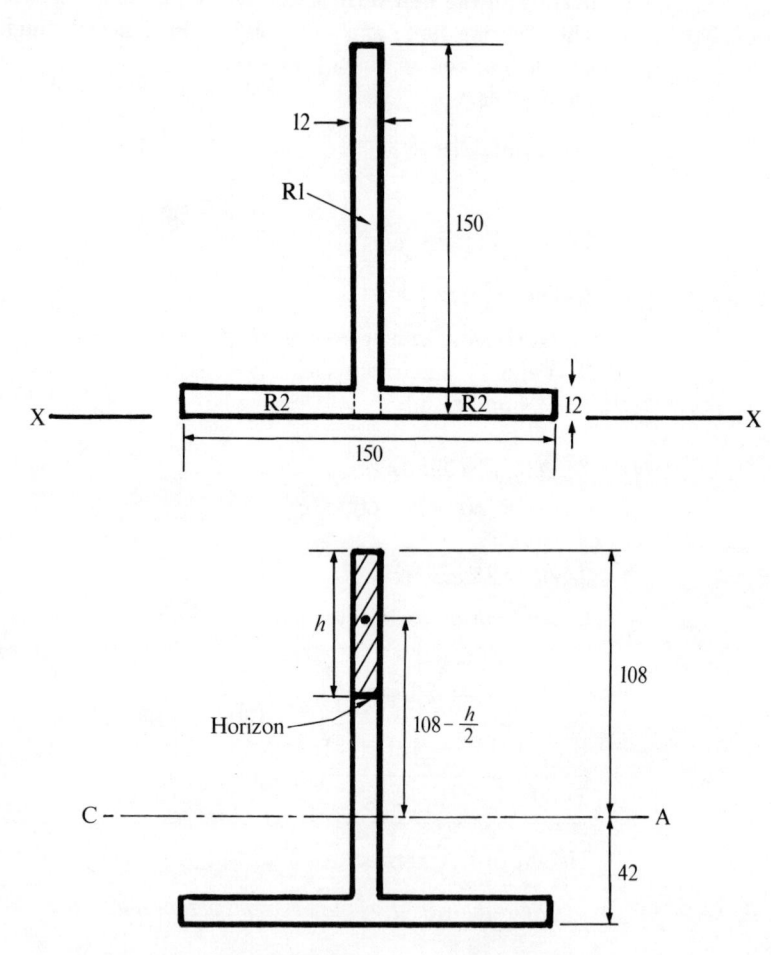

First moment of CHA about CA

	CHA	Moment arm about CA	Moment (m) about CA
R3	$12h$	$(108 - \tfrac{1}{2}h)$	$1296h - 6h^2$

368

E *Evaluation:*

$$\frac{S}{I} = \frac{\tau t}{m}$$

$$\frac{60 \times 10^3}{7\cdot 5 \times 10^6} = \frac{12\tau}{1296h - 6h^2}$$

$$\tau = 0\cdot 667 \times 10^{-3}(1296h\ 6h^2)\ \text{N/mm}^2$$

If τ is to be a maximum, then $d\tau/dh = 0$; $1296 - 12h = 0$. The maximum shearing stress occurs at the centroidal axis. Substitute 108 mm for h in the equation.

S *Solution:*

1. Maximum stress occurs at the centroidal axis.
2. Value of maximum shearing stress is **85 N/mm²**.

Be sure you know what unknowns are required.

Use tabular method to find M and m.

Re-write the information you have been given – the inventory.

15/8

15/8 UNSYMMETRICAL SECTION: Compound girder: Three values of t

A compound girder over a shop-front consists of two I-beams with a flange plate on top. If the maximum shearing force on a section is 150 kN, find the maximum value of the shearing stress. The centroidal axis of the whole section is 186 mm from the lower edge. The second moment of area of one beam about its own centroidal axis is 72×10^6 mm^4, and the cross-sectional area of one beam is 6000 mm^2.

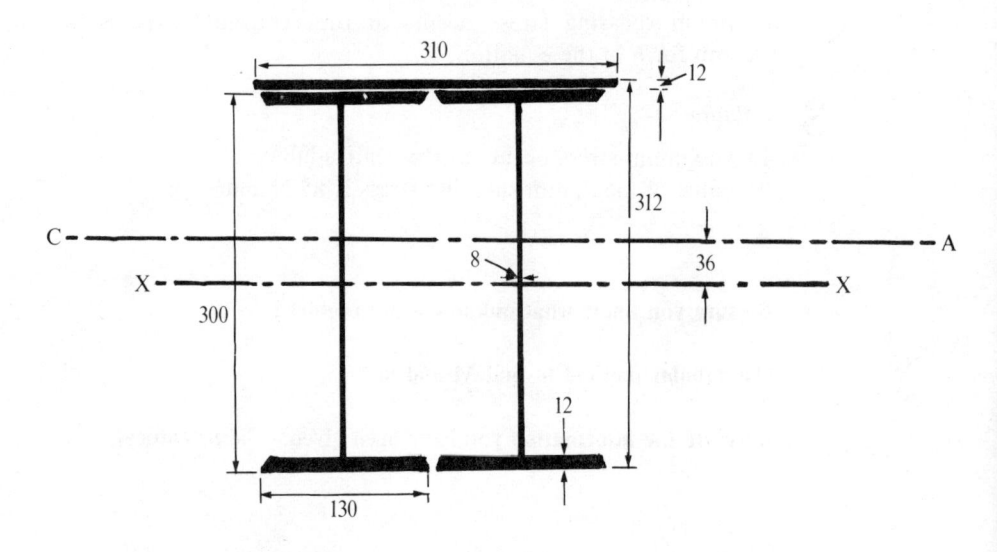

G *Group of Terms:*

$$\frac{S}{I} = \frac{\tau t}{m}$$

U *Unknowns and Units:*

1. Maximum value of the shearing stress.
2. Horizon at which the maximum value occurs.

Use N and mm.

I *Inventory of Information:*

$S = 150 \times 10^3$ N
$I = 72 \times 10^6$ mm^4 for one beam
A: Cross-sectional area of one beam = 6000 mm^2

D *Derived Data:*

Second moment of area of whole beam about its CA:

I_{CA} of two beams $= 2(72 \times 10^6) + 2(Ay^2)$
$\qquad\qquad\qquad\qquad = 144 \times 10^6 + 2(6000 \times 36^2)$

I_{CA} of flange plate $= (310 \times 12^3)/12 + Ay^2$
$\qquad\qquad\qquad\qquad = 310 \times 144 + (310 \times 12)120^2$

I_{CA} of whole girder $= 10^6(144 + 15\cdot6) + 10^6(0\cdot04 + 53\cdot6)$
$\qquad\qquad\qquad\qquad I = 213\cdot2 \times 10^6$ mm^4

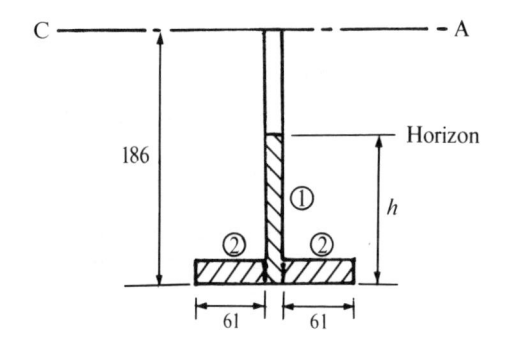

Moment of the cross-hatched area (CHA) about the CA:

Take a distance h from the bottom or from the top. It is easier from the bottom so use that. To make it easier to calculate, take the CHA in two parts as shown in the sketch. The web area is taken right to the bottom which gives its height as a simple h.

Table of first moments

	CHA	Moment arm about CA	m
Area (1)	$8h$	$186 - \tfrac{1}{2}h$	$1488h - 4h^2$
Area (2)	122×12	180	$26\,352$

There are two of these beams, so the total value of m:

$$m = 2\{-4h^2 + 1488h + 26\,352\}$$

The value of t is 8 mm for the selected horizon.

15/9

E *Evaluation:*

$$\frac{S}{I} = \frac{\tau t}{m}$$

$$\frac{150 \times 10^3}{213 \cdot 2 \times 10^6} = \frac{8\tau}{2(-4h^2 + 1488h + 26\,352)}$$

$$\tau = 0 \cdot 705 \times 10^{-3}(-h^2 + 372h + 6588)$$

If τ is to be a maximum, then

$$\frac{d\tau}{dh} = 0 \quad \text{or} \quad -2h + 372 = 0; \quad +h = +186 \text{ mm}$$

The maximum stress occurs at the centroidal axis.
Substitute 186 for h:

$$\text{max. } \tau = 0 \cdot 705 \times 10^{-3}\{+(186)^2 + 2(186)^2 + 6588\}$$

S *Solution:*

1. The maximum value of the shearing stress is **78 N/mm²**.
2. The maximum shearing stress occurs at the centroidal axis.

15/9 UNSYMMETRICAL SECTION: Triangle: Varying m/t

In cross-section, a beam is an equilateral triangle of 200 mm side, the line of loading being perpendicular to one side. If the total shearing force is 50 kN, find the maximum shearing stress and the shearing stress at the neutral axis.

G *Group of Terms:*

$$\frac{S}{I} = \frac{\tau t}{m}$$

U *Unknowns and Units:*

1. Shearing stress at the neutral axis (CA).
2. Location of maximum shearing stress on the section.
3. Maximum shearing stress on the section.

Use N and mm.

I *Inventory of Information:*

S: 50 kN = 50 × 10³ N
b: 200 mm $\qquad d$: 100√3 mm

Section is as shown in sketch.

D *Derived Data:*

Second moment of area of whole section

$$I = \frac{bd^3}{36} \quad \frac{200 \times (100\sqrt{3})^3}{36} = \frac{200\sqrt{3}}{9} \times 10^6 \text{ mm}^4$$

Thickness t

$$t = bh/d$$

from the geometry of the triangles.

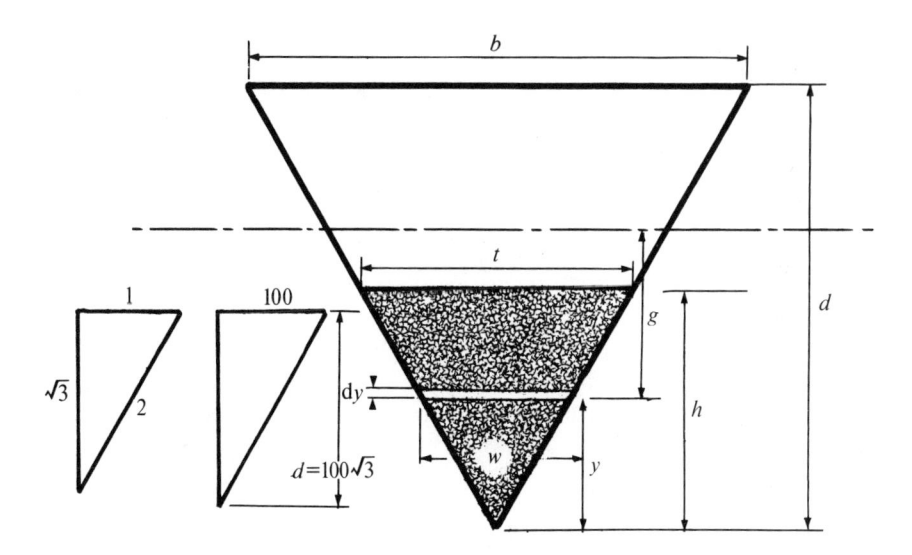

Moment of CHA about CA or NA

The position of the centroidal or neutral axis of a triangle is known to be one-third of the depth from the upper face.

An elementary area (w dy) has a first moment about the neutral axis of (wg dy).

15/9

In terms of the known dimensions, from the geometry of the triangle:

$$w = \frac{by}{d}: \quad g = (\tfrac{2}{3}d - y)$$

$$m = \int_0^h wg\,dy = \int_0^h \frac{by}{d}(\tfrac{2}{3}d - y)\,dy$$

$$= \frac{bh^2}{3}\left(1 - \frac{h}{d}\right) \text{ mm}^3$$

E *Evaluation:*

$$\frac{S}{I} = \frac{\tau t}{m}$$

$$\frac{50 \times 10^3}{\left[\frac{200\sqrt{(3)}}{9}\right] \times 10^6} = \frac{\tau(bh/d)}{(bh^2/3)[1 - (h/d)]}$$

$$\tau = \frac{\sqrt{(3)}hd}{4 \times 10^3}(1 - h)$$

When $h = \tfrac{2}{3}d$, the horizon is at the neutral axis

$$\tau = \frac{2d^2\sqrt{(3)}}{12 \times 10^3}\left(1 - \frac{2}{3}\right) = 2\cdot 9 \text{ N/mm}^2$$

$$\frac{d\tau}{dh} = 0$$

Thus,

$$d - 2h = 0 \quad \text{or} \quad h = \tfrac{1}{2}d$$

Thus the maximum shearing stress on the section occurs when the horizon at which the stress is measured is at half the value of the depth.

At this horizon:

$$\tau = \frac{\sqrt{(3)}hd}{4 \times 10^3}\left(1 - \frac{h}{d}\right) = \frac{\sqrt{(3)}d^2}{16 \times 10^3} = \frac{\sqrt{(3)}[100\sqrt{(3)}]}{16 \times 10^3} \text{ N/mm}^2$$

S *Solution:*

1. At the neutral axis the shearing stress is **2·9 N/mm²**.
2. The maximum value of shearing stress occurs at a horizon half the depth of the section from the horizontal upper face.
3. The value of the maximum shearing stress is **3·3 N/mm²**.

Part II: Group E

16 Torsion

When a bar of circular section is twisted, it can be imagined to consist of a very large number of circular slices, each of which rotates infinitesimally on the one next to it, as the twisting couples are applied. Hold a pencil at each end and twist slightly, and this effect can be visualized. Whereas, in bending, the couples are applied in the plane of the beam, for torsion the couples are applied in planes at right angles to the plane containing the axis of the member. As for bending, there must be a couple at each end in order to produce the stress and deformation which you try to calculate. The deformation of the twisted member over a short length is very small, since it is one of the assumptions that the material remains elastic and returns to its original shape when the couples are released.

16.1 The relationship in torsion

The group of terms relating to stress and deformation of a beam under torsional couples takes the same form as do the other three:

$$\frac{T}{J} = \frac{\tau}{r} = \frac{C\theta}{l}$$

As in the study of the other types of loading, *the first term* refers to the applied loading and a property of the section which, in this instance, is the *polar second moment of area*. *The second term* refers to stress, and, as for bending, the stress increases as the distance from the axis increases. If this were not so, the equation between the first two terms could not be maintained. So long as the torque or twisting couple, and the size of the section remain unaltered, their ratio remains unaltered, and the ratio in the second term must also be maintained at the same figure. Thus if the radius measured out from the axis increases, the shearing or torsional stress must also increase until it is at a maximum at the outer fibres. The *third term* refers, as before, to the deformation, and contains, as in the other groups, the *modulus* – the property of the material which

16.2

defines its sensitivity to deformation. This is the *modulus of rigidity* (C) or *shear modulus*, which is usually smaller than the *modulus of elasticity*. The combined term θ/l represents a type of strain as did d/l for direct stress. Just as d/l represents the rate of deformation, or deformation per unit length, the ratio θ/l represents the rate of twist per unit of length. Once again, for a constant torque and constant diameter of cross-section, the value of the ratio remains constant. Thus if the length of the twisted bar increases, the angle through which one end twists relative to the other also increases in order to maintain the constant ratio. The angle is very small in order to keep within the elastic limit of the material, and is measured in radians, or as a tangent, which, for small angles gives the same figure.

16.2 The influences of variations in individual quantities

As with the other types of loading, it is well if you study the changes which must take place in these three ratios if one of the seven quantities changes. Suppose, for example, that the angle θ is increased, and at the same time the torque T is also increased. Provided they increase in the same ratio, the equality between the first and third terms remains, and the bar is merely twisted to a greater extent than before. But, in these circumstances, the middle term must also increase in value by the same proportion. This means that the stress must increase. Again, assume that the rod is made smaller in diameter. What happens (with the value of torque remaining constant)? A decrease in diameter not only affects the central (stress) term, but causes a considerable reduction in the *second polar moment of area*, since the reduction is not linear, but to the fourth power. The first ratio, therefore increases by a large amount, which means that the other two terms must also do so. Stress, therefore increases to keep the middle term to the correct value, and either θ, the angle of twist, must increase, or the length twisted must decrease. Another method to maintain the equality of all the ratios, if the angle of twist and length remain the same, is to increase C. This means that the new material must be very much stiffer and more resistant to twisting if the diameter is decreased. You are advised to run over similar arguments until you can visualize the twisted rod and how it behaves. Try, for example to work out what happens for four separate changes in each of the three ratios (all must remain equal). In turn, what happens when

(*a*) torque doubles in value, or

16.3

(*b*) stress must be halved, or

(*c*) the angle of twist is to be decreased by one-third, or

(*d*) a much stiffer material is to be used, and simultaneously the shearing stress must be reduced?

As for the other types of loading, the procedure you must adopt in solving the problems is first to clear your mind of what is required by way of answer, secondly what explicit data you have been provided with, thirdly what further data you must derive from the conditions of the problem. The aim is to obtain the seven values for substitution in the terms, and *all must be in consistent units*.

There are seven quantities which influence each other in the torsional problem. Know what they mean as symbols.

Always give the solution of the problem explicitly even if you have worked it out previously. A final statement helps the examiner.

16/1

16.3 THE PROBLEMS

16/1 SOLID SHAFT: Diameter and safe torque

If a circular shaft is twisted until the angle of twist is 1° per metre of length, what should be the diameter of the shaft if the allowable shearing stress is 62 N/mm² and the modulus of rigidity 83 × 10³ MN/m²? In these limiting circumstances, what is the maximum torque which can be applied?

G *Group of Terms:*

$$\frac{T}{J} = \frac{\tau}{r} = \frac{C\theta}{l}$$

U *Unknowns and Units:*
1. The diameter of the shaft.
2. The maximum torque allowed.
Use N and mm.

I *Inventory of Information:*

$\tau = 62$ N/mm² $\quad C = 83 \times 10^3$ MN/m² or N/mm²
$\theta = 1°$ or $\pi/180$ $\quad l = 1$ m $= 1000$ mm

D *Derived Data:*
Polar second moment of area: $J = (\pi r^4/2)$ mm⁴

E *Evaluation:*

$$\frac{2\tau}{\pi r^4} = \frac{62}{r} = \frac{83 \times 10^3 \pi}{180 \times 10^3}$$

$$r = \frac{62 \times 180}{83\pi} = 42.7 \text{ mm}$$

$$T = \frac{62\pi r^3}{2} = 7.6 \times 10^6 \text{ N mm}$$

S *Solutions:*

The answers are:
1. Diameter should be **86 mm**.
2. Torque allowed **7·6 × 10³ N m**.

16/2 SOLID SHAFT: Diameter and safe torque?

In a solid shaft under torsion, the maximum stress must not exceed 33·5 N/mm². The twist permitted is 1° per 1910 mm of length of shaft. If the modulus of rigidity of the material of the shaft is 77 × 10³ MN/m² what is the required diameter of the shaft and the maximum torque permitted?

G *Group of Terms:*
$$\frac{T}{J} = \frac{\tau}{r} = \frac{C\theta}{l}$$

U *Unknowns and Units:*
1. Diameter of shaft.
2. Maximum torque permitted.
Use N and mm.

I *Inventory of Information:*
$\tau \not> 33\cdot5$ N/mm²
$C = 77 \times 10^3$ MN/m² or N/mm²
$\theta/l = 1°$ per 1910 mm $= 9\cdot14 \times 10^{-6}$ radians per mm

D *Derived Data:*
$J = \pi r^4/2$
$C\theta/l = 77 \times 10^3 \times 9\cdot14 \times 10^{-6} = 0\cdot704$ N/mm²

E *Evaluation:*

(1) $\quad \dfrac{\tau}{r} = \dfrac{C\theta}{l} = 0\cdot704; \quad r = \dfrac{33\cdot5}{0\cdot704} = 48$ mm

(2) $\quad \dfrac{T}{J} = 0\cdot704; \quad T = \dfrac{\pi r^4}{2} \times 0\cdot704 = \dfrac{\pi}{2} \times 5\,308\,416 \times 0\cdot704$
$\qquad\qquad\qquad\qquad\qquad\qquad\qquad\qquad = 5\cdot87 \times 10^6$ N mm

S *Solutions:*
The answers required are:
1. Diameter, **96 mm**.
2. Maximum torque **5·87 kN m**.

16/3

16/3 SOLID SHAFT: Angle of twist and maximum stress?

The allowable torque on a solid shaft is 1885 N m. The shaft has a diameter of 50 mm, and a modulus of rigidity of 77×10^3 MN/m². What twist is developed in the shaft when it has the maximum allowable torque applied? What is the maximum stress?

G *Group of Terms:*

$$\frac{T}{J} = \frac{\tau}{r} = \frac{C\theta}{l}$$

U *Unknowns and Units:*
1. Value of twist developed.
2. Stress developed at the outer face of the shaft.
Use n and mm.

I *Inventory of Information:*
$T = 1885$ N m $= 1885 \times 10^3$ N mm
Radius of shaft is half diameter: $r = 25$ mm
$C = 77 \times 10^3$ MN/m² or N/mm²

D *Derived Data:*
$J = \pi r^4/2 = 614 \times 10^3$ mm⁴

E *Evaluation:*

$$\frac{1885 \times 10^3}{614 \times 10^3} = \frac{77 \times 10^3 \theta}{l} \qquad \frac{3 \cdot 07}{77 \times 10^3} = \frac{\theta}{l} = 0 \cdot 04 \text{ radians}$$

$$3 \cdot 07 = \frac{\tau}{25}$$

$$\tau = 25 \times 3 \cdot 07 = 76 \cdot 8 \text{ N/mm}^2$$

S *Solutions:*

The answers to the two questions asked are:
1. Twist per metre of length of shaft **0·04 rad**.
2. Maximum stress **76·8 N/mm²**.

16/4 SOLID SHAFT: Modulus of rigidity and diameter?

A solid straight shaft of circular cross-section is twisted about its longitudinal axis. The shaft twists through an angle of 1° in a length of 1 m, and the maximum permissible stress is 61·8 N/mm². If the torque producing this stress is 7·75 × 10³ N m, what is the modulus of rigidity and the required diameter of the shaft?

G *Group of Terms:*
$$\frac{T}{J} = \frac{\tau}{r} = \frac{C\theta}{l}$$

U *Unknowns and Units:*
1. Modulus of rigidity of the material.
2. Diameter of the shaft.
Use N and mm.

I *Inventory of Information:*
$\theta/l = 1°$ in 1 m $= 0.0175 \times 10^{-3}$ radians/mm
$\tau = 61.8$ N/mm²
$T = 7.75 \times 10^6$ N mm

D *Derived Data:*
$J = \pi r^4/2$ mm⁴

E *Evaluation:*
$$\frac{T}{J} = \frac{\tau}{r}: \qquad \frac{7.75 \times 10^6 \times 2}{\pi r^4} = \frac{61.8}{r}$$

$$\frac{\tau}{r} = \frac{C\theta}{l}: \qquad \frac{61.8}{r} = C \times 0.0175 \times 10^{-3}$$

S *Solutions:*
1. The modulus of rigidity of the material is **82 × 10³ N/mm²**
2. The required diameter of the shaft is **86 mm**.

16/5

16/5 SOLID SHAFT: Diameter and angle of twist?

What is the angle of twist on 3 m of a shaft subjected to 12 350 N m of torque? The stress developed in shear should not be greater than 55 N/mm². The material has a modulus of rigidity of 77×10^3 MN/m². In such conditions what should be the diameter of the shaft?

G *Group of Terms:*
$$\frac{T}{J} = \frac{\tau}{r} = \frac{C\theta}{l}$$

U *Unknowns and Units:*
1. Angle of twist on 3 metres of length.
2. Radius and diameter of shaft.

Use N and mm.

I *Inventory of Information:*
$T = 12\,350$ N m $= 12\cdot35 = 10^6$ N mm
$C = 77 \times 10^3$ MN/m² or N/mm²
$\tau = 55$ MN/m² or N/mm²

D *Derived Data:*
$J = \pi r^4/2$

E *Evaluation:*
$$\frac{12\cdot35 \times 10^6 \times 2}{\pi r^4} = \frac{55}{r} = \frac{77 \times 10^3 \theta}{3 \times 10^3}$$

$$r^3 = \frac{12\cdot35 \times 10^6 \times 2}{55\pi}; \qquad r^3 = 0\cdot143 \times 10^6$$

$$r = 52\cdot3 \text{ mm}$$

$$\theta = \frac{55 \times 3}{77r} = 0\cdot041 \text{ radius}$$

S *Solutions:*

The two answers are:
1. Angle of twist on 3 m is **2·4°**.
2. Diameter of shaft should be at least **105 mm**.

16/6 SOLID SHAFT: Diameter and design of coupling?

If a solid circular shaft, under the action of a torque of 0·037 MN m, is to be coupled to another length by a flange coupling, find the diameter of the shaft and the diameter of the circle of bolts in the coupling flange. There are 12 bolts of 30 mm diameter. The maximum allowable shearing stress in the material of the shaft is 62 MN/m² and in the bolts of the coupling 46 MN/m².

G *Group of Terms:*

There is no mention in the problem of a study of deformation, so the third term of the group is not required.

$$\frac{T}{J} = \frac{\tau}{r}$$

U *Unknowns and Units:*

1. Diameter of the shaft.
2. Diameter of the ring of bolts.

Use MN and m.

I *Inventory of Information:*

$T = 0\cdot037$ MN m $= 37 \times 10^3$ N m
$\tau_s = 62$ MN/m² or N/mm²
$\tau_b = 46$ N/mm²
Number of bolts, 12

D *Derived Data:*

Polar second moment of area: $J = \pi r^4/2$ m⁴

E *Evaluation:*

$$\frac{T}{J} = \frac{\tau_s}{r} \qquad \frac{0\cdot037 \times 2}{\pi r^4} = \frac{62}{r} \qquad r = 0\cdot072 \text{ m}$$

The allowable shearing force in one bolt multiplied by the radius of the ring of bolts gives the torque supplied by one bolt. The total torque is 12 times this value.

Shearing force in one bolt:

$$46\pi \times 15^2 \text{ N} = 0\cdot0326 \text{ MN}$$

16/7

Total torque, which must equal 0·037 MN m for equilibrium, gives rise to equation:

$$\Sigma \text{Torque} = 0$$
$$+12 \times 0{\cdot}0326 \times R - 0{\cdot}037 = 0$$
$$R = 0{\cdot}095 \text{ m}$$

This is the minimum theoretical radius. For the purpose of inserting the bolts, the radius may have to be made larger.

S *Solutions:*

The answers required are:
1. Diameter of shaft **144 mm**.
2. Diameter of ring of bolts **190 mm minimum**.

16/7 HOLLOW SHAFT: Outer and inner diameters?

A hollow shaft when submitted to torsion must not exceed 1° of twist in 3 m of length. The working stress allowed in shear is 46 N/mm². C is 83 × 10³ MN/m². Find the dimensions of the shaft section. The applied torque is 47 kN m. In hollow sections the external radius is R and the internal is r.

G *Group of Terms:*

$$\frac{T}{J} = \frac{\tau}{r} = \frac{C\theta}{l}$$

U *Unknowns and Units:*

1. The outer diameter of the shaft.
2. The inner diameter of the shaft.

Use N and mm.

I *Inventory of Information:*

$T = 47 \text{ kN m} = 47 \times 10^6 \text{ N mm}$
$\tau = 46 \text{ N/mm}^2$
$C = 83 \times 10^3 \text{ N/mm}^2$
$\theta/l = \pi/(180 \times 3 \times 10^3) = 0{\cdot}00583 \times 10^{-3} \text{ rad/m}$

D *Derived Data:*
$$J = \frac{\pi}{2}(R^4 - r^4)$$

E *Evaluation:*
$$\frac{T}{J} = \frac{\tau}{R} = \frac{C\theta}{l}: \quad \frac{2 \cdot 2 \times 47 \times 10^6}{\pi(R^4 - r^4)} = \frac{46}{R} = 83 \times 10^6 \times 0.0058 \times 10^{-3}$$

$$\frac{46}{R} = 0.482: \quad \frac{94 \times 10^6}{\pi(96^4 - r^4)} = 0.482$$

$$R = 96 \text{ mm} \quad\quad r = 69 \text{ mm}$$

S *Solution:*
1. The outer diameter of the hollow shaft: **192 mm**.
2. The inner diameter of the hollow shaft: **138 mm**.

Two differences between the solid-shaft problem and the hollow-shaft problem should be noted:
(*a*) the value of *J* is different;
(*b*) the internal diameter makes no difference (apart from its effect on *J*) on the maximum stress, which is controlled by the value of the external radius *R*.

16/8 **HOLLOW SHAFT: Maximum torque and maximum twist?**

A hollow shaft has an internal diameter which is half of the external diameter, which is 200 mm. The allowable shearing stress is 64 N/mm². Find the maximum permissible torque and the accompanying degree of twist on a length of 3 m. Modulus of rigidity: 80×10^3 N/mm².

G *Group of Terms:*
$$\frac{T}{J} = \frac{\tau}{r} = \frac{C\theta}{l}$$

U *Unknowns and Units:*
1. Maximum permissible torque.
2. Accompanying degree of twist when stress is the maximum permitted.
Use N and mm.

16/9

I   *Inventory of Information:*
$$R = 2r$$
$$\tau = 64 \text{ N/mm}^2$$
$$R = 100 \text{ mm}$$

D   *Derived Data:*
$$J = \frac{\pi}{2}(100^4 - 50^4) = \pi(46\cdot9 \times 10^6) \text{ mm}^4$$

E   *Evaluation:*

$$\frac{T}{J} = \frac{\tau}{R}: \qquad T = \frac{\pi(46\cdot9 \times 10^6) \times 64}{100} = 94\cdot3 \times 10^6 \text{ N mm}$$

$$\frac{T}{R} = \frac{C\theta}{l}: \qquad \frac{\theta}{l} = \frac{\tau}{CR} = \frac{64}{80 \times 10^3 \times 100} = 8 \times 10^{-6} \text{ rad/mm}$$

S   *Solution:*
1. Maximum permissible torque is **94·3 kN m**.
2. Twist on a length of 3 m $8 \times 10^{-6} \times 3 \times 10^3$ is 24×10^{-3} **radians** or **1·4 degrees of angle**.

16/9   HOLLOW SHAFT: Incompatibility of limiting conditions

The specification for a mild-steel hollow circular shaft gives a maximum shearing stress and a maximum allowable twist. The external diameter of the shaft is 75 mm and the internal diameter is 50 mm. Maximum shearing stress is 64 N/mm² and the modulus of rigidity is 80×10^3 N/mm². Maximum permissible twist is 0·0011 rad/m. Explain why the two limiting conditions cannot be reached at the same time.

G   *Group of Terms:*
$$\frac{T}{J} = \frac{\tau}{r} = \frac{C\theta}{l}$$

U   *Unknowns and Units:*
1. Why are the limiting conditions incompatible?
2. What change in design would make the shaft more efficient?

I Inventory of Information:
$$R = 37.5 \text{ mm}$$
$$r = 25.0 \text{ mm}$$
$$\tau = 64 \text{ N/mm}^2$$
$$C = 80 \times 10^3 \text{ N/mm}^2$$
$$\theta/l = 0.0011 \text{ rad/m} = 0.0011 \times 10^{-3} \text{ rad/mm}$$

D Derived Data:
$$J = \frac{\pi}{2}(37.5^4 - 25^4) = 2.49 \times 10^6 \text{ mm}^4$$

E Evaluation:
$$T = \frac{\tau J}{R} = \frac{64 \times 2.49 \times 10^6}{37.5} = 4.25 \times 10^6 \text{ N mm}$$

$$T = \frac{C\theta J}{2} = 80 \times 10^3 \times 0.0011 \times 10^{-3} \times 2.49 \times 10^6 = 0.22 \times 10^6 \text{ N mm}$$

S Solution:
1. Using the specified stress, the allowable torque is about 19 times that permitted by the quoted degree of twist.
2. C, or degree of twist must be increased if the two conditions are to be compatible.

16/10 SOLID AND HOLLOW: Weight reduction: Torsional stiffness?

A solid mild steel shaft is to be replaced by a hollow nickel steel shaft of the same outside diameter. Both materials have the same density and modulus of rigidity. The working stress in mild steel is 75 N/mm², and in nickel steel, is 120 N/mm². What is the percentage reduction in weight effected by the change, and what are the relative torsional stiffnesses of the two shafts?

G Group of Terms:
$$\frac{T}{J} = \frac{\tau}{r} = \frac{C\theta}{l}$$

U Unknowns and Units:
1. Percentage reduction in weight by changing from solid to hollow shaft.
2. Relative torsional stiffness of the two shafts.

Use N and mm.

16/10

I   *Inventory of Information:*

$\tau_s = 75 \text{ N/mm}^2$
$\tau_h = 120 \text{ N/mm}^2$
T is constant
R is same for both shafts
r is unknown

D   *Derived Data:*

$$J_s = \frac{\pi D^4}{64} \text{ mm}^4$$

$$J_h = \frac{\pi}{64}(D^4 - d^4) \text{ mm}^4$$

Area of cross-sections:

$$A_s = \frac{\pi D^2}{4} \text{ mm}^2; \quad A_h = \frac{\pi}{4}(D^2 - d^2) \text{ mm}^2$$

E   *Evaluation:*

$$\frac{T}{J} = \frac{\tau}{D/2}$$

But T and D are constant, and τJ must therefore also be constant. Thus

$$\tau_s J_s = \tau_h J_h \qquad \frac{J_h}{J_s} = \frac{\tau_s}{\tau_h}$$

S   *Solution:*

$$75\left(\frac{\pi D^4}{64}\right) = 120\left\{\frac{\pi}{64}(D^4 - d^4)\right\}$$

d(internal diameter) $= 0.78D$ (external diameter)

$$\% \text{ reduction in weight} = \frac{\frac{\pi D^2}{4} - \frac{\pi}{4}(D^2 - d^2)}{\frac{\pi D^2}{4}} \times 100 = \mathbf{61.2\%}$$

Torsional stiffness is the amount of torque carried for a given twist

or
$$\frac{T}{\theta} = \frac{CJ}{l} = \text{Stiffness}$$

Since C and l are constant in this instance, the stiffness varies as the value of J, or inversely as the value of stress

$$\frac{\text{Stiffness }(h)}{\text{Stiffness }(s)} = \frac{45}{120} = \frac{5}{8}$$

S *Solution:*
1. % reduction in weight: **61·2%**
2. Torsional stiffness of hollow shaft is **62·5%** of that of solid shaft.

Investigate how power can be transmitted by a rotating shaft under torsion.

Always define the unknowns you are expected to evaluate, and then give the figures explicitly at the end in the same terms as the definition of the unknowns.

RECAPITULATION AND FORECAST

There are four sets of Problems in Group E and four in Group F to follow. Know what they all represent.

A beam is a bent structure. Couples causing bending act in the plane in which the member lies.

Couples causing torsion or twisting act in a plane at right angles to the member concerned.

The first term of the fundamental relationship expresses the loading and the type of section resisting it.

The second term defines the stresses produced.

The third term of the fundamental relationship defines the deformation produced.

Stresses from the second term are calculated in

Unit 13 Direct stress.
Unit 14 Bending stress.
Unit 14 Shearing stress from direct shear.
Unit 16 Shearing stress from torsion.

Deformations from the third term are calculated in

Unit 13 Direct stress.
Unit 16 Twist caused by torsion.
Unit 17 Deformation caused by bending.

Combinations of loadings produce combinations of stress and deformation, and two fundamental relationships must be used together. (Units 18 and 19.)

Be sure you know your way about these last two Groups. Solve problems from other textbooks by the methods given here and compare the results and the labour involved.

The last Unit deals with the method of using the energy stored in a deformed body to determine the extent of the deformation. Sometimes this is too elaborate—as for direct stress (20/19)—but usually, for bending, the method is effective, and can easily be applied to more advanced problems in Statically Indeterminate Structures.*

---

* Cassie: *Structural Analysis* (Longman).

GROUP F
Synthesis of Stress and Deformation

Part II: Group F

17 Deformation caused by bending

In this Unit of Study, the deformation caused by bending is discussed. The methods covered are those which have developed from the 'Flexure Formula' as it is called. In unit 14 this 'formula' was used to determine the stresses due to bending. Now a second relationship is developed. The first term of the expression relates to the applied loading (in this instance, bending moment), the second to the stresses developed (Unit 14) and the third to the deformation under bending moment.

$$\frac{M}{I} = \sigma = \frac{E}{R}$$

The part of this relationship we shall use is, of course, that which relates the loading to the deformation produced.

$$\frac{M}{I} = \frac{E}{R}$$

This can be written in various ways, but the most useful in the determination of deformation due to bending is

$$\frac{M}{EI} = \frac{I}{R}$$

On the left-hand side of this equation the loading (bending moment) and the properties of the section – material (E) and geometrical properties (I) – are represented. On the right-hand side is the *curvature*, which is the inverse of the radius (large radius (R), small curvature ($1/R$); small radius, large curvature). It is thus simple to determine the radius of the bent beam at any point where the bending moment is known. Also, if E and I remain constant over the length of the beam (quite a common condition) the curvature at any point is directly proportional to the bending moment applied. This is what you would expect in qualitative terms – the greater the bending moment, the more you expect the beam to bend (small radius; large curvature). The expression used relates in the above form only to one point at a time, but if the bending moment remains constant over a section of the beam, then the radius of curvature remains constant and the beam bends in a circle over that length. In general, this method of determining the deformation of a beam is clumsy and inconvenient.

17.1 Deflection and deflected shapes

The statement of deformation which is looked for by designers is in terms, not of radius of curvature, but of the linear deflection of the beam at any point from its original unloaded state. The maximum deflection suffered is often a good guide as to how well the beam fulfils its purpose and is, therefore, one of the quantities sought. The deflection of a beam under load can be visualized and sketched from personal experience, and such a pictorial representation of the deformation under bending is valuable in guiding the calculations which must be made to find the numerical values required.

It is not difficult to learn to draw *deflected shapes*, freehand. If you take the trouble to acquire this skill, you will find the numerical calculations easier and more meaningful. A well-sketched *deflected shape* shows in a qualitative way, the slopes of various parts of the beam, the deflections in a relative form, and the region in which the maximum deflection is likely to occur.

17.2 Sketching the deflected shapes of bent beams

As a preliminary to any problem on deflections of bent beams, you ought to sketch the deflected shape in order to give you a mental image of the problem before you. The simplest and commonest type of beam is horizontal, and the commonest type of loading is vertical. This section concentrates on such a beam. Other types of beam and loading – beams inclined to the horizontal, and loads inclined to the vertical – are included in the problems, the same techniques of sketching being applied.

The following guide-lines form the antidote to many of the mistakes made by students in sketching deflected shapes:

(*a*) The deflected curves must be smooth, and consist of two types of curve, concave-up or concave-down. There must be no sudden changes of slope; curves and straight portions merge into each other tangentially.

17.2

(b) The sketched deviation from the original state of the beam (usually straight) must be small if the deflected shape is to assume any semblance to reality – do not exaggerate the curves.

(c) There are always salient fixed points through which the deflected shapes *must* pass (e.g., the supports). Locate these fixed points and visualize and sketch the slopes of the beam through these points.

(d) Remember that – even with vertical downward forces acting – the beam may deflect upwards.

As examples of the technique, take a sheet of squared paper, and sketch again the following beams in their deflected form. The skill comes largely in the physical representation by sketching with a pencil; looking at the diagrams will not be sufficient.

Start by indicating possible slopes of the beam at the fixed points through which the beam must pass whether unloaded or bent. Use common knowledge and experience. For example, if a heavy load is close to one end of a beam it is highly likely that the slopes and deflections are likely to be greater there (larger curvature) than on an unloaded portion.

Beam A This has a uniformly distributed load which is symmetrically placed. The slopes at each end must be equal, and the maximum deflection must occur in the centre. The shape of the beam follows the shape of the bending moment diagram.

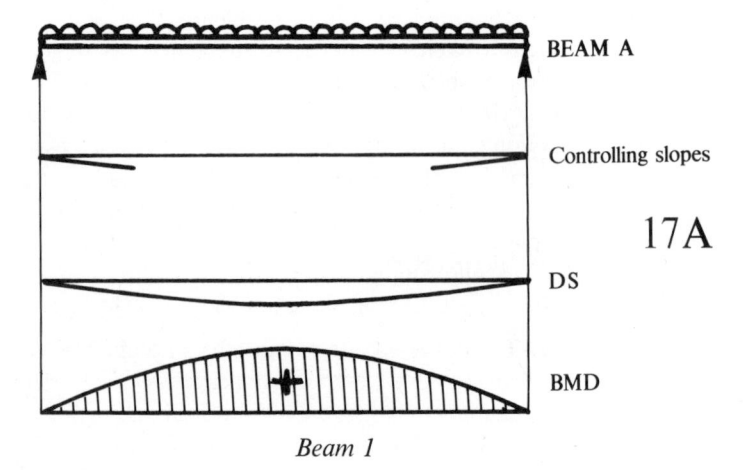

BEAM A

Controlling slopes

17A

DS

BMD

Beam 1

17.2

Beam B This has a heavy concentrated load near one end. From the two supports (fixed points) the possible slopes are sketched. The slope at the left-hand end is shown as somewhat steeper than that at the right-hand end, and your pencil is almost forced to show a larger deflection towards the left-hand end. The maximum deflection is not necessarily under the load.

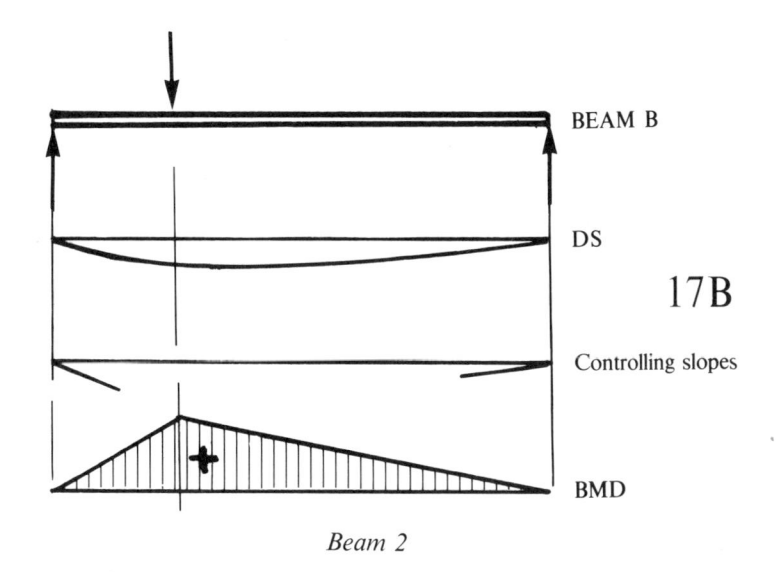

BEAM B

DS

17B

Controlling slopes

BMD

Beam 2

Beam C This beam has an unloaded cantilever at one end, and an unsymmetrical distributed load on the portion between the supports. The bending moment diagram (which it is always well to sketch) shows that there is no bending on the cantilever. When there is no bending the beam is not bent – apparently self-evident, but not always clear to students who are looking for some complexity which does not exist. Thus, where there is no bending moment, the beam remains straight, and at the slope dictated by the last loaded portion of the beam.

Beam D The loaded cantilevers bend in negative bending, and between B and A the beam remains straight since there is no bending moment on that length. For the rest, sketching shows that the beam must bend up in the middle if a smooth curve is to be drawn. The bending is entirely negative, as is shown by the bending moment diagram.

17.2

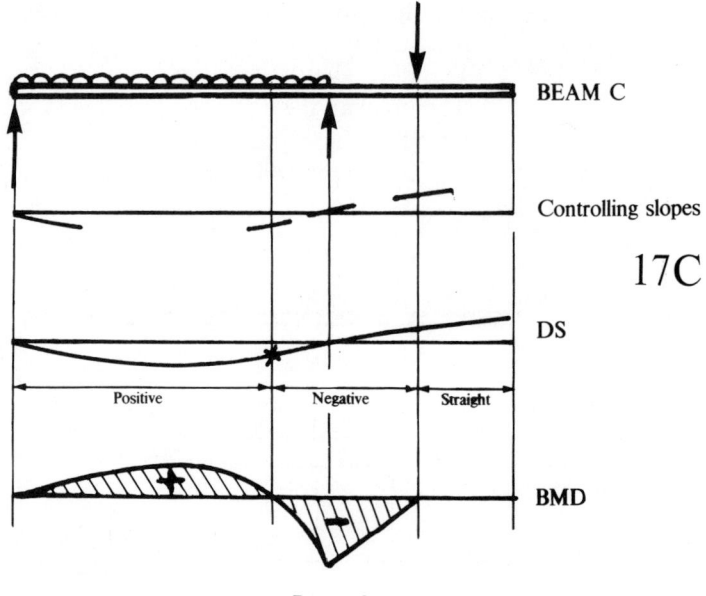

Beam 3

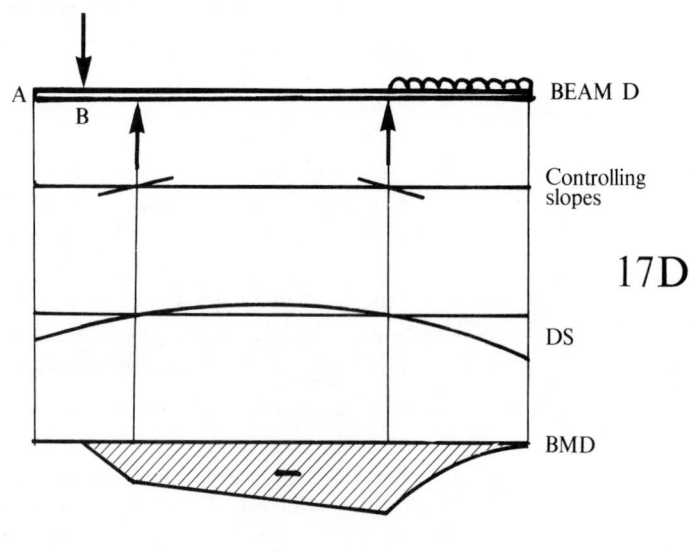

Beam 4

17.2

Points of contraflexure

A comparison of the bending moments and the deflected shapes shows that a bending moment which is entirely above the baseline (positive) or entirely below the baseline (negative) produces a deflected shape which (apart from a few straight portions) is curved in one direction. The positive bending gives a concave-up shape, while negative bending gives a concave-down shape. In the above four examples this relationship between bending moment and curvature is illustrated.

When, however, the bending moment changes from positive to negative (or vice versa) within the length of one beam, the deflected shape must change from one type of curvature to the other. During this change, smooth though the curve may be, there is one point where the deflected shape passes instantaneously through a point of zero bending. At this tangential point, the curves change from concave-up to concave-down or vice versa. Such points are known as *points of contraflexure* and are of great importance in the design of beams. They should always be marked with a cross as in the illustration below of Beam E.

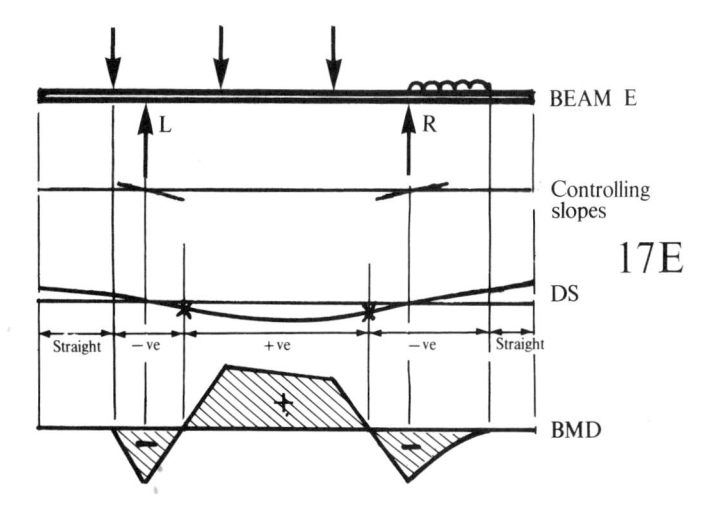

17E

Beam E. There are heavy loads between L and R and light loads on the two cantilevers. At the supports, therefore, the slopes are likely to be downwards towards the middle of the beam. The unloaded portions of the cantilevers remain straight at slopes determined by the inner conditions of loading. If the portion between L and R is to be in positive bending, and the cantilevers, as always, in negative bending, there must be a point at each end of LR where the concave-up changes to the concave-down. These *points of contraflexure* are not at L and R, but directly in line with the points at which the bending moment changes sign – where the outline of the BMD passes through the baseline.

17.3 Relationship between curvature and deflection

Knowing the bending moment at all points on the beam from a bending diagram, you can calculate the radius of curvature (R) at all these points. With some labour you might, with these figures, be able to draw a representation of the bent beam and measure off the values of the deflections. However, this is not a reasonable method of attack, nor is it accurate enough. The deflected shape as obtained qualitatively by sketching, is sufficient guide to allow of accurate further calculation.

The translation of the easily-obtained curvature into linear deflections from the original state is not simple, but various methods of achieving this transformation have been devised. Three of these are discussed here, and a fourth is covered in Unit 20 (strain energy).

In order to understand clearly how the change is made from curvature to deflection, it is important that you should be able to relate the following quantities to each other, and, particularly, be clear in your mind about their signs – whether positive or negative.

The related quantities are:

(*a*) the distance measured along the original beam before bending (x);
(*b*) the displacement of the beam from its original state (y);
(*c*) the slope of the beam at any point;
(*d*) the rate of change of slope of the beam at any point;
(*e*) the radius of curvature and the curvature.

You have been accustomed, if you have worked through Part I, to considering positive directions as upwards and to the right. This can be used again, and the convention, therefore is:

(*a*) *Distance* measured along the beam from an origin, O, is *positive* if it is measured *towards the right* (x).

(*b*) *Vertical deflection* of the beam from its original position at any point is *positive* if it is measured *upwards* (y).

17.3

(c) A *positive slope* is one in which the distance and deflection are both positive or both negative ($+(y/x)$ or $+(dy/dx)$).

A *negative slope* is one in which the deflection or the distance along the beam is negative ($-(y/x)$ or $-(dy/dx)$).

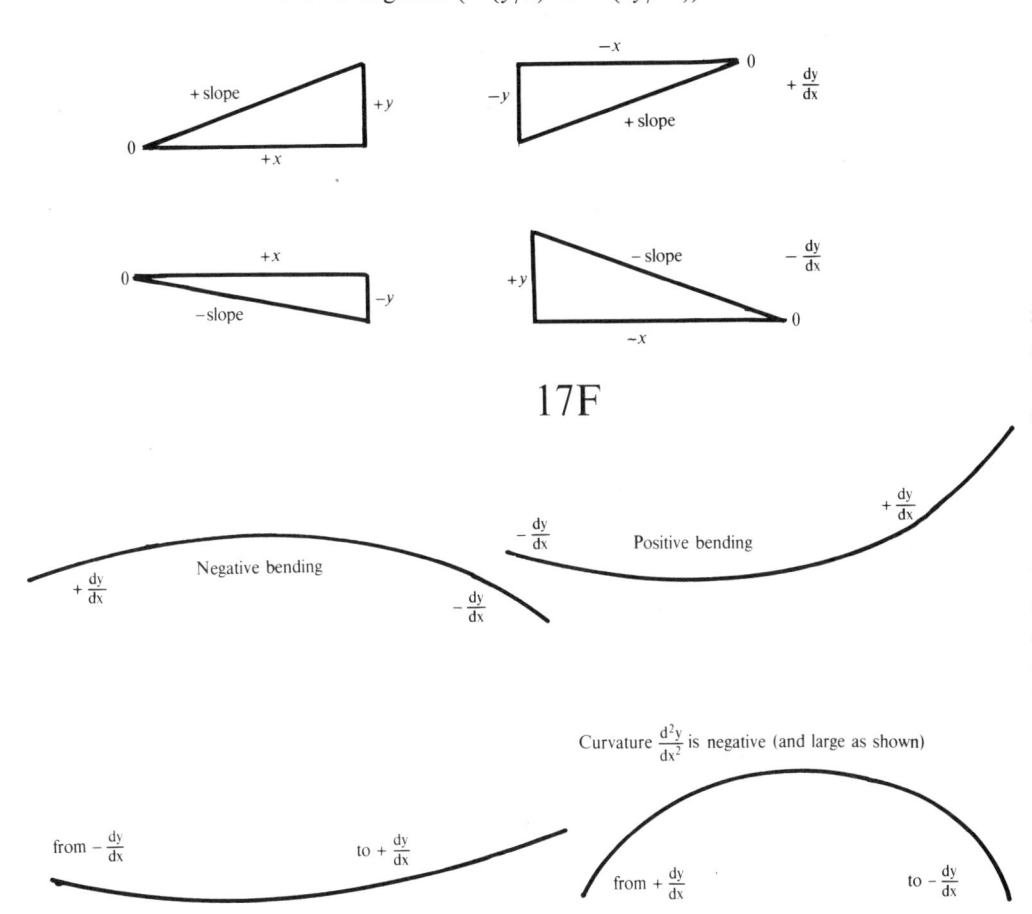

17F

Negative bending

Positive bending

Curvature $\dfrac{d^2y}{dx^2}$ is negative (and large as shown)

Curvature $\dfrac{d^2y}{dx^2}$ is positive (and small as shown)

(d) The sign of the rate of change of slope needs a little discussion. If the slope of the beam, as x increases in the positive direction, changes from positive $[+(dy/dx)]$ to negative $[-(dy/dx)]$, the change of slope is clearly in the *negative* direction. The result is *concave-down*.

399

17.4

If the sign of the slope starts at zero x as negative, and changes to positive, the change of slope is *positive* and *concave-up*. These two curves have already had signs applied to them as reflecting the type of bending (part I). Thus *positive bending gives positive change in slope*, and *negative bending gives negative change in slope*.

If the change of slope is slow in relation to the increase in x, the curvature is small and the radius large. A rapid change in slope gives a larger curvature and a smaller radius. Thus *rate of change of slope* which is (d^2y/dx^2) is related to *curvature*.

(e) The curvature, which you have already found is $1/R$, the reciprocal of the radius, has already had signs allocated to it in (d). You have found in (d) and here, in (e) that the word 'curvature' has been used both for the rate of change of slope and for the reciprocal of the radius. These two terms are, in fact, not perfectly equal, but the difference between them is so very small that they may be equated. The mathematical justification for this may be found in most textbooks on 'strength of materials'.

The relationship for which we have been searching, between the distance along a horizontal beam, and its vertical deflection from its original position thus comes from the curvature. You may write

$$\frac{I}{R} = \frac{d^2y}{dx^2} \text{ which also equals } \frac{M}{EI}$$

Thus,

$$+ EI \frac{d^2y}{dx^2} = M$$

17.4 The double-integration method of finding deflections

There are now only two simple steps to reach the direct relationship between x and y, or the equation to the deflected shape. These are two integrations which give the values of slope and the values of deflection.

$$+ EI \frac{dy}{dx} = + EI \int_0^1 \frac{d^2y}{dx^2} dx = + \int_0^1 M dx \text{ or } \frac{dy}{dx} = \int_0^1 \frac{M}{EI} dx$$

and $$+ EIy = + EI \iint_0^1 M dx \text{ or } y = \int_0^1 \frac{Mx}{EI} dx$$

The *constants of integration* which appear in these operations are evaluated by using known values of the relationship between the distance from the origin and such values as slope or deflection at control points. When the beam is symmetrical, the point where the slope is zero (horizontal) is likely to be obvious, and there are usually two points at which you know the deflection to be zero – at the supports. If E or I varies with x it must appear within the integration, but they are usually constant.

These integrations apply only over that length of the beam for which the expression for M, the bending moment, holds true.

17.5 Macaulay's variation

This method of finding deflection is a variation of the double integration method, with the object of making the integration easier. It is particularly applicable to beams with concentrated loads although, by various 'dodges' it can be used where there are uniformly distributed loads. The beam is necessarily divided, between the concentrated loads, into regions. The rules of procedure for Macaulay's variations of the double-integration method are:

1. Assume an origin at the left-hand end of the beam.
2. Write down the bending moment expression for the region at the extreme right-hand end of the beam.
3. Integrate expressions such as $(x - a)$ in the form $(x - a)^2/2$ *without splitting up the brackets.*
4. The resulting expressions for slope (dy/dx) and deflection (y) are then applicable to any part of the beam, provided that any terms in which the quantity within the brackets is negative, is omitted.

A number of problems illustrate this variation.

17.6 Method of Area-moments

For those who have a visual imagination, a method of using a sketch of the bending moment diagram in order to find the values of the deflections of a beam probably has more attraction than does the double-integration method or Macaulay's variation of it. The *area-moment*

17.6

method is based on the fundamental relationships which lie behind both the other methods. You have already calculated, in section 17.4:

$$\frac{dy}{dx} = \int_0^l \frac{M}{EI} dx \quad \text{and} \quad y = \int_0^l \frac{Mx}{EI} dx$$

The generalized M/EI diagram shown in sketch 17/G illustrates how the diagram is linked to the expressions for slope and deflection. M/EI is the ordinate of the diagram at any point, and dx is an elementary length. The product $M\, dx/EI$ is thus an elementary area and the integral $\int_0^l M\, dx/EI$ is the area of the whole M/EI diagram. Similarly $\int_0^l Mx\, dx/EI$ is the moment of the whole M/EI diagram about O. Thus, if you can assess the values of simple areas and their moments about a given point you can obtain, quite quickly, the values of slopes and deflections.

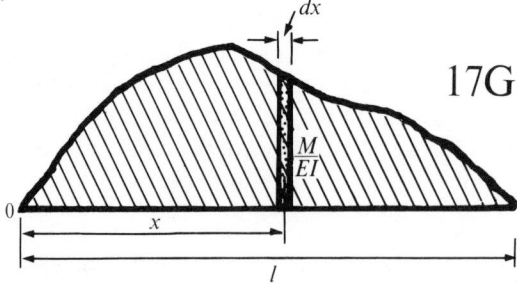

17G

Rules of procedure

A close theoretical analysis is outside the scope of this book which concentrates on the solution of problems, but the mathematical background can be found well developed in most textbooks on the subject. The transference from the sketch of the bending-moment-diagram divided-by-*EI* to values of deflection is not quite so facile as the brief hints given above might lead you to suppose. Mistakes are often made in this method through misconceptions. It is important, therefore, that you should commit to memory the following statements and always use sketches to illustrate them in specific instances. The statements are:

Rule 1. The angle between the tangents to a bent beam at points A and B is equal, numerically, to the area of the M/EI diagram between A and B.

Rule 2. In a bent beam, the displacement of A from the tangent at B is equal, numerically, to the first moment of the M/EI diagram about A.

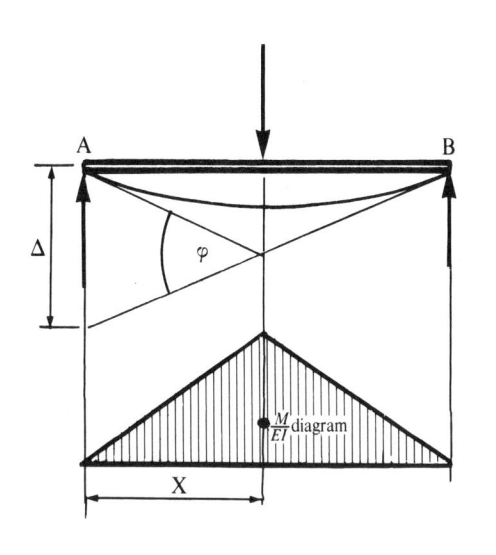

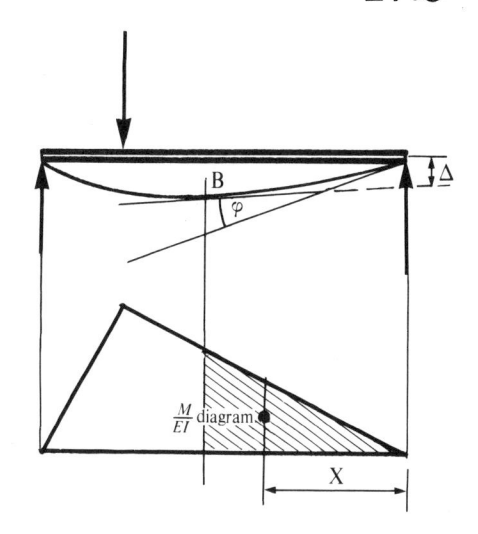

17.6

17H

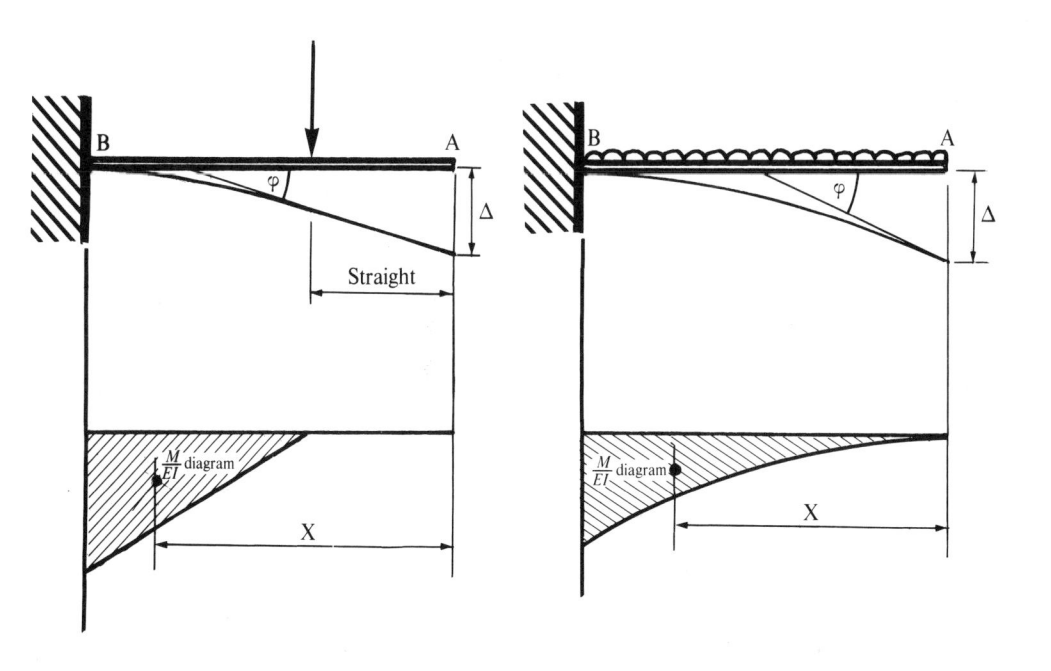

17.6

Two common mistakes are made by students of this method. They often leap to the mistaken conclusion that the area of the M/EI diagram represents a slope and that the moment of the M/EI diagram represents a deflection. Study sketch 17/H carefully, and realise that the area represents a *change in slope* between two points on the beam. The moment of the area represents a displacement, not from the original position of the beam (usually, horizontal) – which is what we mean by a deflection – but *from a tangent*. As the beam deforms, the tangent moves to produce the displacement. The Greek letter ϕ defines the change in slope, and Δ is the displacement (*phi* and *delta*).

If EI is constant over the length of the beam, the M/EI diagram is similar to the M diagram but to different units. If E or I changes along the beam the M/EI diagram may be quite different in shape from the bending moment diagram.

Watch the signs of direction, slope, bending and curvature.

Use the INDEX.

17.7 THE PROBLEMS

17/1 CENTRAL CONCENTRATED LOAD: Maximum deflection?

A simply supported beam is 10 m long and carries a load, concentrated at the centre, of 12 kN. What is the maximum deflection? *EI* is constant over the length of the beam.

G *Governing Statement:*
Displacement of Q from tangent at R is moment of M/EI area about Q.

U *Unknowns and Units:*
1. Deflection of centre, since beam and loading symmetrical, is the maximum deflection.
Use kN and m.

I *Inventory of Information:*
Given in sketch.

D *Derived Data:*
Sketch a tangent which at the centre gives displacement of Q from tangent at R as equal to the deflection of R from the horizontal.

E *Evaluation:*
Area of cross-hatched M/EI diagram is

$$+\left(\frac{12 \times 10}{4} \times \tfrac{1}{2} \times 5\right)\frac{1}{EI} = \frac{75}{EI} \text{ (ratio; dimensionless)}$$

Moment of CHA about Q:

$$-\frac{1}{EI}(75 \times \tfrac{2}{3} \times 5) = -\frac{250}{EI}\text{m (deflection of R from the horizontal)}$$

S *Solution:*
1. The maximum deflection of R from the horizontal is the same as the displacement of Q from the tangent at R and is

$$-\frac{250}{EI}\text{m,}$$

where E is in kN/m² and I in m⁴

17/2

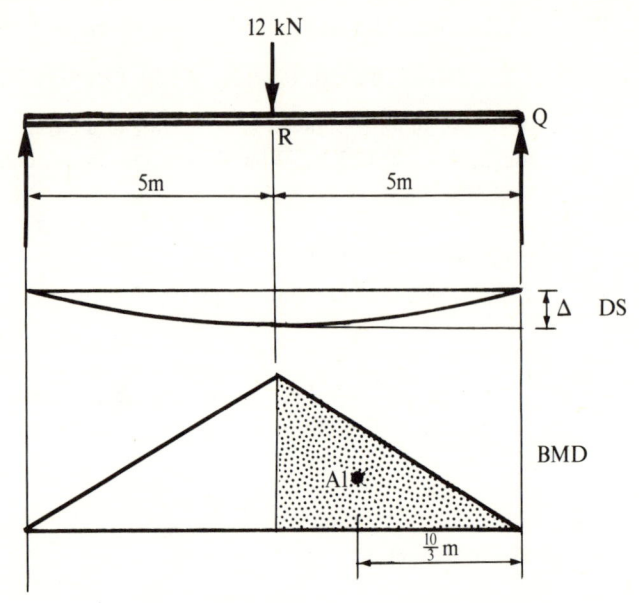

[Figure: Simply supported beam of span 10 m (5 m + 5 m) with 12 kN point load at R at mid-span, deflected shape showing Δ at DS, and BMD triangle with centroid A at 10/3 m]

17/2 CONCENTRATED LOAD: Maximum deflection?: Slope under load?

A simply supported beam carries a point load of 12 kN at 3 m from the left-hand end. The beam has a span of 9 m. Find the maximum deflection and the slope of the beam under the load. *EI* is constant throughout.

G *Guide:*

$$EI\frac{d^2y}{dx^2} = M$$

U *Unknowns and Units:*
  1. Position of maximum deflection.
  2. Value of maximum deflection.
  3. Slope of beam under the load.

I *Information:*
  Span: 9 m
  Origins at *O* (left-hand end) and at R (right-hand end).
  Point load of 12 kN at 3 m from O.

17/2

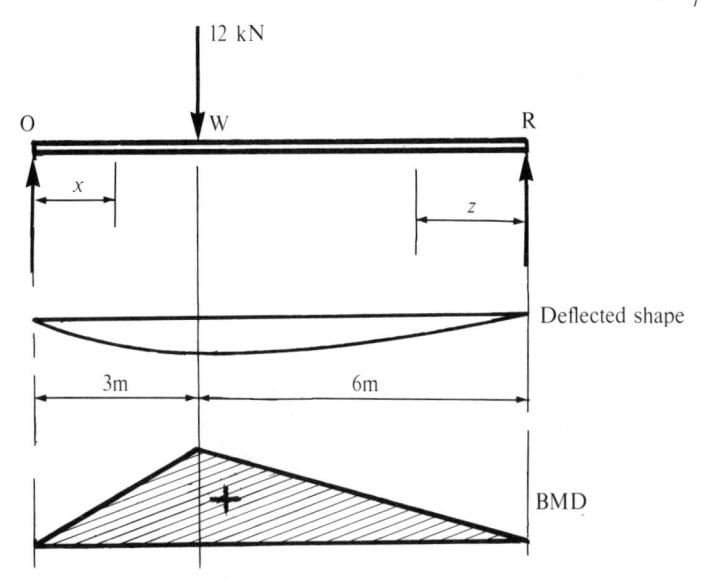

D *Derived Data:*

Left-hand support, 8 kN; Right-hand support, 4 kN.

When x is from 0 to 3, bending moment is $+8x$ kN m.

For the other portion of the beam the expression for bending moment changes and can be obtained by extending x from 3 m to 9 m, or by starting at the other end with a distance z measured in the negative direction. In order to show how signs are handled, take the latter.

When z is from 0 to 6, measured from R, the expression for bending moment includes a negative distance $-z$. Bending moment is $-4z$ kN m.

E *Evaluation:*

From O
$$EI\frac{d^2y}{dx^2} = +8x \qquad EI\frac{dy}{dx} = +4x^2 + A$$

$$EIy_1 = \times \frac{8}{6}x^3 + Ax + B$$

From R
$$EI\frac{d^2y}{d^2z} = -4z \qquad EI\frac{dy}{dz} = -2z^2 + C$$

$$EIy = -\tfrac{2}{3}z^3 + Cz + D$$

407

17/2

S *Substitutions; Solutions:*

	When	Then	Substitute in equation for	Value required
(a)	$x = 0$	$y = 0$	EIy_1	$B = 0$
(b)	$z = 0$	$y = 0$	EIy_2	$D = 0$
(c)	$x = 3$	$y_1 = y_w$	EIy_1	y_w
(d)	$z = -6$	$y_2 = y_w$	EIy_2	y_w
(e)	$x = 3$	$dy/dx = \text{Slope}_w$	$EI[dy/dx]_1$	Slope_w
(f)	$z = -6$	$dy/dx = \text{Slope}_w$	$EI[dy/dx]_2$	Slope_w

Whether the deflection and slope at the load (W) is measured from origin O or origin R, the result must be the same. Thus
Deflection at W from the left (y_w) = y_w from the right.

$$+\tfrac{8}{6}(+3)^3 + A(+3) = -\tfrac{2}{3}(-6)^3 + C(-6)$$

Slope at W, measured from the left $[dy/dx]_w$ = Slope $[dy/dx]_w$ from the right

$$+4(+3)^2 + A = -2(-6)^2 + C$$

Note that the lengths measured from these two origins take their signs with them. The solution of these simultaneous equations gives the values of the constants as

$$A = -60 \quad \text{and} \quad C = +48$$

Now obtain the values of the three unknowns listed under U, by substituting values of x, z, A, and C, being careful to take the signs with the values.

Unknown 1. The maximum deflection occurs where the slope is zero or horizontal. From the sketch of the deflected shape, this appears to lie in the right portion of the beam. Use the slope equation as determined from origin **R**:

$$-2z^2 + C = 0 \quad -2z^2 + 48 = 0 \quad z = \pm 4 \cdot 9 \text{ m}$$

Since the value of z is obtained by taking a square root, its sign may be positive or negative. A glance at the sketch – you must *always* make sketches – shows that it must be negative for it is measured towards the left.

$$\mathbf{z = -4 \cdot 9 \text{ m}}$$

Unknown 2. The value of the maximum deflection is found by substituting $z = -4.9$ in the deflection equation as found from origin R.

$$EIy = -\tfrac{2}{3}z^3 + Cz = -\tfrac{2}{3}(-4.9)^3 + 48(-4.9) = -156$$

$$y_{max} = -(156/EI) \text{ m} - \textbf{a downward deflection}$$

Unknown 3. The slope of the beam under the load appears to be negative. A check on this and on the whole calculation can be made by using the slope equation from origin $O(x = +3)$ and also the slope equation from origin $R(z = -6)$;

$$+4x^2 + A = +4(+3)^2 - 60 = -24$$
or
$$-2z^2 + C = -2(-6)^2 + 48 = -24$$

Slope under the load = $-(24/EI)$ **radians or gradient** which agrees with the down-to-the-right slope in the deflected shape.

17/3 CONCENTRATED LOAD: Maximum deflection?

A beam, of 5-metre span, is supported at its extreme ends and carries one concentrated load of 60 kN, at 1·5 m from the right-hand support. The second moment of area is given as 16×10^3 cm^4, and the modulus of elasticity as 200×10^6 kN/m^2. What are the values of the quantities defining maximum deflection of this horizontal beam?

G *Group of Terms:*

$$EI\frac{d^2y}{dx^2} = M$$

U *Unknowns and Units:*

1. Position of maximum deflection.
2. Value of maximum deflection.
Use kN and m.

I *Inventory of Information:*

Span: 5 m
E: 200×10^6 kN/m^2
I: 16×10^3 cm^4 = $16 \times 10^3 \times 10^{-8}$ m^4 = 0.16×10^{-3} m^4
Load: 60 kN at 1·5 m from right-hand end
EI: $200 \times 10^6 \times 0.16 \times 10^{-3}$ kN m^2 = 32×10^3 kN m^2

17/3

[Beam diagram: span with L at left, R at right; 3·5m from L to load point A, 1·5m from A to R; 60 kN load at A downward; $R_L = 18$ kN upward at L, R_R upward at R; distance 2·75m marked; x measured from L; deflected shape shown below with note "Horizontal $\frac{dy}{dx} = 0$"]

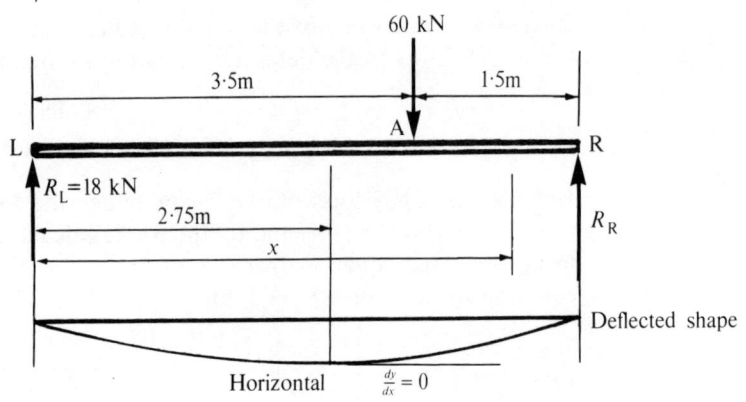

Deflected shape

D *Derived Data:*

The left-hand reaction is 18 kN.

Using Macaulay's method the equation for bending moment must be written for the extreme right-hand region between the load and the right-hand reaction.

$$EI\frac{d^2y}{dx^2} = +R_L x - 60(x - 3·5)$$

E *Evaluation:*

$$EI\frac{dy}{dx} = R_L \frac{x^2}{2} - 30(x - 3·5)^2 + A$$

$$EIy = +R_L \frac{x^3}{6} - 10(x - 3·5)^3 + Ax + B$$

$y = 0$ (second term omitted, as negative) $B = 0$
$y = l = 5$ m (second term included); $A = -68·25$

From the deflected shape, the maximum deflection is likely to be in the left-hand region LA. The final equation for deflection in that region is:

$$EIy = +3x^3 - 68·25x$$

Solving by a tabular method, it is possible to find the position of the maximum deflection directly. It is more efficient, however, to go back to the slope, remembering that the slope is horizontal, or zero, where the deflection is a maximum.

$$EI\frac{d^2y}{dx^2} = +9x^2 - 68·25$$

Equating this to zero,

$$+9x^2 = +68\cdot 25 \quad +x = +2\cdot 75 \text{ m}$$

Maximum deflection occurs at 2·75 from the left-hand end. Deflection at this point:

$$y = \frac{1}{EI}(+3x^3 - 68\cdot 25x) = \frac{1}{32 \times 10^3}(+3(2\cdot 75)^3 - 68\cdot 25 \times 2\cdot 75)$$

S *Solution:*

1. At **2·75 m** from L.
2. Maximum deflection is $-0\cdot 00\,392$ m or **−4 mm**.

The negative sign shows that the deflection is downwards as you would expect.

17/4 UNIFORMLY DISTRIBUTED LOAD: Slope?: Maximum deflection?

A simply supported beam carries a uniformly distributed load over its whole length of 10 m. The load is 16 kN/m of span. Its *EI* is constant, what is the slope at the left-hand end? Find the position and value of the maximum deflection.

G *Guide:*

$$EI\frac{d^2y}{dx^2} = M$$

U *Unknowns and Units:*

1. Slope at left-hand end.
2. Position of maximum deflection.
3. Value of maximum deflection.

Use kN and m.

I *Information:*

Span: 10 m
U.D.L.: 16 kN/m
Origin at left-hand end. Positive *x*.

17/4

D *Derived Data:*

Reaction at O is 80 kN.
Bending moment expression for whole span:
$$+80x - 8x^2 \text{ kN m}$$

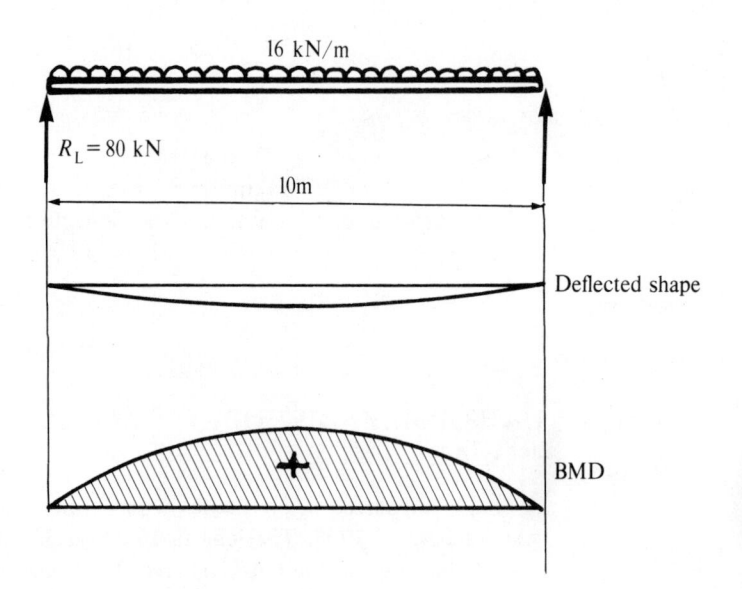

16 kN/m

$R_L = 80$ kN

10m

Deflected shape

BMD

E *Evaluation:*

$$EI\frac{d^2y}{dx^2} = 450x - 8x^2 + EI\frac{dy}{dx} = -40x^2 - \frac{8}{3}x^3 + A$$

$$EIy = -\frac{40}{3}x^3 - \frac{8}{12}x^4 + Ax + B$$

S *Substitution and Solution*

Try to carry out the substitutions and the evaluation of the equations neatly and carefully, being particularly careful about signs of numerical values. The answers agree with the deflected shape in showing a negative slope at O, and a negative (downwards) deflection in the centre of the beam.

17/5

	When	Then	Substitute in equation for	Value required
(a)	$x = +5$	$(dy/dx) = 0$	$EI(dy/dx)$	$A = -667$
	from symmetry of beam and load
(b)	$x = 0$	$(dy/dx) = ?$	$EI(dy/dx)$	$(dy/dx) = -667/EI$
(c)	$x = 0$	$y = 0$	EIy	$B = 0$
(d)	$x = +5$	y is maximum	EIy	$y_{max} = -2084/EI$

1. $-667/EI$
2. **At centre of beam (from symmetry)**
3. $-2084/EI$

The full evaluation of these depends on the values of E and I for the particular beam. These must be in units of kN/m^2 and m^4, respectively.

17/5 CONCENTRATED LOADS: Slope at reaction?: Deflection at load?

A beam as shown in the sketch is loaded with three concentrated loads. The value of *EI* is constant over the whole length. Find, in terms of *EI*, the slope at the left-hand reaction and the deflection under the central load.

G *Group of Terms:*

$$EI \frac{d^2y}{dx^2} = M$$

U *Unknowns and Units:*
1. Slope at the left-hand reaction.
2. Deflection under central load.
Use kN and m.

I *Inventory of Information:*
Loads and dimensions given in the sketch.
EI, constant throughout.

17/5

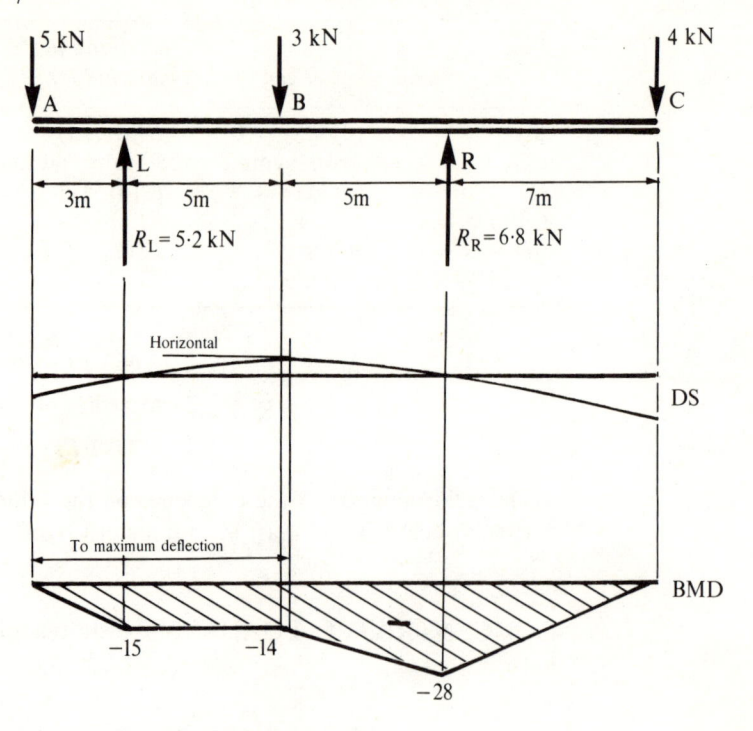

D *Derived Data:*

By usual methods, the left-hand reaction = +5·2 kN and the right-hand reaction = +6·8 kN.

Using Macaulay's method, the origin should be taken at A, and the moment equation written for the region between R and C.

$$M = 5x + 5·2(x - 3) - 3(x - 8) + 6·8(x - 13)$$

E *Evaluation:*

$$EI\frac{dy}{dx} = -\frac{5x^2}{2} + \frac{5·2}{2}(x - 3)^2 - \frac{3}{2}(x - 8)^2 + \frac{6·8}{2}(x - 13)^2 + A$$

$$EIy = -\frac{5x^3}{6} + \frac{5·2}{6}(x - 3)^3 - \frac{3}{6}(x - 8)^3 + \frac{6·8}{6}(x - 13)^3 + Ax + B$$

When $x = +3$, $y = 0$; when $x = +13$, $y = 0$:

$$+B = -278·4; \quad +A = +100·3$$

When $x = +3$

$$\frac{dy}{dx} = +\frac{1}{EI}(-2 \cdot 5 \times 9 + 100 \cdot 3) = +\frac{78}{EI} \text{(dimensionless gradient)}$$

Terms omitted when bracketed quantity is zero or negative. Positive sign should slope up to the right.

When $x = +8$

$$y = +\frac{1}{EI}\left(-\frac{5 \times 512}{6} + \frac{2 \cdot 6 \times 125}{3} + 100 \cdot 3 \times 8 - 278 \cdot 4\right) = +\frac{206}{EI} \text{ m}$$

Positive sign shows an upward deflection

S *Solution:*

1. Slope at L = **+78/*EI* up to the right**.
2. Deflection at B = **+206/*EI* m upwards**.

17/6 TRIANGULAR LOAD: Maximum deflection?

A simply-supported beam is of 20 m span, and carries a distributed load of triangular shape. The apex of the triangle is on the centre line of the beam. The total load is 16 kN and *EI* is constant. Find the position and value of the maximum deflection, and the values of the slopes at the ends of the beam.

G *Guide:*

$$EI \frac{d^2 y}{dx^2} = M$$

U *Unknowns and Units:*

1. Position of maximum deflection.
2. Value of maximum deflection.
3. Slopes at ends of beam.

Use kN and m.

I *Information:*

Span: 20 m; Total load: 16 kN
Horizontal distances (x) measured positively from the left-hand end.

17/6

D *Derived Data:*

Sketch the diagrams of bending moment and deflected shape.
Intensity of loading: since the total load is 16 kN, and the span is 20 m, the mean intensity of loading is 0·8 kN/m of span. But the intensity at the centre of the triangular load is twice this (from geometry of the triangle) or 1·6 kN/m of the span. This is the height of the triangle in the centre of the beam when the loading is sketched.

E *Evaluation:*

The area of triangular loading diagram from the left-hand end to XX, at a distance of x from the left-hand end: from the geometry of the load this is

$$\frac{x}{10} \times 1\cdot 6 \times \frac{x}{2} = 0\cdot 08 x^2$$

The moment of this load about XX is the area multiplied by $x/3$

$$= -0\cdot 08 \frac{x^3}{3}$$

The moment of the whole loading about XX brings in the moment of the reaction (8 kN).

$$M = +8x - 0\cdot 08 \frac{x^3}{3}$$

S *Substitution and Solution:*

Change in slope between the left-hand end and the centre:
$\quad$ = area of M/EI diagram between left-hand end and centre.

$$= \frac{1}{EI} \int_0^{10} M \, dx = \frac{1}{EI} \int_0^{10} (+8x - 0\cdot 08 \frac{x^3}{3}) \, dx$$

$$= \frac{333}{EI} \text{ gradient}$$

Since the slope at the centre (from symmetry) is zero or horizontal, this change in slope also equals the value of the slope at the left-hand end (and the right-hand end also, since the loading is symmetrical).

Displacement of the left-hand end from the tangent at the centre is the moment of the area of the M/EI diagram between the left-hand end and the centre about the left-hand end:

$$\Delta = \frac{1}{EI}\int_0^{10} Mx\,dx = \frac{1}{EI}\int_0^{10}\left(+8x - 0{\cdot}08\frac{x^3}{3}\right)x\,dx$$

$$\Delta = \frac{2134}{EI}\,m$$

Since the tangent at the centre is horizontal, this value also represents the vertical deflection of the centre of the beam, which is the figure sought.

1. The maximum deflection occurs at the centre.

2. The value of the maximum deflection is $\qquad -\dfrac{2134}{EI}\,m$

3. The slopes of the beam at left- and right-hand ends are equal in value, negative at the left-hand end and positive at the right-hand end

$$\frac{333}{EI}\text{ gradient}$$

17/7 MACAULAY'S VARIATION: Partial uniformly distributed load

The only loading on this beam is from O to Q. When using Macaulay's variation with UDL it is necessary to have it covering the whole span. Thus from Q to R the load is extended to do this, and to bring the loading back to UDL on OQ only, an upward loading is applied from Q to R, which restores the status quo. Find the deflection at Q.

G *Governing Statements:*

(a) The length x must be taken into the extreme right-hand portion of the loaded beam.

(b) The load acting under the beam is assumed to be directed upwards.

(c) Double integration of the bending moment gives the deflection required.

17/7

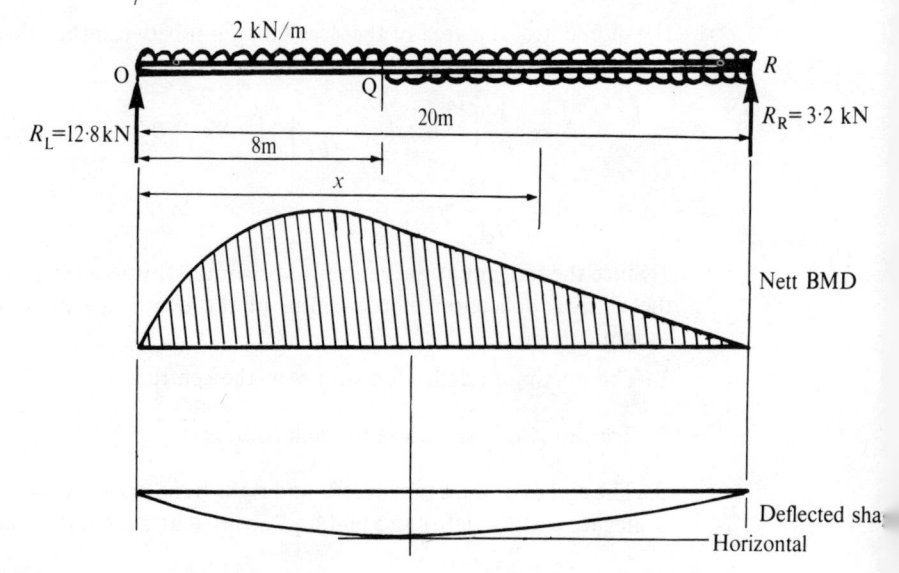

(Figure: Beam with 2 kN/m UDL over 20 m span, support reactions $R_L = 12.8$ kN at O and $R_R = 3.2$ kN at R; point Q at 8 m from O; distance x measured from O. Nett BMD and deflected shape shown below, with horizontal reference.)

U *Unknowns and Units:*

1. Equation to the BM in the extreme right-hand portion of the beam.
2. Result of double integration of this expression.

Use kN and m.

I *Inventory of Information:*

The reactions are obtained as is described in Part I and are found to the 12·8 kN and 3·2 kN. From these and the loading, the equation to the BM can be found.

D *Derived Data:*

Bending moment at end of the length x:

$$+12 \cdot 8x - \frac{2x^2}{2} + \frac{2(x-8)^2}{2}$$

If the quantity in the bracket becomes negative during the calculations, the term including the bracket is omitted from the calculation.

E *Evaluation:*

$$EI \frac{d^2 y}{dx^2} = +12 \cdot 8x - x^2 + (x-8)^2$$

418

$$EI\frac{dy}{dx} = +6\cdot 4x^2 - \frac{x^3}{3} + \frac{(x-8)^3}{3} + A$$

$$EIy = +6\cdot 4\,\frac{x^3}{3} - \frac{x^4}{12} + \frac{(x-8)^4}{4\times 3} + Ax + B$$

When $x = 0$: $y = 0$ (omit third term): $B = 0$.
When $x = 0$: $y = 0$ (use all terms since the third is now positive):
$$A = -273.$$

When $x = 8$

$$EIy = +6\cdot 4 \times \frac{512}{3} - \frac{4096}{12} - 273 \times 8$$

S *Solution:*
The deflection of the beam at the point Q is

$$-\frac{1430}{EI}\,\text{m}$$

where E is in kN/m² and I in m⁴.

By AREA MOMENTS:

D *Derived Data:* A_1 is M/EI diagram $O-Q$: A_2 is M/EI diagram $Q-R$

	M/EI area	Moment of M/EI area about Q	Moment of M/EI area about O
A_1	$\dfrac{I}{EI}\displaystyle\int_0^8 (12\cdot 8x - x^2)\,dx$	—	$\dfrac{I}{EI}\displaystyle\int_0^8 (12\cdot 8x - x^2)x\,dx$
A_2	$(3\cdot 2 \times 12)\dfrac{12}{2} = 230$	$230 \times \dfrac{12}{3}$	$230\left(8 + \dfrac{12}{3}\right)$

The moment of the M/EI areas A_1 and A_2 about O represent the displacement of O from the tangent at R. The moment of the M/EI area A_2 about Q represents the displacement of Q from the tangent at R.

17/8

The relationship of these displacements is

> Deflection at Q + Displacement at Q = 12/20 of the displacement at O

The result should be the same as found above.

17/8 DOUBLE INTEGRATION: Deflection under own weight: Variable I

A cantilever has a horizontal upper surface which is 100 mm wide. The section of the cantilever is rectangular but the depth varies from 200 mm at the fixed end to zero at the free end. If the length of the cantilever is 1·5 m and its mass density is 2·55 Mg/m³, what is the deflection of the free end under its own weight? Some types of problem, of which this is one, are best solved first in general terms.

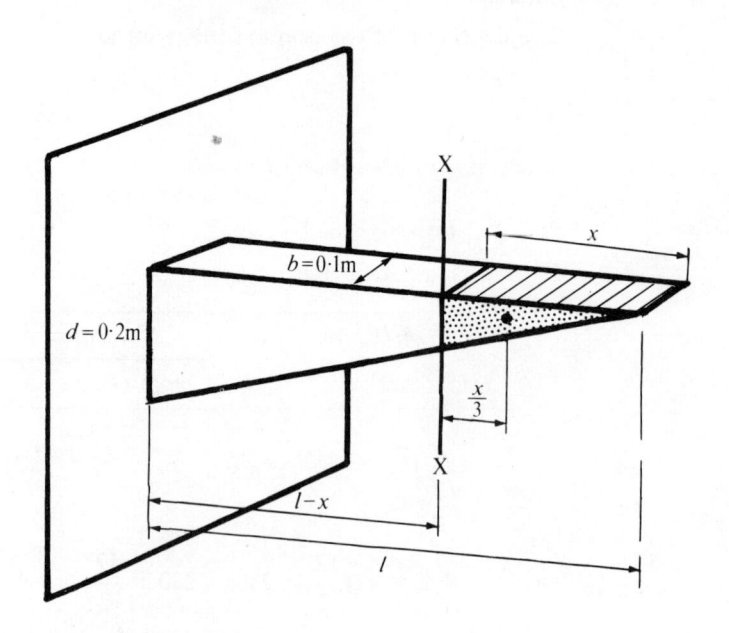

G *Group of Terms:*

$$E \frac{d^2 y}{dx^2} = \frac{M}{I}$$

U *Unknowns and Units:*
1. The deflection of the free end of the cantilever under its own weight, in general terms.
2. The deflection in numerical terms.
Use kN and m.

I *Inventory of Information:*
$l = 1\cdot5$ m
$\sigma = 2\cdot55$ Mg/m^3
$d = 200$ mm $= 0\cdot2$ m
$b = 100$ mm $= 0\cdot1$ m

D *Derived Data:*
Weight of the material calculated from its mass:
$= 2\cdot55$ Mg/m$^3 \times 9\cdot81$
$= 25$ kN/m$^3 = w$

Height of the section at XX $\qquad h = d\dfrac{x}{l}$

Volume of the shaded portion: $\qquad \dfrac{bxh}{2} = \dfrac{bdx^2}{2l}$

Weight of the shaded portion: $\qquad \dfrac{wbdx^2}{2l}$

(M) Moment of the shaded portion about XX – the bending moment
$$\dfrac{wbdx^2}{2l} \cdot \dfrac{x}{3} = \dfrac{wbdx^3}{6l}$$

(I) Second moment of area at XX:
$$\dfrac{bh^3}{12} = \dfrac{b}{12}\left(d\dfrac{x}{l}\right)^3 = \dfrac{bd^3x^3}{12l^3}$$

E *Evaluation:*
$$E\dfrac{d^2y}{dx^2} = \dfrac{M}{I} = +\dfrac{2wl^2}{d^2}$$

17/8

$$E \frac{dy}{dx} = + \frac{2wl^2}{d^2} x + A$$

$$Ey = + \frac{wl^2 x^2}{d^2} + Ax + B$$

When $x = -l$, both dy/dx and y are zero. Thus,

$$+A = + \frac{2wl^3}{d^2} \text{ and } +B = - \frac{wl^4}{d^2}$$

$$Ey = + \frac{wl^2}{d^2}(x^2 + 2lx - l^2)$$

Deflection at the end under its own weight is obtained by equating x to zero:

$$+y_E = - \frac{wl^4}{d^2 E}$$

Substituting the given values, and the value of E for steel (207×10^6 kN/m²)

$$+y_E = - \frac{25 \times 1 \cdot 5^4}{0 \cdot 2^2 \times 207 \times 10^6} = - 0 \cdot 015 \times 10^{-3} \text{ m}$$

Notice that the value of b is not significant, the same deflection would be obtained with any width.

S *Solution:*

The deflection of the extreme end under its own weight, if the cantilever is of steel, is $\qquad -0 \cdot 015$ mm.

The general expression is: Deflection of the free end is: $\qquad -\dfrac{wl^4}{d^2 E}$

17/9

17/9 **DOUBLE INTEGRATION: Horizontal Beam: Varying distributed load**

A simply-supported beam, 10 m span, carries no load on the right half, but on the left half the load varies linearly from 1 kN/m at the left end, to zero at the centre. Find the deflection under the load, and the position and magnitude of the maximum deflection. *EI* is constant.

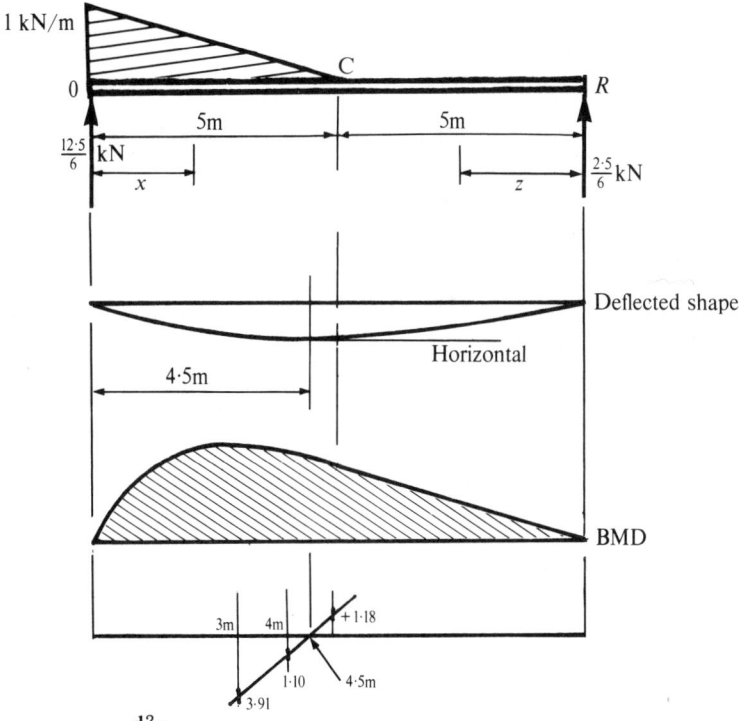

G *Guide:* $EI \dfrac{d^2y}{dx^2} = M$

U *Unknowns and Units:*
1. Deflection under the centre of the beam.
2. Position of the maximum deflection.
3. Value of the maximum deflection.
Use kN and m

I *Information:*
Span: 10 m; Load on left half varies.
The value of *x* is positive and *z* negative.

17/9

D *Derived Data:*

Deflected-shape and bending-moment diagrams must be drawn. The only difficulty which might develop from this requirement stems from any doubt you may have in finding the expression for bending moment in the left half of the beam. When that is found, the procedure is not difficult. Re-read Units 9 and 10 if you are in any doubt as to how the following expressions were obtained:

Support at the left-hand end (O) $(12 \cdot 5)/6$ kN

Support at the right-hand end (R) $(2 \cdot 5)/6$ kN

Expression for bending moment at x, anywhere between O and C is found by covering the beam to the right of x and dividing the trapezium formed by the load into two triangles each of which applies its own moment at the edge of the covering card.

$$Mx = +(12 \cdot 5x)/6 - 1 \cdot \frac{x}{2} \cdot \frac{2x}{3} - \frac{5-x}{5} \cdot \frac{x}{2} \cdot \frac{x}{3} \cdot 1$$

$$= +(12 \cdot 5x)/6 - \frac{x^2}{2} + \frac{x^3}{30}$$

Expression for bending moment anywhere between R and C

$$Mx = -(2 \cdot 5z)/6$$

E *Evaluation:*

From O towards the right:

$$EI \frac{d^2y}{dx^2} = + \frac{12 \cdot 5x}{6} - \frac{x^2}{2} - \frac{x^3}{30}$$

$$EI \frac{dy}{dx} = + \frac{12 \cdot 5x^2}{12} - \frac{x^3}{6} - \frac{x^4}{120} + A$$

$$EIy = + \frac{12 \cdot 5x^3}{36} - \frac{x^4}{24} - \frac{x^5}{600} + Ax + B$$

From R towards the left:

$$EI \frac{d^2y}{dx^2} = - \frac{2 \cdot 5z}{6} \qquad EI \frac{dy}{dx} = - \frac{2 \cdot 5z^2}{12} + C$$

$$EIy = - \frac{2 \cdot 5z^3}{36} + Cz + D$$

S Substitution and Solution

Substitute critical values of x and z, and so find expressions containing the unknown constants of integration. Equating slopes and deflections working from the left and from the right, the constants can be found. Use a tabular method, which allows you to do one thing at a time. Work down each column, and not along each line.

	When	Then	Substitute in equation for	Values determined
(a)	$x = 0$	$y = 0$	yEI (from O)	$B = 0$
(b)	$z = 0$	$y = 0$	yEI (from R)	$D = 0$
(c)	$x = 5$	$y = y_c$	yEI (from O)	$\left. \begin{array}{l} y_c \\ y_c \end{array} \right\}$ Equal
(d)	$z = 5$	$y = y_c$	yEI (from R)	
(e)	$x = 5$	$dy/dx = \text{Slope}_c$	EI (from O)	$\left. \begin{array}{l} \text{Slope}_c \\ \text{Slope}_c \end{array} \right\}$ Equal
(f)	$z = 5$	$dy/dx = \text{Slope}_c$	EI (from R)	

Whether deflection and slope at the centre are derived from equations originating from O or from R, the results must be the same. This equivalence may be used to evaluate the constants A and C.

$$\text{Deflection at centre (from } O) = \text{Deflection at centre (from } R)$$
$$\text{Substitution } (c) = \text{Substitution } (d)$$

By substituting $x = +5$ and $z = -5$ in the appropriate deflection equations, the following expression is obtained.

$$+5A + 5C + 140 = 0$$

Using the values of slope in the same way,

$$\text{Substitution } (e) \quad \text{Substitution } (f)$$

and thus
$$+A - C + 15 \cdot 6 = 0$$

is the expression linking the two values of slope, when $x = +5$ and $z = -5$ m.

Solving these simultaneously, $+C = +6 \cdot 4$ and $+A = -9 \cdot 2$

The Unknowns can now be found by substituting these values in the equations for deflection and slope and substituting also the appropriate values of x and z, remembering signs, x being positive, and z negative. By this means you should obtain the following results:

425

17/9

Unknown (1): The *deflection at the centre* can be found either by substituting $x = +5$ in the equation from O, or $z = -5$ in the equation from R.

You should find both results to agree at a value of $-\dfrac{23 \cdot 3}{EI}$ m.

The negative sign indicates a downward deflection.

Unknown (2): The *position of the maximum deflection* is at the point where the slope is horizontal or zero. From the sketched deflected shape, the maximum deflection appears to occur just inside the left half, near the centre. For a trial, therefore, use the slope equation as found from the left-hand (O), and equate the value of slope to zero.

$$+\frac{12 \cdot 5 x^2}{12} - \frac{x^3}{6} + \frac{x^4}{120} - 9 \cdot 2 = 0$$

Using trial values of x you can interpolate the value at which slope is zero, either numerically or graphically. The graphical solution is shown in the sketch.

The zero slope, and thus the maximum deflection occurs at

4·5 m from 0

Unknown (3): The *value of the maximum deflection* is obtained by substituting $x = +4 \cdot 5$ m in the equation for deflection obtained from the left-hand end O.

You should find this to be $-\dfrac{23 \cdot 8}{EI}$ m.

which is very little different from the downward deflection estimated at the centre. This close agreement was predicted by the deflected shape.

Deflection goes through a 'maximum' at zero slope, but it may not be the maximum deflection on the structure.

17/10

17/10 **VARYING CROSS-SECTION: Double integration: Cantilever**

A cantilever is 6 m long and carries 10 kN at the free end. In plan, the shape of the cantilever is a triangle, being of zero width at the free end, and 1 m wide at the fixed end. It is uniformly thick throughout at 0·1 m. The cantilever has a built-in zero slope at the fixed end. Find the deflection at the end and at half the length.

G *Guide:* $E \dfrac{d^2 y}{dx^2} = \dfrac{M}{I}$ (I varies)

U *Unknowns and Units:*
1. Deflection at the free end.
2. Deflection at 3 m from the free end.
Use kN and m.

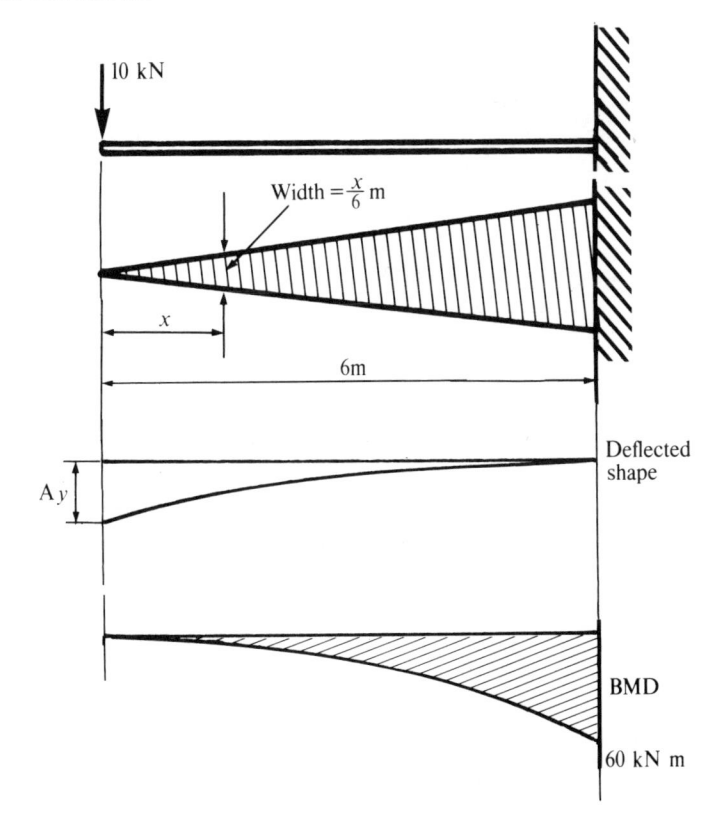

10 kN

Width = $\dfrac{x}{6}$ m

x

6m

A y

Deflected shape

BMD

60 kN m

427

17/10

I   *Information:*

Length of cantilever: 6 m.
Width at free end: zero: Width at fixed end: 1 m.
Thickness: 0·1 m.
Concentrated load at free end: 10 kN.

D   *Derived Data:*

Width of beam at x from the free end is $x/6$ m.

Second moment of area $(bd^3/12)$ is $\dfrac{0 \cdot 1 x}{7200}$ m^4.

Since this is a term in x it must be taken into the integration.
Bending moment at any point in the length is negative and of value $-10x$ kN m

E   *Evaluation:* $E\dfrac{d^2y}{dx^2} = \dfrac{M}{I} = -\dfrac{10x \times 7200}{0 \cdot 1 x}$

$$E\frac{dy}{dx} = -72 \times 10^4 x + A; \qquad Ey = -36 \times 10^4 x^2 + Ax + B$$

S   *Substitution and Solution:*

When	Then	Substitute in equation for	Value determined
(a) $x = +6$	$\dfrac{dy}{dx} = 0$	$\dfrac{dy}{dx}$	$A = -432 \times 10^4$
(b) $x = +6$	$y = 0$	y	$B = -1296 \times 10^4$

Substitution in the equation for y the values $x = 0$ and $x = +3$ gives the deflection at the end and at half-length.

$$\text{Deflection at the end} \quad -\frac{1296 \times 10^4}{E}$$

$$\text{Deflection at centre} \quad -\frac{324 \times 10^4}{E}$$

The negative sign indicates that deflection is downward, as was expected.

17/11

17/11 AREA-MOMENTS: Cantilever: Constant EI: Two types of load

A horizontal cantilever, of uniform *EI*, carries a uniformly distributed load and a concentrated load, as shown in the sketch. Find the slope and deflection of the free end.

G *Governing Statements:*
  (a) Change in slope from A to B is area of *M/EI* diagram.
  (b) Displacement of A from tangent at B is moment of *M/EI* diagram about A.
  (c) Since B is held horizontally, the slope at A is equal to the change in slope (a) and the deflection at A is equal to the displacement (b).

U *Unknowns and Units:*
  1. Area of *M/EI* diagram between A and B.
  2. Moment of *M/EI* diagram between A and B, about A.
  Use kN and m.

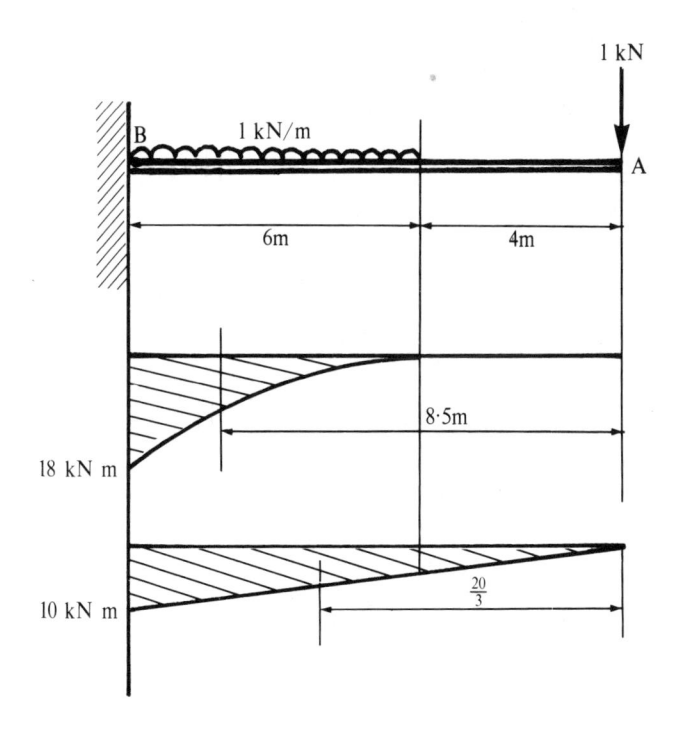

429

17/11

I   *Inventory of Information:*

Loading shows that the bending moment at B due to the uniformly distributed load is 18 kN m.

The concentrated load gives a BM at B of 10 kN m.

The BMD for the UDL is parabolic, and for the concentrated load, is triangular. These areas A_1 and A_2 are taken separately for convenience. Superposition is not required.

D   *Derived Data:*

Area of A_1: $\frac{1}{3} \times 18 \times 6 = 36$ kN m².
Moment of A_1 about A: $36 \times 8 \cdot 5 = 306$ kN m³.
Area of A_2: $\frac{1}{2} \times 10 \times 10 = 50$ kN m².
Moment of A_2 about A: $50 \times \frac{2}{3} \times 10 = 333$ kN m³.

E   *Evaluation:*

Area of M/EI diagram between A and B $= -(36 + 50)$
$\qquad\qquad\qquad\qquad\qquad\qquad\qquad\qquad = -86$ kN m².
Moment of M/EI diagram between A and B $= -306 - 333$
$\qquad\qquad\qquad\qquad\qquad\qquad\qquad\qquad = -339$ kN m³.

For a numerical solution, E must be in kN/m² and I in m⁴ (EI is in kN m²).

S   *Solution:*

1. Change in slope between A and B equals the slope at A, since the slope at B is zero:

$$= -\frac{86}{EI} \text{ (ratio)}$$

2. Displacement of A from tangent at B equals the deflection of A, since the tangent at B is horizontal:

$$= -\frac{339}{EI} \text{ m}$$

17/12 AREA-MOMENTS: Symmetrical beam: Concentrated loads

The beam shown in the sketch is symmetrical and symmetrically loaded, and has a constant *EI*. The maximum deflection must occur at the centre of the span. Find the value of this deflection if E is 207×10^6 kN/m² and I is 73×10^{-6} m⁴.

G *Governing Statements:*

(*a*) The slope of the beam at S is horizontal. Thus the displacement of P from the tangent at S is equal to the desired maximum deflection.

(*b*) Displacement of P from tangent at S is moment of M/EI diagram between S and P about P.

U *Unknowns and Units:*

1. Moment of M/EI diagram between S and P about P.
Use kN and m.

I *Inventory of Information:*

The BMD for the whole beam shows a uniform bending in the central length BC.

The value of the BM in this region is $5{\cdot}77 \times 4 = 23{\cdot}2$ kN m.

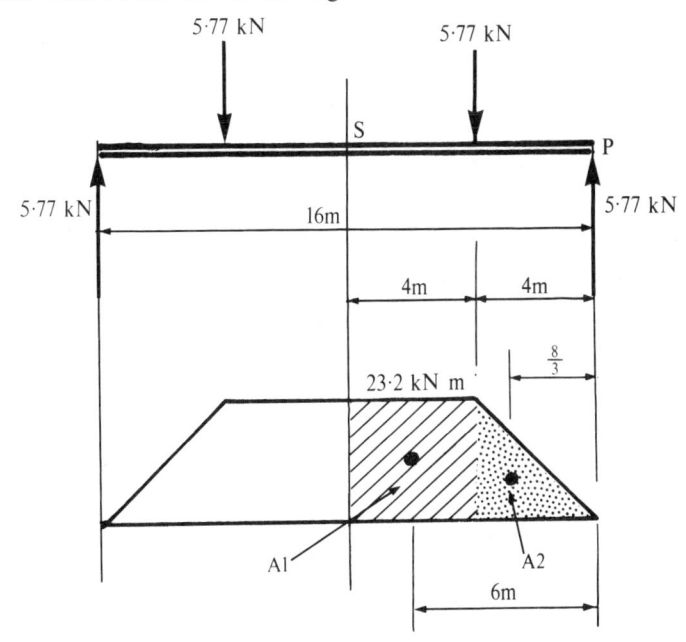

431

D *Derived Data:*

Only the CHA (cross-hatched area) needs be considered, as it represents the BMD between the points S and P.

Area of A_1: $\qquad +(23\cdot 2 \times 4) = +92\cdot 8$ kN m².
Moment of A_1 about P: $-(92\cdot 8 \times 6) = -557$ kN m³.
Area of A_2: $\qquad +(23\cdot 2 \times \tfrac{1}{2} \times 4) = +46\cdot 4$ kN m².
Moment of A_2 about P: $-(46\cdot 4 \times \tfrac{2}{3} \times 4) = -124$ kN m³.

E *Evaluation:*

Moment of the M/EI diagram between S and P about P is

$$-\frac{1}{EI}(557 + 124) = -\frac{681}{EI}$$

S *Solution:*

1. Maximum deflection of S from horizontal equals the displacement of P from the horizontal tangent at S.

$$= -\frac{681}{EI} = -\frac{681}{207 \times 10^6 \times 73 \times 10^{-6}} = -0\cdot 045 \text{ m (down)}$$

17/13 AREA-MOMENTS: Symmetrical beam: Uniform loading

The beam shown in the sketch is symmetrical, symmetrically loaded, and has a constant *EI*. The maximum deflection is, therefore, at the centre. Find the value of this maximum deflection in terms of *EI*, and also the value of the slope of the beam at the outer ends.

G *Governing Statements:*

(a) Change in slope from E to D is the area of the M/EI diagram between E and D.
(b) Displacement of C from the tangent at E is the moment of the M/EI diagram between E and C about C.
(c) Since the beam is symmetrical, the change (a) denotes the slope of the beam at D, and the displacement (b) denotes the maximum deflection (at E). Tangent at E is horizontal.

17/13

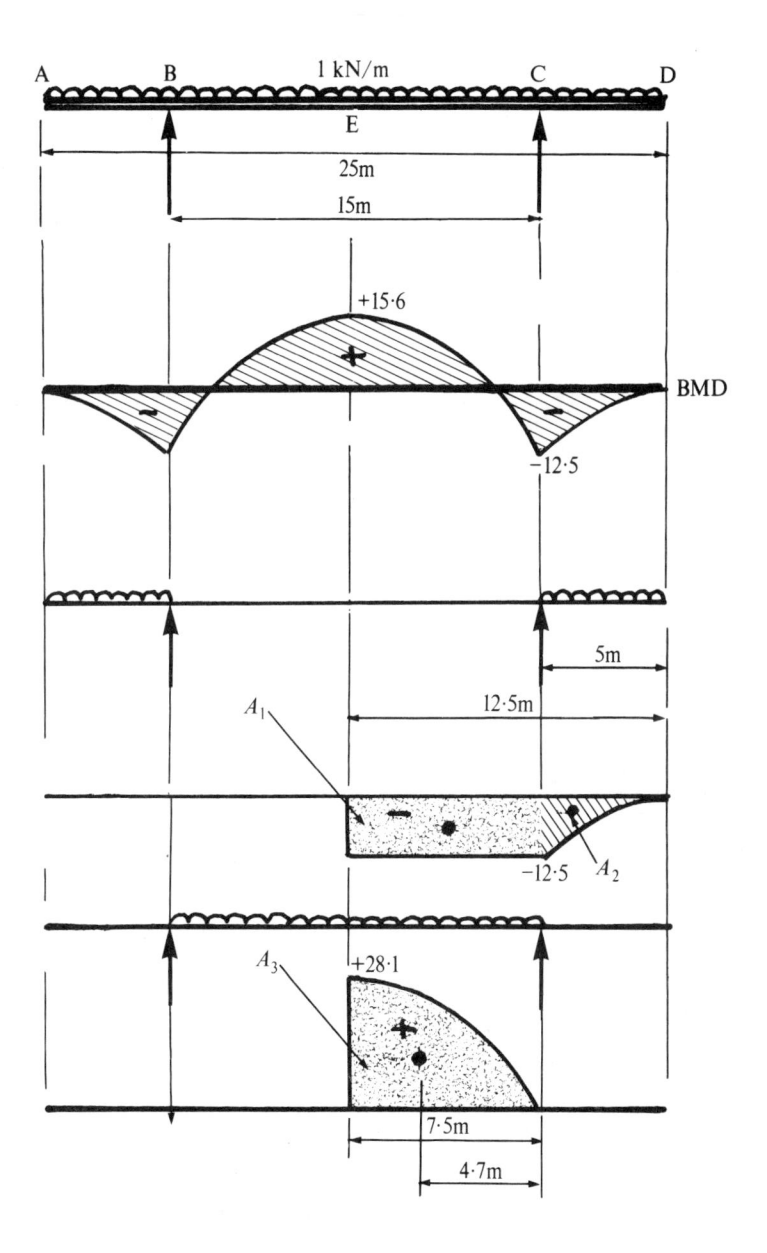

17/13

U *Unknowns and Units:*
  1. Area of M/EI diagram between E and D.
  2. Moment of M/EI diagram between E and C, about C.
  Use kN and m.

I *Inventory of Information:*
Where the areas of the BMD may be complex, and their moments about a point even more complex, it is helpful to divide the loading into several portions, and sum the effects.

   Diagram (1) shows the BMD for the whole beam.

   Diagram (3) shows half the BMD for loading on AB and CD.

   Diagram (5) shows half the BMD for loading on BC.

Superposition of (5) and (3) gives the same result as (1) but are much more amenable to calculation.

D *Derived Data:*
Area of A_1: $\quad -(12 \cdot 5 \times 7 \cdot 5) = -94$ kN m².
Area of A_2: $\quad -(\frac{1}{3} \times 12 \cdot 5 \times 5) = -21$ kN m².
Area of A_3: $\quad +(\frac{2}{3} \times 28 \cdot 1 \times 7 \cdot 5) = +141$ kN m².
Moment of A_1 about C: $-(94 \times \frac{1}{2} \times 7 \cdot 5) = -353$ kN m³.
Moment of A_3 about C: $+(141 \times \frac{5}{8} \times 7 \cdot 5) = +661$ kN m³.

E *Evaluation:*

Moment of M/EI diagram between E and C about C $= \dfrac{1}{EI}(+661 - 353)$.

Area of M/EI diagram between E and D $= \dfrac{1}{EI}(+141 - 113)$.

S *Solution:*
  1. Slope of beam at D is $+28/EI$ **(up to the right)**.
  2. Deflection of beam at E from horizontal is $+308/EI$ **m (down)**.

17/14

17/14 AREA-MOMENTS: Simply supported beam: Unsymmetrical loading

This beam is unsymmetrically loaded. Determine the vertical deflection at the centre of the beam. *EI* is constant.

G *Governing Statements:*

(a) Displacement of T from the tangent at E is the moment of the M/EI diagram between T and E about T. This is represented on the sketch by δ_T.

(b) Displacement of L from the tangent at E is the moment of the M/EI diagram between L and E about L. This is represented on the sketch by δ_L.

U *Unknowns and Units:*

1. Deflection of the beam at T, the central point. This deflection is represented on the sketch by Δ.

Use kN and m.

I *Inventory of Information:*

The BMD for the single load is a triangle, with a maximum ordinate under the load of 20 kN m.

D *Derived Data:*

The whole area A_w of the BMD: $\frac{1}{2} \times 20 \times 9 = 90$ kN m²

The moment of A_w about L: $90 \times 4 = 360$ kN m³

The right-hand area A_r: $\frac{1}{2} \times 15 \times 4\cdot5 = 34$ kN m².

The moment of the right-hand area about T: $34 \times 1\cdot5 = 51$ kN m³.

E *Evaluation:*

Moment of area of M/EI diagram between T and E about T is

$$\frac{51}{EI} \text{ m} = \delta_T$$

Moment of area of M/EI diagram between L and E about L is

$$\frac{360}{EI} \text{ m} = \delta_L$$

From the geometry of triangle XRS, $2(\Delta + \delta_T) = \delta_L$

17/14

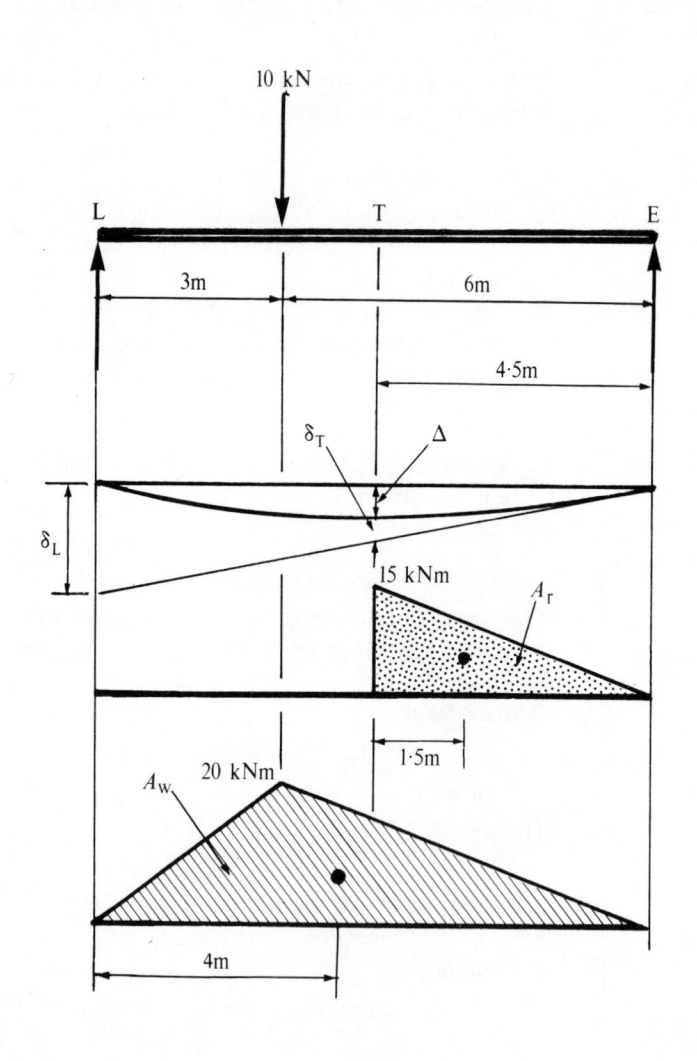

Solution:
Deflection of the centre, T, from the horizontal

$$\Delta = -\frac{129}{EI}\,\text{m}$$

17/15 AREA-MOMENTS: Vertical mast with stay

A vertical steel mast is 30 m high, and has a constant flexural rigidity, *EI*. It is firmly built into the ground and is supported by a stay at 45° to the horizontal. The stay is adjusted in length so that the point B remains undeflected and in the original unloaded position. Find the value of the tension in the stay when a load of 10 kN is applied at the top, horizontally. What is the deflection of the top of the mast? (A, top: M, centre: B, bottom).

G *Governing Statements:*

(a) Horizontal displacement of M from the vertical tangent at B is the moment of the M/EI diagram between M and B about M. This is stated to be zero.

(b) Horizontal displacement of A from the vertical tangent at B is the moment of the M/EI diagram between A and B about A.

U *Unknowns and Units:*

1. The value of P (by equating (a) above to zero).
2. The displacement or deflection of A.

Use kN and m.

I *Inventory of Information:*

The BMD can be broken into two; the BMD caused by the force at the top of the mast, which can be considered as producing positive bending, and the BMD caused by the horizontal component of P, which gives negative bending.

D *Derived Data:*

Area A_1: $\tfrac{1}{2} \times 150 \times 15 = 1125$ kN m²   Area A_3: $\dfrac{225P}{2\sqrt{2}}$ kN m²

Area A_2: $\tfrac{1}{2} \times 300 \times 15 = 2250$ kN m²   Area A_4: 4500 kN m²

Moment of A_1 about M: $1125 \times 5 = 5625$ kN m³

Moment of A_2 about M: $2250 \times 10 = 22\,500$ kN m³

Moment of A_3 about M: $\dfrac{225P}{2\sqrt{2}} \times 10 = \dfrac{1125P}{\sqrt{2}}$ kN m³

Moment of A_4 about A: $4500 \times 20 = 90\,000$ kN m³

437

17/15

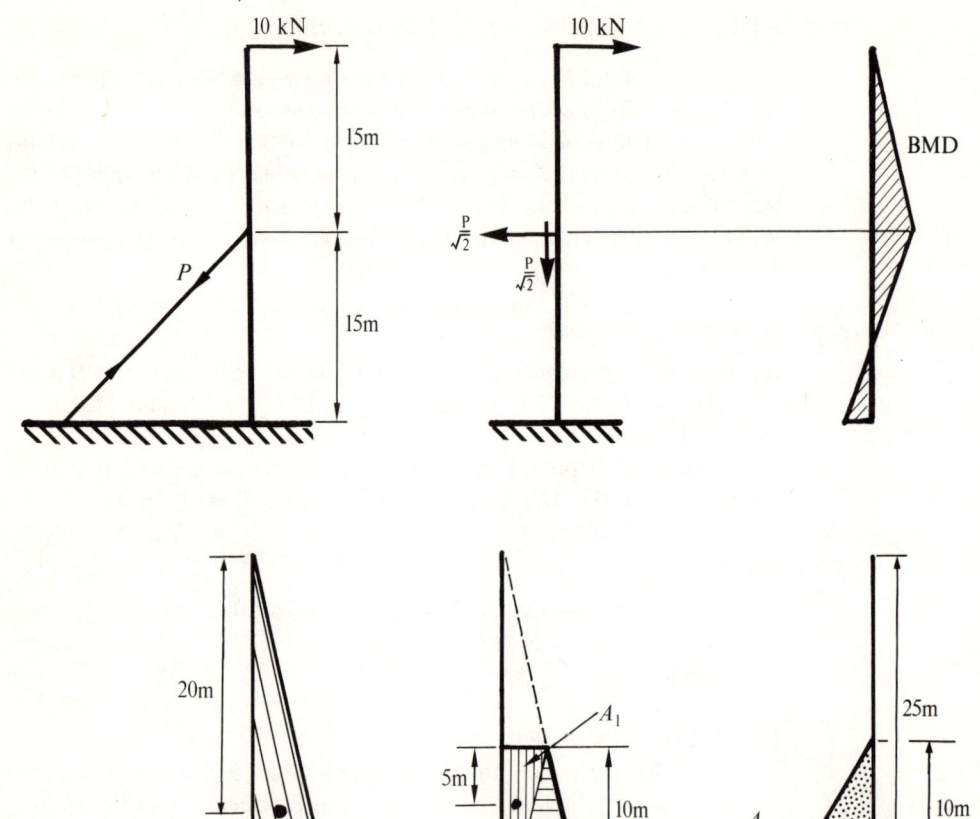

E Moment of A_3 about A: $\dfrac{225P}{2\sqrt{2}} \times 25$ (with P evaluated)

Evaluation:

Displacement of M from tangent at B is zero (see governing statement).

$$\frac{1}{EI}\left(5625 - 22\,500 - \frac{1125P}{\sqrt{2}}\right) = 0$$

Displacement of A from tangent at B is moment of M/EI diagram between A and B about A:

$$\frac{1}{EI}\left(90\,000 - \frac{225 \times 25\sqrt{2}}{2\sqrt{2}} \times 25\right)$$

S *Solution:*
1. The value of P when M is held in its original position is

$$25\sqrt{2}\text{ kN}$$

2. In these circumstances, the horizontal deflection of A is

$$\frac{19\,700}{EI}\text{ m}$$

For Area-moment determinations always draw lots of diagrams, and mark on them the various quantities given or calculated.

Compare the methods of evaluating displacements by solving the same problem by different methods. Decide which method is suitable for various categories of problem.

17/16

17/16 **AREA-MOMENTS: Bent cantilever: Varying second moment of area**

The cantilever shown in the sketch carries a vertical load at the extreme end. The displacement of this point O from the tangent at Q is also the vertical deflection of the load. Find this vertical deflection.

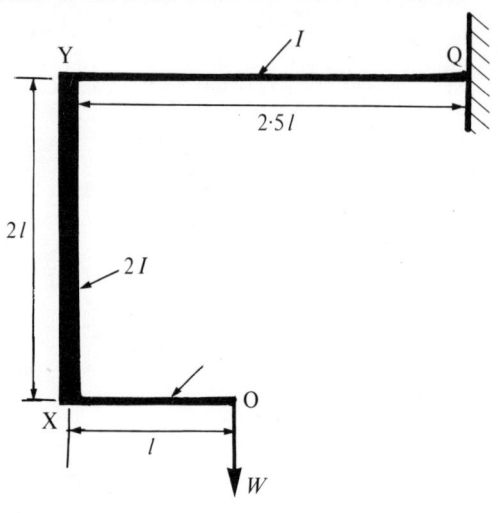

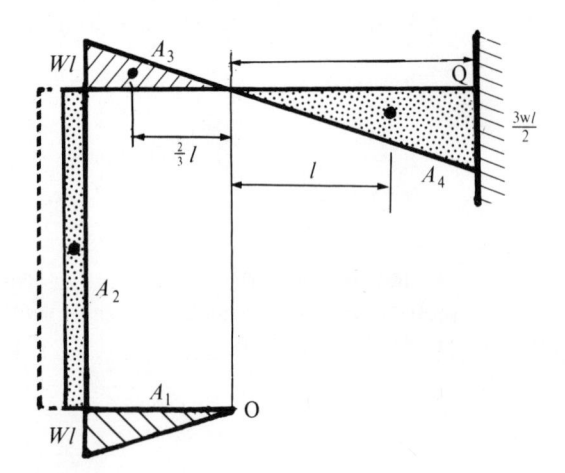

G *Governing Statements:*

(a) Displacement of O from tangent at Q is the moment of the areas A_1, A_2, A_3, A_4, about O.

440

U *Unknowns and Units:*
  1. Values of the four M/EI diagrams.
  2. Moments of the four M/EI diagrams about O.
  Use kN and m.

I *Inventory of Information:*
  The second moment of area of the vertical member, XY is twice that of the others. E is in kN/m² and I in m⁴.

D *Derived Data:*
  The M/EI diagrams and their moments about O can be recorded in a table:

Area of M/EI diagram	Moment Arm	Moment about O
$A_1 \quad +\dfrac{1}{EI} \times \dfrac{1}{2} Wl^2$	$+\tfrac{2}{3}l$	$+\dfrac{Wl^3}{3EI}$
$A_2 \quad +\dfrac{1}{2EI} \times 2Wl^2$	$+l$	$+\dfrac{Wl^3}{3EI}$
$A_3 \quad +\dfrac{1}{EI} \times \dfrac{1}{2} Wl^2$	$+\tfrac{2}{3}l$	$+\dfrac{Wl^3}{3EI}$
$A_4 \quad -\dfrac{1}{EI} \times \dfrac{9Wl^2}{8}$	$-l$	$+\dfrac{9}{8}Wl^3$

E *Evaluation:*
  The total moment of the areas about O is
  $$\frac{67Wl^3}{24EI}$$

S *Solution:*
  The vertical deflection of O is $\qquad 2\cdot79 \, \dfrac{Wl^3}{EI} \, \text{m}$

441

17/17

17/17 AREA-MOMENTS: Eccentric U.D.L.: Varying second moment of area

The left half of the beam in the sketch has a second moment of area of cross-section which is one-half of that in the right half of the beam. Find the deflection of the beam at a point 5 m from A. (Point P.)

G Governing Statements:

(a) Displacement of P from tangent at B is the moment of the M/EI diagram between P and B about P.

(b) Displacement of A from tangent at B is the moment of the M/EI diagram between A and B about A.

U Unknowns and Units:

1. Displacements δ_A and δ_P.
2. Value of Δ from relationship $\Delta + \delta_P = \tfrac{3}{4}\delta_A$.

Use kN and m.

I Inventory of Information:

Here, there is no easy way of finding the area and the moment of the M/EI diagram from simple geometrical shapes. The best method, and one which is often quicker than those used in earlier problems is by integrating the value of the BM.

$$\text{Area of } \frac{M}{EI} \text{ diagram} = \frac{1}{EI}\int M\,dx: \text{ Moment} = \frac{1}{EI}\int Mx\,dx$$

For the left half of the beam there is a trapezium A_M which could be split into two triangles as before, but whose area and moment can also be found easily by integration. For the right half, integration is the only simple way.

D Derived Data:

Area of A_R:

$$\frac{1}{EL}\int_0^{10} EL\,(15x - x^2)\,dx = \frac{417}{EI} \text{ (ratio)}$$

Moment of A_R about B:

$$\frac{1}{EI}\int_0^{10}(15x - x^2)x\,dx = \frac{2500}{EI}\,\text{m}$$

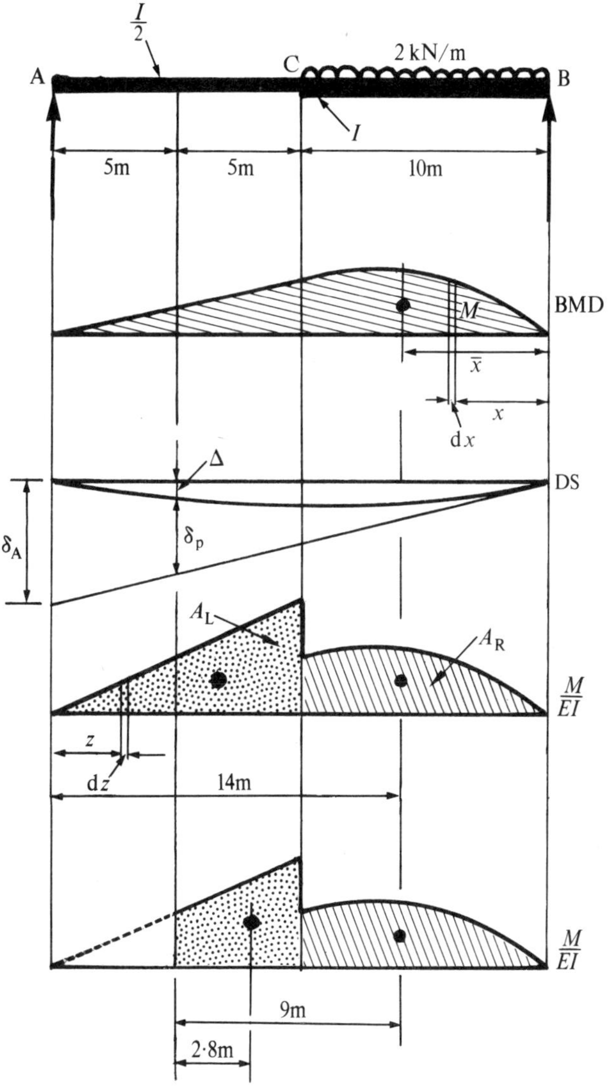

I/2
2 kN/m
A
C
B
I
5m
5m
10m
BMD
M
x̄
x
dx
Δ
DS
δA
δp
AL
AR
M/EI
z
dz
14m
M/EI
9m
2·8m

17/17

Value of $\bar{x} = 2500/417 = 6$ m and thus the value of l is 14 m.

Moment of A_L about A:

$$\frac{1}{(EI)/2} \int_0^{10} 5x^2 \, dx = \frac{3333}{EI} \text{ m}$$

Area of Trapezium (A_M):

$$\frac{1}{(EI)/2} \int_5^{10} 5x \, dx = \frac{375}{EI} \text{ (ratio)}$$

Moment of A_M about A:

$$\frac{2}{EI} \int_5^{10} 5x^2 \, dx = \frac{2917}{EI} \text{ m}$$

Moment of A_M about P:

$$\frac{375}{EI}(7 \cdot 8 - 5) = \frac{1050}{EI} \text{ m}$$

E *Evaluation:*

Moment of M/EI diagram about A:

$$\delta_A = \frac{1}{EI}(417 \times 14 + 3333) = \frac{9171}{EI} \text{ m}$$

Moment of M/EI diagram about P:

$$\delta_P = \frac{1}{EI}(417 \times 9 + 1050) = \frac{4803}{EI} \text{ m}$$

S *Solution:*

The displacement δ_P together with the required deflection Δ represents a length which is three-quarters of the displacement δ_A.

Thus

$$\Delta + \frac{4803}{EI} = \frac{3}{4} \times \frac{9171}{EI}$$

$$\Delta = \frac{2075}{EI} \text{ m (down)}$$

17/18

With a value of E of say, 200×10^6 kN/m² and a second moment of area, of the right-hand half of the beam, of 60×10^{-6} m⁴ (30×10^{-6} on the left), the deflection Δ would be

$$\frac{2075}{12\,000} = 0.17 \text{ m or } \frac{1}{116} \text{ of the span.}$$

If the whole beam had a uniform EI equal to that of the right half in this question, the beam would be stiffer. Find the deflection Δ which should be about $\hfill 1360/EI$.

17/18   AREA-MOMENTS: Cantilever: Varying second moment of area

The cantilever shown in the sketch has a second moment of area of I over the outer third, a value of $2I$ in the middle third, and one of $3I$ in the third nearest the support. Find the slope of the cantilever to the horizontal at the extreme end, and the deflection of that end. How do these compare with the corresponding values when EI is constant? Fixed end, K: free end, N.

G   *Governing Statements:*

(a) The change in slope between K and N is the area of the M/EI diagram between K and N.

(b) The displacement of N from the tangent at K is the moment of the M/EI diagram between K and N about N.

(c) Since the tangent to the deflected shape at K is horizontal (slope zero) the values in (b) and (c) are those asked for in the problem.

U   *Unknowns and Units:*

1. The area of the M/EI diagram for the whole cantilever when there are three values of second moment of area.
2. The area of the M/EI diagram for the whole cantilever when EI is constant.
3. The moment of the M/EI diagram about N when there are three values of second moment of area.
4. The moment of the M/EI diagram about N when EI is constant over the whole length.

Use kN and m.

I   *Inventory of Information:*

The maximum bending moment is 90 kN m; this remains true for both cases. The changes in I have no effect on the bending moment.

The M/EI diagram does not keep the same shape as the BMD in this

17/18

problem; all values of BM must be divided by the appropriate I to give the M/I diagram, the value of E remaining constant.

D *Derived Data:*

This is best worked out in tabular form:

Area No.	Area × EI	Distance to N from centroids	Moment of M/EI diagram
1	−45	8	−360
2	−30	7	−210
3	−45	5	−225
4	−23	4	−90
5	−45	2	−90
Total	−188		−975
6	−405	6	−2430

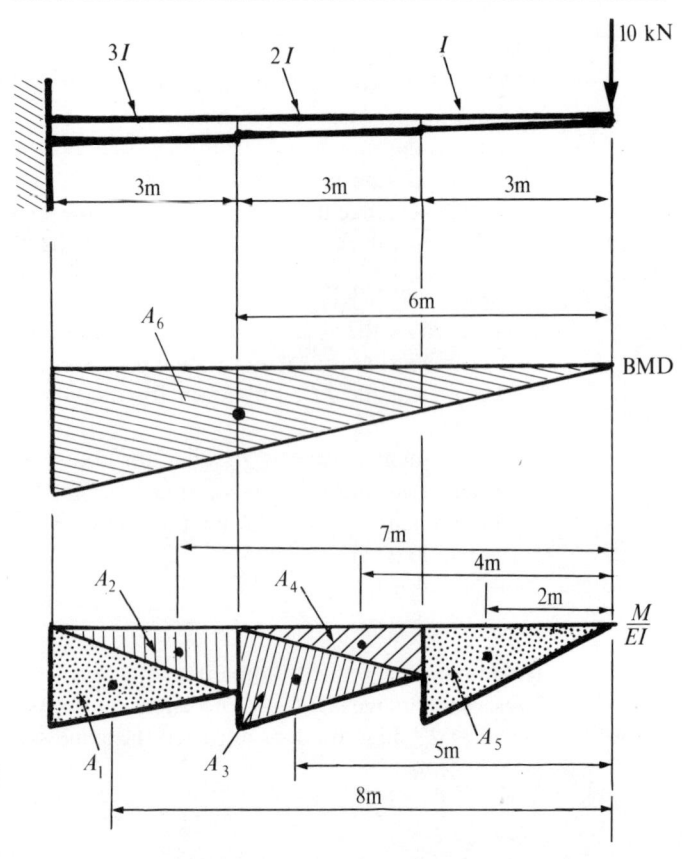

E *Evaluate:*

The slope of the beam at N is the sum of the areas of the M/EI diagrams, or $-188/EI$ **gradient**.

The deflection of the beam at N is the sum of the moments of the M/EI diagrams about N, or $-975/EI$ **m**.

Similar figures for a constant I are $-405/EI$ and $-2430/EI$.

S *Solution:*

The stiffening of the cantilever caused by the increasing second moment of area towards the support has decreased the slope at N by a ratio of 405/188 or **2·15**.

The stiffening has also decreased the deflection of the end N by a ratio of 2430/975 or **2·49**.

17/19 AREA-MOMENTS: Straight beam: Simple supports: Applied couple

This beam is hinged at both ends so that it can take an upward or a downward reaction at both A and C. A couple applies a moment of 1200 N m at C. Find the position of the maximum deflection from the original straight length, and the value of this maximum deflection.

G *Governing Statements:*

(a) When the maximum deflection is reached, the tangent at that point is horizontal, and, therefore, the displacements δ_A and δ_C must be equal.

(b) The displacement of A from the tangent at D is the moment of the M/EI diagram between A and D about A.

(c) The displacement of C from the tangent at D is the moment of the M/EI diagram between D and C about C.

U *Unknowns and Units:*

1. Lengths d and e such that δ_A and δ_C are equal. Length d, measured from A, is the distance to the point of maximum deflection.
2. Value of maximum deflection.

Use kN and m.

17/19

I *Inventory of Information:*

The FSD and BMD are found from the methods of part I, and the BMD can be divided into three areas for calculating the displacements.

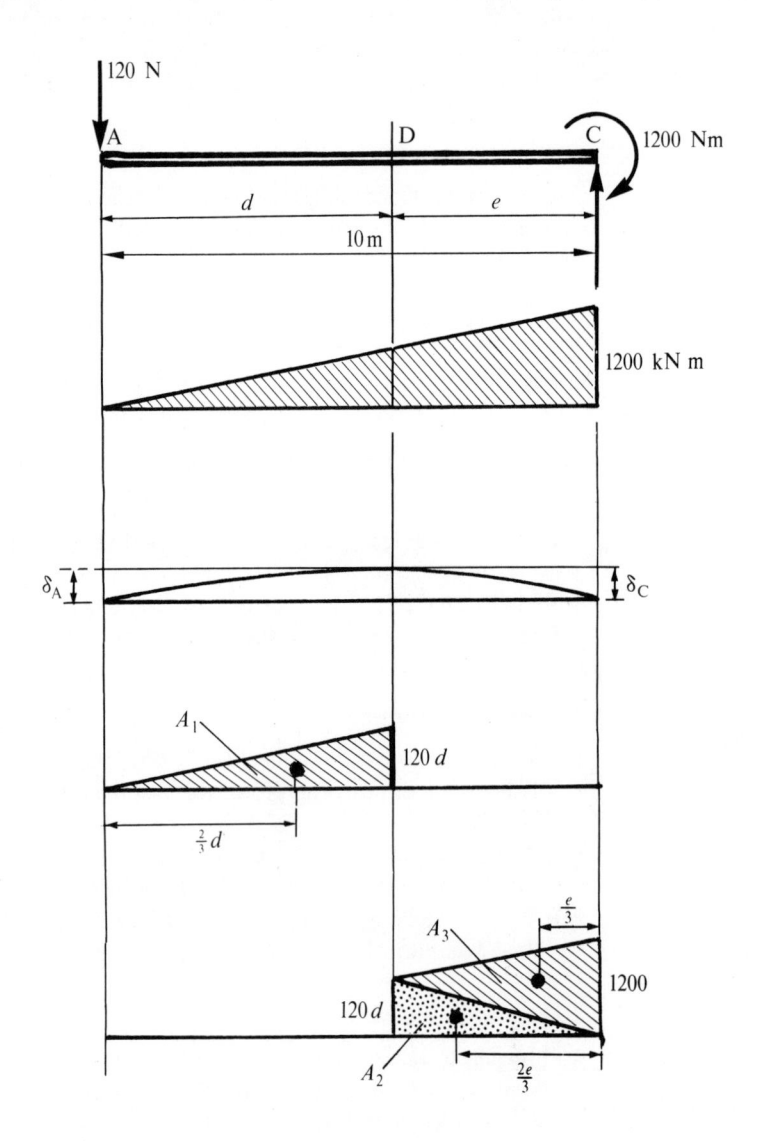

448

D *Derived Data:*

Area of A_1: $\quad \frac{1}{2} \times 120d \times d = 60d^2$ N m²

Moment of A_1 about A: $60d^2 \times \frac{2}{3}d = 40d^3$ N m³

Area of A_2: $\quad \frac{1}{2} \times 120d \times e = 60de$ N m²

Moment of A_2 about C: $60de \times \frac{2}{3}e = 40de^2$ N m³

Area of A_3: $\quad \frac{1}{2} \times 1200 \times e = 600e$ N m²

Moment of A_3 about C: $600e \times l/3 = 200e^2$ N m³

E *Evaluation:*

$$\delta_A = \delta_C$$
$$+40d^3 = +40de^2 + 200e^2$$

But $e = (10 - d)$

$$+40d^3 = +40d(10 - d)^2 + 200(10 - d)^2$$
$$+d^3 = +d(10 - d)^2 + 5(10 - d)^2$$

Whence $\quad d = 5 \cdot 8$ m

S *Solution:*

1. Maximum deflection occurs at **5·8 m from A**.
2. Value of maximum deflection is

$$\frac{40d^3}{EI} = \frac{7670}{EI} \text{ m}$$

17/20 AREA-MOMENTS: Beam carrying both load and couple

The beam (AB) in the sketch carries a 1 kN load on a bracket. The flexural rigidity is constant along the whole length (EI). Find the vertical deflection of the central point of the beam. (C)

G *Governing Statements:*

(a) The load on the bracket applies a central vertical force of 1 kN and also a couple of 2 kN m.

(b) Displacement of C (and A) from tangent at B is the moment of the M/EI diagram between B and C (or A) about C (or A).

449

17/20

U *Unknowns and Units:*
1. Displacement at C and displacement at A from tangent at B.
2. Value of Δ by geometrical relationship.

I *Inventory of Information:*
Two BMD's can be used, one for the load, and one for the couple. The deflected shape shows positive bending on the right half of the beam and negative on the left half.

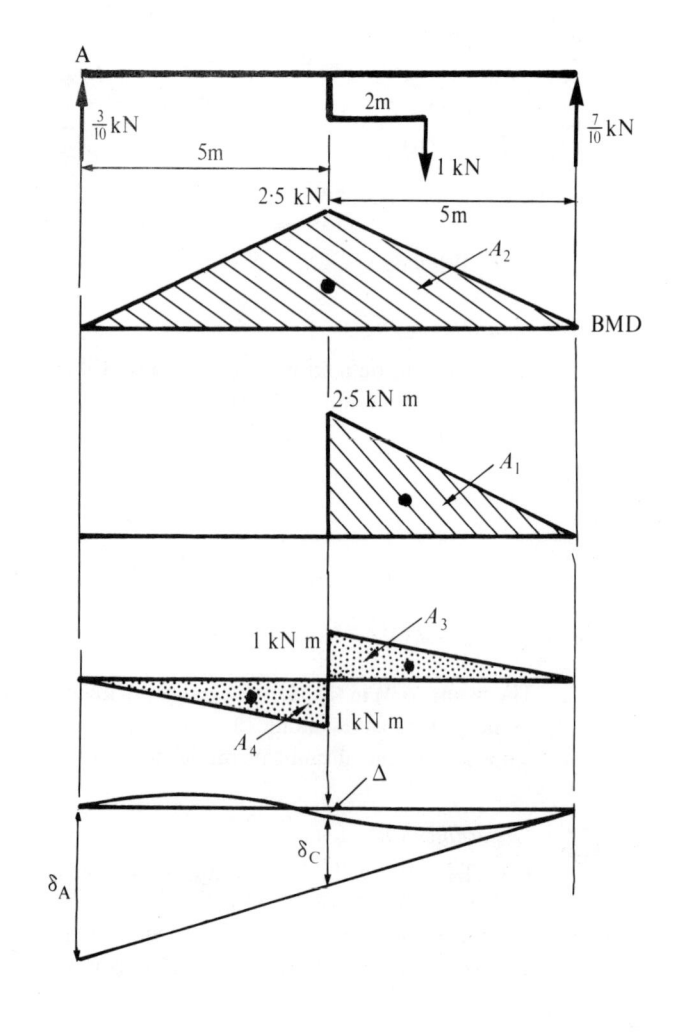

450

D *Derived Data:*

Area A_1: $\quad +(\frac{1}{2} \times 2 \cdot 5 \times 5) = +6 \cdot 3 \text{ kN m}^2$

Area A_2: $\quad +(2 \times A_1) = +12 \cdot 5 \text{ kN m}^2$

Area A_3: $\quad +(\frac{1}{2} \times 1 \times 5) = +2 \cdot 5 \text{ kN m}^2$

Area A_4: $\quad\quad\quad\quad\quad\quad\quad\quad -2 \cdot 5 \text{ kN m}^2$

Moment of A_1 about C: $+(6 \cdot 3 \times \frac{5}{3}) = +10 \cdot 4 \text{ kN m}^3$

Moment of A_3 about C: $+(2 \cdot 5 \times \frac{5}{3}) = +4 \cdot 2 \text{ kN m}^3$

Moment of A_2 about A: $+(12 \cdot 5 \times 5) = +62 \cdot 5 \text{ kN m}^3$

Moment of A_3 about A: $+(2 \cdot 5 \times \frac{20}{3}) = +16 \cdot 7 \text{ kN m}^3$

Moment of A_4 about A: $-(2 \cdot 5 \times \frac{10}{3}) = -8 \cdot 3 \text{ kN m}^3$

E *Evaluation:*

Displacement of C from tangent at B:

Moment ($\div EI$) of A_1 and A_3 about C:

$$\delta_C = \frac{1}{EI}(+10 \cdot 4 + 4 \cdot 2) = +\frac{15}{EI} \text{ m}$$

Displacement of A from tangent at B:

Moment of A_2, A_3, and A_4 ($\div EI$) about A:

$$\delta_A = \frac{1}{EI}(+62 \cdot 5 + 16 \cdot 7 - 8 \cdot 3)$$

$$= +\frac{71}{EI} \text{ m}$$

From the geometry of the original horizontal line of the beam, the deflected shape and the displacements calculated,

$$\Delta + \delta_C = \frac{1}{2}\delta_A$$

$$\Delta + \frac{15}{EI} = \frac{71}{2EI}$$

S *Solution:*

The vertical deflection of the centre of the beam is

$$\Delta = +\frac{41}{2EI} \text{ m}$$

17/21

17/21 VARYING CROSS-SECTION: Beam loaded by couples

A plate spring is tested as shown in the sketch. When the equal loads, applied at the ends, reach the value of 3·6 kN, describe the curvature of the plates at the ends and in the central portion. The plate is 12 mm thick throughout.

G *Guide:* $\dfrac{M}{I} = \dfrac{E}{R}$

U *Unknowns and Units:*
  1. Curvature of the triangular ends.
  2. Curvature of the central portion.
Use kN and mm.

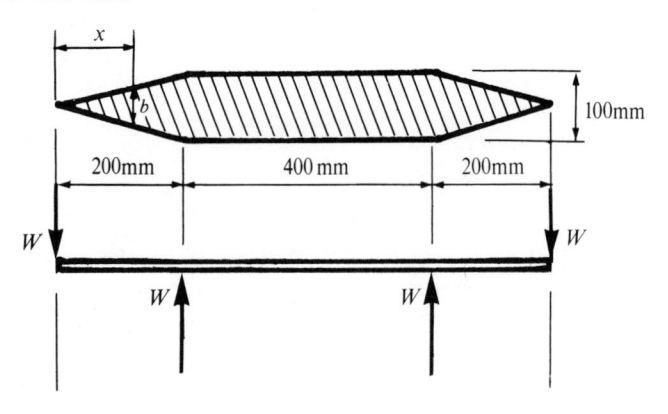

I *Information:*
W in the sketch is 3·6 kN.
Thickness, t, is 12 mm.

D *Derived Data:*
Width of plate (b) in the triangular portion is $x/2$ mm.
Second moment of area in triangular portion:
$$\frac{bt^3}{12} = 72x \text{ mm}^4$$
Second moment of area in parallel portion:
$$\frac{100 \times 12^3}{12} = 144 \times 10^2 \text{ mm}^4$$
Bending moment in triangular portion: $= Wx = 3\cdot 6x$ kN mm.
Bending moment in central portion: $= 3\cdot 6 \times 200 = 720$ kN mm.

E *Evaluation:*

Triangular portion: $\dfrac{M_1}{I_1} = \dfrac{E}{R_1}$

$\dfrac{3\cdot 6x}{72x} = \dfrac{E}{R_1}$ $\hspace{4em}$ $R_1 = 20E$ mm

Parallel portion: $\dfrac{M_2}{I_2} = \dfrac{E}{R_2}$

$\dfrac{720}{14400} = \dfrac{E}{R_2}$ $\hspace{4em}$ $R_2 = 20E$ mm

S *Solution:*

The plate bends into a uniform radius over its whole length. For the conditions given, the radius is **$20E$ mm, when E is in kN/mm².**

Attempt many problems from other textbooks by the methods shown here.

Use the mnemonics throughout this book. One step at a time is easy, and they add up to a complete solution.

Part II: Group F

18 Direct and bending stress combined

Few structures are exposed to only one type of loading. The stresses described in Group E were those caused separately by direct force, bending couples, transverse shear, and torsional couples. There are many different ways in which the stresses caused by these simple loadings can be combined. Sometimes the stresses caused by two different types of loading are of the same type and occur on the same plane, and therefore, may be added algebraically. If they are of different types, such as a direct stress and a shearing stress, or act on different planes passing through one point, their combination is more complex and is discussed in Unit 19.

In this present Unit of Study examples are given of the addition of two stresses which occur on the same plane and are of the same type. Other examples are possible, and the general condition is that if the stresses are of the same type (both direct (σ) or both shear (τ)) and act on the same plane, they can be calculated separately and added together.

With a skill in adding like stresses, and in finding the effects of unlike stresses acting together (see Unit 19), you can consider yourself well fitted for solving the more elementary problems concerning states of stress in a loaded material.

18.1 Bending combined with direct force

Direct stress is caused by a direct force acting on the area of the cross-section. Direct stress is uniform over the whole cross section. Bending stress is caused by the action of a couple which produces varying stress – from tension to compression – across the cross-section. The bending stress is of like kind to the direct stress, and can be added algebraically to the simultaneous direct stress.

For example, if you examine the silhouette of a cross-section on which have been sketched the distributions of direct stress and of bending stress, the combined effect is obtained by adding arrows of the same

18.1

18 A

Direct load | Couple | Combined effect

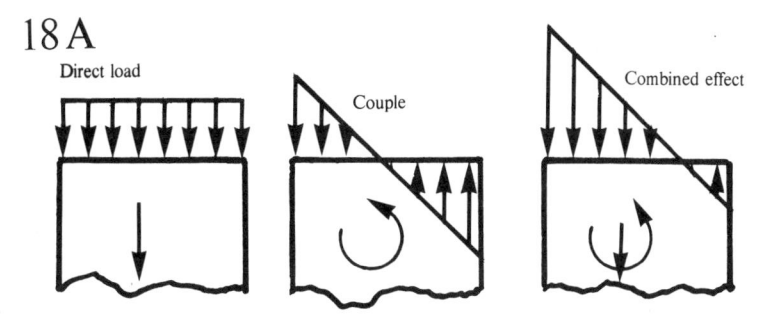

sense, and subtracting arrows of opposite senses. At L, for example, the combined stresses cause a final stress greater than either of the original ones. At R the couple produces a tensile stress which counteracts the compressive stress caused by the direct force. According to the ratio of the two effects, three different profiles to the stress distribution are possible (sketch 18B).

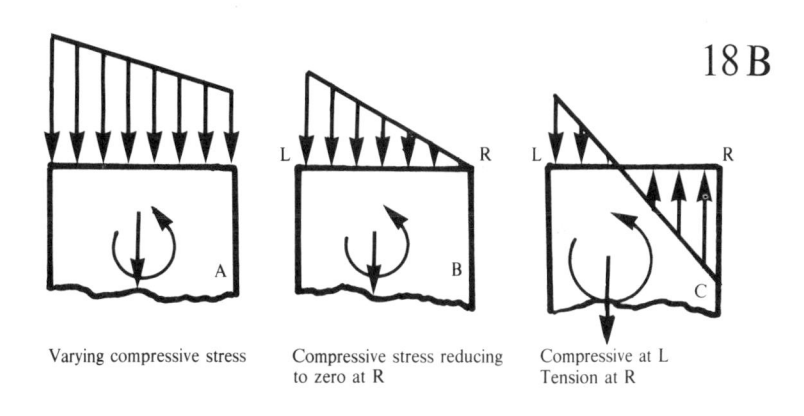

18 B

Varying compressive stress | Compressive stress reducing to zero at R | Compressive at L Tension at R

If the direct stress is dominant and the bending stress of less value, the condition at A is likely. As the bending stress increases in proportion to the direct stress there comes a point when, at R, tension is just about to appear (B). This is an important condition when the material of which the structure is composed is concrete or jointed masonry or other material weak in tension. The designer's efforts are directed to ensuring that the condition in B is not overstepped. If the effect of the bending couple overcomes that of the direct load, then condition C will appear in which there are both compressive and tensile stresses.

18.2

18.2 Overturning moments

When a couple is applied to a cross-section already under direct stress, even condition C may be overstepped if the couple is large enough. This limiting condition is often the subject of examination papers. The sketches of 18C show what happens. Here, a wall of some weight

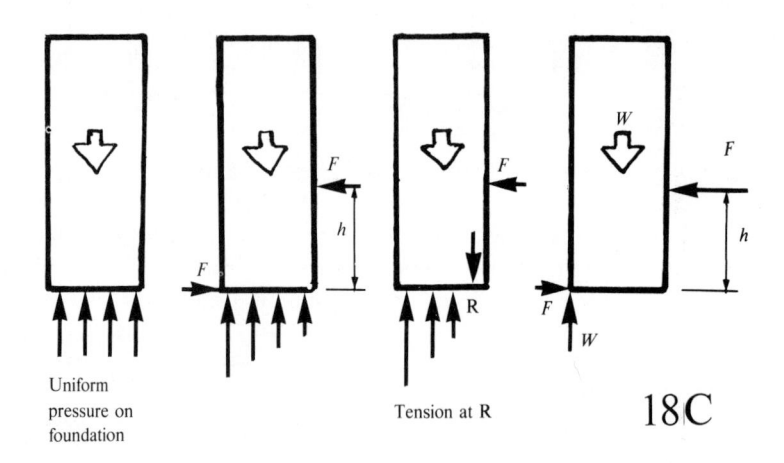

Uniform pressure on foundation

Tension at R

18C

applies, from its self-weight, a direct stress on the ground. If a couple is applied some of the direct compressive stress is counteracted. (The couple is formed by the applied direct force from wind or earth pressure and the frictional force on the base.) As the value of F increases and the couple increases, even the small compressive force at R can be counteracted and tension developed. Continuing increase in F (assuming the wall holds together in one piece) can produce a condition when the couple has counteracted the whole of the compressive stress, and lifted the wall off its base. Overturning takes place if Fh is greater than Wd. The value of Wd is the restoring couple, which can always be called into play to counteract overturning couples. d is half-width of base.

In most problems on combined stress the state of overturning does not occur and some of the intermediate stress diagrams (combined from direct and bending stresses) are in question. It should be noted that this type of problem contains two sets of forces and stresses which are quite independent. The overturning couple can have a value independent of the direct stress applied by the weight of the wall, as in the above example. In the next type of combined stress, the two loads – the direct force and the bending couple – are not independent.

18.3 Eccentric loading

Another way in which a couple may be applied in combination with a direct force, is for the force to be applied eccentrically – not at the centroid of the section, as is usually assumed for direct-force conditions. Suppose you imagine a tie-bar of any section pulled by a force F which passes down the axis of the rod (i.e., through the centroid of the area). The stress is then uniform across the section, and this kind of stress is dealt with in Unit 13. If F is not acting through the centroid of the cross-section, but at some other point on its area, it is acting with an *eccentricity*. This value of eccentricity is measured by the distance between the line of F and the centroid of the section. The eccentricity is measured in linear units (mm) and is often given the symbol e.

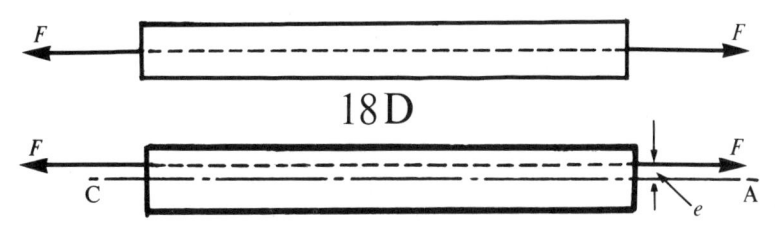

18D

To understand how this effect can be estimated in terms of stress, look at an enlarged section in silhouette. At (*a*) you have the given eccentric loading for which you are required to calculate the stress distribution. If you add, as at (*b*), two forces of value F but exactly opposing each other, they cancel each other out, and the effect is still the same as at (*a*). The three F's at (*b*) can now be split into a couple with

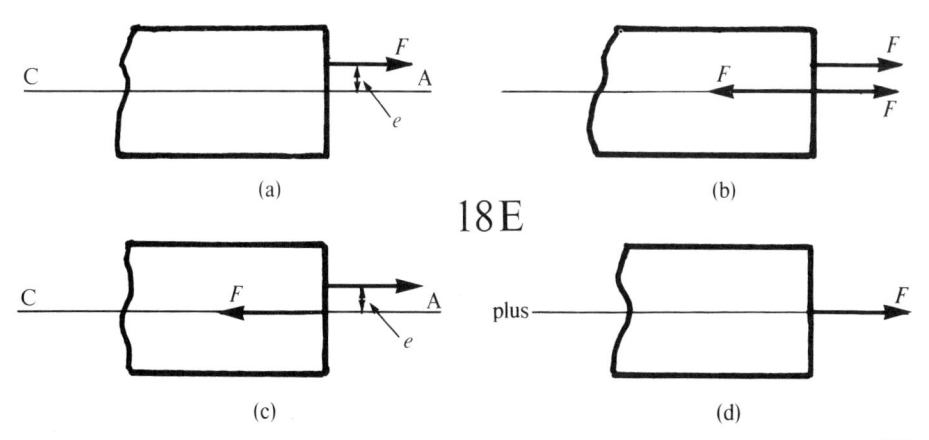

18E

18.4

a moment of *Fe*, producing varying stress, plus a direct force *F* acting through the centroid and producing uniform stress (*d*). This condition is the same as at (*a*).

Thus, if a force is eccentric on a section its effect is two-fold. There is first the uniform effect of *F* on the area *A*, and, secondly, there is the bending effect and the stress produced by the moment (*Fe*) of the couple. In this type of problem, the moment applied by the couple is related to the value of the direct stress and the location of the direct force on the cross-section.

18.4 Quantities and equations

For the estimation of the values of combined stress, therefore, both the expression for direct stress, and that for bending stress must be used:

$$\frac{F}{A} = \sigma_d : \qquad \sigma_d = \frac{F}{A}$$

$$\frac{M}{I} = \frac{\sigma_b}{h} : \qquad \sigma_b = \frac{Mh}{I} = \frac{M}{z}$$

The bending stress may be added to or subtracted from the direct stress. The direct stress is easily obtained by using *F* and *A*. The bending stress requires the calculation of the second moment of area, and the distance *h* from the centroidal axis to the outer fibres of the cross-section. This part of the calculation is made more readily by the use of the *section modulus*, which is merely the second moment of area divided by the depth *h* from CA to the outer fibre. The section modulus is measured in m³ or mm³ and is given the symbol *z*

$$z = \frac{I}{h} \text{ mm}^3$$

Thus the combined effect, giving the combined stress is:

$$\sigma_c = \sigma_d + \sigma_b = +\frac{F}{A} \pm \frac{M}{z}$$

The problems which follow are divided into *eccentric compression*, *eccentric tension*, and *dead load plus overturning moment*. These cover the usual problems of this kind. The final problem concerns a foundation.

18.5 THE PROBLEMS

18/1 ECCENTRIC TENSION: Rectangular section: One load

The line of pull on a tie bar of rectangular section does not pass through the centroid of the section, but at Y which is 3 mm from one centre line and exactly on the other, as shown in the sketch. The tension exerted is 240 kN. The dimensions of the cross-section are 100 × 40 mm. Find the maximum and minimum intensities of tensile stress in the section.

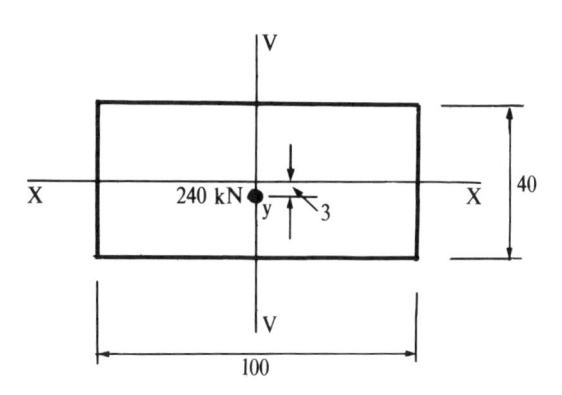

G *Groups of Terms:*

$$\frac{F}{A} = \sigma_d \qquad \frac{M}{I} = \frac{\sigma_b}{h}$$

$$+ \sigma_d \pm \sigma_b = + \frac{F}{A} \pm \frac{M}{z}$$

U *Unknowns and Units:*
1. Maximum tensile stress.
2. Minimum tensile stress.
Use kN and mm.

I *Inventory of Information:*

$F = 240$ kN $\qquad M = 240 \times 3$ kN mm

D *Derived Data:*

(a) Bending occurs on the centre line VV and about the centre line XX. There is no bending in other direction.

459

18/2

(b) *Modulus of section* for a rectangle is

$$z = \frac{100 \times 40^2}{6} = 26 \cdot 7 \times 10^3 \text{ mm}^3$$

(c) *Area of section:* 4000 mm²

E *Evaluation:*

$$\frac{F}{A} + \frac{M}{z} = \frac{240}{4000} + \frac{240 \times 3}{26 \cdot 7 \times 10^3} = 0 \cdot 087 \text{ kM/mm}^2$$

$$\frac{F}{A} - \frac{M}{z} = 0 \cdot 060 - 0 \cdot 27 = 0 \cdot 033 \text{ kN/mm}^2$$

S *Solution:*

1. Maximum tensile stress: 87 N/mm².
2. Minimum tensile stress: 33 N/mm².

18/2 ECCENTRIC TENSION: Circular section: Solid

A circular tie rod, 50 mm in diameter, was originally subjected to axial tension of 200 kN. The rod has corroded more on one side than on the other, and is now only 46 mm in diameter, as shown in the sketch. Calculate the maximum tensile stress in the original bar, and the percentage increase in stress due to the reduction in area through corrosion. The point of application of the load is still at the centre of the original circle.

G *Groups of Terms:*

$$\frac{F}{A} = \sigma_d \qquad \frac{M}{I} = \frac{\sigma_b}{h}$$

$$+\sigma_d \pm \sigma_b = +\frac{F}{A} \pm \frac{M}{z}$$

U *Unknowns and Units:*

1. Original direct stress.
2. Maximum value of present stress.
3. Percentage increase in stress.

Use kN and mm.

18/2

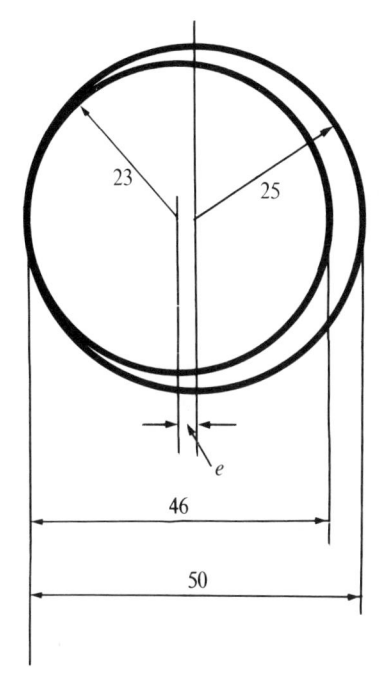

I *Inventory of Information*

$F = 200$ kN $\quad D = 50$ mm $\quad d = 46$ mm $\quad h = r = 23$ mm

F is now eccentric by a distance of $\frac{1}{2}(50 - 46) = 2$ mm.

D *Derived Data:*

(a) *Areas:* original $= 25^2$; present $= 23^2$

(b) *Second moment of area* of present section which is still circular ($r = 23$ mm):

$$I = \frac{\pi r^4}{4} = 0{\cdot}22 \times 10^6 \text{ mm}^4$$

(c) *Modulus of Section:*

$$z = \frac{I}{h} = 0{\cdot}0096 \times 10^6 \text{ mm}^3$$

(d) *Moment applied*

$$M = 200 \times 2 = 400 \text{ kN mm}$$

461

18/3

E *Evaluation:*

Original stress

$$\frac{F}{A} = +\frac{200}{25^2} = 0\cdot 102 \text{ kN/mm}^2$$

New stress:

$$\frac{F}{A} + \frac{M}{z} \qquad \frac{+200}{23^2} + \frac{400}{9\cdot 6 \times 10^3} = 0\cdot 162 \text{ kN/mm}^2$$

S *Solution:*

1. **102 N/mm²**
2. **162 N/mm²**
3. **59% increase.**

18/3 ECCENTRIC TENSION: Hollow circular bar: Concentric void

A hollow circular bar carries a tensile eccentric load, parallel to the axis of the bar. The outside diameter is 80 mm, and the internal diameter is 50 mm. The two circles are concentric. The maximum stress must not exceed the mean stress by more than 25%. What is the allowable eccentricity of the load?

G *Groups of Terms:*

$$\frac{F}{A} = \sigma_d \qquad \frac{M}{I} = \frac{\sigma_b}{h}$$

or

$$+\sigma_d \pm \sigma_b = +\frac{F}{A} \pm \frac{M}{z}$$

U *Unknowns and Units:*

1. Allowable eccentricity of load.

Use N and mm.

I *Inventory of Information:*

$R = 40$ mm $\qquad r = 25$ mm (radii of circles)

Mean stress is F/A. Therefore, M/z must not be greater than $0\cdot 25\ F/A$.

D *Derived Data:*

(a) Moment (M) is force times eccentricity $= Fe$ (N mm).
(b) Value of h is equal to the outer radius, or 40 mm.
(c) Second moment of area of a hollow circular tube

$$I_{NA} = \frac{\pi}{4}(R^4 - r^4) = 1.71 \times 10^6 \text{ mm}^4$$

(d) Area of the metal of the cross-section $= \pi(R^2 - r^2) = 3060 \text{ mm}^2$.
(e) Modulus of section

$$\frac{I}{h} = \frac{1.71 \times 10^6}{40} = 42.8 \times 10^3 \text{ mm}^3$$

E *Evaluation:*

$$\frac{M}{z} = 0.25 \frac{F}{A} \text{ at the limit}$$

$$\frac{Fe}{z} = 0.25 \frac{F}{A} \qquad e = \frac{0.25z}{A}$$

$$= 3.5 \text{ mm}$$

S *Solution:*

1. Thus maximum eccentricity allowable is $\qquad e = \mathbf{3.5 \text{ mm}}$.

18/4 ECCENTRIC COMPRESSION: I Section: One load

A short column, whose cross-sectional area is **2240 mm²**, consists of an I or H section whose depth is **150 mm**. The second moment of area, taken from published steel tables is $8.4 \times 10^6 \text{ mm}^4$. The load acts on the axis of the web, but is at a distance of **50 mm** from the centroidal axis. What are the maximum and minimum stresses on the column if the load is **100 kN**.

G *Groups of Terms:*

$$\frac{F}{A} = \sigma_d \qquad \frac{M}{I} = \frac{\sigma_b}{h}$$

$$+\sigma_d \pm \sigma_b = +\frac{F}{A} \pm \frac{M}{z}$$

463

18/4

U *Unknowns and Units:*
1. Maximum stress on the section.
2. Minimum stress on the section.
Use kN and mm.

I *Inventory and Information:*

Depth of section = 150 mm $\quad h = 75$ mm
$I = 8\cdot4 \times 10^6$ mm^4 $\quad z = 0\cdot112 \times 10^6$ mm^3
$F = 100$ kN $\quad e = 50$ mm
$M = 5000$ kN mm $\quad$ Area $= 2240$ mm^2

D *Derived Data:*
None required.

E *Evaluation:*

$$\frac{F}{A} \pm \frac{M}{z} = +\frac{100}{2240} \pm \frac{5000}{0\cdot112 \times 10^6}$$

$$= +0\cdot092 \text{ kN/mm}^2$$

or

$$-0\cdot002 \text{ kN/m}^2$$

S *Solution:*
1. The maximum compressive stress in the column is **92 N/mm²**.
2. The minimum stress in the column is negative which means the column has to be held down; the stress is a tension; **2 N/mm²**.

Compressive bending stress occurs on the side of the eccentricity; tensile bending stress on the side away from the eccentricity, when the load is compressive.

18/5

18/5 ECCENTRIC COMPRESSION: Circular section: Hollow

A cylindrical cast-iron column has an external diameter of 300 mm and an internal diameter of 200 mm. Because of faulty casting, the core is 12 mm off centre. Where must the load be applied so that no bending stresses are induced?

G *Groups of Terms:*

If no bending stresses are to occur, the stress must be uniform, so the load must be applied at the centroid of the whole section.

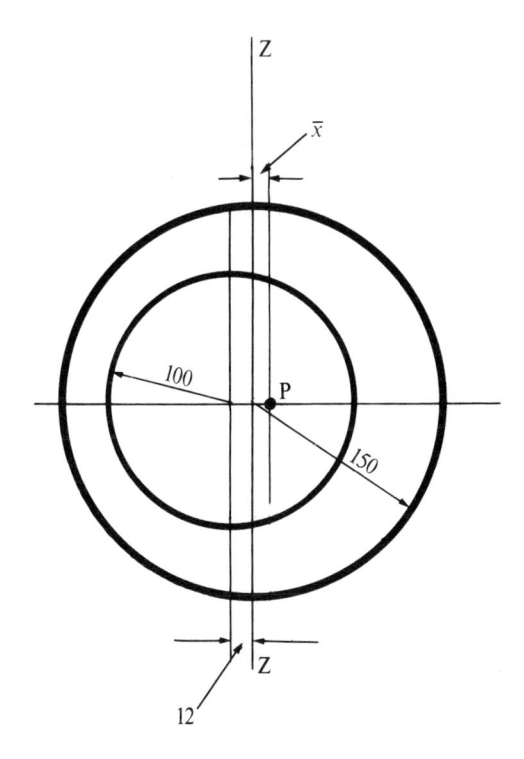

U *Unknowns and Units:*

1. Where must load be applied so that there are no bending stresses? It is thus a question of where is the centroid? (value of $\bar{x}$ in the sketch). Use mm.

465

18/5

I *Inventory of Information:*

When two circles are not concentric, their eccentricity can be in one direction only, for it is always possible to draw a diameter through the two centres. The centroid thus lies on a line joining the two centres.

$$D = 300 \text{ mm} \qquad d = 200 \text{ mm} \qquad \text{Eccentricity} = 12 \text{ mm}$$

D *Derived Data:*

(a) *Centroid:* The reference axis is ZZ which passes through the centre of the large circle. The point P is the centroid of the perforated column.

E *Evaluation:*

	Area (mm²)	Distance (mm) to ZZ (x)	Moment Ax (mm³)
Large circle	$150^2\pi$	0	0
Small circle	$100^2\pi$	-12	$-12 \times 100^2\pi$
Whole area	$(150^2 - 100^2)\pi$	$+\bar{x}$	$+\bar{x}\pi(150^2 - 100^2)$
Total			0

$$+\bar{x}\pi(150^2 - 100^2) - 12 \times 100^2\pi = 0$$
$$\bar{x} = 9{\cdot}6 \text{ mm}$$

S *Solution:*

1. Load should be applied over a point **9·6 mm** on the opposite side of ZZ from the centre of the smaller circle.

The Second Moment of Area divided by the distance from centroidal axis to the outside fibre is the Section Modulus.

18/6 ECCENTRIC COMPRESSION: Circular section: Hollow

A cast-iron vertical pillar is 150 mm in diameter. It has a hole, 25 mm in radius, drilled along its axis. The centre of the hole is 38 mm distant from the centre of the pillar (P). Find the values of the maximum and minimum stresses on the section.

G *Groups of Terms:*

$$\frac{F}{A} = \sigma_d \qquad \frac{M}{I} = \frac{\sigma_b}{h}$$

$$+\sigma_d \pm \sigma_b = +\frac{F}{A} \pm \frac{M}{z}$$

U *Unknowns and Units:*
1. Maximum compressive stress.
2. Minimum stress (tension or compression?).

Use kN and mm.

I *Inventory of Information:*

Diameter of pillar = 150 mm
Eccentricity of hole = 38 mm
h_R = 79·35 mm

Diameter of hole = 50 mm
F = 1000 kN
h_L = 70·65 mm

467

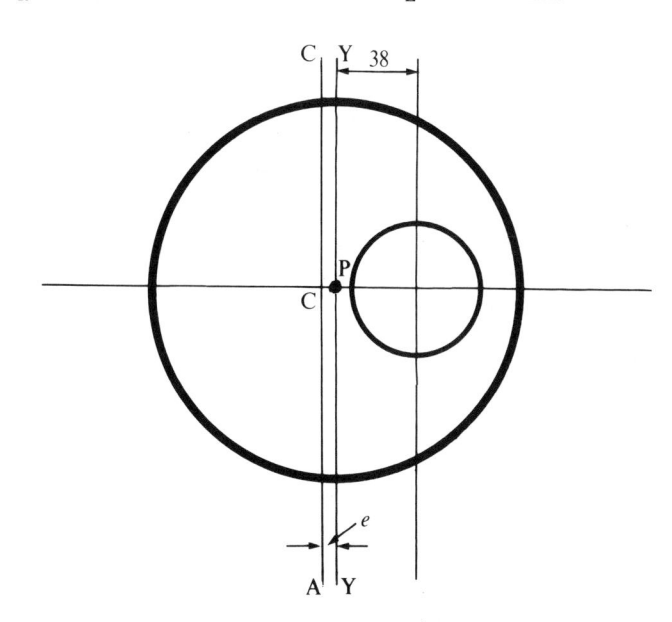

18/6

D *Derived Data:*

(a) *Distance from YY* (through centre of circle) to the centroidal axis (N.A) is 4·35 mm which is also the eccentricity of the load.

(b) *Second moment of area* about N.A.

$$I = 29 \times 10^6 \text{ mm}^4$$

(c) *Modulus of section:*

$$z_R = \frac{29 \times 10^6}{79 \cdot 35} = 366 \times 10^3 \text{ mm}^3$$

$$z_L = \frac{29 \times 10^6}{70 \cdot 65} = 410 \times 10^3 \text{ mm}^3$$

(d) *Eccentricity* of load from the neutral axis of the whole section is 4·35 mm.

E *Evaluation:*

$$\frac{F}{A} \pm \frac{M}{z}$$

$$= \frac{1000}{5000\pi} + \frac{(1000 \times 4 \cdot 35)}{366 \times 10^3} = 0 \cdot 0756 \text{ kN/mm}^2$$

or

$$= \frac{1000}{5000\pi} - \frac{(1000 \times 4 \cdot 35)}{410 \times 10^3} = 0 \cdot 0531 \text{ kN/mm}^2$$

S *Solution:*

1. Maximum stress is compressive: **76 N/mm².**
2. Minimum stress is also compressive: **53 N/mm².**

18/7 ECCENTRIC COMPRESSION: I Section: Two loads

The sketch shows the simplified outline of the cross-section of a stanchion. The loading is 3000 kN uniformly distributed over the cross-section, together with the concentrated loads shown. Calculate the maximum and minimum stresses produced in the stanchion if the section is symmetrical about the centroidal axis.

G *Groups of Terms:*

$$\frac{F}{A} = \sigma_d \qquad \frac{M}{I} = \frac{\sigma_b}{h}$$

$$+\sigma_d \pm \sigma_b = +\frac{F}{A} \pm \frac{M}{z}$$

U *Unknowns and Units:*
1. Maximum stress on the cross-section.
2. Minimum stress on the cross-section.
Use kN and mm.

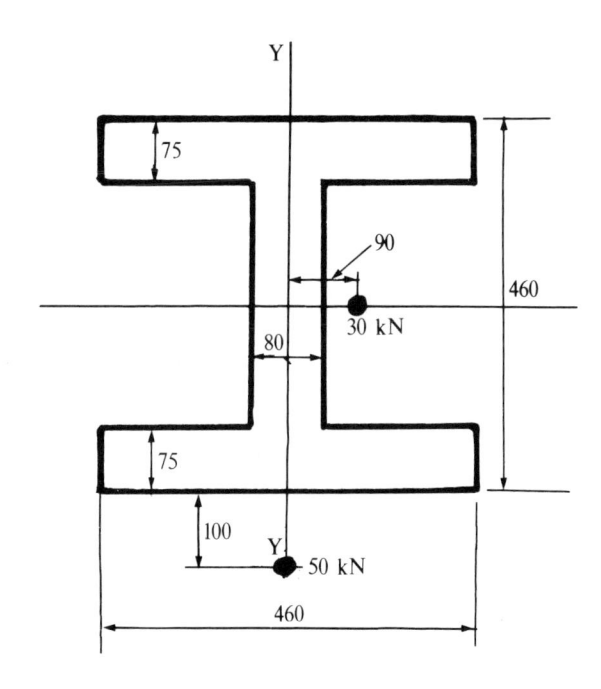

18/7

I   *Inventory of Information:*

$F = 3000 + 30 + 50 = 3080 \text{ kN}$ $A = 93{\cdot}8 \times 10^3 \text{ mm}^2$
$P_1 = 30 \text{ kN}$ $P_2 = 50 \text{ kN}$
$e_1 = 90 \text{ mm}$ $e_2 = 330 \text{ mm}$
$h = 230 \text{ mm}$ for both XX and YY

D   *Derived Data:*

(a) Second moment of area about XX and about YY:

$$I_{XX} = 226{\cdot}4 \times 10^6 \text{ mm}^4$$
$$I_{YY} = 73{\cdot}05 \times 10^6 \text{ mm}^4$$

(b) Section modulus about XX and about YY:

$$z_{XX} = 985 \times 10^3 \text{ mm}^3$$
$$z_{YY} = 317 \times 10^3 \text{ mm}^3$$

E   *Evaluation:*

$$+\frac{F}{A} \pm \frac{Pe}{z} = \frac{3080}{33{\cdot}8 \times 10^3} \pm \frac{30 \times 90}{317 \times 10^3} \pm \frac{50 \times 330}{985 \times 10^3}$$

$$= 0{\cdot}0329 \pm 0{\cdot}008\,52 \pm 0{\cdot}0168$$

$$= 0{\cdot}0582 \text{ or } 0{\cdot}0076 \text{ kN/mm}^2$$

S   *Solution:*

1. Maximum stress on cross-section: **58·2 N/mm².**
2. Minimum stress on cross section: **7·6 N/mm².**

When there are several eccentric loads on the same cross-section, add their effects to obtain the final stresses.

18/8

18/8 ECCENTRIC COMPRESSION: Rectangular section: Three loads

A short rectangular column supports an axial load and two eccentric loads as shown in the sketch. Find the greatest and least stress in the section.

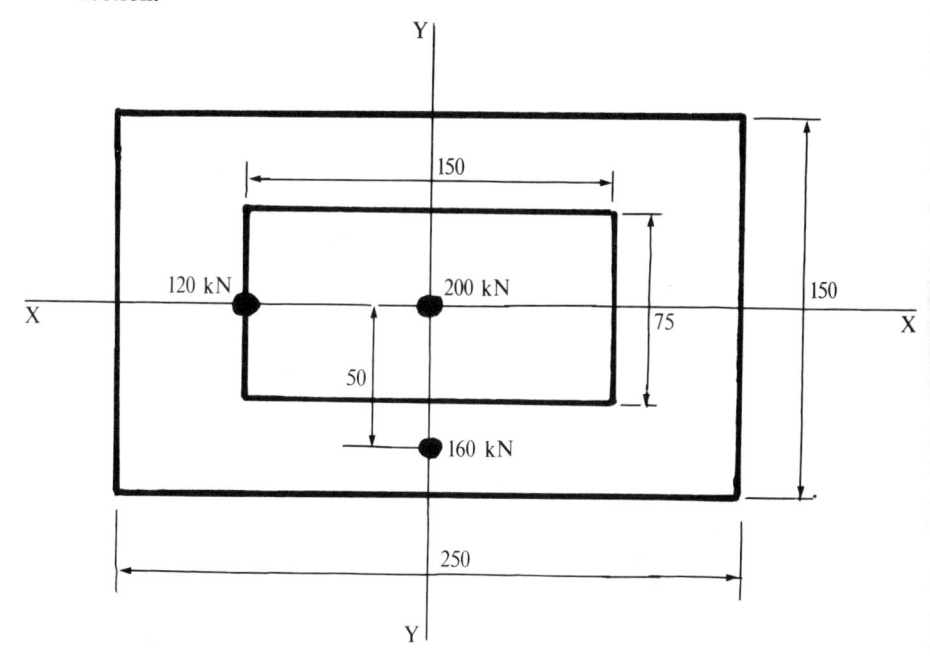

[Sketch: Rectangular section with X-X and Y-Y axes. Overall dimensions 150 (half-height above X-X) and 250 (half-width to right of Y-Y — full width therefore 500, full height 300). Inner rectangle drawn; 150 marked from left inner edge to centre (Y-axis), 75 marked from X-axis down to inner edge, 50 marked from X-axis down to 160 kN load. Loads: 120 kN at left (on X-axis, at inner rectangle left edge), 200 kN at centre (origin), 160 kN below centre at 50 mm below X-axis.]

G *Groups of Terms:*

$$\frac{F}{A} = \sigma_d \qquad \frac{M}{I} = \frac{\sigma_b}{h}$$

$$+\sigma_d \pm \sigma_b = +\frac{F}{A} \pm \frac{M_1}{z_W} \pm \frac{M_2}{z_{XX}}$$

U *Unknowns and Units:*

1. Greatest stress in the column at A.
2. Least stress in the column at B.

The positions where greatest and least stress occur in such an eccentrically loaded column can usually be estimated, or at the most, two possible points selected. In this instance the loading is all towards corner A and away from corner B.

Use kN and mm.

18/8

I *Inventory of Information:*

$F = 200 + 120 + 160 = 480$ kN
$A = (250 \times 150) - (75 \times 150) = 26 \cdot 25 \times 10^3$ mm^2
$M_1 = 120 \times 75 = 9000$ kN mm
$M_2 = 160 \times 50 = 8000$ kN mm
h (at right angles to XX) $= 75$ mm for both A and B.
h (at right angles to YY) $= 125$ mm for both A and B.

D *Derived Data:*

The second moment of area is required about the centroidal or neutral axis in both directions – about both XX and YY – since there is bending in both directions. These two values must be found quite separately. This problem is worked out in Unit 12.

(a) $I_{XX} = 75 \cdot 5 \times 10^6$ mm^4 $I_{YY} = 216 \cdot 3 \times 10^6$ mm^4

(b) $z_{XX} = \dfrac{75 \cdot 5 \times 10^6}{75}$ mm^3 $z_{YY} = \dfrac{216 \cdot 3 \times 10^6}{125}$ mm^3

$z_{XX} = 10^6$ mm^3 $z_{YY} = 1 \cdot 73 \times 10^6$ mm^3

E *Evaluation:*

$$\dfrac{F}{A} \pm \dfrac{M_1}{z_{YY}} \pm \dfrac{M_2}{z_{XY}}$$

Point A: $\dfrac{450}{26 \cdot 25 \times 10^3} + \dfrac{9000}{1 \cdot 73 \times 10^6} + \dfrac{5000}{10^6} = 0 \cdot 0315$ kN/mm^2

Point B: $\dfrac{450}{26 \cdot 25 \times 10^3} - \dfrac{9000}{1 \cdot 73 \times 10^6} - \dfrac{5000}{10^6} = 0 \cdot 0051$ kN/mm^2

S *Solution:*

1. Greatest stress in the column (at A): **31·5 N/mm^2**
2. Least stress in the column (at B): **5·1 N/mm^2**

18/9 DEAD LOAD: Wind pressure

A concrete chimney is 15 m high. The chimney is 1·8 m square and has a square central flue of 1·2 m side. Concrete has a mass of 2500 kg/m³.

Wind pressure (p N/m²) is distributed uniformly over the windward face. On the leeward side there is a suction which amounts to half of the wind pressure ($p/2$ N/m²) at the top of the chimney, and reduces to zero at the bottom. The value of the maximum wind pressure to be expected at the site is 2·4 kN/m². Discuss the safety of the chimney in these circumstances.

G Groups of Terms:

$$\frac{F}{A} = \sigma_d \qquad \frac{M}{I} = \frac{\sigma_b}{h}$$

$$+\sigma_d \pm \sigma_b = +\frac{F}{A} \pm \frac{M}{z}$$

U Unknowns and Units:

There are two stages at which safety is imperilled: when tension appears on the windward face, and when the chimney begins to overturn. The wind pressures at which these occur are thus of importance.

1. Wind pressure at which tension develops on the windward side.
2. Wind pressure at which chimney begins to overturn.

These pressures, when calculated, can be compared with the expected wind pressure and conclusions drawn.

Use N and m.

I Inventory of Information:

Height of chimney: 15 m

Area of concrete in a cross-section: $1·8^2 - 1·2^2 = 1·8$ m².

Concrete has mass of 2500 kg/m³

Expected wind pressure (to be used for comparison with calculated pressures): $2·4 \times 10^3$ N/m².

D Derived Data:

(a) *Centroid* of the section is at its centre

$$h = 0·9 \text{ m}$$

(b) *Second moment of area*

$$I_{NA} = \tfrac{1}{12}(1 \cdot 8^4 - 1 \cdot 2^4) = 0 \cdot 702 \text{ m}^4$$

(c) *Modulus of section*

$$z = I/h = 0 \cdot 702/0 \cdot 90 = 0 \cdot 780 \text{ m}^3$$

(d) *Vertical pressure* due to weight of chimney:

For each cubic metre the mass is 2500 kg.

This represents a force due to gravity of $2500 \times 9 \cdot 81$ N for every metre of depth, on each square metre of area.

For 15 m of depth the pressure is 15 times the unit value.

Thus, pressure due to weight of chimney

$$\frac{F}{A} = (2500 \times 9 \cdot 81)15 = 367 \cdot 9 \times 10^3 \text{ N/m}^2$$

This pressure is acting over the area of the concrete of the chimney. Thus, the total weight of the chimney is

$$F = (367 \cdot 9 \times 10^3 \text{ N/m}^2) \times 1 \cdot 8 \text{ m}^2$$

which represents a downward force of

$$F = 662 \cdot 4 \times 10^3 \text{ N}$$

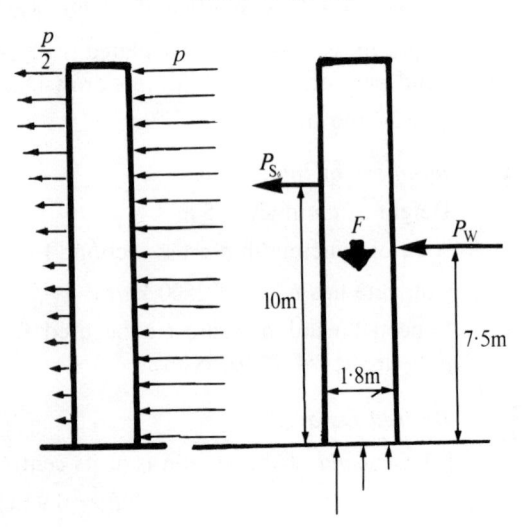

(e) *Force exerted on the chimney* by wind pressure and suction.

The *pressure* on the windward side is uniform, so the total force is the pressure multiplied by the area of one face of the chimney

$$P_W = p \times 1.8 \times 15 = 27p \text{ N}$$

This force acts at halfway up the chimney, for the purposes of calculating moments.

The *suction* on the leeward side varies from half the wind pressure to zero. The resultant of this distributed force acts at the centroid of the area of pressure – at 10 m from the ground. The value of the suction is

$$P_S = p/4 \times 1.8 \times 15 = 6.75p \text{ N}$$

(f) *The moment of wind pressure and suction* about the base of the chimney where the greatest moment occurs

$$\begin{aligned} M &= P_W \times 7.5 + P_S \times 10 \\ &= 27p \times 7.5 + 6.75p \times 10 \\ &= 202.5 + 67.5p \\ M &= 270p \text{ Nm} \end{aligned}$$

E *Evaluation:*

The terms of the two groups are:

$$\frac{F}{A} = 367.9 \times 10^3 \text{ N/m}^2$$

$$M = 270p \text{ N m}$$

$$z = 0.78 \text{ m}^3$$

1. When the pressure due to bending is equal to that due to direct force, the compressive stress due to the weight of the chimney is exactly balanced by the negative pressure caused by the wind moment.

$$\sigma_d - \sigma_b = \frac{F}{A} - \frac{M}{z} = 0$$

$$367.9 \times 10^3 - \frac{270p}{0.78} = 0$$

18/9

When this is solved, the wind pressure at which the first tension cracks appear in the wall of the chimney is

$$p_1 = 1\cdot06 \times 10^3 \text{ N/m}^2$$
$$p_1 = 1\cdot06 \text{ kN/m}^2$$

2. The wind pressure at which overturning of the chimney takes place is that which causes the moment of the weight of the chimney about the leeward edge (point L) to be counteracted by the overturning moment caused by the wind and suction pressures.

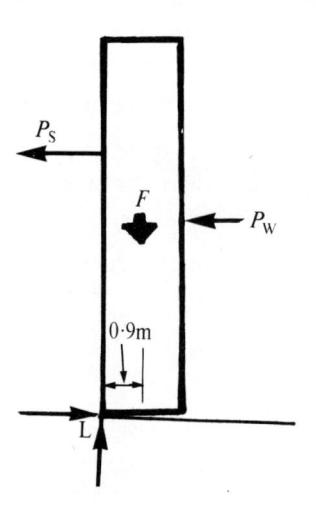

Moments about L of all forces

$$+F \times 0\cdot9 - P_w \times 7\cdot5 - P_s \times 10 = 0$$
$$+0\cdot9F - 270p = 0 \qquad p_2 = 2\cdot2 \text{ kN/m}^2$$

S *Solution:*

The conclusion to be drawn from these figures is that a wind pressure of 2·4 kN/m², which is expected, will not only cause cracking in the windward face of the chimney, but would cause overturning.

18/10

18/10 DEAD LOAD: Foundation pressure

A foundation block is 1·5 m wide by 1·2 m deep (at right angles to the plane of the paper). It is loaded with 6 kN (eccentric 0·25 m) and a clockwise couple whose moment is 11 kN m. The foundation material should be loaded in compression at all points of the base. Check the validity of the design of the base for these conditions.

G *Group of Terms:*

$$+\sigma_d \pm \sigma_b = +\frac{F}{A} \pm \frac{M}{z}$$

U *Unknowns and Units:*

1. Bearing pressure on the soil at L.
2. Bearing pressure on the soil at R.

Use kN and m.

I *Inventory of Information:*

Direct load eccentrically placed: 6·0 kN
Applied couple has moment of: 11·0 kN m
Eccentricity of 6 kN: 0·25 m
Width of foundation block: 1·5 m
Depth of foundation block: 1·2 m

Thickness of block is immaterial; it is assumed that it will be made strong enough.

D *Derived Data:*

Couple formed by the eccentrically placed 6 kN has a moment of $-6 \times 0.25 = -1.5$ kN m.

Couple giving 11 kN of moment is positive.

Area of base: $1.2 \times 1.5 = 1.8$ m².

Nett moment applied by eccentric load and applied couple is $+11 - 1.5 = +9.5$ kN m

Section modulus for a rectangle:

$$\frac{bd^2}{6} = \frac{1.2 \times 1.5^2}{6} = 0.45 \text{ m}^3$$

477

18/10

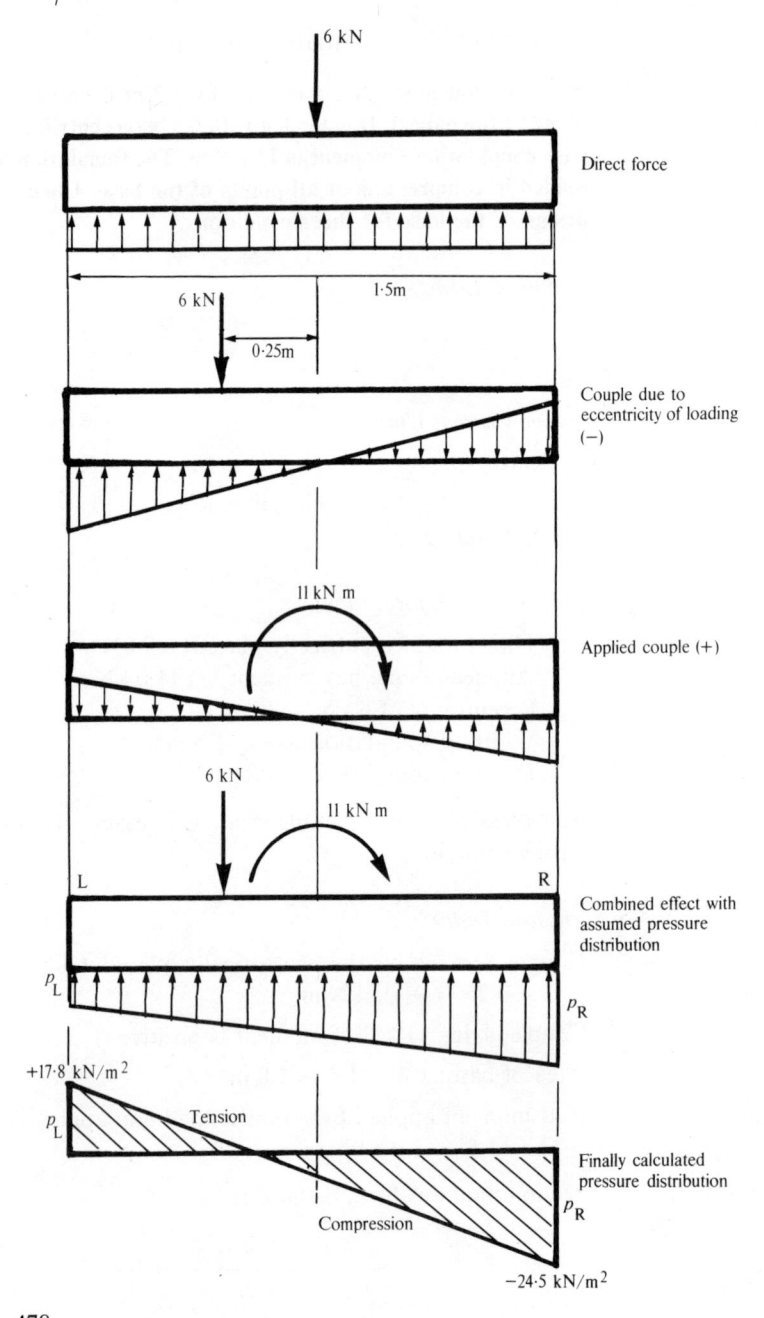

6 kN

Direct force

1·5m

6 kN

0·25m

Couple due to eccentricity of loading (−)

11 kN m

Applied couple (+)

6 kN

11 kN m

L   R

Combined effect with assumed pressure distribution

p_L p_R

+17·8 kN/m²

p_L Tension

Compression

p_R

−24·5 kN/m²

Finally calculated pressure distribution

478

E *Evaluation:*

A positive moment applies compression at R and tension at L, so p_L will be less than p_R.

$$+\frac{F}{A} \pm \frac{M}{z} = +\frac{6}{1\cdot 8} + \frac{9\cdot 5}{0\cdot 45} \quad \text{or} \quad +\frac{6}{1\cdot 8} - \frac{9\cdot 5}{0\cdot 45}$$

$$= +24\cdot 5(p_R) \quad \text{or} \quad -17\cdot 8(p_L)$$

The negative sign for p_L means that there would be tension at L if the block were held down. The real distribution of stress is different from that assumed.

S *Solution:*

1. The bearing pressure at L would be a tension of **17·8 kN/m²** if this could be resisted. As conditions are, the foundation block would lift off from the soil.
2. The bearing pressure on the soil at R is **24·5 kN/m²** if it could be supported.
3. The conditions given required that the bearing pressure should be compression over the base of the block. The conditions you have found show that this does not take place. The block must be enlarged or the loading reduced or changed in position in order to ensure compression loading over the whole area, as was assumed.

A centrally applied couple on a symmetrical section gives equal tensile and compressive stresses at the extreme edges of the section.

Part II: Group F

19 Direct and shearing stress combined

So far, in Units 13 to 16, the problems have been concerned with the determination of stresses for particular types of loading – direct force, shearing force, bending couples, and torsional couples. In unit 18 it is shown that both a direct force and a bending couple produce direct stresses – the first uniform over the section and the second varying – and these can be combined. In the same way, transverse shear and torsional shear can be combined algebraically. There are few instances, however, where a transverse shear and a torsional shear exist together without a bending couple being present, and this type of combination (and no doubt, others which could be thought of) is not important at this stage of your studies.

A type of combination of stress conditions which is vitally important in the design of structures is the simultaneous application of direct and shearing forces on several planes through the stressed body. An inclined stress on any plane can be resolved into a direct stress and a shearing stress. As the plane is rotated about a point, the values of the normal and shearing stresses on it change in response to the forces acting externally on the material. The problems with which this Unit is concerned are those of defining the effects of the combination of applied direct and shearing stresses at a point in a body. This combination can be described as the 'state of stress' at the given point. The Unit is chiefly concerned with a graphical method of calculation to elucidate the complexities which often puzzle the student of this subject.

The loadings which cause simultaneous application of direct and shearing stresses in a structure are:

a transverse shearing force combined with a direct force

a transverse shearing force combined with a bending couple

a torsional shearing force combined with a direct force

a torsional shearing force combined with a bending couple

and variations of these.

It is assumed that you know how to obtain the simple values of

19.1

direct stress, bending stress (a type of direct stress) and shearing stresses, separately calculated. This unit deals only with how these are compounded after their values are determined. The two types of stress – direct and shearing – are not acting in line as were the stresses of Unit 18. The shearing stress acts across the face of the section, and the direct stress normal (at right angles) to the section. The direct stress when at right angles to the section is usually called the 'normal' stress.

19.1 The stressed block

A simple example of how these stresses act in combination is shown by magnifying a very small block of matter in the end-section of a simply-supported beam (sketch 19A). Near the lower face of the beam the bending couple produces a normal stress (direct tension) in the fibres of the beam. The shearing force (see Part I) develops a positive shearing stress on the same elementary stressed block, which is an enlargement of the small stressed particle under a state of stress.

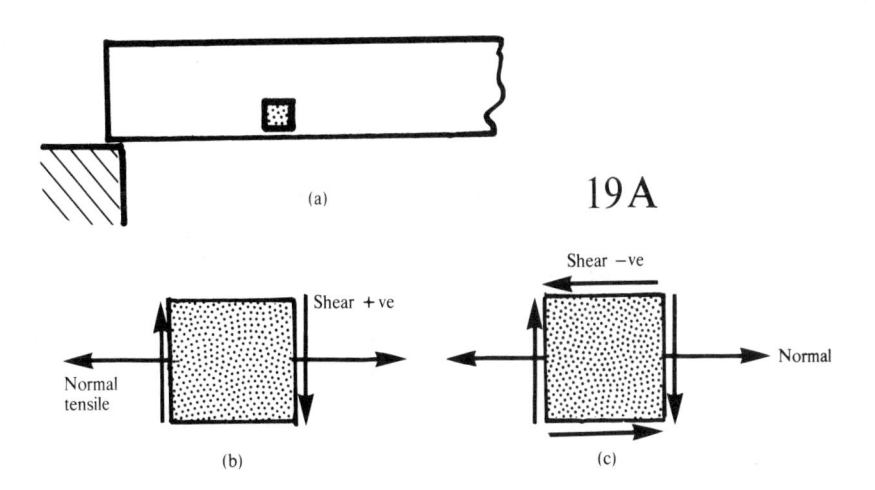

This block (b) has been cut away and removed from the beam to form a free-structure diagram. You can see that the block, as it is shown, cannot be in equilibrium. It would rotate under the couple formed by the shearing stress. As, in the beam, it does not rotate, it must be held in equilibrium by an opposing couple so that the moment equation of equilibrium, discussed in Part I, can be satisfied.

19.2

This example, simple though it is, illustrates the fundamental principle that one shearing stress (+ or −) cannot exist alone if equilibrium is to be maintained. A shearing stress of the opposite sign must exist at right angles to the first, just as a direct force of one sign must be balanced by an equal and opposite force for equilibrium. Textbooks on strength of materials show that the shearing stresses which exist at right angles must be of the same magnitude regardless of the values of the normal stresses acting. Remember, then:

A shearing stress acting on one plane through the point being studied causes an equal shearing stress of opposite sign to act on a plane at right angles to the first.

You start, therefore, on any problem of this unit thinking of a small, stressed block, usually considered to be square in cross-section. This block carries normal stresses and also shearing stresses on both sets of faces, the shears being equal, and of opposite sign. If the two planes chosen are not at right angles, other conditions apply, which will be clear when you carry out the solution of the problems. Remember positive shear shows the arrows acting so that they form a clockwise couple. Positive normal stress is tensile. Some of the possibilities are shown in sketch 19B.

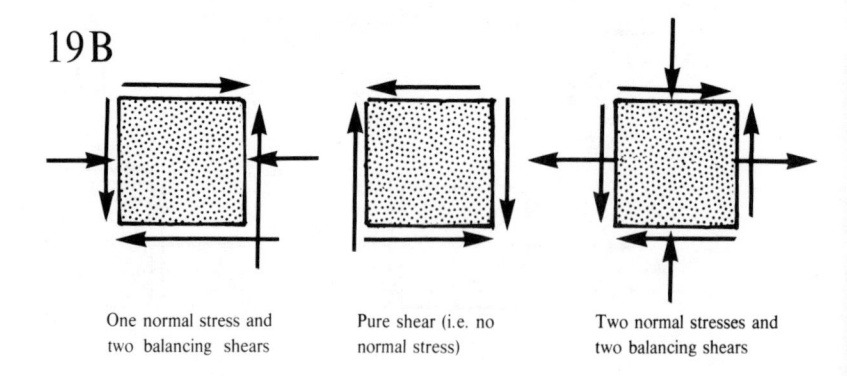

19B

One normal stress and two balancing shears

Pure shear (i.e. no normal stress)

Two normal stresses and two balancing shears

19.2 Principal stresses

The stressed block is very small, and you should think of the two opposing faces of the block as one plane with normal or shearing stresses opposing each other on the two sides. The expansion into a physical block is made in order to allow of clear thought on how the

19.2

stresses vary on the infinite number of planes which can pass through the chosen point. Try this experiment:

Hold a piece of card vertically in front of you, and push against it with the index finger of the other hand, making the angle between the finger and the card 90°. You are applying a *normal force*. Now rotate the card slowly so that it begins to slope away from or towards you, still maintaining the push of the finger. You will find that your finger has a tendency to slip. This shows that you have, on the sloping card, been applying a *shearing force* to the surface in addition to the normal force. As the plane rotates through a chosen point, the normal stress originally acting, changes to variations of both normal and shearing stresses with the change in angle.

When the stress applied to the plane is normal only, and there is no shearing stress, it is called a *principal stress*. The plane on which it acts can be called a *principal plane*. Other planes through the same point are stressed by both normal and shearing stresses. For a planar condition, such as you study in this unit, there are two planes on which principal stresses act, and these planes are at right angles to each other. One of

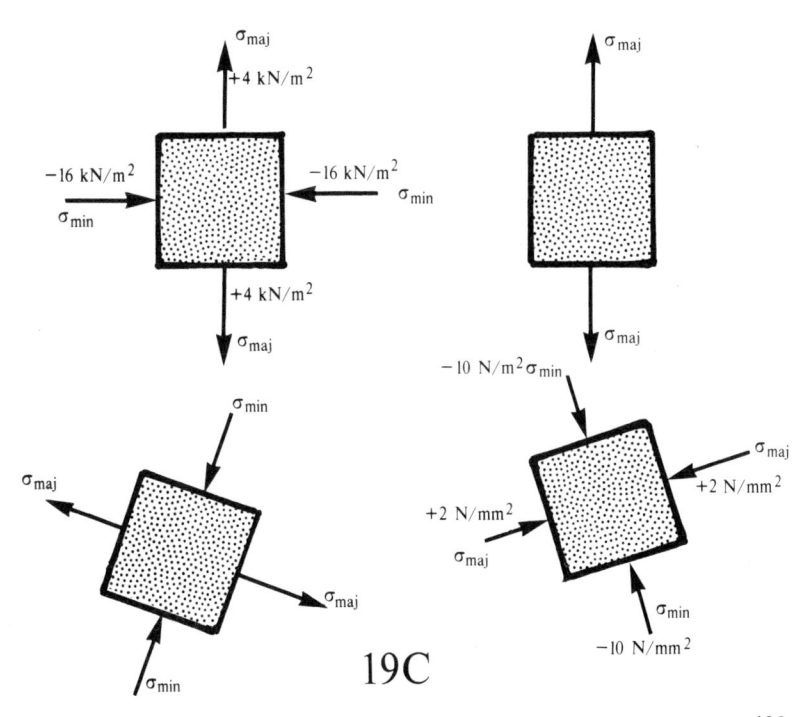

19C

Principal stresses

the principal stresses is the greatest normal stress which can occur at this point – the *major principal stress* (σ_{maj}). The other principal stress is the least normal stress occuring at the point and is the *minor principal stress* (σ_{min}).

There can be various combinations of principal stresses (p.s.) and some of these are shown in sketch 19C, but remember that these principal stresses occur only on planes where there is zero shear, the whole stress being normal to the plane concerned. A tensile p.s. is positive, and a compressive p.s. is negative. Thus, when both a tensile and a compressive principal stress act, the tensile stress is always the major p.s. regardless of the numerical magnitude of the values. Similarly, if two compressive p.s. act, the smaller of these in terms of magnitude is the 'major' p.s. (because they are both negative).

19.3 Graphical calculation by the Mohr's Circle

The argument so far advanced in this Unit is that through a point in a stressed body planes can be defined on each of which the stress can be resolved into a normal and a shearing stress. On two of the planes the shearing stress is zero and the planes carry *principal stresses*. The principal stresses are the greatest and the least normal stress at the stressed point. Similarly, there may be two planes not necessarily at right angles to each other, on which the normal stresses are zero, and only shearing stresses act – the planes of *pure shear*. But, whereas at any point in a stressed planar body there are always two principal planes and two principal stresses, pure shear occurs only if one of the principal stresses is positive and one negative. On all the other planes passing through the stressed point and making various angles with the plane carrying the major principal stress, there are two components of the stress on the plane – a normal stress and a shearing stress.

The relationships between principal stresses, normal stresses, shearing stresses, and the directions of the planes on which they act can be calculated by considering the equilibrium of an elementary stressed block, by the methods of Part I. The results are given as mathematical expressions in Section 19.7. Substituting values in these 'formulae' can give the numerical results required, but not necessarily a clear understanding of the 'state of stress'. However, the mathematician Mohr recognized that the expressions which connect the values of the various stresses can be graphically displayed by the lengths and angles related through the geometry of a circle. The demonstration of the state of stress within a body by means of the *Mohr's Circle* gives, not only the numerical 'answers' but allows you to obtain a much clearer appreciation of the complex conditions on the many planes passing through a point in a stressed body.

The problems in this unit deal with the combination of normal and shearing stresses by means of the Mohr's Circle. Since this circle is a method of calculation, and its dimensions give the values required in the solution of problems, the circle must be drawn accurately. It is no longer sufficient to sketch approximately to scale, and proper drawing techniques must be used for accurate and quick solutions.

19.4 Lengths in the Mohr's Circle diagram

Before attempting to draw the Mohr's Circle (MC) be sure that you understand how stresses are set out in the diagram from the zero point O (sketch 19D). There are two axes, XX and YY, the normal stresses being measured (from O as origin) along the XX axis in both the positive (right) and negative (left) directions. The shearing stresses are set off to scale parallel to YY, measuring upwards from the XX axis for positive shear, and downwards for negative shear. The lengths shown in sketch 19D represent:

ON_1: positive (tensile) normal stress: N_1S_1: negative shearing stress

ON_2: positive (tensile) normal stress: N_2S_2: positive shearing stress

ON_3: negative (compressive) normal stress: N_3S_3: positive shearing stress

ON_4: negative (compressive) normal stress: N_4S_4: negative shearing stress

Each S point represents the state of stress on a given plane in relation to the zero point O. After you have set out various states of stress on several planes passing through the chosen point in a stressed structure you will find that all the S points lie on a circle. This is the Mohr's Circle (MC), and as long as the S values all lie on the circumference, they all belong to the same state of stress at the chosen point. Each Mohr's Circle applies only to one point in a stressed body.

In sketch 19E, the components ON_2, N_2S_2: ON_6, N_6S_6 and ON_3, N_3S_3 all belong to the same stressed point and the same state of stress. The difference in their values and signs originate from the angles at

19.5

19D

+ve shear

−ve shear

or

or

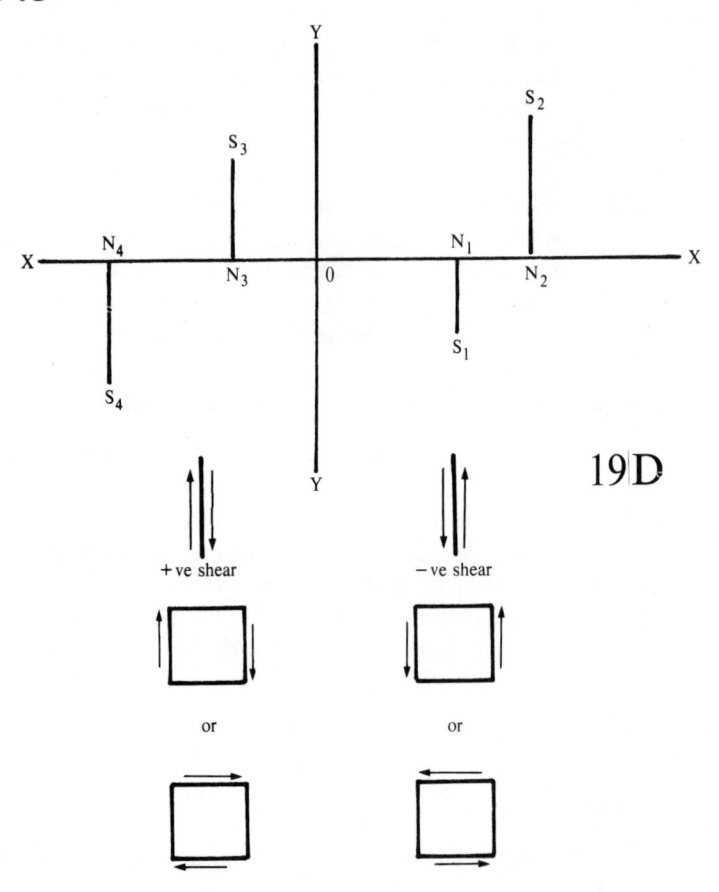

which the various planes lie in relation to each other. The 1, 4, and 5 points do not belong to the state of stress represented by the MC. Indeed, these three do not all belong to the stress condition at any other single point in the stressed body, for it is not possible to draw a single circle through S_1, S_4, and S_5 with its centre on OX. These three thus represent some of the stresses at at least two other points.

19.5 Angles in the Mohr's Circle diagram

You now know how to find, by measurement, the values of the components of stress on various planes passing through a point (normal and shearing stress) and also the special cases of principal and pure shearing stresses. But, so far, the MC has given you no indication of the directions of the various planes on which these stresses act. This is the second significant measurement you must make.

The MC's function, in showing the directions of the planes, overlaps its other purpose of giving magnitudes of stresses. Remember the circle is a technique of calculation and not a physical representation of the state of stress. Do not, therefore be confused by the double use of the MC: treat the two functions separately (sketch 19F).

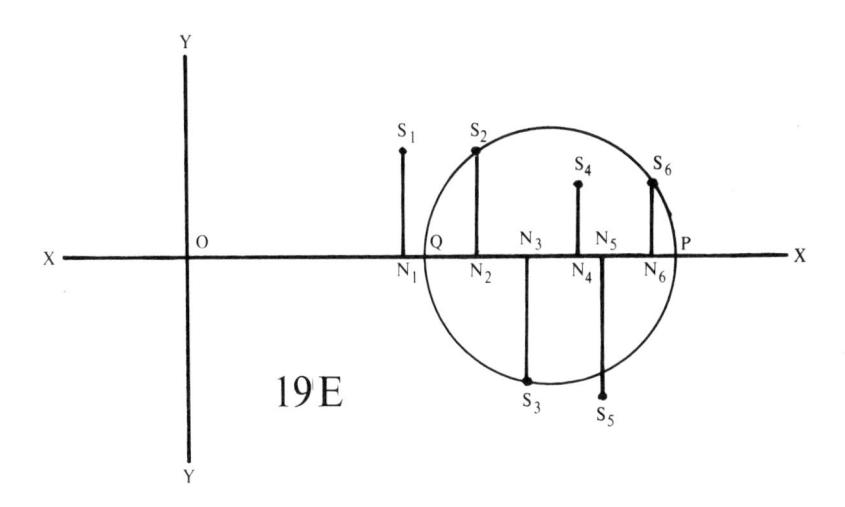

The plane on which the major p.s. acts is supposed to lie in the direction XX or QP. As the pencil point moves round to S_3, S_2, and S_1 so the angle of the plane on which each of these stresses act, is measured anticlockwise from the line QP (the direction of the plane carrying σ_{maj}). This the angle θ_1. (Sketch 19E for position of Q)

When your pencil point reaches Q, the plane is parallel to YY and tangent to the MC at Q. OQ is thus the minor p.s. acting on a plane at right angles to XX. As your pencil point passes through Q and continues anticlockwise round the circle, the angle of the plane concerned, for any S point, is defined by θ_2 measured all the way round from QP.

The direction of the plane on which the major p.s. acts is shown as horizontal merely for the convenience of printing. It can lie at any angle to the horizontal. It is a simple matter, however, after drawing the MC, to swing the whole diagram round until XX is parallel to the plane

19.6

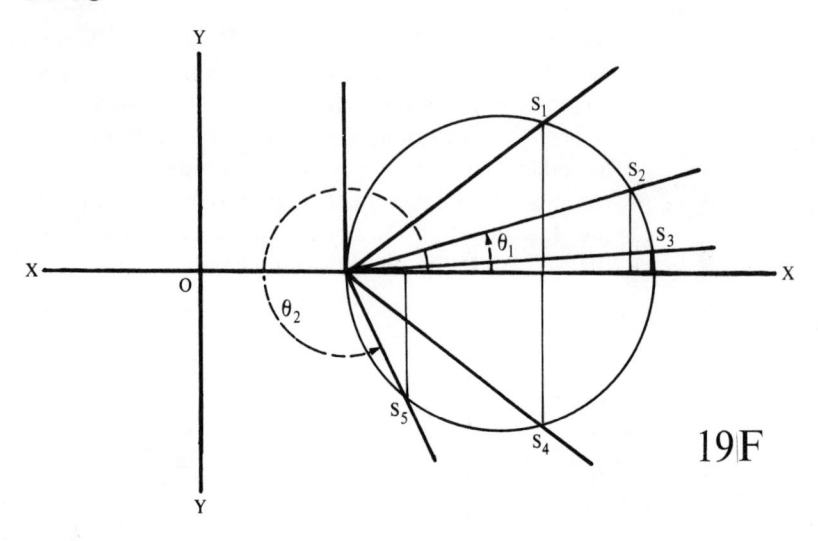

19F

carrying the major p.s. (if you know its direction). This, perhaps, facilitates understanding the technique.

19.6 Drawing the Mohr's Circle

Usually, when a problem is presented, you will be looking for the worst conditions of stress for the material in question. You always start with some information and some data of loading. From the data of the loads acting, you can calculate from the methods of earlier Units in Part II, the simple stresses developed. The combination of these might then lead to a more critical stress situation.

For example, a cylinder of clay in compression, but unsupported on the sides, has a vertical negative p.s. applied to its horizontal cross-section. The major p.s. is represented by the lateral stress on the sides of the sample, and that value is zero. The MC for this condition is, therefore, a circle with centre on XX, and touching YY at point O. The circle lies wholly to the left of YY, OQ measured across the diameter, representing the minor p.s. The maximum shearing stress which can occur within this compressed cylinder is represented by the maximum height of the circumference above or below XX – the radius. Join Q to the end of this radius and it is clear that the maximum shearing stress is equal to half the applied compressive stress and acts at an angle of 45° to the plane carrying the major p.s. (in this instance

Guide table **19.6**

Data provided		I	II	III	IV	V	VI	VII
D1	Two principal stresses (OP and OQ)	×						
D2	One principal stress (OP or OQ) one shearing stress and one related normal		×					
D3	Two shearing stresses (N_1S_1 and N_2S_2) with related normal stress			×	×	×	×	
D4	The two shears are numerically equal			×	×			
D5	The two shears are numerically unequal					×	×	
D6	The two shears are of the same sign				×		×	
D7	The two shears are of opposite signs			×		×		
D8	Combinations of shears and normals and directions of planes on which they act							×
Operation to draw Mohr's Circle		↓	↓	↓	↘	↓	↙	↓
Step 1	CHOOSE appropriate operation from the category number	O1	O2	O3	O4			O5
Step 2	SET OFF, measuring from O, (+ to right; − to left)	OP and OQ	OP or OQ	ON_1 ON_2	ON_1 ON_2			Angles and values
Step 3	SET OFF, measuring from N, (+ shear up; − shear down)	—	ON_1 N_1S_1	N_1S_1 N_2S_2	N_1S_1 N_2S_2			of shears and normals
Step 4	BISECT, at B the line (between points on the diagram)	PQ	PS_1 or QS_1	S_1S_2	S_1S_2			S_1S_2
Step 5	ERECT perpendicular at B (meeting PQ in C) to line	(B and C coincide)	PS_1 or QS_1	S_1S_2	S_1S_2			S_1S_2
Step 6	MOHR'S CIRCLE has centre C and radius	CP	CS_1	CS_1	CS_1			CS_1

19.7

that is vertical). As the compressive stress is increased in a test, the circle grows bigger, but still touches YY at O. When the material fails, it is found that one portion of the cylinder slides over the other (or shears) at an angle of 45° to the vertical. This occurs because clay is weak in shear. Dip into any book on soil mechanics and you are likely to find an illustration of this confirmation of the accuracy of Mohr's Circle derivations.

Examiners can present the candidate with values of stresses and with angles of planes in many different combinations, but, basically, the drawing of the MC depends on knowing the position of two S points (P and Q being included). The GUIDE TABLE should be used initially, but soon you will be able to discard it if you work sufficient examples.

(*a*) Enter the table horizontally at the various data levels D. You may have information on several of these conditions.
(*b*) Find the category I, II, ..., VIII for which all the data given are included.
(*c*) Run your finger down the category column to find the operation O1, O2, O3, etc. required.
(*d*) Carry out the steps of setting off, bisecting etc., shown in the column below the operation number.
(*e*) With step 6 you can draw the Mohr's Circle from which measurements of lengths and angles give the values of stresses and the directions of planes. The stresses and directions can be displayed in stressed-block diagrams as is shown in the problems.

19.7 Relationships between stresses and directions of planes

σ_{maj}: Major principal stress

σ_{min}: Minor principal stress

σ_A: Normal stress on plane A

σ_B: Normal stress on plane B, at right angles to plane A

τ_A: τ_B: Shearing stress on planes A and B

θ: Angle between major principal plane and plane A (measured anticlockwise)

OC: Distance from zero of co-ordinates to centre of Mohr's Circle

τ_{max}: Radius of Mohr's Circle and maximum shearing stress.

19.8

Location of centre of MC and the value of the radius

$$OC = +\sigma_{min} + \tfrac{1}{2}(\sigma_{maj} - \sigma_{min})$$
$$\tau_{max} = \tfrac{1}{2}(\sigma_{maj} - \sigma_{min})$$

Normal and shearing stresses from principal stresses

$$\sigma_A = +\sigma_{maj}\cos^2\theta + \sigma_{min}\sin^2\theta$$
$$\tau_A = +\tfrac{1}{2}(\sigma_{maj} - \sigma_{min})\sin 2\theta \text{ or } +(\sigma_{maj} - \sigma_{min})\cos\theta\sin\theta$$

Principal stresses from normal and shearing stresses on two planes at right angles

$$\sigma_{maj} \text{ and } \sigma_{min} = +\tfrac{1}{2}(\sigma_A + \sigma_B) \pm \sqrt{[\tfrac{1}{4}(+\sigma_A - \sigma_B)^2 + \tau_A^2]}$$

Angle between major principal plane and plane A, measured anticlockwise

$$\tan 2\theta = +\frac{2\tau_A}{\sigma_A - \sigma_B}$$

These have been written for positive stresses. If principal, normal or shearing stresses are negative, they must carry their sign with them.

Reminders:

$$\sin 2\theta = 2\cos\theta\sin\theta \qquad \cos^2\theta = \tfrac{1}{2}(1 + \cos 2\theta)$$
$$\tan 2\theta = \frac{2\tan\theta}{1 - \tan\theta} \qquad \sin^2\theta = \tfrac{1}{2}(1 - \cos 2\theta)$$

19/1

19.8 THE PROBLEMS

The problems in this unit cover most of the variations given in examination papers. It is not enough, however, to obtain the required results by mere substitution in the 'formulae' of section 19.7. The Mohr's Circles should be drawn each time, and the results checked by calculation. Never merely read through the worked examples. Do them again, preferably without reference to the text, and then compare your solution with the one given here. Have you managed the problem more efficiently; what time did you take over it? Do better with the next one. Look up other textbooks and work out their problems by the methods given here. Pay close attention to signs and never write down a quantity without its sign.

19/1 PRINCIPAL STRESSES GIVEN: Find pure shear

If the two principal stresses at a point in a stressed body are $+1\cdot0$ N/mm² and $-4\cdot0$ N/mm², what are the normal stresses on planes where the shearing stresses are $+2\cdot3$ N/mm² and $-2\cdot3$ N/mm²? At what angles do these planes lie? What is the stress on the plane carrying pure shear, and at what angle does this plane lie?

G *Guide Table:*

Enter the table at D1: this gives **Category I**.

U *Unknowns and Units:*

1. Normal stresses on defined planes, where shear is known.
2. Shearing stress on planes carrying pure shear.
3. Angles that these planes make with the plane carrying major principal stress.

Use N and mm.

I *Inventory of Information:*

Use Operation OI to draw the Mohr Circle.

D *Derived Data:*

(i) Draw two lines parallel to PQ and at $2\cdot3$ N/mm² both above and below PQ. These intersect the circle in four points, S_1, S_2, S_3, and S_4.

(ii) Drop perpendiculars to PQ from the S points to define corresponding N points.

(iii) The plane of pure shear is the one on which there is no normal stress. On such a plane, O and N must coincide. The value of the pure shear is thus OS_p both positive and negative.

(iv) Angles to the planes on which these stresses occur are obtained by taking angles counterclockwise from QP to the various S points.

E Evaluation:

The normal stresses are obtained by measuring ON_1, ON_2, ON_3, and ON_4. These are, by measurement, -2.0 N/mm² and -0.9 N/mm².

The value of pure shear is OS_p, which is 1.9 N/mm² as measured from the diagram.

The angles required are measured by protractor counter-clockwise from QP as PQS. Those points S below QP show an angle of over 180° to the plane on which the major p.s. acts. The plane on which ON_3 and SN_3 act, for example, is at 38° to the plane carrying the major p.s.

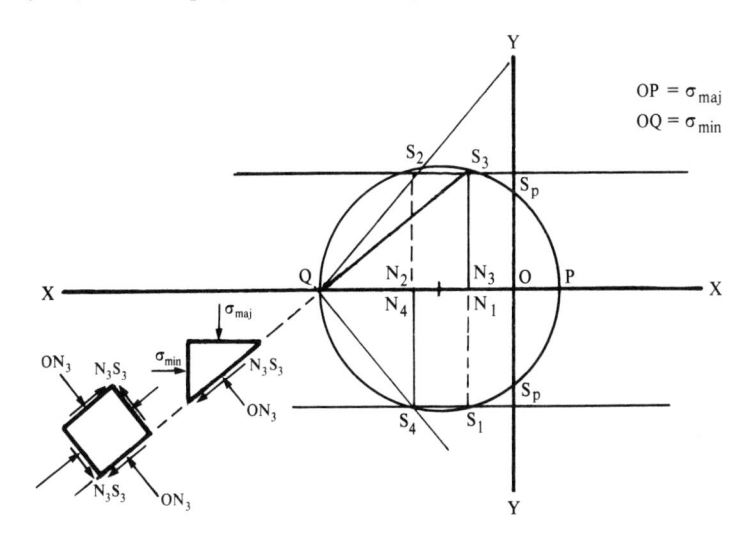

S Solution:

1. -2.0 N/mm: -0.9 N/mm
2. 1.9 N/mm (both + and −)
3. Measure as required.

19/2

19/2 **PRINCIPAL STRESSES GIVEN: Find maximum shear**

At a point in a stressed material the principal stresses are $-8\cdot0$ N/mm^2 and $-17\cdot3$ N/mm^2. Find the maximum value of shearing stress at the point, and the normal and shearing stresses on a plane at 346° to the plane carrying the major principal stress.

G *Guide Table:*

Enter at D1: this is Category I, requiring Operation O1.

U *Unknowns and Units:*

1. Maximum shearing stress.
2. Stresses on a plane at 346° to XX.

Use N and mm.

I *Inventory of Information:*

With two values of compression as principal stresses follow operation O1, and draw the Mohr Circle.

D *Derived Data:*

(i) Measure maximum shearing stress as CS_m (half the difference of the principal stresses which is $12\cdot6$ N/mm^2).

(ii) Rotate QP through 346° counterclockwise to position QS_2 and measure ON_2 ($-8\cdot7$ N/mm^2) and N_2S_2 ($-2\cdot0$ N/mm^2).

E *Evaluation:*

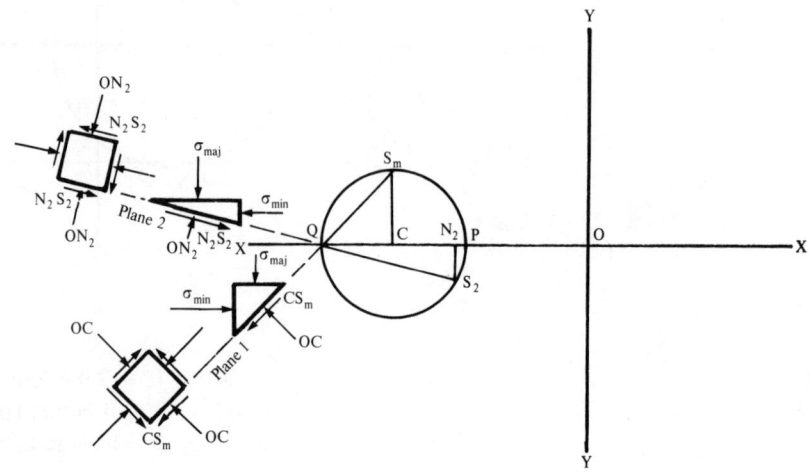

No further evaluation is required, except perhaps that a check might be made from the theory to make sure that you are drawing the circle correctly, and that the measurements give values close to the correct ones.

S *Solution:*
1. The maximum shearing stress is about **13 N/mm²** (plane 1).
2. The normal and shearing stresses on plane 2 are approximately -9 N/mm² and -2 N/mm² respectively.

19/3 PRINCIPAL STRESSES GIVEN WITH ANGLE TO HORIZONTAL: Draw Mohr Circle

At a point in a stressed body, the major principal stress acts at 135° to the horizontal. If the principal stresses are $+205$ N/mm² and $+30$ N/mm² at what angle to the horizontal is a plane on which the normal stress is $+145$ N/mm²? What is the shearing stress on this plane assuming it is positive?

G *Guide Table:*
Enter at D1: this is Category I, requiring Operation O1.

U *Unknowns and Units:*
1. Angle of plane on which normal stress is known.
2. Shearing stress on that plane.
Use N and mm.

I *Inventory of Information:*
The major principal stress always acts along the direction YY, and the plane on which it acts is in the direction XX. These two must be drawn at the angle to the horizontal defined above.
   Set off two tensile principal stresses and follow Operation O1 to draw the Mohr Circle.

D *Derived Data:*
Set off ON_1 as $+145$ N/mm², and erect N_1S_1 on the positive side.
   Measure angle S_1QP which is the angle of the plane on which the 145 N/mm² stress acts, relative to the plane on which the major p.s. acts.

19/3

Measure angle S_1QH which is the corresponding angle to the horizontal.

Measure N_1S_1, which is the required shearing stress.

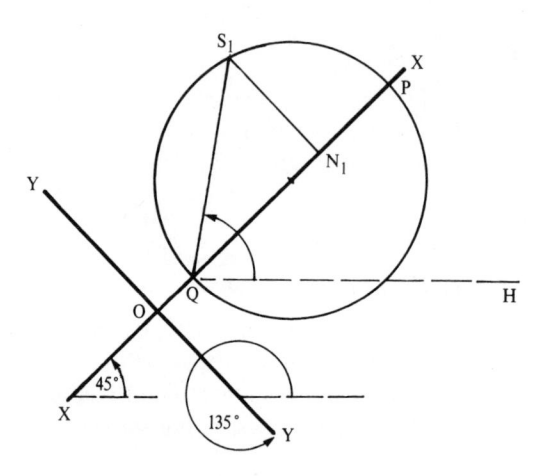

E Evaluation:

Check value of N_1S_1 against theory.

S Solution:

As measured from the diagram,
1. The plane concerned is at 37° to the plane carrying the major principal stress, and at **82° to the horizontal (HQS₁)**.
2. The shearing stress on the defined plane is **83 N/mm²**.

OX is the direction of the plane on which the major principal stress acts.

The minimum principal stress may be greater numerically, than the min.

The minor principal stress may be greater, numerically, than the major principal stress if the major is tensile and the minor is compressive (negative).

19/4

19/4 **ANGLE OF MAJOR PRINCIPAL STRESS GIVEN:** Find angles of maximum shear

At a point in a stressed body, the direction of major principal stress is at 43° to the horizontal (measured counter-clockwise). At what angle to the horizontal is the plane on which the maximum shearing force acts, for each of the combinations?

Major p.s.	70 N/mm²	Minor p.s.	15 N/mm²
Major p.s.	60 N/mm²	Minor p.s.	15 N/mm²
Major p.s.	60 N/mm²	Minor p.s.	25 N/mm²

G *Guide Table:*

Enter at D1: category I.

U *Unknowns and Units:*

1. Angle of the plane on which maximum shearing force acts. Use N and mm.

I *Inventory of Information:*

From the facts given, XX must be drawn at 43° to the horizontal. Use Operation O1 to draw the Mohr Circles.

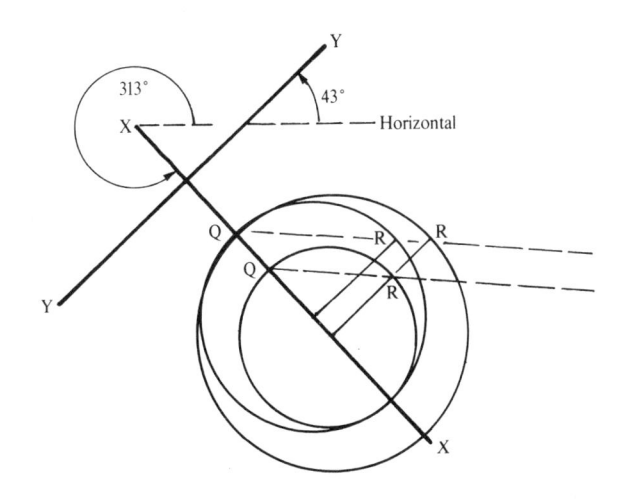

497

19/5

D *Derived Data:*

There are three values of CR, which is the radius of the circle and therefore the maximum shearing stress. By chance, two of the circles have the same centre.

Join Q to R to find the angle of the plane on which the maximum shearing stress acts for each combination.

E *Evaluation:*

Since the plane on which the maximum shearing stress acts is always 45° to the plane carrying the major principal stress, the various lines QR are parallel, and the same angle is produced for all combinations of principal stresses.

S *Solution:*

1. The angle made by the plane on which maximum shearing stress acts with the plane carrying major p.s. is the same for all combinations, and lies at **2° below the horizontal**.

19/5 PRINCIPAL STRESSES EQUAL TO MAXIMUM SHEARING STRESS: Draw Mohr Circle

The two principal stresses at a point in a stressed body are equal in numerical value and also equal in value to OR, the maximum shearing stress (36 N/mm². What are the stresses on planes at 20° and 290° measured counter-clockwise from the plane on which the major principal stress acts? Illustrate the result with stressed-block diagrams.

G *Guide Table:*

Enter at D1: this is category I.

U *Unknowns and Units:*

1. Normal and shearing stresses on plane at 20° to XX.
2. Normal and shearing stresses on plane at 290° to XX.
Use N and mm.

I *Inventory of Information:*

OP cannot be numerically equal to OQ unless one is tensile (+) and the other is compressive (−). The value of each is 36 N/mm², which is thus the radius of the MC and equal to the maximum shearing stress. Draw the Mohr Circle.

D Derived Data:

Rotate PQ on Q in a counterclockwise sense until S_1QP is 20° and continue until S_2QP is 290°.

Produce line S_2Q and show on plane 1-2 a compressive stress and a negative shear (since N_2S_2 is below XX). Place the right thumb on plane 1-2 and forefinger to the right, and turn them counterclockwise. This shows the direction of the shear on plane 1-2.

Draw two short lines intersecting in 3 showing the planes on which the p.s. act. The major p.s. is tensile and the minor, compressive.

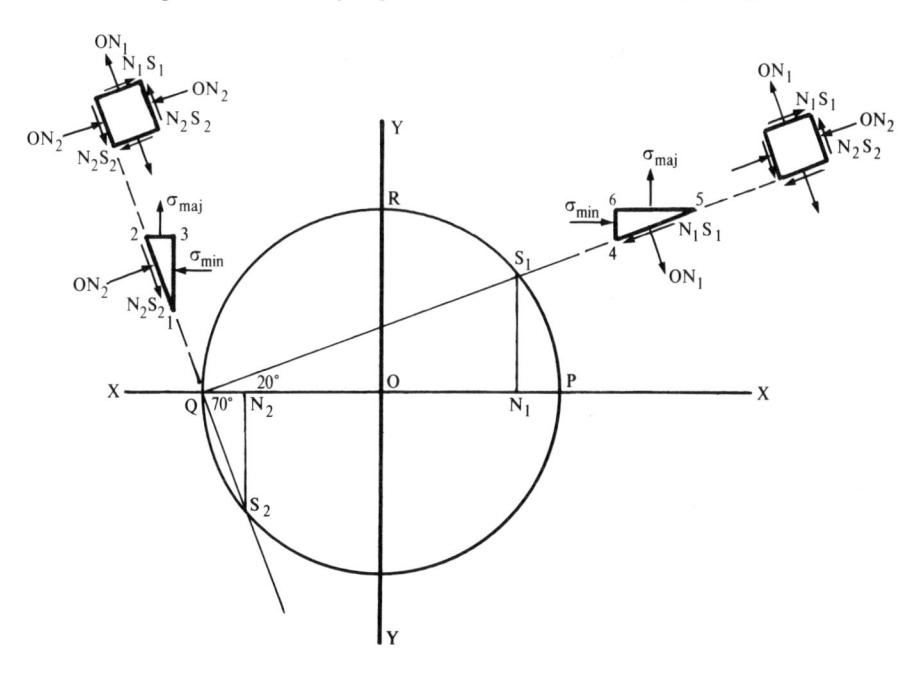

Repeat plane 1-2 and erect on it a square, on which the stresses ON_1, N_1S_1, ON_2, and N_2S_2 can be marked.

Carry out a similar step for plane 4-5.

E Evaluate:

The stresses measured are – on the 20° plane 4-5,

Normal $ON_1 = +28$ N/mm²   Shear $N_1S_1 = +23$ N/mm²

Similarly the stresses on the 290° plane can be measured.

S *Solution:*

1. Normal: $+28$ N/mm^2: Shearing: $+23$ N/mm^2.
2. Normal: -28 N/mm^2: Shearing: -23 N/mm^2.

Note that both sets of stresses can be shown on both stressed-block diagrams because the two planes are at right angles to each other. Normally only one set of stresses can be shown unless calculations for a plane at right angles to the first are carried out.

19/6 PRINCIPAL, NORMAL, and SHEARING STRESSES GIVEN

At a point in a stressed body the maximum p.s. is $+134$ N/mm^2 (tension) and on another plane passing through the point, the normal stress is $+93$ N/mm^2, with a shearing stress of -39 N/mm^2. Find the value of the minor p.s., and also the normal and shearing stresses on a plane at 26° to the plane carrying the major principal stress. What is the value of the maximum shearing stress in the body at this point? Draw stressed-block diagrams.

G *Guide Table:*

Enter at D2: this is Category II. The major p.s. is used, so the point P is used in the Operation to draw the MC.

U *Unknowns and Units:*

1. The minor principal stress.
2. Normal and shearing stresses on a plane at 26° to the plane carrying the major p.s.
3. Maximum shearing stress.
Use N and mm.

I *Inventory of Information:*

The Operation to determine the MC is O2.
The point P is used and not Q.
Draw the Mohr Circle; C and N_1 are coincident.

D *Derived Data:*

(i) Draw the angle PQS$_2$ at 26°. This gives S_2 and N_2.
(ii) Measure OQ to obtain value of minor p.s.

19/6

(*iii*) Value of the maximum shearing stress is the radius of the circle.

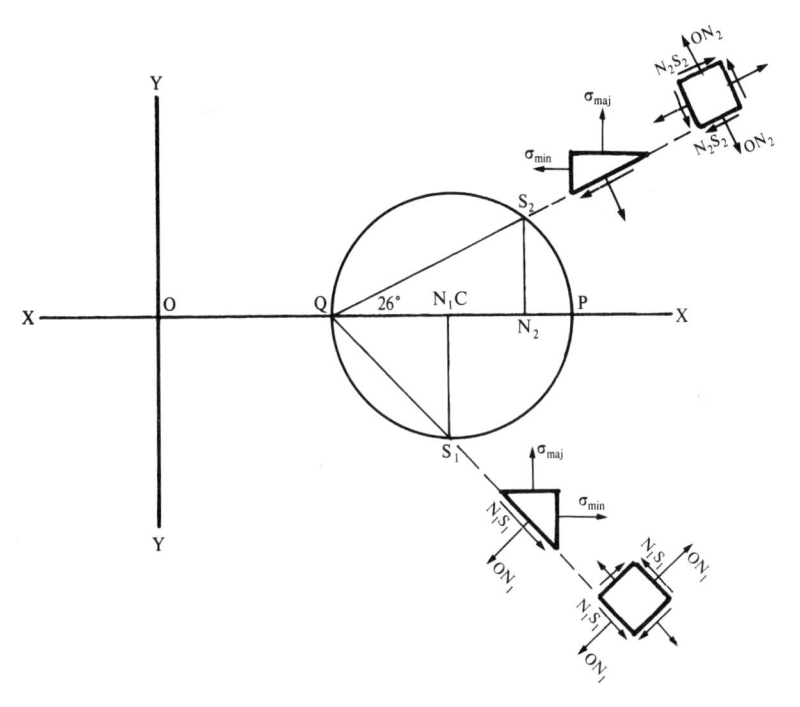

E *Evaluation:*

OQ is measured as +56 N/mm². Thus the maximum shearing stress is half the difference of 134 and 56 which is 39 N/mm². This explains why C and N_1 were coincident.

S *Solution:*

1. **+56 N/mm²** is the value of the minor p.s.
2. Normal stress: **+118 N/mm²**: Shearing stress: **+30 N/mm²**.
3. Maximum shearing stress is **39 N/mm²**, (+ and −).
   Stress-block diagrams are drawn for planes 1 and 2 showing negative shearing on plane 1 and positive on plane 2.

19/7

19/7 **NORMAL AND SHEARING STRESSES GIVEN:** What are the principal stresses?

On two planes at right angles to each other, the forces acting are -60 N/mm^2 and -188 N/mm^2 normal stress, together with a shearing stress of 95 N/mm^2 (negative with the -60 N/mm^2 normal stress). What are the principal stresses? What change in minimum p.s. would cause the major p.s. to be zero? Would the shearing stress on the two planes increase or decrease? It is assumed that the same centre C would be used for the MC.

G *Guide Table:*
Enter at D3, D4, and D7: this is Category III.

U *Unknowns and Units:*
1. Values of major and minor p.s.
2. Change of minimum p.s. to cause major p.s. to be zero.
3. Would the shearing stress increase or decrease?
Use N and mm.

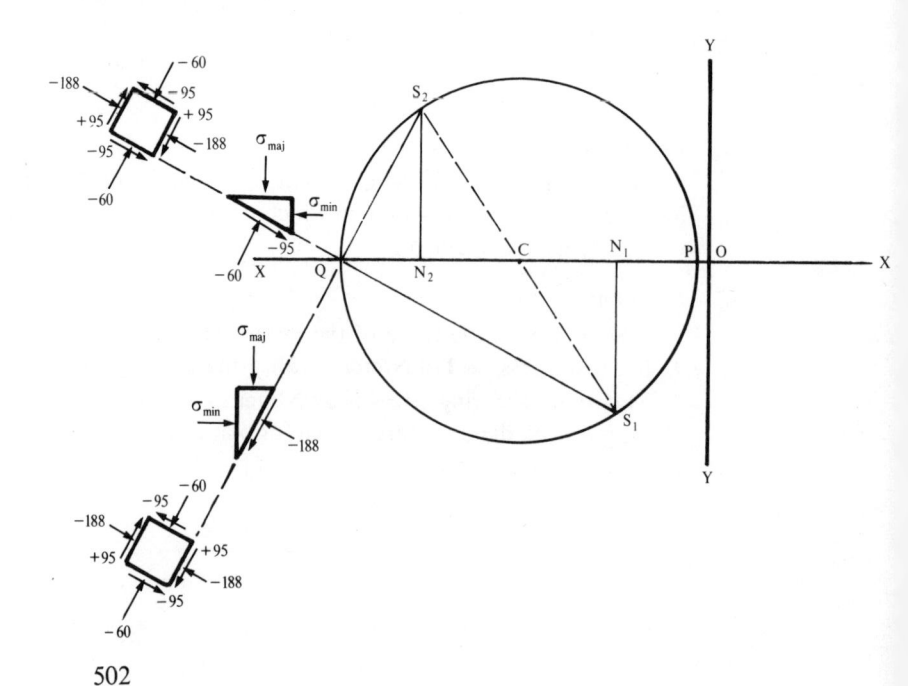

502

I *Inventory of Information:*
The table shows that the Operation for drawing the Mohr Circle is O3. Draw the circle.

D *Derived Data:*
(*i*) The major p.s. is OP, and the minor OQ. Numerically OP is less than OQ, but both are negative, so OP is larger than OQ as usual.
(*ii*) If the same centre is used, OP can increase to zero from its present negative value by an increase in the diameter of the circle.
(*iii*) The shearing stresses are all measured vertically from PQ, thus a larger circle means a larger shearing stress for the same normal stress, and this is true for all the normal stresses.

E *Evaluation:*
Measure OP and OQ: p.s. are -8 and -238 N/mm^2

If OP increases from -8 to zero N/mm^2 with the same centre to the circle, then the minor p.s. (which is OQ) decreases from -238 to -246 N/mm^2.

S *Solution:*
1. Major p.s. is **-8 N/mm^2**; minor p.s. is **-238 N/mm^2**.
2. Minor p.s. decreases to **-246 N/mm^2** if major p.s. increases to zero.
3. All shearing stresses would increase for given normal stress.

SQ gives the direction of the plane on which stress OS acts.

19/8

19/8 **NORMAL AND SHEARING STRESSES ON TWO PLANES:**
Find maximum-shear plane

If the plane on which the major horizontal stress acts lies at 25° to the horizontal (measured counter-clockwise), at what angle does the plane carrying the maximum shear lie? The stress on a plane E are $+19$ N/mm^2 normal and -12 N/mm^2 shear. On a plane F, the normal stress is $+51$ N/mm^2 and the shearing stress is $+12$ N/mm^2.

G *Guide Table:*
Enter at D3, D4, and D7: this is category III.

U *Unknowns and Units:*
1. Angle of plane carrying maximum shear.
Use N and mm.

I *Inventory of Information:*
The table shows that the Operation for drawing the Mohr Circle is O3. Draw the circle.

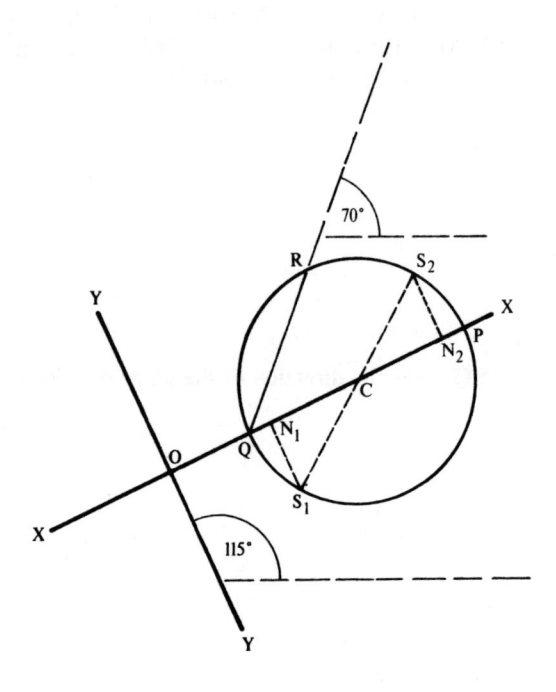

D *Derived Data:*

(*i*) The maximum shear is represented by CR, at right angles to PQ.

(*ii*) The direction of the plane on which the maximum shear acts is obtained by joining QR. The angle RQP represents the angle between the plane carrying the major p.s. and the plane carrying maximum shear.

E *Evaluation:*

Measurement of the major and minor p.s. shows $+55$ and $+15$ N/mm². The maximum shear is half the difference and is thus 20 N/mm² either positive or negative.

The plane on which the positive maximum shear acts lies at an angle RQP to the plane carrying major p.s. This angle is always 45°, so the plane carrying maximum shear is at $+45°$, $+25°$, or $+70°$ to the horizontal.

No circle is required for measurement, for the answer could be given immediately by adding 45 and 25. The determination of the p.s. is not required in order to obtain the angle asked for. The same answer occurs each time (see problem 19/4).

S *Solution:*

The plane on which the maximum shearing stress acts lies at **70° to the horizontal.**

19/9 STRESSES AND ANGLES GIVEN: Find principal stresses

The normal stress on a plane at 13° to the plane carrying the major p.s. is -24 N/mm² (compression). On a plane at 283° to XX, the normal stress is -61 N/mm² and the shearing stress is -9 N/mm². Find the values of the principal stresses.

G *Guide Table:*

Enter at D8: this is category VII.

U *Unknowns and Units:*

1. Values of the principal stresses.
Use N and mm.

19/9

I *Inventory of Information:*
Operation O5.
  (*i*) Note that the two planes (13° and 283°) are at right angles to each other. The shearing stresses on each must then be equal and of opposite sign.
  (*ii*) This means that the shearing stress on the plane at 13° is $+9$ N/mm², and the data concerning the other plane can be discarded as having served its purpose. The value of 61 N/mm² is not required.
  (*iii*) Set off ON_1 equal to -24 N/mm².
  Set up N_1S_1 equal to $+9$ N/mm² (above PQ).
  (*iv*) Draw a line, at 13° to the horizontal, through S_1. This cuts the line XX in Q. Bisect QS_1 at B and drop perpendicular on to XX. This is the centre C.

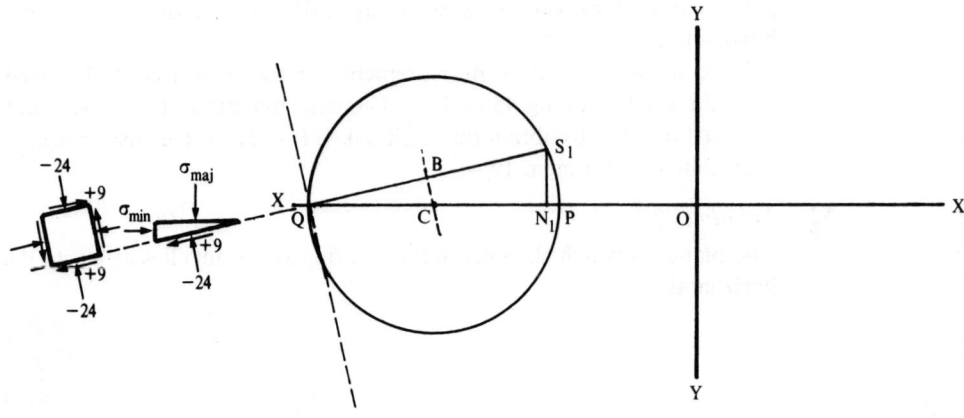

D *Derived data:*
Draw Mohr Circle centre C and radius CS_1.

E *Evaluation:*
Care must be taken in drawing, preferably on a large scale, so that Q can be found correctly.
  Measurement shows the values of OP and OQ, the principal stresses.

S *Solution:*
  1. The principal stresses are measured off as
     Major: -22 N/mm² (OP)   Minor: -63 N/mm² (OQ).

506

19/10 NORMAL AND SHEARING STRESSES GIVEN: Find principal stresses

On a plane A in a stressed material, the normal stress is tensile (54 N/mm^2) and the shearing stress is positive (25 N/mm^2). On another plane B passing through the same point, the normal stress is tensile (14 N/mm^2) and the shearing stress is negative (34 N/mm^2). Find the values of the principal stresses. Draw stressed-block diagrams for the state of maximum shear.

G *Guide Table:*

Enter at D3, D5, and D7: this is Category VII.

U *Unknowns and Units:*

1. Values of the principal stresses.
Use N and mm.

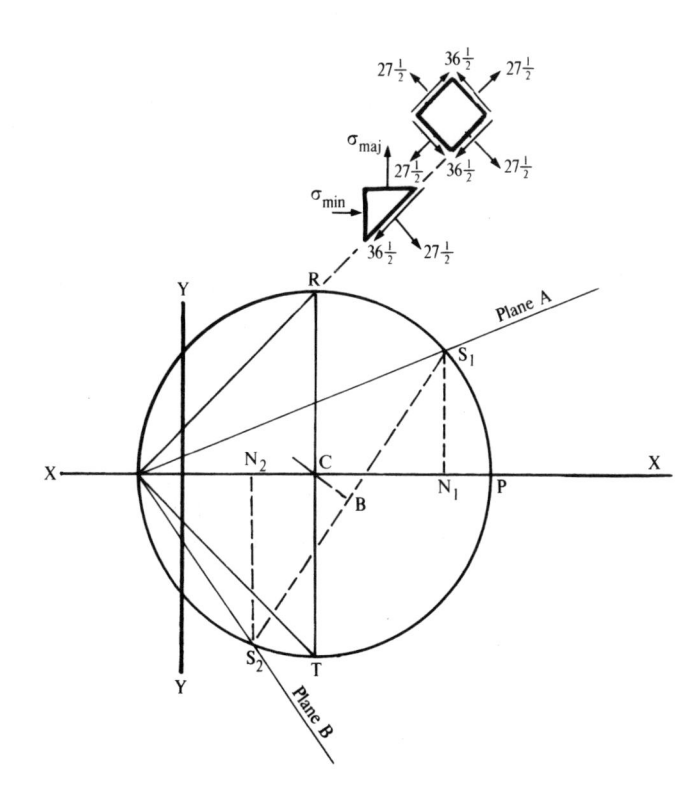

507

19/10

I   *Inventory of Information:*

The Operation for the drawing of the Mohr Circle is shown by the table to be O4. Draw the Mohr Circle.

D   *Derived Data:*

(*i*) Join QR to find the plane on which the positive maximum shear acts.

(*ii*) Stressed-block diagrams for this are drawn as follows:

Mark length 1-2 and draw lines intersecting in 3, parallel to the planes on which the principal stresses act.

Draw the length 1-2 again and complete the square 1-2-3-4.

Note that, since QT represents the other plane on which maximum shearing stress acts, the normal stresses on each of the two planes QR and QT are equal. The maximum shearing stress is positive (clockwise) on QR and negative on QT.

E   *Evaluation:*

Measurement of OQ and OP give the values required.

The normal stress on all faces of the square 1-2-3-4 is OC or $27\frac{1}{2}$ N/mm². The value of the maximum shearing stress is CR = CT = $36\frac{1}{2}$ N/mm². Check this by theory.

S   *Solution:*

1. The principal stresses are $+64$ N/mm² and -9 N/mm².

The *obliquity* of a stress is the angle its direction makes to the normal to the plane on which it acts. This is angle SOP.

19/11 NORMAL AND SHEARING STRESSES GIVEN: Find principal stresses

On two planes passing through a stressed point, the normal and shearing stresses are:

	Normal	Shearing
Plane Z	+1·5	+1·9 N/mm²
Plane W	+3·5	+1·4 N/mm²

Find the values of the principal stresses, and the angles of the planes on which the shearing stress is $-1·8$ N/mm².

G *Guide Table:*
Enter at D3, D5, and D6: this is Category VI.

U *Unknowns and Units:*
1. Values of the principal stresses.
2. Planes on which shearing stress is $-1·8$ N/mm².
Use N and mm.

I *Inventory of Information:*
The table shows that the Operation for drawing the Mohr Circle is O4. Draw the Mohr Circle.

D *Derived Data:*
(*i*) The drawing of the Mohr Circle locates OP and OQ to determine the p.s.
(*ii*) Draw a horizontal line S_3S_4 at a level of shearing stress of $-1·8$ N/mm².
(*iii*) Angles at which the planes carrying this shear are measured are shown by S_3QP and S_4QP.

E *Evaluation:*
Measure OP. Measure angles S_3QP and S_4QP.

S *Solution:*
1. The principal stresses are **4·0 and zero N/mm²**.
2. The angles concerned **302° and 329°** measured counterclockwise from the plane carrying the major principal stress.

This state of stress represents a simple tension under a force F. The sketches of a rod being pulled in this way shows how the force F is

19/11

translated into shearing and normal stresses. At other angles of section, other values of shearing and normal stresses will be developed.

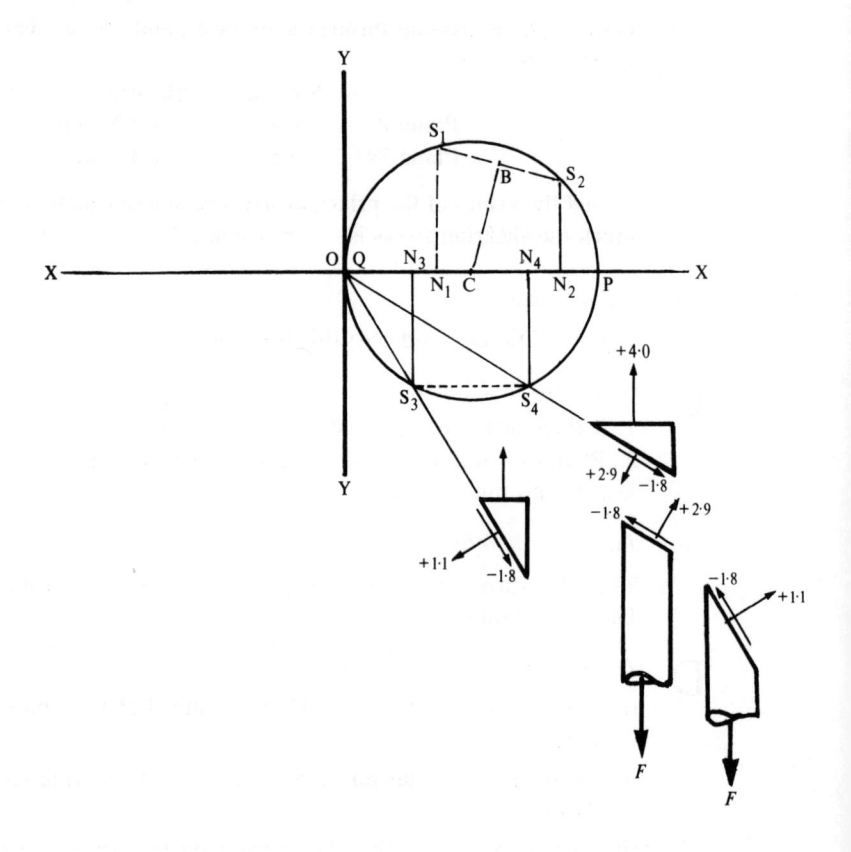

The stress acting on a plane defined by the point S has a numerical value of OS. The normal component is ON and the shearing component is NS – all in stress units.

Part II: Group F

20 Strain energy

When deformation takes place in a structure with the production of stress, and the structure is maintained under load, a store of energy is retained within the material of construction. When the load is removed, the structure returns to its original shape (since you are considering only structures loaded in the elastic range) and releases the energy which was supplied in order to deform it. The example best appreciated from personal experience is what happens when you deform a spring. A spring is a structure constructed in such a way as to suffer considerable deformation, and to absorb energy. So long as you continue to pull steadily on the spring of a spring balance, nothing happens – the structure is in equilibrium under the loading. But, when you release the load, the spring follows your hand and returns to its unstressed condition. That considerable energy is stored in the motionless spring under steady tension is evident if you release your hand suddenly, when there is a violent, and even destructive movement. The unwinding of a watch spring slowly, drives the hands, but a sudden release of the energy in the spring is a disaster.

Springs form obvious and demonstrable stores of energy, but every structure which deforms elastically and returns to its original shape on the release of the load, also stores energy. There is a logical relationship between the amount of energy stored and the elastic deformation required to store it. This relationship allows displacement of various parts of loaded structures from their original positions to be calculated, often with much less complexity than was required in the problems of Unit 17.

20.1 Representation of the magnitude of the energy stored

In earlier units you found that a series of *groups of terms* similar in arrangement, allowed of an investigation of stresses and deformations under the basic loadings. Similar conditions apply for calculating the energy stored by deforming and stressing a structure. For the basic

20.1

types of loading groups of terms, similar to those used earlier, give the values of strain energy stored in a stressed material. As you learned, the deformation under transverse shear is very small in comparison with the deformation under the bending which usually accompanies it, so, for this unit, only the three loadings; *Direct Forces*, *Bending*, and *Torsion*, will be considered. The GUIDE TABLE shows how the strain energy

GUIDE TABLE

LOADING		STRAIN ENERGY			
Type of loading	Conditions of Loading	in terms of the loading (force or couple)	in terms of the maximum stress produced by the loading	in terms of the deformation produced by the loading	Ref. Letter
Direct force	F constant	$\dfrac{F^2 l}{2AE}$ =	$\dfrac{\sigma^2 A l}{2E}$ =	$\dfrac{EAd^2}{2l}$	A
Bending Couple	M constant over length l	$\dfrac{M^2 l}{2EI}$ =	$\dfrac{\sigma^2 I l}{2Eh^2}$ =	$\dfrac{EI\theta^2}{2l}$	B
	M varying over length	$\displaystyle\int_0^l \dfrac{M^2}{2EI}\, dl$	—	—	C
Torsional Couple	T constant over length l	$\dfrac{T^2 l}{2CJ}$ =	$\dfrac{r^2 J l}{2Cr^2}$ =	$\dfrac{CJ\theta^2}{2l}$	D
	T varying over length	$\displaystyle\int_0^l \dfrac{T^2 dl}{CJ}$	—	—	E

stored can be expressed in terms of loading, stress, and deformation. There are eleven possibilities, but, as you will soon appreciate, many of these possibilities are not of great value, and the problem of using

20.1

strain energy to determine deformation results in the use of only one or two of these terms for the majority of problems encountered.

The chief object of studying strain energy in the present context is to find a simpler method of determining displacement in structures, where the methods already used are clumsy or of limited application. Under some loadings, displacements or deformations are easily found by the methods of Units in group E and although they could also be determined by strain energy methods, no advantage is obtained. One or two problems given at the end of this unit show that there is little to be gained by finding deformations by strain energy methods for structures under such loadings as are shown in lines A and D of the guide table.

Further, at this stage of your studies you will find problems of deformation under *direct force* or *torsional couples* presented in terms of a constant value of the force or couple. Neither the force nor the couple varies along the length of the member concerned. In the central section of the guide table, however, it is the exception to find lengths of beams where the bending couple remains constant. This occurs under certain types of symmetrical loading, but, in general, the moment caused by the bending couple (the bending moment) varies along the length of the beam.

By this elimination, therefore, the important study is of deformation under bending, as indicated by line C of the guide table, and it is to this determination that the problems are mainly directed. The earlier methods you practised in Unit 17 were complex and time-consuming, and became unworkable if the problem was not a simple one.

The categories of problem encountered in this field are usually as follows:

(*i*) Problems dealing with the deformation and deflection of beams and frames under bending.

(*ii*) Problems where several different types of loading act simultaneously. If a structure is loaded in bending and torsion or in bending, torsion, and direct force, then it is more convenient to use strain energy methods to determine the combined effect, and lines A, C, and E of the guide table may be used.

(*iii*) Problems dealing with the deformation of pin-jointed frames composed of members under tension or compression. Although it is just as simple to calculate the deformation of a single member under direct force by the methods of Unit 13, the joint effect of

20.3

many members under different values of force is more manageable by strain energy than by any other method.

20.2 The use of strain energy in the determination of deformation

You should remind yourself from Unit 17, that the difficulty encountered there in determining the deformation of bent structures (beams) is that of being able to translate the deformed radius of curvature (R) into terms of linear displacement. However, in 1879, Alberto Castigliano elucidated a theorem which allows of linear displacement being derived directly from energy stored in a deformed structure. An advantage of Castigliano's method is that you can calculate displacements of points on complete structures loaded by a multiplicity of loads – a step not easily taken by the methods of Units 13, 16, and 17.

The theorem which Castigliano put forward as a means of determining linear displacement from the value of strain energy, is:

For any elastic system on unyielding supports, the first partial derivative of the strain energy stored in the whole system, with respect to any of the external forces of the system, is equal to the displacement of the point of application of that force, in the direction of the line of action of that force.

LEARN THIS AND REMEMBER IT

20.3 Calculation of displacement under bending couples

When the bending moment varies along the length of a beam (l), the strain energy (U) is

$$U = \int_0^l \frac{M^2 dl}{2EI}$$

If both E and I are constant over the length (l) this can be written as

$$U = \frac{1}{2EI} \int_0^l M^2 dl$$

Sometimes the *second moment of area* varies along the length (although E usually remains constant). Then, the expression becomes:

$$U = \frac{1}{2E} \int_0^l \frac{M^2 dl}{I}$$

When you use Castigliano's theorem to find displacement, the first partial derivative of the strain energy, with respect to a chosen load must also be defined, and is:

$$\frac{\partial U}{\partial p} = \frac{1}{EI}\int_0^l M \frac{\partial M}{\partial p} dl \quad \text{or} \quad \frac{1}{E}\int_0^l \frac{M}{I}\frac{\partial M}{\partial p} dl$$

This expression, when developed, gives the displacement of the load P (one of the loads causing the bending moments M) in the direction of its line of action. It is very important, as with any application of theoretical principles, to avoid straying outside the limits of applicability of the theory. Accordingly the following points must be remembered:

(*i*) There must be a concentrated force P acting on the beam at the point where the displacement is required, and in the direction of the required displacement. This condition is not always directly fulfilled in all problems as they are presented, but is a basic requirement, as can be understood from the statement of Castigliano's theorem. This condition must be achieved, even if some artifice must be used, and that artifice must not influence the final result.

(*ii*) Remember the displacement obtained is always that of the movement of the point of application of P *in the direction of its action*. If the deflection required is horizontal or inclined, a vertical load P will not give the required result.

(*iii*) If there is no concentrated load P at the point where the deflection is required, a 'dummy' load P must be inserted and, later, equated to zero. If there is a load at the selected point, and in the correct direction, but it has a numerical value, it should be given the value P, and only later changed back to its numerical value. The use of a *unit load* for P has its advantages as is described below.

(*iv*) The integration must be taken over the *whole* structure. Students often fail to realize this.

20.4 Laying out the calculation for displacement due to bending

You must be thoroughly familiar with the methods of stating bending moments in general terms for any part of a loaded structure. If you are not certain about about this, go back and study Units 9 and 10.

20.4

Step 1: Break the structure up into lengths, AB, BC, CD, ... over each of which the expression for bending moment remains the same.

Step 2: Write down the general expression for bending moment for the lengths in question.

Step 3: Obtain the first partial derivative of each bending moment expression with respect to the concentrated load *P* which you have made sure is at the point where the displacement is required and and acting in the direction of the required displacement ($\partial M/\partial P$).

Step 4: Multiply the bending moment by the partial derivative and integrate between the appropriate limits, including *E* and *I* either as constants or variables.

Step 5: Add the values obtained for all the lengths, so that the *whole structure* is included.

These steps are most effectively accomplished by using a tabular method. An example shows what should be done when there is a concentrated load at the point where the displacement is required:

A beam of 20 m span carries a load, concentrated at the centre of 10 kN. What is the deflection at the centre? (sketch 20A).

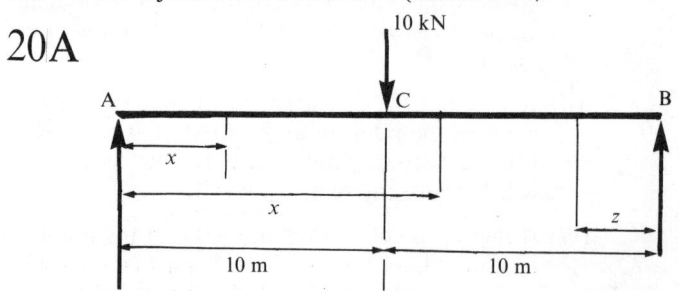

20A

First divide the structure into lengths AC and CB on which the bending moments are $+5x$ and $\{+5x - 10(x - 10)\}$. This will make a clumsy calculation for it is clear that the bending moment on BC can be expressed as $-5z$, which is similar to the $+5x$ of the other end. Secondly, change the 10 kN to *P* kN, and construct the table:

Length	Bending moment (M)	$\partial M/\partial p$	$M \, \partial M/\partial p$	Limits
AC	$+\dfrac{Px}{2}$	$+\dfrac{x}{2}$	$+\dfrac{Px^2}{4}$	$0 - \dfrac{l}{2}$
CB	$+\dfrac{Px}{2} - P\left(x - \dfrac{l}{2}\right)$	$+\left(\dfrac{x}{2} - x\right)$	$+\dfrac{P}{4}(x^2 - lx)$	$\dfrac{l}{2} - l$

516

20.4

This is clumsy and it is better to take the easier path, so the second line of the table can be:

| BC | $-\dfrac{Pz}{2}$ | $-\dfrac{z}{2}$ | $+\dfrac{Pz^2}{4}$ | $0 - \dfrac{l}{2}$ |

Using x again instead of z, measuring from B, the integration is:

$$\Delta = \frac{1}{EI} \int_0^l M \frac{\partial M}{\partial P} dx = \frac{2}{EI} \int_0^{l/2} \left(+ \frac{Px^2}{4} \right) dx = \frac{Pl^3}{48EI}$$

When $P = 10$ kN and $l = 20$ m, $\qquad \Delta = \mathbf{1667}/\boldsymbol{EI}$ **m**.

As a second example, if the beam carries a U.D.L. and no concentrated load, a concentrated load must be imposed. The sketch 20B shows the

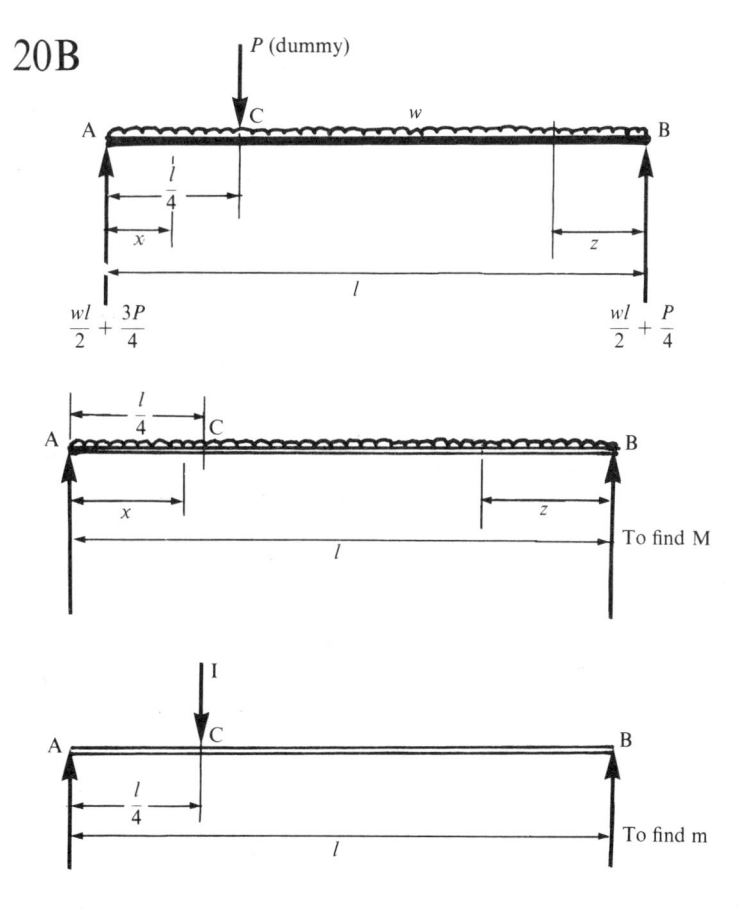

20B

517

20.5

result. *P* is placed where the displacement is required – say at the quarter-point. The same procedure of writing down the expression for bending moment and obtaining the derivative with respect to *P* is followed, but, this time, *P* is not given a value but is equated to zero. The table for this calculation is:

Length	Bending moment (M)	$\partial M/\partial P$	$M\partial M/\partial P$ (when P = 0)	Limits
AC	$\left(+\dfrac{wl}{2} + \dfrac{3P}{4}\right)x - \dfrac{wx^2}{2}$	$+\dfrac{3x}{4}$	$\left(+\dfrac{3wlx^2}{8} - \dfrac{3wx^3}{8}\right)$	$0 - \dfrac{l}{4}$
BC	$\left(-\dfrac{wl}{2} - \dfrac{P}{4}\right)z + \dfrac{wz^2}{2}$	$-\dfrac{z}{4}$	$\left(+\dfrac{wlz^2}{8} - \dfrac{wz^3}{8}\right)$	$0 - \dfrac{3l}{4}$

at this point *P* can be equated to zero, since it was merely put there as a dummy load.

The integration is:

$$\Delta_{AC} = \frac{3w}{8EI} \int_0^{l/4} (+lx^2 - x^3)\,dx = \frac{wl^4}{630EI}$$

$$\Delta_{BC} = \frac{w}{8EI} \int_0^{3l/4} (+lz^2 - z^3)\,dz = \frac{wl^4}{130EI}$$

$$\Delta = \Delta_{AC} + \Delta_{BC} = \frac{wl^4}{108EI}$$

which is the vertical deflection of a beam carrying a uniformly distributed load (*w*), as measured at the quarter point.

The general expression for the deflection of such a beam at a point *a* from the left-hand end, and *b* from the right-hand end is

$$\Delta = \frac{wab}{EI}\left(\frac{a^2 + b^2}{6} - \frac{a^3 + b^3}{8l}\right)$$

Substitute *l*/4 for *a* and 3*l*/4 for *b*, and check the above calculation.

20.5 The unit-load variation

An examination of the quantity $\partial M/\partial P$ in the two examples given in the last Section shows you that this quantity represents the bending moment which would be acting on the beam if there were a unit load on the beam at the point where the deflection is required, and no other loading was present. A simple method of using Castigliano's theorem, then, is to have two sketches for each problem. One of these shows the loading as given. The other shows the same beam or frame, but with a unit load acting at the point where the deflection is required and in the direction of that deflection. The bending moment expression obtained from the first is M and that obtained from the second is $\partial M/\partial P$, but more usually, when the device of a unit load is used, given the symbol m. In the second example of the last section the table would then appear as follows:

Length	Bending moment (M)	Bending moment (m)	Mm	Limits
AC	$+\dfrac{wlx}{2} - \dfrac{wx^2}{2}$	$+\dfrac{3x}{4}$		As before
BC	$-\dfrac{wlz}{2} + \dfrac{wz^2}{2}$	$-\dfrac{z}{4}$		As before

20.6 The unit-load method applied to the deformation of pin-jointed frames

If you wish to find the displacement of a joint on a pin-jointed frame you must first be thoroughly familiar with the techniques of calculating the force in each of the members of the frame under a given loading. Since you will be using two diagrams – one for the given loading and one for a unit load applied where the deflection is required – this calculation must be done rapidly if the deformation or deflection is to be calculated effectively.

The method of attack is to draw the two sketches and calculate:

(*i*) The force in each member under the given loading – symbol F.

(*ii*) The force in each member under a unit load only, placed at the joint whose deflection is required and acting in the direction in which the displacement or deflection is to be measured – symbol u.

In addition to these, the cross-sectional area and length of each member must be known, and the modulus of elasticity of the material of which the members are made. The first line (A) of the guide table is then used.

The strain energy stored in each member under direct force is:

$$\frac{F^2 l}{2AE}$$

20.7

Using Castigliano's theorem in a similar way to its development in the study of bending, you have

$$\Delta = \frac{\partial U}{\partial F} = \frac{Flu}{AE}$$

The work is best done in a tabular form, and as the members are not linked to each other in the simple mathematical relationship you found when studying the displacement due to bending, addition is used, the effect of each member being summed to obtain the combined effect on the structure. Thus, the displacement of the chosen joint on the structure is found by:

$$\sum \frac{Ful}{AE}$$

The later problems at the end of the unit show the method.

Anyone can do these problems if given unlimited time; your aim should be to do them fast. This needs a great deal of practice and a clear knowledge of the method to be followed.

The problems towards the end of the book are not so fully explained as at the beginning. You will find you have to work them out afresh; this is the aim of the text. You must not be able merely to read the problem; you must work at it.

20.7 THE PROBLEMS
20/1 UNIFORMLY DISTRIBUTED LOADS: Simple supports

This is the same problem as 17/4 where a simply-supported beam is loaded over its whole length with a uniformly distributed load. The problem is to find the maximum deflection which is clearly at the centre.

G *Guides:*

Sketch 17/4 again and also a second sketch showing the beam unloaded except for a unit load at the centre, where the deflection is required. See Section 20.5 above for explanations.

U *Unknowns and Units:*

1. Vertical deflection of point C at the centre of the beam.
Use kN and m.

I *Inventory of Information:*

EI is constant and can be taken out of the integration as shown in section 20.3.

$$R_L = R_R = 80 \text{ kN (reactions from the UDL)}$$
$$r_L = r_R = \tfrac{1}{2} \text{ kN (reactions from the unit load)}$$

Load is uniformly distributed over whole length.

D *Derived Data:*

Bending moment from UDL, at a section x metres from the left-hand end, is
$$M_L = +80x - 8x^2$$
and from right-hand end
$$M_R = -80z + 8z^2$$

Bending moment at a section x metres from left-hand end, from unit load
$$m = +\tfrac{1}{2}x$$
and from right-hand end
$$m = -\tfrac{1}{2}z$$

E *Evaluation:*

Integration must be taken over whole length.

Length	M	m	Limits
OC	$+80x - 8x^2$	$+x/z$	0 – 5
BC	$-80z + 8z^2$	$-z/z$	0 – 5

$$\Delta = \frac{2}{EI}\int_0^5 Mm\,dx = \frac{8}{EI}\int_0^5 (+10x^2 - x^3)\,dx = +\frac{2083}{EI} \text{ m}$$

20/2

S *Solution:*
1. Vertical deflection at the centre: $+2083/EI$ m
In problem 17/4, the deflection was found to be negative or downwards ($+$up). By strain energy, the displacement is always in the direction or sense of the unit load (or concentrated load).

20/2 CONCENTRATED LOAD: Deflection under load?

This is the same problem as in 17/2, but you will find instead of the maximum deflection, the deflection under the load. The 12 kN should be changed to P, and only given a value later on.

G *Guides:*
The sketch of 17/2 can be used with P instead of 12 kN.
$$U = \int_0^l \frac{M^2}{2EI} \qquad \Delta = \int_0^l \frac{M \, \partial M/\partial P}{EI} \, dl$$

U *Unknowns and Units:*
1. Vertical deflection under load.
Use kN and m.

I *Inventory of Information:*
EI is constant: $R_L = \frac{6}{9}P$ kN $\quad R_R = \frac{3}{9}P$ kN

D *Derived Data:*
$$M_L = +\tfrac{6}{9}Px \text{ on left-hand portion}$$
$$M_R = -\tfrac{3}{9}Pz \text{ on right-hand portion}$$

E *Evaluation:*

Length	M	$\partial M/\partial P$	Limits
LB (left 3 m)	$+\tfrac{6}{9}Px$	$+\tfrac{6}{9}x$	0–3
RB (right 9 m)	$-\tfrac{3}{9}Pz$	$-\tfrac{3}{9}z$	0–6

$$\Delta_L = \frac{1}{EI}\int_0^3 M\frac{\partial M}{\partial P}\,dx = \frac{36}{81EI}\int_0^3 Px^2\,dx = +\frac{4P}{EI}$$

$$\Delta_R = \frac{1}{EI}\int_0^6 M\frac{\partial M}{\partial P}\,dz = \frac{9}{81EI}\int_0^6 Pz^2\,dz = +\frac{8P}{EI}$$

$$\Delta = \Delta_L + \Delta_R = +\frac{12P}{EI}$$

S *Solution:*
When P is 12 kN, vertical deflection under the load is **$144/EI$ m**, EI being in kN m².

20/3 CONCENTRATED LOAD: Maximum deflection?

This is the same problem as 17/2 when it was required to find the maximum deflection. This occurs, as was shown in 17/2, at 4.1 m from the LHE, and 4.9 m from the RHE.

G *Guides:*

$$\text{Guides: } \Delta = \int_0^l \frac{M \dfrac{\partial M}{\partial P} \, dl}{EI} = \frac{1}{EI} \int_0^l Mm \, dl$$

Remember that the symbol m represents the bending moment at any point caused by a unit load at the point where the deflection is required. Draw a sketch of the beam loaded only by a unit load at 4·1 m from the left-hand end.

U *Unknowns and Units:*
1. Maximum deflection at 4·1 m from the left-hand end.
Use kN and m.

I *Inventory of Information:*
EI is constant.

$$R_L = \frac{6}{9} \times 12 = 8 \text{ kN:} \quad R_R = 4 \text{ kN}$$

From the beam carrying the unit load at 4·1 m from LHE:

$$r_L = \frac{4\cdot 9}{9} \text{ kN:} \quad r_R = \frac{4\cdot 1}{9} \text{ kN}$$

D *Derived Data:*
Since the point where the deflection is required is not at the loaded point, the beam must be divided into three lengths LB, RD, and DB. (L is the left end, R the right end, D the point of maximum deflection, B the position of the load).

523

20/4

E *Evaluation:*

Length	Moment (M)	m(unit load at D)	Limits
LB	$+8x$	$+\dfrac{4\cdot9}{9}x$	0–3
RD	$-4z$	$-\dfrac{4\cdot1}{9}z$	0–4·9
BD	$+8x - 12(x-3)$	$+\dfrac{4\cdot9}{9}x$	3–4·1

$$A_1 = \frac{1}{EI}\int_0^3 Mm\,dx = \frac{1}{EI}\int_0^3 \frac{39\cdot2}{9}x^2\,dz = +\frac{39}{EI}$$

$$A_2 = \frac{1}{EI}\int_0^{4\cdot9} Mm\,dz = \frac{1}{EI}\int_0^{4\cdot9} \frac{16\cdot4}{9}z^2\,dz = +\frac{72}{EI}$$

$$A_3 = \frac{1}{EI}\int_3^{4\cdot1} Mm\,dx = \frac{1}{EI}\int_3^{4\cdot1} (-4x+3)\frac{4\cdot9}{9}x\,dx = +\frac{46}{EI}$$

S *Solution:*
Total deflection at point of maximum deflection:

$$(+39 + 72 + 46)\frac{1}{EI} = \frac{157}{EI}\text{ m}$$

20/4 VARYING SECOND MOMENT OF AREA

This is the same problem as 17/10 where the second moment of area varies along the length.

G *Guides:*

$$\Delta = \frac{1}{E}\int_0^l \frac{M}{I}\frac{\partial M}{\partial P}\,dl \quad\text{or}\quad \Delta = \frac{1}{E}\int_0^l \frac{M}{I}m\,dl$$

U *Unknowns and Units:*
1. Deflection at the free end.
2. Deflection at half length.
Use kN and m.

I *Inventory of Information:*

$$I = \frac{x}{72 \times 10^3} \text{ m}^4 \text{ (from 17/4)}$$

$$P = 10 \text{ kN} \qquad l = 6 \text{ m}$$

D *Derived Data:*
Draw two cantilevers, one with a unit load at the half point. The other has a concentrated load at the end, which is the sketch of 17/4.

$$M = -Px \quad \text{or} \quad 10x$$

For calculation of the deflection at the end,

$$\frac{\partial M}{\partial P} = -x$$

For calculation of the deflection at half-point find m for the unit load at this point.

On the left half, $m = 0$

On the right half, $m = (x - 3)$ kN m

E *Evaluation:*
For Δ at the free end:

Length	M	$\partial M/\partial P$	$M(\partial M/\partial P)$	I	Limits
Whole	$-Px$	$-x$	$+Px^2$	$\dfrac{x}{72 \times 10^3}$	0–6

Since I contains the term x and is thus not constant, it must appear in the integration:

$$\Delta = \frac{1}{E} \int_0^6 \frac{(+Px^2)}{I} \, dx = \frac{72 \times 10^3}{E} \int_0^6 Px \, dx$$

When $P = 10$ kN, $\Delta = \dfrac{1296 \times 10^4}{E}$ m

20/5

For Δ at mid-point:

Length	M	m	I	Limits
Right half	$-10x$	$-(x-3)$	$\dfrac{x}{72 \times 10^3}$	3–6
Left half	$-10x$	0	$\dfrac{x}{72 \times 10^3}$	0–3

$$\Delta = \frac{1}{E}\int_3^6 \frac{10x(x-3)}{I}\,dx = \frac{72 \times 10^3}{E}\int_0^6 (+10x - 30)\,dx$$

When $P = 10$ kN, $\Delta = \dfrac{324 \times 10^4}{EI}$ m

Note that, since you were not taking the partial derivative, the concentrated load could be used as 10 kN.

S Solution:

1. The deflection at the free end is $\dfrac{1296 \times 10^4}{E}$ m.

2. The deflection at the mid-point is $\dfrac{324 \times 10^4}{E}$ m.

20/5 **PARTIAL U.D.L.**: Deflection at selected point

This is the same problem as 17/7.

G Guides:

$$\Delta = \int_d^l \frac{Mm\,dl}{EI}$$

A unit load must be placed at Q, 8 m from O. Draw a separate sketch showing the beam unloaded except for the unit load.

D Derived Data:

$$R_L = 12\cdot 8 \text{ kN}: \quad R_R = 3\cdot 2 \text{ kN}$$
$$r_L = 0\cdot 6 \text{ kN}: \quad r_R = 0\cdot 4 \text{ kN}$$

from a unit load at Q.

E *Evaluation:*

Length	M	m	Limits
OQ	$+12 \cdot 8x - x^2$	$+0 \cdot 6x$	0–8
RQ	$-3 \cdot 2z$	$-0 \cdot 4z$	0–12

$$\Delta_1 = \frac{1}{EI} \int_0^8 (7 \cdot 7 x^2 - 0 \cdot 6 x^3) \, dx = + \frac{697}{EI}$$

$$\Delta_2 = \frac{1}{EI} \int_0^{12} (1 \cdot 3 x^2) \, dx = + \frac{737}{EI}$$

S *Solution:*
Deflection at Q in a vertical direction is $+1434/EI$ m (downwards).

20/6 DEFLECTION UNDER OWN WEIGHT
This is the same problem as 17/8.

G *Guides:*

$$\Delta = \int_0^l \frac{Mm \, dl}{EI}$$

A unit load must be placed at the end of an otherwise unloaded cantilever to develop the bending moments m. Draw a separate sketch.

I *Inventory of Information:*
$l = 1 \cdot 5$ m
$d = 0 \cdot 2$ m
$I = \dfrac{bd^3 x^3}{12 l^3}$
$w = 25$ kN/m^3
$E = 207 \times 10^6$ kN/m^2
(see sketch 17/8)

D *Derived Data:*

$$\frac{M}{I} = \frac{2wl^2}{d^2}$$

527

E *Evaluation:*

Length	M/I	m	(M/I)m	Limits
Whole length	$+\dfrac{2wl^2}{d^2}$	$+x$	$+\dfrac{2wl^2}{d^2}x$	0–l

$$\Delta = \frac{1}{E}\int_0^l \frac{M}{I} m\, dx = \frac{1}{E}\int_0^l \frac{2wl^2}{d^2} x = \frac{wl^4}{Ed^2}$$

S *Solution:*
Substituting values in the expression for deflection, the vertical deflection of the end of the cantilever is

$$\Delta = +0.015 \times 10^{-3}\ \text{m} = \mathbf{0.015\ mm}$$

20/7 CENTRAL CONCENTRATED LOAD
This is the same problem as 17/1.

G *Guides:*

Call the 12 kN load 'P' for the present. No unit-load diagram is required, as there is a concentrated load already acting (as part of the specified loading) at the point where the deflection is required and in the direction of the required deflection (vertical).

E *Evaluation:*

Length	M	∂M/∂P	Limits
AR	$+\dfrac{P}{2}x$	$+\dfrac{x}{2}$	0–5
QR	$-\dfrac{P}{2}z$	$-\dfrac{z}{2}$	0–5

Both lines are the same, so use x for both integrations.

$$\Delta = \frac{2}{EI}\int_0^5 \frac{Px^2}{4}\, dx = \frac{125P}{6EI}$$

S *Solution:*
When P is 12 kN, the vertical deflection at the centre of the beam is
$$\mathbf{250/EI\ m.}$$

528

20/8 VERTICAL MAST WITH STAY

This is the problem of 17/15.

G *Guides:*

$$\Delta = \int_0^l \frac{M}{EI} \frac{\partial M}{\partial P} \, dl$$

Draw a sketch of the mast as in the second sketch of 17/15. The vertical load $P/\sqrt{2}$ which is a component of the pull in the stay, has no effect on the horizontal deflection and can be omitted. Only two horizontal loads act: 10 kN and $P/\sqrt{2}$. The displacement you find in the evaluation will be in the direction of the horizontal component $P/\sqrt{2}$.

E *Evaluation:*

Length	M	$\partial M/\partial P$	Limits
AM	$+10x$	0	—
	(x measured from A)		
MB	$+10(x+15) - Px/\sqrt{2}$	$-x/\sqrt{2}$	0–15
	(x measured from M)		(length MB)

$$\Delta = \frac{1}{EI} \int_0^{15} \left(-\frac{10x^2}{\sqrt{2}} - \frac{150x}{\sqrt{2}} + \frac{Px^2}{2} \right) dx = -19\,900 + 563P$$

S *Solution:*

If this displacement is zero, as is stated in the problem, then

$$+563P = +19\,900 \qquad\qquad +P = +35\cdot4 \text{ kN}$$

20/9 BENT CANTILEVER: Varying I

This the problem of 17/16.

G *Guides:*

$$\Delta = \int_0^l \frac{M}{EI} \frac{\partial M}{\partial P} \, dl$$

There is no need to use a unit-load diagram since there is already a concentrated load W at the point where the deflection is required, and acting in the direction of the desired deflection.

529

20/10

E *Evaluation:*

Length	M	$\partial M/\partial W$	$M(\partial M/\partial W)$	EI	Limits
OX	$+Wx$	$+x$	$+Wx^2$	EI	0–1
XY	$+W$	$+1$	$+W$	$2EI$	0–2
YE	$+Wx$	$+x$	$+Wx^2$	EI	0–1
EQ	$-Wx$	$-x$	$+Wx^2$	EI	0–1·5

$$\Delta_1 = \frac{2}{EI}\int_0^1 Wx^2\,dx = \frac{2W}{3EI}; \quad \Delta_2 = \frac{1}{2EI}\int_0^2 W\,dx = \frac{W}{EI}$$

$$\Delta_3 = \frac{1}{EI}\int_0^{1\cdot 5} Wx^2\,dx = \frac{3\cdot 4 W}{3EI}; \quad \text{Total} = \frac{8\cdot 4 W}{3EI}$$

S *Solution:*

If the load W is 10 kN, this deflection becomes $\Delta = \dfrac{28}{EI}$ m

20/10 ECCENTRIC U.D.L.: Varying I

This is the problem of 17/17.

G *Guides:*

$$\Delta = \int_0^1 \frac{Mm\,dl}{EI}$$

Deflection is to be found at 5 m from A. Sketch an unloaded beam and place a unit load at this point. Calculate m from this diagram. Re-draw the diagram of the loaded beam from 17/17.

E *Evaluation:*

The integration must be taken in three parts, since changes in the expressions for M or for m take place at both P and C.

Length	M	m	EI	Limits
AP	$+5x$	$+\frac{3}{4}x$	$\frac{1}{2}EI$	0–5
PC	$+5x$	$+\frac{3}{4}x - (x-5)$	$\frac{1}{2}EI$	5–10
BC	$-15z + z^2$	$-\frac{1}{4}z$	EI	0–10

x is measured from the left-hand end; z from the right-hand end.

$$\Delta_1 = \frac{1}{\frac{1}{2}EI} \int_0^5 + \frac{15x^2}{4} \, dx = \frac{313}{EI}$$

$$\Delta_2 = \frac{1}{\frac{1}{2}EI} \int_5^{10} + 5x \left(5 - \frac{x}{4}\right) dx = +\frac{1146}{EI}$$

$$\Delta_3 = \frac{1}{EI} \int_0^{10} \left(+ \frac{15z^2}{4} - \frac{z^3}{4}\right) dz = +\frac{625}{EI}$$

S *Solution:*

Deflection of a point at 5 m from A is obtained by adding the values found in the evaluation:

$$\Delta = \frac{2084}{EI} \text{ m}$$

This is of the right order, the difference from the result in 17/17 being due to small approximations being made.

20/11 **CANTILEVER: Varying I**

This is the problem of 17/18.

G *Guides:*

$$\Delta = \int_0^l \frac{M}{EI} \frac{\partial M}{\partial P} \, dl$$

Since the deflection of the free end is in question, the concentrated load there can be called P in the meantime, and the first derivative calculated from it. The free end is A and the fixed end, D.

E *Evaluation:*

Length	M	m	Mm	EI	Limits
AB	$+Px$	$+x$	$+Px^2$	EI	0–3
BC	$+Px$	$+x$	$+Px^2$	$2EI$	3–6
CD	$+Px$	$+x$	$+Px^2$	$3EI$	6–9

531

20/12

$$\Delta_1 = \frac{1}{EI}\int_0^3 (+Px^2)\,dx = +\frac{\partial P}{EI}$$

$$\Delta_2 = \frac{1}{2EI}\int_3^6 (+Px^2)\,dx = +\frac{31\cdot 5P}{EI}$$

$$\Delta_3 = \frac{1}{3EI}\int_6^9 (+Px^2)\,dx = +\frac{57P}{EI}$$

S *Solution:*
The vertical deflection of the free end of this cantilever with three different second moments of area, when $P = 10$ kN, is

$$\Delta = \frac{975}{EI}\,\text{m}$$

20/12 **APPLIED COUPLES: Simple supports**
This is the problem of 17/19.

G *Guides:*

$$\Delta = \int_0^l \frac{Mm\,dl}{EI}$$

The maximum deflection takes place at 5·8 m from A. Find this maximum deflection. A unit load must be placed at the point where the maximum deflection takes place and in the direction (vertical) of the deflection desired. The sense can be taken at random – say downward as usual. Draw a sketch which shows an upward reaction at A of 0·42 kN, and a downward load of unity at D.

E *Evaluation:*

Length	M	m	Mm	Limits
AD	$-120x$	$+0\cdot 42x$	$-(120 \times 0\cdot 42)x^2$	0–5·8
CD	$+1200 - 120z$	$-0\cdot 58z$	$-(1200 - 120z)0\cdot 58z$	0–4·2

x is measured from A and z from C.

$$\Delta_1 = \frac{1}{EI} \int_0^{5\cdot 8} (-50\cdot 4x^2)\, dx = -\frac{3278}{EI}$$

$$\Delta_2 = \frac{1}{EI} \int_0^{4\cdot 2} (-696z + 69\cdot 6z^2) = -\frac{4420}{EI}$$

Use reference tables for squares and cubes.

S *Solution:*
The vertical deflection at the point of maximum deflection (D) is

$$\Delta = -\frac{7698}{EI}\, \text{m}$$

The negative sign shows that the sense was wrongly chosen. The true deflection is vertically upwards and not downwards. This is clear from the deflected shape, but there is no need to puzzle over complex displacements. Use a random sense carefully observe signs of all moments, and the correctness of the sense is shown in the final answer.

20/13 APPLIED COUPLES
This is the problem of 17/20.

G *Guides:*

$$\Delta = \int_0^l \frac{Mm\, dl}{EI}$$

A second drawing must be made of the beam without the loading bracket but with a unit load acting vertically downward at C. From this sketch, the values of the bending moment m are found.

E *Evaluation:*

Length	M	m	Mm	Limits
AC	$+0\cdot 3x$	$+0\cdot 5x$	$+0\cdot 15x^2$	0–5
BC	$-0\cdot 7z$	$-0\cdot 5z$	$+0\cdot 35z^2$	0–5

533

20/14

$$\Delta = \frac{1}{EI}\int_0^5 (+0{\cdot}15x^2 + 0{\cdot}35x^2)\,dx = \frac{125}{6EI}$$

S *Solution:*
The vertical displacement of the beam at C is

$$+\Delta = +\frac{20{\cdot}8}{EI}\,\text{m}$$

20/14 VARYING I: Concentrated load

A beam of 12-metre span carries a load of 3 kN at one third of its length from the left-hand support. The beam is simply supported. Find the deflection under the load is the second moment of area of the right part of the beam is one-third that of the left part.

G *Guides:*

$$\Delta = \int_0^l M\,\frac{\partial M}{\partial P}\,dl$$

Draw only one sketch, making the load have a value of P kN, for the moment.

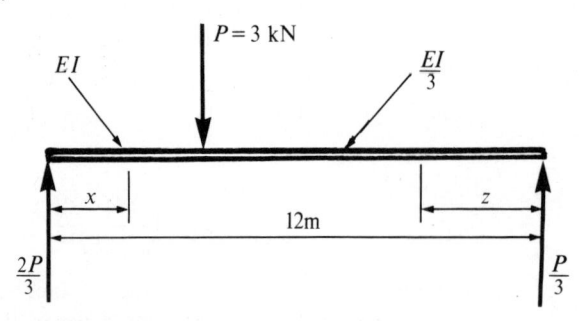

U *Unknowns and Units:*
1. Deflection under the load.
Use kN and m.

I *Inventory of Information:*
Distance from left-hand end to the load is 4 m. Span is 12 m. Clockwise moments positive,

D *Derived Units:*

$$R_L = \tfrac{2}{3}P\text{ kn} \qquad R_R = \tfrac{1}{3}P\text{ kn}$$

534

E *Evaluation:*

Length	M	$\partial M/\partial P$	$M(\partial M/\partial P)$	EI	Limits
Left-hand third	$+\tfrac{2}{3}Px$	$+\tfrac{2}{3}x$	$+\tfrac{4}{9}Px^2$	EI	0–4
Right-hand two-thirds	$-\tfrac{1}{3}Pz$	$-\tfrac{1}{3}z$	$+\tfrac{1}{9}Pz^2$	$\tfrac{1}{3}EI$	0–8

$$\Delta_L = \frac{1}{EI}\int_0^4 (+\tfrac{4}{9}Px^2)\,dx = +\frac{256P}{27EI}$$

$$\Delta_R = \frac{3}{EI}\int_0^8 (+\tfrac{1}{9}Pz^2)\,dz = +\frac{512P}{9EI}$$

S *Solution:*

The vertical deflection under the load is $\dfrac{1792}{27EI}P$

which becomes

$$\frac{1792}{9EI}\text{ m when }P = 3\text{ kN}.$$

20/15 APPLIED COUPLES: Simple supports

Find the central deflection of a beam which is unloaded except for a clockwise couple applied at one end and a counter-clockwise couple at the other end.

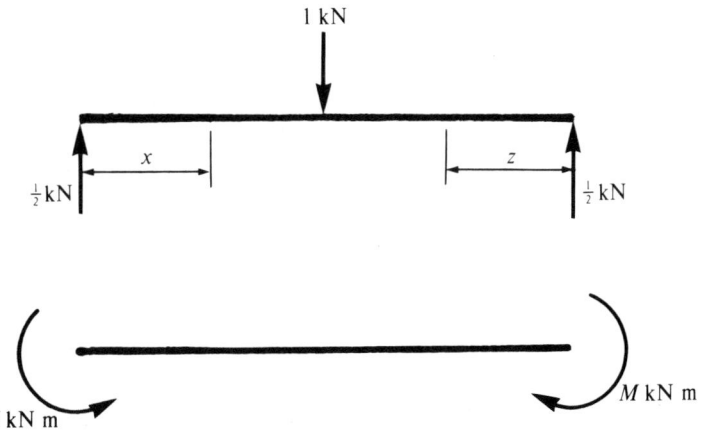

20/15

G *Guides:*
For equilibrium, the two couples must have the same moment, of opposite senses.

$$\Delta = \int_0^l \frac{Mm}{EI} dl$$

A unit load must be placed at the centre of the unloaded beam so that the moments m can be determined.

U *Unknowns and Units:*
1. Vertical deflection of the centre of a horizontal beam, loaded as shown.
Use kN and m.

I *Inventory of Information:*
Symbols will be used throughout:
  Length: l (m)
  Second moment of area: I (m^4)
  Modulus of elasticity: E (kN/m^2)
  Moment of the couples: M (kN m)

D:E *Evaluation:*

Length	M	m	Limits
Left-hand half	$-M$	$+x/2$	$0 - l/2$
Right-hand half	$+M$	$-z/2$	$0 - l/2$

$$\Delta = \frac{2}{EI} \int_0^{l/2} -M \frac{x}{2} dx = -\frac{2}{EI}\left(\frac{Ml^2}{16}\right) = -\frac{Ml^2}{8EI}$$

S *Solution:*

The central deflection of the beam is $-\dfrac{Ml^2}{8EI}$ m.

The negative sign shows that the sense of the deflection is opposite to that shown in the sketch. The true deflection is upwards.

20/16 CANTILEVERED PORTAL FRAME

A rectangular portal frame is fixed at one end and free at the other as shown in the sketch. It is loaded with a concentrated 5 kN load. Find both the vertical and the horizontal movements of the free end. The leg DE has an *EI* value $2\frac{1}{2}$ times that of the *EI* of AB and BD.

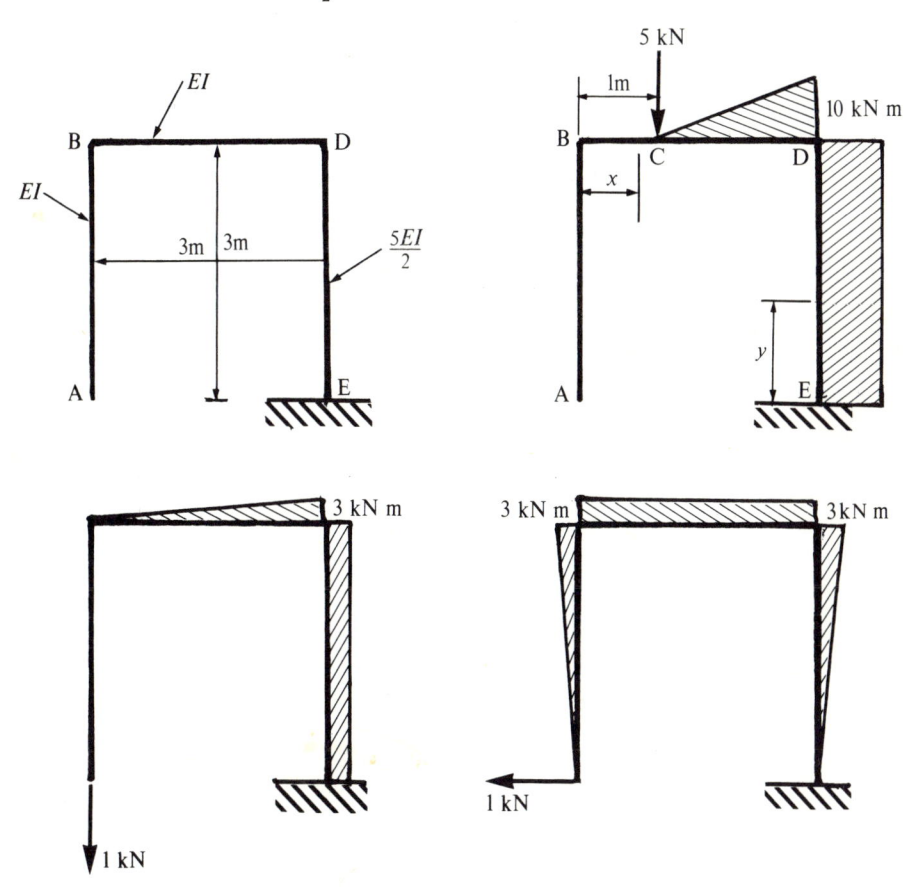

G *Guides:*

A unit load must be placed at the point where the deflection is required and in the direction of the desired deflection. The sense (as opposed to direction) can be chosen at random:

$$\Delta = \int_0^l \frac{Mm}{EI} \, dl$$

537

20/16

U *Unknowns and Units:*
1. Deflection of the end of the portal in a vertical direction.
2. Deflection of the end of the portal in a horizontal direction.
Use kN and m.

I *Inventory of Information:*
Stiffness of DE is $2\tfrac{1}{2}EI$. Stiffness of AB and BD is EI. Lengths are shown in the sketch.

D *Derived Data:*
The three bending moment diagrams are for the loading, for a unit load acting vertically, and for a unit load acting horizontally. Equations for BM taken from these.

E *Evaluation:*
For vertical deflection: Vertical unit load as shown

Length	M	m	Limits and EI
AB	—	—	—
BC	—	$-x$	—
CD	$-5(x-1)$	$-x$	1–3 (EI)
DE	-10	-3	0–3 ($2 \cdot 5EI$)

$$\Delta = \frac{1}{EI}\int_1^3 -5x(x-1)\,dx + \frac{1}{2\cdot5EI}\int_0^3 +30\,dy = +\frac{59\cdot3}{EI}\,\text{m}$$

For horizontal deflection: Horizontal unit load as shown

Length	M	m	Limits and EI
AB	0	$+y$	—
BC	0	$+3$	—
CD	$-5(x-1)$	$+3$	1–3 (EI)
DE	-10	$+y$	0–3 ($2\cdot5EI$)

$$\Delta = \frac{1}{EI}\int_1^3 \{-5(x-1)\}3\,dx + \frac{2}{5EI}\int_0^3 (-10y)\,dy = -\frac{48}{EI}\,\text{m}$$

S *Solution:*

1. The vertical deflection of the end of the frame is $+\dfrac{59\cdot 3}{EI}$ in the *direction* and *sense* shown.

The horizontal deflection of the end of the frame is $-\dfrac{48}{EI}$ in the horizontal *direction* but the negative sign indicates that the *sense* is wrong, and the deflection takes place inwards.

20/17 VARYING I: Varying E: Varying load

A cantilever has a cross-section with a second moment of area of *I* on its outer half and a value of 2*I* on its inner half. The outer half has a modulus of elasticity of 207×10^6 kN/m^2 (steel) and the inner half, firmly attached to the outer half, has a value of *E* of 115×10^6 kN/m^2. If *I* is $63\cdot 3 \times 10^6$ mm^4 what is the deflection of the free end of the cantilever?

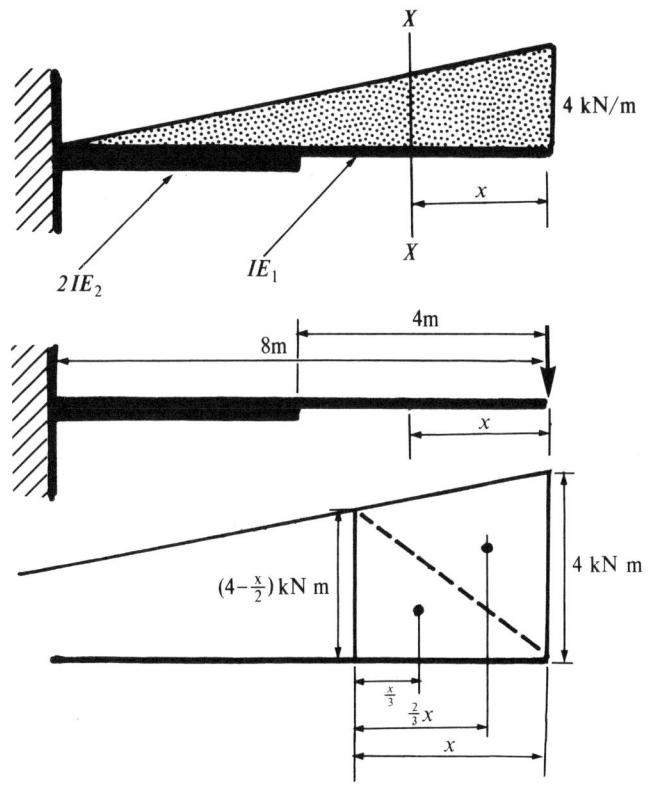

539

20/17

G *Guides:*
Both E and I have different values on different parts of the cantilever
The value of M varies along the beam also. The deflection is thus

$$\Delta = \frac{1}{E_1 I_1} \int_0^{l/2} Mm \, dx + \frac{1}{E_2 I_2} \int_{l/2}^{l} Mm \, dx$$

U *Unknowns and Units:*
1. Vertical displacement of the end of the cantilever. A unit load acting vertically is required at the end in order to find the values of m. Use kN and m.

I *Inventory of Information:*
The distributed load is not uniform and increases in value with x
Length is 8 m: I of outer half is $63 \cdot 3 \times 10^{-6}$ m^4
I of inner half is $126 \cdot 6 \times 10^{-6}$ m^4
E of outer half is 207×10^6 kN/m^2
E of inner half is 115×10^6 kN/m^2

D *Derived Data:*
The loading decreases from 4 kN/m at the outer end to zero at the inner end. The decrease is at the rate of $\frac{1}{2}$ kN/m.

Intensity of loading at distance x from the free end is

$$4 - \frac{x}{2} \text{ kN/m}$$

Moment of loading from free end over length about the point X is

$$+4 \frac{x}{2} \cdot \frac{2x}{3} + \left(4 - \frac{x}{2}\right) \frac{x}{2} \cdot \frac{x}{3} = 2x^2 - \frac{x^3}{12} \text{ kN m}$$

E *Evaluation:*

Length	M (kN m)	m (kN m)	E (kN/m^2)	I (m^4)	Limits (m)
AC	$+2x^2 - \dfrac{x^3}{12}$	$+x$	207×10^6	63×10^{-6}	0–4
CB	$+2x^2 - \dfrac{x^3}{12}$	$+x$	115×10^6	126×10^{-6}	4–8

$$\Delta_1 = \frac{1}{207 \times 63} \int_0^4 \left(2x^2 - \frac{x^3}{12}\right) x \, dx = \frac{1}{13\,041} \left[\frac{x^4}{2} - \frac{x^5}{60}\right]_0^4 = 0 \cdot 0085 \text{ m}$$

540

$$\Delta_2 = \frac{1}{115 \times 126} \int_4^8 \left(2x^2 - \frac{x^3}{12}\right) x \, dx = \frac{1}{14\,490} \left[\frac{x^4}{2} - \frac{x^5}{60}\right]_4^8 = 0.0960 \text{ m}$$

S *Solution:*

The vertical deflection of this very badly designed beam is, at the extreme end:

$$+0.0085 + 0.0960 = +0.1045 \text{ m or } 104.5 \text{ mm}$$

This combination of non-uniformly distributed load with varying E and I is certainly as complex a problem as you are likely to encounter at this stage in your studies. It is, of course, quite unreal, but gives exercise in the procedures.

20/18 DEFLECTION INTO THE THIRD DIMENSION

A rod of circular cross-section is bent to form the quadrant of a circle, and is fixed horizontally at one end. It is loaded at right angles to its plane by a horizontal load **P**. What is the horizontal displacement of the free end if the radius of the circle is 300 mm and the diameter of the rod is 20 mm?

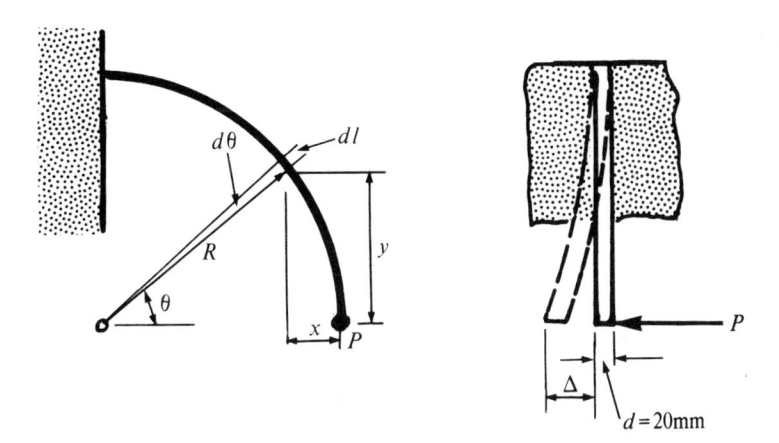

G *Guides:*

The load P causes both bending and torsion. The bending moment caused by P is p multiplied by the vertical distance to the section (since P is horizontal). This distance is y. The torque caused by p is P multiplied by the horizontal distance to the section (x). The plane of torque is at right angles to the plane of bending, for the same force.

Be careful to observe, in the statements of problems involving circles, whether terms are given as 'radius' or 'diameter'. Many students go wrong at this point.

$$\Delta = \frac{1}{EI} \int_0^l M \frac{\partial M}{\partial P} dl + \frac{1}{CJ} \int_0^l T \frac{\partial T}{\partial P} dl$$

U *Unknowns and Units:*
1. Horizontal displacement of *P* out of the original vertical plane. Use kN and m.

I *Inventory of Information:*
   $R = 300$ mm
   $d = 20$ mm
   $P = 0.2$ kN

D *Derived Data:*
Second moment of area
$$I = \frac{\pi d^4}{64} = 7855 \text{ mm}^4$$
Polar second moment of area
$$J = \frac{\pi d^4}{32} = 15\,710 \text{ mm}^4$$
If the rod is made of aluminium,
   $E = 72 \times 10^6$ kN/m^2
   $C = 27 \times 10^6$ kN/m^2
   $EI = 72 \times 10^6 \times 7855 \times 10^{-12}$ kN m^2 = **0·565 kN m²**
   $CJ = 27 \times 10^6 \times 15\,710 \times 10^{-12}$ kN m^2 = **0·424 kN m²**
(mm^4 = $(10^3)^4$ m^4)

E *Evaluation:*

Length	M	$\partial M/\partial P$	T	$\partial T/\partial P$	Limits (linear)
A–B along curve	$+Py$	$+y$	$+Px$	$+x$	0–1/4 circumference

It is not convenient to use these cartesian co-ordinates directly, since the integration must be taken along the length of the member – round the curve of the quadrant. The terms must be changed to polar co-ordinates.

$$y = R \sin \theta \quad x = R(1 - \cos \theta) \quad dl = R \, d\theta$$

Length	M	$\partial M/\partial P$	T	$\partial T/\partial P$	Limits (angular)
A–B along curve	$+PR \sin \theta$	$+R \sin \theta$	$+PR(1 - \cos \theta)$	$+R(1 - \cos \theta)$	$0 - \pi/2$

$$\Delta_1 = \int_0^{\pi/2} \frac{PR^3}{EI} \sin^2 \theta (R \, d\theta) = \frac{PR^3}{EI} \int_0^{\pi/2} \sin^2 \theta \, d\theta =$$

$$\frac{PR^3}{EI} \left[\frac{\theta}{2} - \frac{1}{4} \sin 2\theta \right]_0^{\pi/2} = \frac{PR^3 \pi}{4EI}$$

$$\Delta_2 = \int_0^{\pi/2} \frac{PR^2}{CJ} (1 - \cos \theta)^2 (R \, d\theta) = \frac{PR^3}{CJ} \left[\theta - 2 \sin \theta + \frac{\theta}{2} + \frac{1}{4} \sin 2\theta \right]_0^{\pi/2} =$$

$$\frac{PR^3}{CJ} \left(\frac{3\pi}{4} - 2 \right)$$

$$\Delta = \Delta_1 + \Delta_2 = \frac{PR^3 \pi}{4} \left\{ \frac{1}{EI} + \frac{1}{CJ} (3\pi - 8) \right\}$$

S *Solution:*

1. Substituting values from *I* and D, above, the deflection of the rod out of its plane in a horizontal direction is (kN and m):

$$\Delta = \frac{0 \cdot 2(0 \cdot 3)^3 \pi}{4} \left\{ \frac{1}{0 \cdot 565} + \frac{1}{0 \cdot 424} (3\pi - 8) \right\} = 0 \cdot 022 \text{ m} = \mathbf{22 \text{ mm}}$$

20/19

20/19 ROD UNDER TENSION

A steel bar of square cross-section (6 × 6 mm) and length 2 m, is loaded axially with a force of F kN. What is the extension of the bar if F is 72 kN?

G *Guides:*
This is the application of strain energy to the determination of the extension of a single rod. In the following problems the effect of connecting a number of these extended rods together with pin-jointed ends shows how this simple technique can be extended to full frames. (Refer to line A in the guide table of this unit.)

$$\bar{U} = \frac{F^2 l}{2AE}$$

U *Unknowns and Units:*
1. Extension of the rod.
Use kN and m.

I *Inventory of Data:*
$A = 6 \times 6 = 36$ mm² $\qquad l = 2$ m $= 2000$ mm
$E = 207$ kN/m² $\qquad F = 72$ kN

D *Derived Data:*
Partial derivative

$$\frac{\partial \bar{U}}{\partial F} = \frac{Fl}{AE} = \text{deformation}$$

E *Evaluation:*
Fl/AE is the same expression as was obtained in Unit 13 and the same result can be obtained readily without the use of strain energy.

$$\frac{Fl}{AE} = \frac{72 \times 2000}{36 \times 207} = 19 \text{ mm}$$

S *Solution:*
1. The extension of the rod is **19 mm**.

20/20 CLOSE-COILED SPRING

A close-coiled helical spring carries a load W. The radius of the spring is R mm and the diameter of the rod forming the spring is d mm. The length of the spring is represented by n complete turns, and the modulus of rigidity, or torsion modulus is C kN/mm^2. Find what extension of the spring takes place. It starts by being tightly set with each turn of the rod in contact with those above and below.

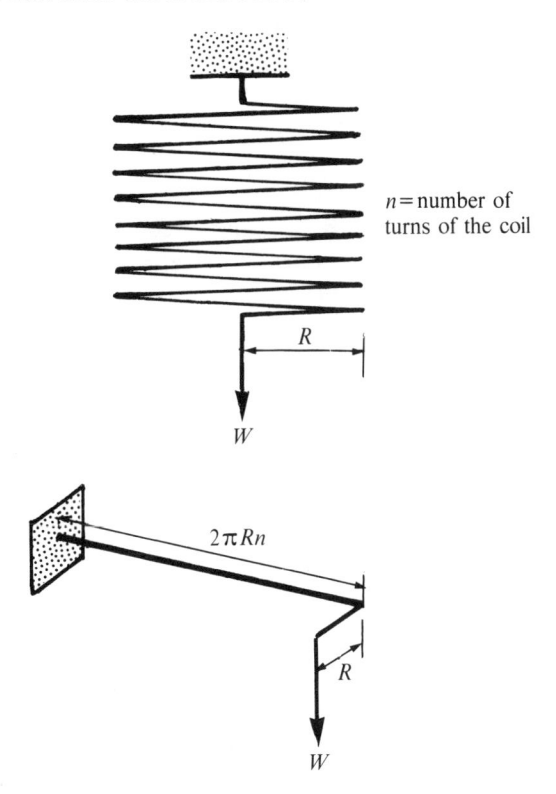

n = number of turns of the coil

G Guides:

This is a torsion problem. The spring can be considered as a long rod, coiled for convenience. The torque applied is represented by the load multiplied by the radius of the coil – assuming the load to hang in the centre of the spring. The extension is thus the first partial derivative of the stored strain energy – line D of the guide table of this unit.

$$\bar{U} = \frac{2T^2 l}{CJ} \qquad \Delta = \frac{\partial \bar{U}}{\partial W} = \frac{T}{CJ} \frac{\partial T}{\partial W} l$$

545

20/20

U   *Unknowns and Units:*
1. Extension of the coil, equivalent to the deformation caused by the torque.

Use kN and m.

I   *Inventory of Information:*

n = number of turns of the coil   Radius of coil = R mm
Diameter of rod = d mm   Load carried = W kN
Moldulus of rigidity = C kN/mm²
l is the length of the rod as developed from the number of complete circular turns = $n2\pi R$

D   *Derived Data:*

Torque, $T = W \times$ radius of coil $= WR$ kN mm.
Polar second moment of area $= \pi d^4/32$.
$\partial T/\partial W = R$ mm.

E   *Evaluation:*

$$\Delta = \frac{T}{CJ}\frac{\partial T}{\partial W}l = \frac{WR}{CJ} \times R \times n2\pi R = \frac{2WR^3 n\pi}{CJ}$$

$$= \frac{2WR^3 n\pi}{C} \cdot \frac{32}{\pi d^4} = \frac{64 WR^3 n}{Cd^4} \text{ mm}$$

S   *Solution:*

Knowing the load applied, the radius of the coil, the number of complete turns of the coil, the modulus of rigidity of the material and the diameter of the rod forming the spring, you can determine the extension of any close-coiled spring under load.

These problems become very heavy unless you have practised determining the forces in the members of a pin-jointed frame by the methods of Part I.

20/21 PIN-JOINTED TRUSS: Cantilever

Determine the deflection vertically of the loaded point A in the cantilever truss shown. The areas of the cross-sections of members are such that when the load is applied, members carrying tension (ties) are loaded to 90 N/mm², while compression members (struts) carry 61 N/mm². Modulus of elasticity is 207×10^3 N/mm².

Member	F/A (N/mm²)	u	l (m) (convert to mm)	Ful/A
AB	−61	−1	5	305×10^3
BC	−61	−2	5	610×10^3
BD	+90	$+\sqrt{2}$	$5\sqrt{2}$	900×10^3
DE	+90	+1	5	450×10^3
EA	+90	$+\sqrt{2}$	$5\sqrt{2}$	900×10^3
BE	−61	−1	5	305×10^3
Total				3470×10^3 N/mm

$$\Delta = \frac{3470 \times 10^3}{E} = \frac{3470 \times 10^3}{207 \times 10^3} = 17 \text{ mm vertical deflection of } A$$

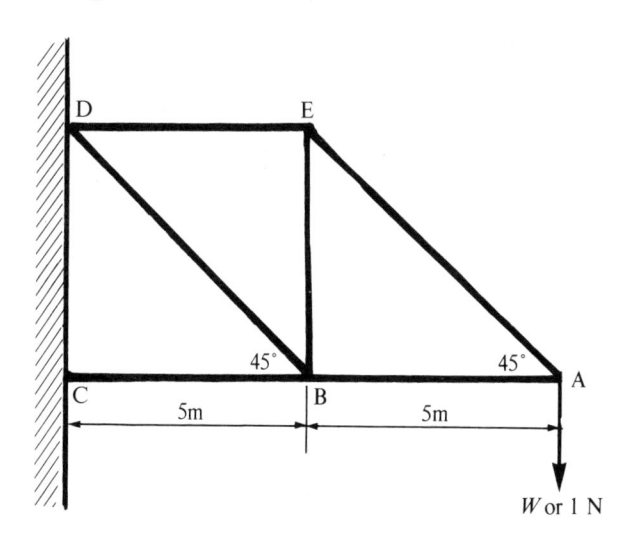

20/22

20/22 PIN-JOINTED TRUSS: Simple supports

The N-girder shown has four bays, each of 6 m in length. The height of the truss is 4 m. The area of each member is the same, the cross-section being 7750 mm². If E is 207×10^3 N/mm² find the deflection of the central point H.

In this problem the F column is found from the first diagram. The (u) forces in all the members must be found by the application of a unit dummy or fictitious load at H. This requires the use of two FSD's so be sure you know how to determine forces in pin-jointed trusses.

Member	F(kN)	u(kN)	l(m)	Ful
AB	−675	−0·75	6·00	3038
BC	−900	−1·50	6·00	8100
CD	−900	−1·50	6·00	8100
DE	−675	−0·75	6·00	3038
FG	0	0	6·00	0
GH	+675	+0·75	6·00	3038
HI	+675	+0·75	6·00	3038
IK	0	0	6·00	0
AG	+815	+0·90	7·21	5289
BH	+272	+0·90	7·21	1765
DH	+272	+0·90	7·21	1765
EI	+815	+0·90	7·21	5289
AF	−450	−0·50	4·00	900
BG	−150	−0·50	4·00	300
CH	0	0	4·00	0
DI	−150	−0·50	4·00	300
EK	−450	−0·50	4·00	900
Total				44 860

So far, everything has been in kN and m. Converting to kN and mm, E 207 kN/mm² and Ful is $44\,860 \times 10^3$ kN mm.

$$\Delta = \frac{Ful}{AE} = \frac{44\,860 \times 10^3}{7750 \times 207} = \textbf{28 mm deflection vertically of joint H}$$

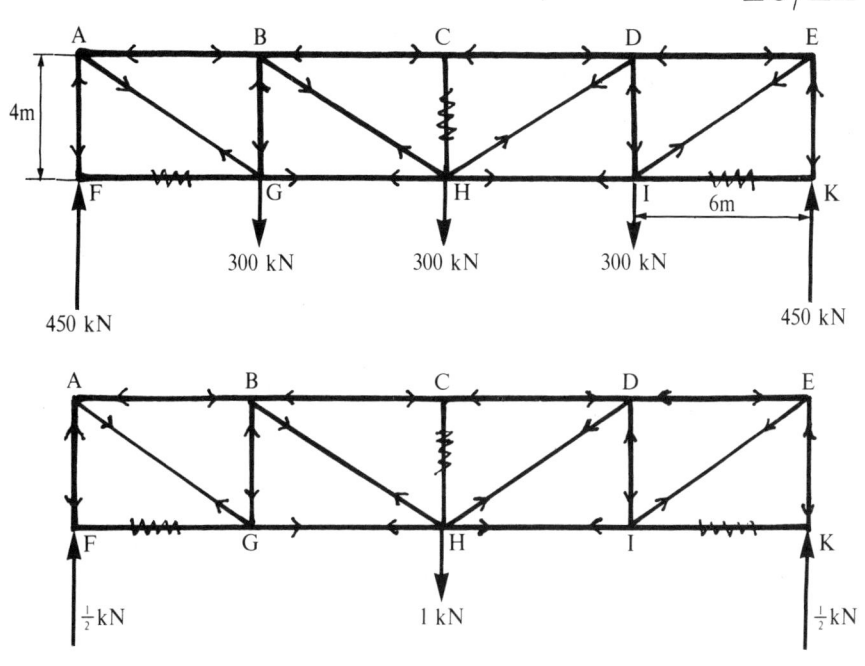

Always mark on each member the arrows showing whether it is in tension or compression. The tensile condition is shown by 'inward' arrows. Use plus for tension and minus for compression.

549

20/23

20/23 PIN-JOINTED CANTILEVER

The cantilever shown is loaded at B with a load of 10 kN. Determine the vertical deflection of B. The cross-sectional area of AB is 150 mm² but the others all have a cross-sectional area of 300 mm².

When there is only one load on the structure, and the deflection of the loaded point is required, it is best to calculate as for a load of W rather than for a numerical load. This allows of the values of u being determined more easily. Tension, positive; compression, negative.

Member	F(kN)	u	A(mm²)	l(m)	Ful/A (W = 10 kN)
AB	$+\sqrt{3}W$	$+\sqrt{3}$	150	$2\sqrt{3}$	693
BC	$-2W$	-2	300	2·0	267
AC	$-W$	-1	300	2·0	67
AD	$+\tfrac{1}{2}W$	$+\tfrac{1}{2}$	300	4·0	33
CD	$-\sqrt{3}\dot{W}$	$-\sqrt{3}$	300	$2\sqrt{3}$	346

Using l in mm, the total is $\qquad$ 1406 kN/mm

$$\Delta = \frac{Ful}{AE} = \frac{1406}{E} \text{ mm, where } E \text{ is in kN/mm}^2$$

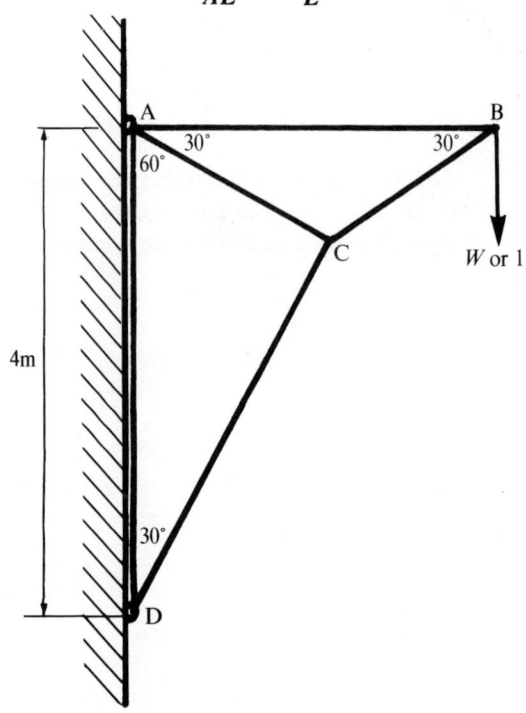

550

20/24 PIN-JOINTED FRAME

Find the vertical deflection of the point F under a loading which produces 124 N/mm² in the tension members, and 93 N/mm² in the compression members. Modulus of elasticity is 207×10^3 N/mm².

The point which you wish to study, is unloaded. A unit load must be placed at this point to obtain the values of u.

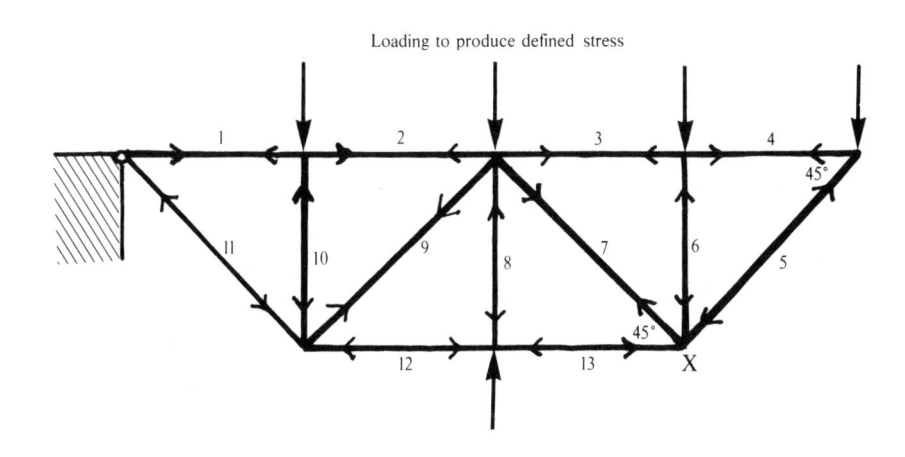

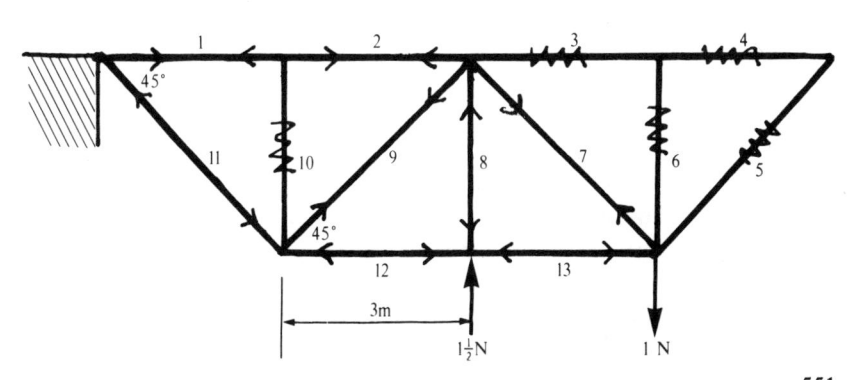

Loading to produce defined stress

551

20/24

Member	FAl (N/mm²)	l(m)	u	$Ful/A(10^3)$
1	+124	3	$+\frac{1}{2}$	186
2	+124	3	$+\frac{1}{2}$	186
3	+124	3	0	0
4	+124	3	0	0
5	−93	$3\sqrt{2}$	0	0
6	−93	3	0	0
7	+124	$3\sqrt{2}$	$+\sqrt{2}$	744
8	−93	3	$-1\frac{1}{2}$	419
9	+124	$3\sqrt{2}$	$1/\sqrt{2}$	372
10	−93	3	0	0
11	−93	$3\sqrt{2}$	$-1/\sqrt{2}$	279
12	−93	3	−1	279
13	−93	3	−1	279
Total				2744×10^3 N/mm

$$\Delta_F = \frac{Ful}{AE} = \frac{2744 \times 10^3}{207 \times 10^3} = \mathbf{13\ mm\ vertical\ deflection\ of\ joint}\ F$$

ENVOI

Writers, in past times, used to give their readers a farewell message at the end of the book. This is not necessary on this occasion, for this book will never be read from beginning to end. It should not be read at all, but worked with as a tool. It is a reference volume, and you may well have turned to 20/24 on opening the volume. Always use pencil and paper. Use the INDEX.

Index

Sections are indicated thus: **12.3, 17.6**
Problems are indicated thus: 19/8, 10/7

Angle of twist, 16/3, 16/5, 16/8, 16/9
Angle section: second moment of area, 11/8
Angles in Mohr's circle, **19.5**, 19/2, 19/3, 19/4, 19/8, 19/9, 19/11
Applied couples
 effect on conditions of equilibrium, **5.4**
 on beam, BMD, 10/23
Arch or bent rod, forces and couples, 9/7, 9/8, 9/11, 9/39, 9/40, 9/44
Area moments, method of, **17.6**, 17/1, 17/6, 17/11, 17/12, 17/13, 17/14, 17/15, 17/16, 17/17, 17/18, 17/19, 17/20

Beam
 horizontal, BMD, 10/12–10/17, 10/19
  with mixed loads, BMD, 10/12–10/14, 10/16, 10/17, 10/19
  SFD, 10/3–10/5
 inclined loads, SFD, TD, 10/7, 10/10
 inclined SFD, TD, BMD, 10/11, 10/21, 10/22
 non-uniform load, BMD, 10/26
 treatment of self-weight, 10/13
 with applied couples, 10/11, 10/23, 10/25
 with cantilever projection, 10/6, 10/10, 10/16, 10/19, 10/28
 with hinges, 10/28, 10/29
 with several spans, ILD, 10/29
 with two rolling loads, ILD, 10/30
Bearing pressure, 18/10
Bending
 deformation caused by, **14.4**, 17/all problems, **17.1**
 formula for, **12.4**
 stress caused by, **14.1, 14.2**
Bending couples
 displacement produced, **20.3**
 effect on structure, **12.3**
Bending moment diagrams, **10.2, 10.6**

horizontal beams, 10/12, 10/13, 10/14, 10/15, 10/16, 10/17, 10/19
Bending of
 circular shaft, 14/2, 14/7
 compound girder, 14/4, 14/5
 rectangular beam, 14/3
 sheet pile, 14/10
 steel channels, 14/5
 steel joist, 14/1, 14/4, 14/5, 14/6, 14/9
 tapered beam, 14/7
 Tee section, 14/8, 14/9
 unsymmetrical section, 14/8, 14/9, 14/10
Bent cantilever, displacement, 17/16, 20/9, 20/16
Body forces, **3.1**
Boiler, pressure in, 13/3
Bulk modulus, hydrostatic pressure, 13/19

Cable suspension, 9/46, 9/47, 9/49
Canopy, BMD, 10/20
Cantilever
 and beam, SFD, BMD, 10/10, 10/16, 10/17
 double, BMD, 10/20
 inclined load and couple, 10/9
 non-uniform load, 10/8
 U.D.L., 10/1
 with self-weight, 10/2
 with varying I, displacement, 17/16, 20/9
Cantilevers
 bent, displacement of, 17/6, 20/9, 20/16
 SFD, 10/6
Centre of Rotation, **8.4**
Centroid
 circular punched plate, 11/6
 perforated area, 11/5, 11/6, 11/7
 symmetrical areas, 11/1, 11/2
 unsymmetrical areas, 11/3, 11/4, 11/6, 11/7
Change in temperature: stress due to, 13/11–13/13

553

Change in volume, **13.4**, 13/15, 13/16
Circle
   of stress, **19.3** et seq
   properties of, **11.6**
Circular
   hollow sections, 18/3, 18/5, 18/6
   punched plate, centroid, 11/6
   section, bending in, 14/2, 14/7
   section, shear in, 15/4, 15/5
   section under torsion, 16/1–16/10
   sections under eccentric load, 18/2, 18/3, 18/5, 18/6
Close-coiled spring, 20/20
Column
   compression of, 13/1
   eccentric load, 10/24
   with cantilevers, 10/20
Combined stress
   (bending and direct), **18.1**
   (direct and shearing), **19.1**
Compound steel section, shear in, 15/8
Compression
   eccentric, 10/24, 18/4–18/9
   of column, 13/1
Conditions of Equilibrium, **5.1, 5.3, 5.5**
Contraflexure, points of, **17.2**, 17/5, 17/13, 17/15, 17/16, 17/20
Controlled strain, **13.5**
Couple combined with inclined loads, 10/9
Couples, **3.5, 20.3**
   and eccentric loads, **10.4**
   as part of a supporting system, **6.5**
   effect of Conditions of Equilibrium, **5.4**
   representation of effects, 10.1
Crane girder, ILD, 10/30
Curvature and deflection, relationship, **17.3**
Cylindrical boiler, stress in, 13/3

Deflected shapes, **10.3, 17.1, 17.2**
Deflection
   and curvature, **17.1** et seq., **20.1** et seq.
   by area-moments, **17.6**, 17/1, 17/11–17/20
   by Macaulay, **17.5**, 17/3, 17/5, 17/6, 17/7
   by strain energy, **20.2**
   of cantilever, 17/8, 17/10, 17/11, 17/15, 17/16, 17/18, 20/4, 20/5, 20/6, 20/8, 20/9, 20/11, 20/16, 20/17
   of curved bar, 20/18
   of horizontal beam, 17/1–17/7, 17/12, 17/13, 17/14, 17/17, 17/19, 17/20, 20/1, 20/2, 20/7, 20/10, 20/12, 20/14, 20/15
   of mast with stay, 17/15, 20/8
   of pin-jointed frames, 20/21–20/24
   of portal frame, 20/16
   of spring, 20/20
   into third dimension, 20/18
   under applied couples, 17/19, 20/12, 20/15
   under concentrated load, 17/1, 17/2, 20/2, 20/3, 20/7, 20/14
   under own weight, 17/8, 20/6
   under uniformly distributed load, 17/4, 17/17, 20/1, 20/5, 20/10
   under varying load, 17/9, 20/17
   under varying modulus of elasticity, 20/17
   under varying second moment of area, 17/10, 17/16, 17/18, 20/4, 20/6, 20/9, 20/11, 20/14, 20/17
Deformation, **12.1**
   due to bending, **14.4, 17.1**
   from direct force, **13.1**, 20/19
   from strain energy, **20.2**
   of pin-jointed frames, **20.6**
Diagrams of bending moment, **10.2**
Differential extension, 13/7
Direct and bending stress combined, **18.1**
Direct and shearing stress combined, **19.1**
Direct force
   effect on structure, **12.3**
   formula for, **12.4, 13.1**
Direction and sense of forces, **3.3**
Displacement
   of pin-jointed frames, **20.6**
   underbending couples, **20.3**
Double integration method, **17.4**, 17/2–17/10

Eccentric
   compression, 18/4–18/9
   load due to wind, 18/9
   load on column, 10/24
   loading, **18.3**
   loads on beam, 10/25
   loads with couples, **10.4**
   tension, 18/1
Elasticity, condition of, **12.2**
Energy, storage of, **12.1**
Equations of Equilibrium, **5.1, 5.3, 5.5**
Equilibrium
   achievement of, **5.1**
   under more than three forces, **5.3**
   under two or three forces, **5.2**

First Moment of Area, **11.3**
Flexure formula, **12.4**
Forces
   and couples in bent rod or arch, 9/7, 9/8, 9/11, 9/39, 9/40, 9/41, 9/44

body and surface, **3.1**
direction and sense, **3.3**
representation of effects, **10.1**
Formulae for direct force, bending, shear and torsion, **12.4**
Foundation
　pressure, 18/10
　slab, 10/18
Frame
　portal, displacement, 20/16
　rigid, 10/20
Frames, deformation of, **20.6**
Free structure, definition of, **6.1** et seq.

Geometry
　of curves, **7.3**
　of the cross-section, **11.1**
　of the structure, **7.1**
Graphical calculation of stress, **19.3** et seq.
Guide table for strain energy, **20.1**

Hexagon, Second Moment of Area, 11/11
Hollow sections,
　circular, 18/3, 18/5, 18/6
　rectangular, 18/8
Hollow shaft
　shear in, 15/6
　torsion of, 16/7–16/10
Horizontal beam,
　BMD 10/12–10/19
　couples and eccentric loads, 10/25
　non-UDL, 10/15
Hydrostatic pressure, bulk modulus, 13/19

Inclined and eccentric loads, 10/7, 10/25
Inclined beams, 10/11, 10/21, 10/22
Inclined loads and couple, 10/9, 10/11
Index, use of, **1.3**
Influence line diagrams, **10.5, 10.6**
Influence lines
　beam with cantilever, 10/28
　beam with hinges, 10/27
　several spans and hinges, 10/29
I section under eccentric load, 18/4, 18/7

Lack of fit, 13/6
Lateral strain, **13.3**, 13/14
　restrained conditions, 13/17, 13/18
　three applied stresses, 13/17, 13/18
Limitations of study for rigid structures, **4.1, 4.4**
Load, eccentric, on column, 10/24
Loading, eccentric, **18.3**, 10/25

Loads
　eccentric and couples, **10.4**
　inclined and eccentric, 10/25
　mixed, on horizontal beam, 10/12, 10/13, 10/14, 10/16, 10/17, 10/19
　non-uniform, 10/26
Longitudinal strain, **13.2**

Macaulay's method, **17.5**, 17/3, 17/5, 17/6, 17/7
Mass and weight relationship, **3.2**
Mast, deflection of, 17/15, 20/8
Method of double integration, **17.4**, 17/2, 17/3, 17/4, 17/5, 17/6, 17/8, 17/9, 17/10
Mixed loads, horizontal beam, 10/12, 10/13, 10/14, 10/16, 10/17, 10/19
Mnemonic
　for deformable structures, **12.5**
　for rigid structures, **8.1**
Modulus, bulk, 13/19
Mohr's circle, **19.3** et seq.
Moment of a couple, **3.5**
Moments, **3.4**
　overturning, **18.2**

Non-uniformly distributed load, 10/8, 10/15, 10/26
Normal stress, **19.4**

Oval link, second moment of area, 11/16
Overturning moments, **18.2**

Parabola, properties of, **11.6**
Perforated areas and plates, 11/5–11/7, 11/13–11/15
Pin-jointed frames, deformations, **20.6**, 20/21–20/24
Points of contraflexure, **17.2**, 17/5, 17/13, 17/15, 17/16, 17/20
Poisson's ratio, **13.3**
Polar Second Moment of Area, **11.5**
Portal frame, displacement, 20/16
Pressure in boiler, 13/3
Principal stress, **19.2**, 19/1, 19/2, 19/3, 19/5, 19/7, 19/9, 19/10, 19/11
Pure shear, 19/1

Quantities
　and symbols used, **2.1**
　changing from one unit to another, **2.4**
　multiplying and dividing, **2.3, 2.4**

Rectangle
　perforated, Second Moment of Area, 11/13, 11/14

Rectangle—contd.
　properties of, **11.6**
　Second Moment of Area, 11/9, 11/10
Rectangular
　hollow sections, 18/8
　section under eccentric load, 18/1, 18/8
Relationships
　in bending, **14.2**
　in transverse shear, **15.1, 15.2**
Restrained strain, **13.5**
　change in temperature, 13/11–13/13
　two materials, 13/13
Reversal of stress, 13/5
Rigid frame, bending moments, 10/20
Rigid structures
　technique of solution, **8.1**
　unknowns
　　one, 9/1, 9/3–9/8, 9/10–9/13
　　two, 9/2, 9/15, 9/33
　　three, 9/9, 9/14, 9/16, 9/23, 9/28–9/32, 9/34, 9/36, 9/38
　　four, 9/17, 9/20, 9/21, 9/22, 9/35, 9/49
　　five, 9/27, 9/37, 9/39–9/44, 9/46, 9/48
　　six, 9/18, 9/19, 9/24, 9/47
　　seven, 9/45
　　eight, 9/25, 9/26
Road or bent arch, forces and couples, 9/7, 9/8, 9/11, 9/39–9/44
Rolling loads, influence lines, 10/30

Second Moment of Area, **11.4**
　angle section, 11/8
　hexagon, 11/11
　oval, 11/16
　rectangle, 11/9, 11/10
　Tee, 11/9
　triangle, 11/10
　truncated square, 11/17
　unsymmetrical section, 11/7–11/10, 11/12
　varying along beam, 17/10, 17/16, 20/4, 20/9
Second Polar Moment of Area, **11.5**
Self-weight, horizontal beam, 10/13
Semi-circle, properties of, **11.6**
Sense and direction of forces, **3.3**
Senses of supports, **6.6**
Shaft, torsion of
　hollow, 16/7–16/10
　solid, 16/1–16/6, 16/10
Shapes, deflected, **10.3**
Shared load
　differential extension, 13/7

lack of fit, 13/6
two materials, 13/8–13/10
Shear, formulae and effects, **12.3, 12.4**
Shearing force diagram, **10.6**
　and couple, 10/9, 10/11
　due to self-weight, 10/2
　inclined loads, 10/7
　non-uniform load, 10/8
Shearing stress, **15.1, 19.1, 19.4**, 19/1, 19/2, 19/5
　in circular section, 15/4, 15/5
　in compound steel section, 15/8
　in hollow circular section, 15/6
　in I section, 15/3
　in rectangular section, 15/1, 15/2
　in Tee section, 15/7
　in triangular section, 15/9
　maximum, 19/2, 19/4, 19/5, 19/8
Slab under foundation load, 10/18
Slope and curvature, signs of, **17.3**
Solid shaft
　shear in, 15/4, 15/5
　Torsion in, 16/1–16/6, 16/10
Spring, close-coiled, 20/20
Square plate
　punched, centroid of, 11/7
　truncated, I of, 11/17
Steel beams, bending of, 14/1, 14/4, 14/5, 14/6, 14/9, 14/10
Steel section, shear in, 15/8
Storage of energy, **12.1**
Strain
　controlled, **13.5**
　definition of, **13.2**
　energy, **12.1, 20.1**
　　and deformation, **20.2**
　　method of calculation, **20.4**
　　unit load method, **20.5, 20.6**
　lateral, **13.3**, 13/14
　longitudinal, **13.2**
Stress, **12.1**
　and lateral strain, 13/17, 13/18
　by Mohr's circle, **19.3** et seq.
　caused by bending, **14.1, 14.2**
　caused by shear, **15.1**
　reversal of, 13/5
Support, types of, **6.3**
Supports, relationship between types, **6.4**
Surface tractions, **3.1**
Suspension cables, 9/46–9/48
Symmetrical area, centroid of, 11/1, 11/2

Table for strain energy, **20.1**
Technique of solution for rigid structures, **8.1**

Tee section
   bending of, 14/8, 14/9
   Second Moment of Area, 11/9
   shear in, 15/7
Temperature, stress due to change, 13/11–13/13
Tension
   eccentric, 18/1–18/3
   of wire, 13/2
   varying along length, 13/4
Third dimension deflection, 20/18
Thrust
   inclined load and couple, 10/9, 10/11
   inclined load on beam, 10/7
Torsion
   formulae for, **12.4**
   of circular sections, **16.1,** 16/1–16/10
Torsional couple, effect on structures, **12.3**
Transverse shear
   formulae for, **12.4**
   stress caused, **15.1**
Triangle
   properties of, **11.6**

Second Moment of Area, 11/10
   shear in, 15/9
Two materials
   restrained strain, 13/13
   shared load, 13/8, 13/9, 13/10

Unit load method, strain energy, **20.5, 20.6**
Unsymmetrical area, centroid, 11/3, 11/4, 11/6, 11/7
Unsymmetrical section under bending, 14/8–14/10

Varying second moment of area, 17/10, 17/16, 20/4, 20/9
Volume, change in, **13.4,** 13/15, 13/16

Weight and mass, relationship, **3.2**
Wind load, eccentric pressure, 18/9
Wire, tension of, 13/2, 13/4
Working pressure in boiler, 13/3